Food-Evoked Emotion, Product Acceptance, Food Preference, Food Choice and Consumption: The New Perspective on Developing Novel and Healthier Products

Food-Evoked Emotion, Product Acceptance, Food Preference, Food Choice and Consumption: The New Perspective on Developing Novel and Healthier Products

Editor

Witoon Prinyawiwatkul

MDPI

Basel • Beijing • Wuhan • Barcelona • Belgrade • Novi Sad • Cluj • Manchester

Editor
Witoon Prinyawiwatkul
Louisiana State University
Agricultural Center
USA

Editorial Office
MDPI
St. Alban-Anlage 66
4052 Basel, Switzerland

This is a reprint of articles from the Special Issue published online in the open access journal *Foods* (ISSN 2304-8158) (available at: https://www.mdpi.com/journal/foods/special_issues/Food_Evoked_Emotion_Acceptance_Preference_Choice_Consumption).

For citation purposes, cite each article independently as indicated on the article page online and as indicated below:

Lastname, A.A.; Lastname, B.B. Article Title. *Journal Name* **Year**, *Volume Number*, Page Range.

ISBN 978-3-0365-7882-8 (Hbk)
ISBN 978-3-0365-7883-5 (PDF)
doi.org/10.3390/books978-3-0365-7883-5

Contents

About the Editor vii

Witoon Prinyawiwatkul
Food-Evoked Emotion, Product Acceptance, Food Preference, Food Choice and Consumption: Some New Perspectives
Reprinted from: *Foods* **2023**, *12*, 2095, doi:10.3390/foods12112095 1

Cristhiam E. Gurdian, Damir D. Torrico, Bin Li, Georgianna Tuuri and Witoon Prinyawiwatkul
Effect of Disclosed Information on Product Liking, Emotional Profile, and Purchase Intent: A Case of Chocolate Brownies Containing Edible-Cricket Protein
Reprinted from: *Foods* **2021**, *10*, 1769, doi:10.3390/foods10081769 7

Michelangelo Serpico, Dominic Rovai, Kristine Wilke, Ruta Lesniauskas, Jeff Garza and Amy Lammert
Studying the Emotional Response to Insects Food Products
Reprinted from: *Foods* **2021**, *10*, 2404, doi:10.3390/foods10102404 31

Ziyang Chen, Cristhiam Gurdian, Chetan Sharma, Witoon Prinyawiwatkul and Damir D. Torrico
Exploring Text Mining for Recent Consumer and Sensory Studies about Alternative Proteins
Reprinted from: *Foods* **2021**, *10*, 2537, doi:10.3390/foods10112537 51

Wang-Chin Tsai and Xuqi Chen
Understanding the Key Factors of Older Adults' Continuance Intention in Congregate Meal Halls
Reprinted from: *Foods* **2021**, *10*, 2638, doi:10.3390/foods10112638 65

Sara R. Jaeger, David Jin, Grace S. Ryan and Joachim J. Schouteten
Emoji for Food and Beverage Research: Pleasure, Arousal and Dominance Meanings and Appropriateness for Use
Reprinted from: *Foods* **2021**, *10*, 2880, doi:10.3390/foods10112880 83

Ana Curutchet, Maite Serantes, Carolina Pontet, Fatima Prisco, Patricia Arcia, Gabriel Barg, et al.
Sensory Features Introduced by Brewery Spent Grain with Impact on Consumers' Motivations and Emotions for Fibre-Enriched Products
Reprinted from: *Foods* **2022**, *11*, 36, doi:10.3390/foods11010036 99

Ammaraporn Pinsuwan, Suntaree Suwonsichon, Penkwan Chompreeda and Witoon Prinyawiwatkul
Sensory Drivers of Consumer Acceptance, Purchase Intent and Emotions toward Brewed Black Coffee
Reprinted from: *Foods* **2022**, *11*, 180, doi:10.3390/foods11020180 115

Cristhiam E. Gurdian, Damir D. Torrico, Bin Li and Witoon Prinyawiwatkul
Effects of Tasting and Ingredient Information Statement on Acceptability, Elicited Emotions, and Willingness to Purchase: A Case of Pita Chips Containing Edible Cricket Protein
Reprinted from: *Foods* **2022**, *11*, 337, doi:10.3390/foods11030337 131

Pannapa Hanmontree, Witoon Prinyawiwatkul and Amporn Sae-Eaw
Emotion and Wellness Profiles of Herbal Drinks Measured Using Different Questionnaire Designs
Reprinted from: *Foods* **2022**, *11*, 348, doi:10.3390/foods11030348 155

Petar Vrgović, Milica Pojić, Nemanja Teslić, Anamarija Mandić, Aleksandra Cvetanović Kljakić, Branimir Pavlić, et al.
Communicating Function and Co-Creating Healthy Food: Designing a Functional Food Product Together with Consumers
Reprinted from: *Foods* **2022**, *11*, 961, doi:10.3390/foods11070961 . . . . **175**

Saara Vanhatalo, Hilkka Liedes and Kyösti Pennanen
Nature Ambience in a Lunch Restaurant Has the Potential to Evoke Positive Emotions, Reduce Stress, and Support Healthy Food Choices and Sustainable Behavior: A Field Experiment among Finnish Customers
Reprinted from: *Foods* **2022**, *11*, 964, doi:10.3390/foods11070964 . . . . **199**

Leidy Montoya, Natalia Quintero, Stevens Ortiz, Juan Lopera, Patricia Millán and Aída Rodríguez-Stouvenel
Inulin as a Fat-Reduction Ingredient in Pork and Chicken Meatballs: Its Effects on Physicochemical Characteristics and Consumer Perceptions
Reprinted from: *Foods* **2022**, *11*, 1066, doi:10.3390/foods11081066 . . . . **217**

Aline Oliveira Penedo, Sophie Bucher Della Torre, Franziska Götze, Thomas A. Brunner and Wolfram Manuel Brück
The Consumption of Insects in Switzerland: University-Based Perspectives of Entomophagy
Reprinted from: *Foods* **2022**, *11*, 2771, doi:10.3390/foods11182771 . . . . **237**

Armand V. Cardello, Fabien Llobell, Davide Giacalone, Sok L. Chheang and Sara R. Jaeger
Consumer Preference Segments for Plant-Based Foods: The Role of Product Category
Reprinted from: *Foods* **2022**, *11*, 3059, doi:10.3390/foods11193059 . . . . **255**

Charles Spence
Behavioural Nudges, Physico-Chemical Solutions, and Sensory Strategies to Reduce People's Salt Consumption
Reprinted from: *Foods* **2022**, *11*, 3092, doi:10.3390/foods11193092 . . . . **285**

Isaac Ho, Adelynn Peterson, Jack Madden, Kylie Wai, Ruta Lesniauskas, Jeff Garza, et al.
The Crick-Eatery: A Novel Approach to Evaluate Cricket (*Acheta domesticus*) Powder Replacement in Food Products through Product Eating Experience and Emotional Response
Reprinted from: *Foods* **2022**, *11*, 4115, doi:10.3390/foods11244115 . . . . **303**

Heng Xu, Mengyun Xiao, Jun Zeng and Huihui Hao
Green-Labelled Rice versus Conventional Rice: Perception and Emotion of Chinese Consumers Based on Review Mining
Reprinted from: *Foods* **2023**, *12*, 87, doi:10.3390/foods12010087 . . . . **337**

Silvia Murillo, Ryan Ardoin and Witoon Prinyawiwatkul
Factors Influencing Consumers' Willingness-to-Try Seafood Byproducts
Reprinted from: *Foods* **2023**, *12*, 1313, doi:10.3390/foods12061313 . . . . **357**

About the Editor

Witoon Prinyawiwatkul

Witoon Prinyawiwatkul holds a PhD degree in Food Science from University of Georgia, USA, and two Honorary Doctorate degrees in Agro-Industrial Product Development from Kasetsart and ChiangMai Universities, Thailand. He is currently a Professor of Food Science at Louisiana State University, LA, USA. His research focuses include sensory sciences, consumer research, and value-added/healthier products. He oversees the Sensory Services Lab. He actively collaborates with scientists worldwide, has published 260+ refereed articles (an h-index of 55 with 10,800+ citations, as of May 2023), 1 edited book, and 5 book chapters, and has presented 360+ scientific presentations. He is a fellow of the Institute of Food Technologists (USA). He has served as Scientific/Academic/Section/Associate Editor for several prestigious journals, including *Foods*.

foods

MDPI

Editorial

Food-Evoked Emotion, Product Acceptance, Food Preference, Food Choice and Consumption: Some New Perspectives

Witoon Prinyawiwatkul

School of Nutrition and Food Sciences, Louisiana State University, Agricultural Center, Baton Rouge, LA 70803, USA; wprinya@lsu.edu; Tel.: +1-225-578-5188

Introduction

Food is more than just a source of nutrients—it is a source of basic pleasure and aesthetic experiences. Consumers are now more health-conscious and more educated about what goes into their foods. Many consumers want healthier versions of retailed food products, such as those with high-fiber or low-sodium content. These consumers are concerned about the health benefit or risk associated with food consumption. Currently, globalization allows consumers to be exposed to various cuisines which can be readily available to them. Additionally, with the world population increasing rapidly, alternative food sources and food production will be needed to support sustainability. Potential alternative protein sources (e.g., from edible insects or plant-based sources), for instance, will likely be needed. By-products from various food processing and manufacturing find their way to be incorporated into food and beverage with additional health benefits. The questions are "would consumers be willing to consume them? How do they feel when consuming them, and do they like them?". Based on many studies, food choice, acceptance, preference, and consumption are affected by many factors, including both intrinsic and extrinsic factors and cues and consumer characteristics.

It is known that food elicits emotions, mainly positive ones. Measuring food-evoked emotions is a topical topic in sensory and consumer sciences. New methods for emotion measurement have been proposed. Emotions are becoming a critical component in designing products that meet consumers' needs and expectations [1]. Emotional profiles and ratings may effectively differentiate products with similar sensory characteristics and hedonic ratings; hence, they may provide additional information that goes beyond traditional hedonic ratings and may provide more insight toward food choice and behavior. Several studies have reported emotional responses to food and their relationships to product acceptability. In addition to the sensory quality of food, food-evoked emotion has been reported to be critical in predicting product acceptance and willingness-to-try, which are, in turn, critical in developing novel products. Appropriate health benefit information has also been reported to impact emotion, purchase decisions, and food choices. Human senses and cues perform an instrumental role in food choice and intake, emotion, and product acceptance; hence, understanding their roles and importance is critical.

This Special Issue aimed to present both original and cutting-edge research contributing to a deeper understanding of food-evoked emotion, food choice, preference, acceptance, and consumption, which is a valuable source of information for research and development, product innovation, and marketing that goes beyond traditional sensory preference and acceptability measurement. In this Special Issue, there are eighteen papers showcasing a diversity of approaches applied to different types of food products as test samples that provide excellent examples of the complex and multidisciplinary nature of the subject matter. A snapshot of these papers is given below.

Several methods and testing conditions for measuring and collecting emotions associated with foods have been developed and reported. Several studies reported in this Special Issue provide additional insights in this area. Different questionnaire designs can

Citation: Prinyawiwatkul, W. Food-Evoked Emotion, Product Acceptance, Food Preference, Food Choice and Consumption: Some New Perspectives. *Foods* **2023**, *12*, 2095. https://doi.org/10.3390/foods12112095

Received: 16 May 2023
Accepted: 18 May 2023
Published: 23 May 2023

influence the results of emotion and wellness assessments. Hanmontree et al. assessed emotion and wellness profiles of herbal drinks using six different questionnaire designs [a questionnaire item written in the form of words vs. full sentences and three types of measuring methods: a rating scale, CATA, and RATA] [2]. They reported that using a full sentence did not provide a clear benefit over using some keywords, especially when using familiar words clearly understood by consumers. All three measuring methods produced similar emotion and wellness profiles. However, the authors cautioned that each method has advantages and limitations that should be carefully considered [2]. For emotion measurement, questionnaires are primarily used and can be implemented using words or emojis. However, it has been questionable if emojis are appropriate to be used in the context of food consumption situations. Jeager et al. used 24 emojis (14 facial and 10 non-facial) characterized using the Pleasure-Arousal-Dominance model and compared the perception of two consumer groups (n = 165 from New Zealand and n = 861 from the UK) [3]. They found that emojis were suitable for cross-cultural research and identified similar groups of emoji as most and least suitable for food-related consumer research [3]. In order to be effective in collecting, sorting, and arranging a large pool of data on consumers' perceptions, Chen et al. introduced novel approaches—text mining and Natural Language Processing (NLP)—to obtain sensory data and identify current consumer trends on alternative proteins [4]. Based on some selected publications between 2018 to 2021, they found that insect- and plant-based proteins were the centers of alternative protein research, and there was no significant association between emotions and alternative protein categories [4].

Several studies have reported that additional product information, along with relevant information, such as health benefits and sustainability, can significantly impact consumers' perceptions. Gurdian et al. [5,6] investigated the effects of ingredient (that is, cricket protein) information, environmental and nutritional-quality claims, and tasting moments [before tasting and after tasting with or without additional statement) on hedonic, emotional, and purchase intent (PI) of pita chips and chocolate brownies. Although expectations, disconfirmations, and product claims may not have a direct effect on product liking, they may still indirectly affect the overall product acceptability via emotional elicitation [5]. Evoking "interested" and "adventurous" emotions performed a significant role as drivers of product liking regardless of the formulation and moment, and PI can be improved through the elicitation of pleasant emotions upon tasting. For chocolate brownies, the disclosed information affected emotional profiles more than formulation. After-tasting emotions "happy" and "satisfied" were critical predictors of PI. Gender also significantly affected perceived emotion and PI [6]. Another study was devoted to understanding how the "green-food label" of rice affected the perception and emotions of Chinese consumers using a text mining approach [7]. Green food is produced in an ecological environment that meets standards of safety, nutrition, and non-pollution. The COVID-19 pandemic increased public health awareness, turning consumers toward safe and healthy foods to strengthen their immunity. Xu et al. compared differences in consumer perceptions and emotions towards green-labeled rice vs. conventional rice [7]. They reported that green-label mistrust and packaging breakage during logistics were the leading causes of negative emotions (e.g., disappointing and doubtful) among consumers of green-labeled rice [7].

Environments, where consumers make food choice decisions and consume food, have effects on their consumption, choices, preference, and product acceptance. This may likely and partly due to different emotional responses induced by repeated exposure to such environments. Vanhatalo et al. designed two environments in the restaurant: 'nature ambiance' to induce a positive emotional response and 'fast food ambiance' to induce a less positive emotional response [8]. They found that during 'nature ambiance' week, consumers more often chose vegetarian dishes and generated less plate waste. This study provided some evidence that ambiance modification in a real-life setting may insert a moderate effect on healthier and more sustainable food behaviors [8]. In another study, Tsai and Chen focused on the "trust" level of older adults that influenced continuance intention in congregate meal halls [9]. Trust is a mental state and a basis of social participation. In this study, trust was classified as "pre-use trust" and "post-use trust" in products or services. The authors found that the perceived service quality was the main factor affecting the perceived satisfaction, which, in turn, affected the post-use trust; hence, the older adults showed a positive continuance intention to participate in the senior meal halls [9].

In addition to methods and testing conditions and environments as described above, the type of test products and consumer characteristics can largely affect acceptance, emotion, food choice, and consumption. Insects have been proposed as a sustainable and alternative food source; however, in the Western world, insects are viewed as disgusting. Serpico et al. attempted to understand the acceptance of insect-based products in the US market by studying emotional responses [10]. Insect food products were positively correlated with the emotions of interest, understanding, daring, adventurous, and worried and negatively correlated with the emotions satisfied, good, pleasant, happiness, calm, warm, nostalgia, and secure. Disgust was found as a barrier to product acceptance [10]. Likewise, Penedo et al. found that the most common reason among young adults (n = 290 in a Swiss university) for not eating insect-containing foods was disgust. In contrast, the most likely reason for eating insect foods was curiosity [11]. Product appropriateness for insect incorporation performs a significant role in the acceptance of insect-containing food. Ho et al. evaluated how product eating experience affected consumer acceptability, emotional response, satiety, and plate waste [12]. Three different dishes (sausage, pasta, and brownies) containing cricket powder (CP) were tested against the controls. Negative terms selected, such as worried, decreased once the products were consumed. Significant correlations were found between appearance liking and satiety and taste liking and plate waste for both the control and CP dishes. They concluded that formulation and serving methods might increase the overall acceptability of food products containing insect powders [12].

Two studies were devoted to evaluating consumer perception of products containing byproducts from seafood [13] and brewing [14] processing. With increasing global demand for seafood, seafood byproducts (SB) utilization can contribute to a more sustainable food supply chain through waste-to-value food product development. Murillo et al. evaluated some factors influencing consumers' willingness-to-try (WTT) seafood byproducts using an online survey with 904 adult seafood consumers [13]. This research was concerned with using consumers' emotional baseline scores during an early stage of the COVID-19 pandemic as predictors of WTT foods containing SB. Consumers feeling more unsafe during the pandemic may have been reluctant to report positive WTT. The SB safety and health claim information cue may have reassured some of these consumers, making "unsafe" emotion a significant predictor of WTT [13]. Brewery spent grain (BSG) is a byproduct from a brewing process of malted barley and a good source of fiber (30–50% dry wt basis). Curutchet et al. evaluated the impact of this fiber enrichment on sensory quality, acceptability, and purchase intention of bread, pasta, and chocolate milk [14]. Ambivalence was seen in the emotions generated by the different fiber-enriched products, depending on whether the fiber was perceived. Consumers felt more warm, bored, active, and less guilty when tasting fiber-enriched pasta and bread, but the opposite was observed for fiber-enriched chocolate milk. The authors concluded that the effect of BSG addition was product-specific and that fiber perception makes consumers feel more confident [14].

Cardello et al. conducted a survey of willingness to consume (WTC) 5 types of plant-based (PB) (Milk, Cheese, Meat (33%PB), Meat (100%PB) and Fish) food in the USA, Australia, Singapore, and India (n = 2494) [15]. In addition to WTC, emotional, conceptual, and situational use characterizations were also evaluated and found to exert significant lifts/penalties on WTC. They reported that positive valence ('enthusiastic', 'comforting', 'easygoing', 'energetic', 'happy', and 'inspiring') significantly impacted positive WTC, while the five terms with negative valence—'boring', 'dissatisfied', 'nervous', 'tense', and 'uninspired'—overwhelmingly reduced WTC. This study demonstrated that consumers are not monolithic in their WTC PB foods and that WTC is often a function of the food category of the PB food [15].

Coffee is one of the most widely consumed beverages in the world, and black coffee drinking is becoming more popular and a new cultural practice among Thai people. The link between coffee aroma/flavor and elicited emotions remains underexplored. Pinsuwan et al. identified key sensory characteristics of brewed black coffee via descriptive sensory analysis that affected acceptance, purchase intent, and emotions for Thai consumers [16]. This information could be useful for creating or modifying the sensory profile of brewed black coffee to increase consumer acceptance [16].

Three papers were devoted to product development and sensory strategies for functional and healthy foods. Vrgović et al. described a systematic process so-called "co-creation" for developing functional food, which directly involves consumers in various stages of the creation process [17]. They used raspberry seeds as a source of natural bioactive compounds to produce healthy, functional food products other than just an extract. Montoya et al. developed and characterized reduced-fat pork and chicken meatballs. Inulin was used as a fat substitute to mimic the properties of fat [18]. There was a greater probability of meatball consumption with a claim of "preservative-free" as compared to other claims "reduced-fat" or "a good source of fiber". Last but not least important, Spence noted the importance of sodium in the human diet while emphasizing the risk of excessive sodium intake [19]. The author provided a narrative historical review, discussing a wide range of sensory approaches that have been experimented with to modulate expected and perceived saltiness in order to reduce salt consumption. The author also highlighted a number of important questions that remains for future research [19].

The editor hopes that the readers will find this Special Issue insightful, interesting, and useful for future research. The diversity of both the content and the methodologies presented in this Special Issue should inspire and encourage future exploration of multidisciplinary research collaboration, which would lead to a better understanding of the complex relationship among emotion, acceptance, preference, choice, and consumption of food.

Funding: This research received no external funding.

Conflicts of Interest: The author declares no conflict of interest.

References

1. Prinyawiwatkul, W. Relationships between Emotion, Acceptance, Food Choice, and Consumption: Some New Perspectives. *Foods* **2020**, *9*, 1573. [CrossRef] [PubMed]
2. Hanmontree, P.; Prinyawiwatkul, W.; Sae-Eaw, A. Emotion and Wellness Profiles of Herbal Drinks Measured Using Different Questionnaire Designs. *Foods* **2022**, *11*, 348. [CrossRef] [PubMed]
3. Jaeger, S.R.; Jin, D.; Ryan, G.S.; Schouteten, J.J. Emoji for Food and Beverage Research: Pleasure, Arousal and Dominance Meanings and Appropriateness for Use. *Foods* **2021**, *10*, 2880. [CrossRef] [PubMed]
4. Chen, Z.; Gurdian, C.; Sharma, C.; Prinyawiwatkul, W.; Torrico, D.D. Exploring Text Mining for Recent Consumer and Sensory Studies about Alternative Proteins. *Foods* **2021**, *10*, 2537. [CrossRef] [PubMed]
5. Gurdian, C.E.; Torrico, D.D.; Li, B.; Prinyawiwatkul, W. Effects of Tasting and Ingredient Information Statement on Acceptability, Elicited Emotions, and Willingness to Purchase: A Case of Pita Chips Containing Edible Cricket Protein. *Foods* **2022**, *11*, 337. [CrossRef] [PubMed]
6. Gurdian, C.E.; Torrico, D.D.; Li, B.; Tuuri, G.; Prinyawiwatkul, W. Effect of Disclosed Information on Product Liking, Emotional Profile, and Purchase Intent: A Case of Chocolate Brownies Containing Edible-Cricket Protein. *Foods* **2021**, *10*, 1769. [CrossRef] [PubMed]
7. Xu, H.; Xiao, M.; Zeng, J.; Hao, H. Green-Labelled Rice versus Conventional Rice: Perception and Emotion of Chinese Consumers Based on Review Mining. *Foods* **2023**, *12*, 87. [CrossRef] [PubMed]
8. Vanhatalo, S.; Liedes, H.; Pennanen, K. Nature Ambience in a Lunch Restaurant Has the Potential to Evoke Positive Emotions, Reduce Stress, and Support Healthy Food Choices and Sustainable Behavior: A Field Experiment among Finnish Customers. *Foods* **2022**, *11*, 964. [CrossRef] [PubMed]
9. Tsai, W.-C.; Chen, X. Understanding the Key Factors of Older Adults' Continuance Intention in Congregate Meal Halls. *Foods* **2021**, *10*, 2638. [CrossRef] [PubMed]
10. Serpico, M.; Rovai, D.; Wilke, K.; Lesniauskas, R.; Garza, J.; Lammert, A. Studying the Emotional Response to Insects Food Products. *Foods* **2021**, *10*, 2404. [CrossRef] [PubMed]

11. Penedo, A.O.; Bucher Della Torre, S.; Götze, F.; Brunner, T.A.; Brück, W.M. The Consumption of Insects in Switzerland: University-Based Perspectives of Entomophagy. *Foods* **2022**, *11*, 2771. [CrossRef] [PubMed]
12. Ho, I.; Peterson, A.; Madden, J.; Wai, K.; Lesniauskas, R.; Garza, J.; Gere, A.; Amin, S.; Lammert, A. The Crick-Eatery: A Novel Approach to Evaluate Cricket (*Acheta domesticus*) Powder Replacement in Food Products through Product Eating Experience and Emotional Response. *Foods* **2022**, *11*, 4115. [CrossRef] [PubMed]
13. Murillo, S.; Ardoin, R.; Prinyawiwatkul, W. Factors Influencing Consumers' Willingness-to-Try Seafood Byproducts. *Foods* **2023**, *12*, 1313. [CrossRef] [PubMed]
14. Curutchet, A.; Serantes, M.; Pontet, C.; Prisco, F.; Arcia, P.; Barg, G.; Menendez, J.A.; Tárrega, A. Sensory Features Introduced by Brewery Spent Grain with Impact on Consumers' Motivations and Emotions for Fibre-Enriched Products. *Foods* **2022**, *11*, 36. [CrossRef] [PubMed]
15. Cardello, A.V.; Llobell, F.; Giacalone, D.; Chheang, S.L.; Jaeger, S.R. Consumer Preference Segments for Plant-Based Foods: The Role of Product Category. *Foods* **2022**, *11*, 3059. [CrossRef] [PubMed]
16. Pinsuwan, A.; Suwonsichon, S.; Chompreeda, P.; Prinyawiwatkul, W. Sensory Drivers of Consumer Acceptance, Purchase Intent and Emotions toward Brewed Black Coffee. *Foods* **2022**, *11*, 180. [CrossRef] [PubMed]
17. Vrgović, P.; Pojić, M.; Teslić, N.; Mandić, A.; Kljakić, A.C.; Pavlić, B.; Stupar, A.; Pestorić, M.; Škrobot, D.; Mišan, A. Communicating Function and Co-Creating Healthy Food: Designing a Functional Food Product Together with Consumers. *Foods* **2022**, *11*, 961. [CrossRef] [PubMed]
18. Montoya, L.; Quintero, N.; Ortiz, S.; Lopera, J.; Millán, P.; Rodríguez-Stouvenel, A. Inulin as a Fat-Reduction Ingredient in Pork and Chicken Meatballs: Its Effects on Physicochemical Characteristics and Consumer Perceptions. *Foods* **2022**, *11*, 1066. [CrossRef] [PubMed]
19. Spence, C. Behavioural Nudges, Physico-Chemical Solutions, and Sensory Strategies to Reduce People's Salt Consumption. *Foods* **2022**, *11*, 3092. [CrossRef] [PubMed]

foods

MDPI

Article

Effect of Disclosed Information on Product Liking, Emotional Profile, and Purchase Intent: A Case of Chocolate Brownies Containing Edible-Cricket Protein

Cristhiam E. Gurdian [1], Damir D. Torrico [2], Bin Li [3], Georgianna Tuuri [1] and Witoon Prinyawiwatkul [1,*]

1 Agricultural Center, School of Nutrition and Food Sciences, Louisiana State University, Baton Rouge, LA 70803, USA; cgurdi3@lsu.edu (C.E.G.); gtuuri@agcenter.lsu.edu (G.T.)
2 Department of Wine, Food and Molecular Biosciences, Faculty of Agriculture and Life Sciences, Lincoln University, Lincoln 7647, New Zealand; damir.torrico@lincoln.ac.nz
3 Agricultural Center, Department of Experimental Statistics, Louisiana State University, Baton Rouge, LA 70803, USA; bli@lsu.edu
* Correspondence: wprinya@lsu.edu

Abstract: Edible insects, a sustainable and nutritious alternative to conventionally derived proteins, are unfamiliar to Westerners and often associated with negative sentiments. Edible-cricket protein (ECP) added to chocolate brownies (CB) [0% ECP = CBWO (without) vs. 6% *w/w* ECP = CBW (with)], and disclosed information [no ECP added = (−) vs. ECP with benefits = (+), ECP− and ECP+, respectively] yielded four CB treatments (CBWO−, CBWO+, CBW−, and CBW+). Subjects (n = 112 female and n = 98 male) rated liking, selected emotions before- and after-tasting, and determined consumption (CI) and purchase intent (PI) after tasting. Likings were analyzed with mixed-effects ANOVA and post hoc Tukey's HSD test. Emotions were evaluated with Cochran's-Q test and correspondence analysis. Emotions driving or inhibiting overall liking (OL) were assessed with penalty-lift analyses using two-sample t-tests. A random forest algorithm was used to predict PI and estimate variables' importance. Female's and male's expected OL were higher for CBWO− than for CBWO+. Females' actual OL was higher for CBWO than for CBW regardless of the disclosed information but males' actual OL was the same across treatments. Females exhibited negative-liking disconfirmation for CBW−. In both tasting conditions, the disclosed information affected treatments' emotional profiles more than formulation. After-tasting emotions "happy" and "satisfied" were critical predictors of PI.

Keywords: sentiment analysis; alternative insect protein; anthropo-entomophagy; consumer behavior; extrinsic-product cue; cognitive information

Citation: Gurdian, C.E.; Torrico, D.D.; Li, B.; Tuuri, G.; Prinyawiwatkul, W. Effect of Disclosed Information on Product Liking, Emotional Profile, and Purchase Intent: A Case of Chocolate Brownies Containing Edible-Cricket Protein. *Foods* **2021**, *10*, 1769. https://doi.org/10.3390/foods10081769

Academic Editor: Zein Kallas

Received: 5 July 2021
Accepted: 28 July 2021
Published: 30 July 2021

1. Introduction

The expected rise in the global population has increased the need for finding more efficient ways to obtain nutrient-dense sustainable foods [1,2]. Presently, protein deficiency is a leading cause of malnutrition for over one billion people worldwide [3]. Thus, investigations of ways to achieve a sustainable protein supply are being conducted, which includes using new technologies and ingredients to produce protein-rich foods [4]. For instance, edible insects can be produced with a higher feed-conversion efficiency, lower spending of environmental resources (e.g., water, land, feed), and less ecosystem pollution than conventional-source-derived proteins (including plant-based and livestock) [5]. Overall, edible insects have a high-quality nutritional profile, which depends on species, developmental stage, diet, sex, and processing among other variables [6–8]. Hence, functional ingredients could be obtained from their protein, fat, and chitin components. Particularly, the incorporation of edible insect protein in foods will be governed by the functionality they can add to the formulations; hence, there is a growing interest in studying their physio-chemical properties and sensory acceptability in different food categories. Although insects represent an eco-friendly and nutritious alternative to conventionally derived protein sources, the available information regarding their safety as a food ingredient is still limited [5]. According to Murefu et al. [9], the potential food safety hazards associated with edible insects are of chemical, biological, and allergenic nature (arginine kinase, α-amylase, tropomyosin, and other proteins also present in crustaceans which are widely known allergens in edible insects). The extent to which contaminants negatively affect the safety of edible insects as foods is determined primarily by the production system, species, developmental stage at harvest, and feeding (including sources) in the rearing process. This suggests that controlled conditions rather than wild harvesting shall be practiced to guarantee adequate food safety standards in edible insects [5].

Acceptable food products containing edible insect ingredients in bakery [8,10–15], energy/protein bars [16], extruded snacks [17], pasta [18], and meat [19,20] categories have been reported. However, there is still a significant reluctance to consume edible insect food products mainly in western cultures, where entomophagy is not a common practice [21]. Such rejection has been associated primarily with disgust [22] and neophobia [23]. These challenges could be addressed or at least counterbalanced by educating consumers about the benefits of edible insects [24], introducing novel food products with a "similarity of tasting" approach (i.e., tastes like another popular product) [19], incorporating edible insects or their nutrient fractions as "invisible ingredients," such as flours, extracts or powders [25] in familiar products [26], and promoting tasting experiences with edible-insect products to improve consumers' familiarity [27]. Different food matrices can be used to study the incorporation of novel ingredients in food products. For instance, chocolate brownies (CB) are familiar to consumers, highly acceptable, and commonly associated with positive feelings, which makes CB an appropriate food model for the incorporation of edible insect products [28–30]. Edible cricket protein (ECP) is a high-quality protein produced in more sustainable conditions than plant or animal-based proteins, but has not been sufficiently explored regarding its acceptability in the US marketplace [31]. Yet, Fischer and Steenbekkers [32] reported that Westerners are more receptive to crickets, mealworms, and grasshoppers than to other edible insects.

Predicting consumers' food choice with models based solely on hedonic information may not yield adequate prediction power compared to more holistic models that incorporate product-elicited emotions information [33]. In sensory studies, the collected data are usually analyzed via multivariate projection techniques such as principal component analysis, to describe the treatments and/or explain the observed differences among them [34]. However, predictive discriminant models can be built on sensory and emotional data to efficiently discriminate among treatments and to provide a measure of variable importance for future sensory analysis applications [35]. Recently, machine learning and data mining have become more popular by providing modeling tools to predict variable outcomes based on ensembles of predictors, such as random forest (RF) and bootstrap-aggregation (bagging) trees, that perform better than their single predictors [36]. To the best of our knowledge, these tools have not been fully explored to model sensory and emotional data together with demographic information. The inclusion of emotions (before- and after-tasting) evoked by CB formulations without and with ECP upon disclosing ECP presence and its benefits to consumers in addition to product acceptance and other demographic and experimental variables may improve the performance of an RF model predicting purchase intent (PI) and aid marketing strategies for the introduction of edible-insect foods into the US marketplace.

The effect of product benefit claims, such as sustainability or high-nutritional value on the PI, emotions, and overall liking (OL) has been widely studied in different products. The effect of the claims varies depending on the food category, implied benefits, and the population being studied [34,37]. Several studies have reported the positive effects of disclosed benefit claims on consumer acceptability, perception, PI, or emotional profiles [38] albeit others have found them irrelevant [39] or not significant for certain demographic groups [40]. To our knowledge, the effect of disclosing the presence of ECP in CB while communicating the sustainability and nutritional-quality benefits derived from its consumption on product acceptability and emotional profiles as they relate to PI has not yet been studied. Therefore, the objective of this study was to investigate whether disclosing ECP presence accompanied by an environmental and nutritional-quality claim affected the expected (before-tasting) and actual (after-tasting) OL, emotional profiles, and/or PI of CB formulations (CBWO and CBW).

2. Materials and Methods

2.1. Chocolate Brownies (CB) Preparation

Chocolate brownies (CB) were prepared with Betty-Crocker fudge batter mix comprising sugar, enriched flour bleached (wheat flour, niacin, iron, thiamin mononitrate, riboflavin, folic acid), cocoa processed with alkali, palm oil, corn syrup, corn starch, and 2% or less of: carob powder, salt, canola oil, and artificial flavor (General Mills Sales, Inc., Minneapolis, MN, USA), USDA grade A large-white eggs (Great Value, Walmart Stores, Inc., Bentonville, AR, USA), and Wesson canola oil (Conagra Brands, Chicago, IL, USA). Edible cricket protein (ECP) commercialized as Griopro 100% cricket powder (All Things Bugs LLC, Midwest City, OK, USA) made of whole crickets (*Acheta domesticus* and *Gryllodes sigillatus*) containing 65% *w/w* protein, 22.5% *w/w* fat, and 5% total carbohydrate (wet basis) was added (6% *w/w*) to the formulation. This concentration of ECP was based on preliminary data from a trial with 25 subjects tasting CB within a range of ECP (3–10% *w/w*) for which 6% *w/w* was the highest percentage before significant taste and aroma rejection occurred due to an earthy off-flavor/aroma and/or a rancid aftertaste. Batches of each CB formulation (without ECP, CBWO, and with 6% *w/w* ECP, CBW) were prepared the day before the consumer study. Briefly, eggs (875 g), water (258 g), canola oil (621 g), batter mix (3128 g), and ECP powder (312 g, only added for CBW) were stirred together in a Globe SP20 commercial food mixer (Globe Food Equipment CO, Dayton, OH, USA) at speed 2 for each batch. The mixture was then placed in a 45.7 cm × 66 cm aluminum tray and baked in a pre-heated OV310G mini rotating rack oven (Baxter Mfg, a Division of ITW FEG, LLC, Orting, WA, USA) at 325 °F for 52 min. Baked CBWO and CBW were stored separately at room temperature in food grade BPA-free polypropylene 2 oz. clear plastic-lidded portion cups (CrystalWare, Lakewood, NJ, USA) overnight until the consumer study was performed.

2.2. Consumer Study

The research protocol was approved by Louisiana State University (LSU) Agricultural Center Institutional Review Board (IRB # HE 18-9 and # HE 18-22). Participants (n = 210 untrained consumers 18 years of age and older; Table 1) were recruited from a pool of faculty, staff, and students at the LSU campus, Baton Rouge, LA, USA. Recruitment criteria included: (1) no self-reported allergy or adverse reactions towards any ingredients of the CB samples or unsalted crackers, (2) willingness to taste samples that contain edible cricket protein (ECP) powder, (3) absence of any physiological or medical conditions that would compromise their performance in the sensory evaluation, and (4) self-reported regular consumption (at least once per month) of CB. Subsequently, subjects agreed with and signed a consent form included in the approved research protocol.

Table 1. Demographic profile of participants from the consumer study.

Demographic Variables	Levels	*n*	%
Gender	Female	112	53.33
	Male	98	46.67
Age group	18–22	93	44.29
	23–29	84	40.00
	30–39	24	11.43
	40–49	5	2.38
	50–59	3	1.43
	≥60	1	0.48
Race	Asian	37	17.62
	Black/African American	27	12.86
	Latino	41	19.52
	White/Caucasian	100	47.62
	Other	5	2.38
Highest education level achieved	College degree	56	26.67
	Graduate or professional degree	74	35.24
	High school or lower degree	80	38.10
High-protein products consumption	Yes	123	58.57
	No	87	41.43
Previously tasted products with edible insects	Yes	117	55.71
	No	93	44.29

2.3. *Questionnaire: Consumer Liking, Emotions, Consumption (CI) and Purchase Intent (PI)*

Qualtrics software (Qualtrics, Provo, UT, USA) was used to administer the computer-based questionnaires given to panelists and to collect their responses. The four CB treatments (Figure 1) were presented together before starting the evaluation. Then, consumers were instructed to evaluate them in a monadic-sequential order as indicated on the screen based on the three-digit sample blinding code, and specific sample information and related benefits of ECP were given for each sample. The full related benefits of ECP statement was as followed "This sample contains ECP. Edible insects are safe to eat and are considered a sustainable source of high-quality protein and other nutrients. Edible insect production has less negative environmental impact than traditional livestock production. An estimated two billion people worldwide consume edible insects" [26].

The experimental design was a completely randomized block design (CRBD) with a factorial arrangement. The formulation and disclosed information factors had 2 levels each (formulation levels: without ECP and with 6% ECP; disclosed information levels: "No ECP added" statement and "Sample contains ECP and benefits" statement). The Qualtrics software instructed the panelists which of the treatments shall be evaluated first and disclosing absence of ECP or alternatively its presence accompanied by its benefits was part of the treatment identity. The presentation order for the four treatments was balanced and randomized, so each of them had the same chance of being present in all four possible positions (half of the consumers evaluated first those treatments that contained the "this sample has ECP and benefits" information whereas the other half evaluated first those treatments that contained the "No ECP was added" information. Specifically, when consumers evaluated CBWO− (treatment 1) first, it involved a CB without ECP accompanied by a "No ECP was added to this sample" statement; when they evaluated

CBWO+ (treatment 2) first, it involved a CB without ECP accompanied by a "This sample contains ECP and the nutritional and environmental benefits from ECP"; when they evaluated CBW− (treatment 3) first, it involved a CB with 6% ECP accompanied by a "No ECP was added to this sample"; finally, when they evaluated CBW+ (treatment 4) first, it involved a CB with 6% ECP accompanied by a "This sample contains ECP and the nutritional and environmental benefits from ECP". The treatments were evaluated in two experimental conditions (before- and after-tasting) in one sensory session.

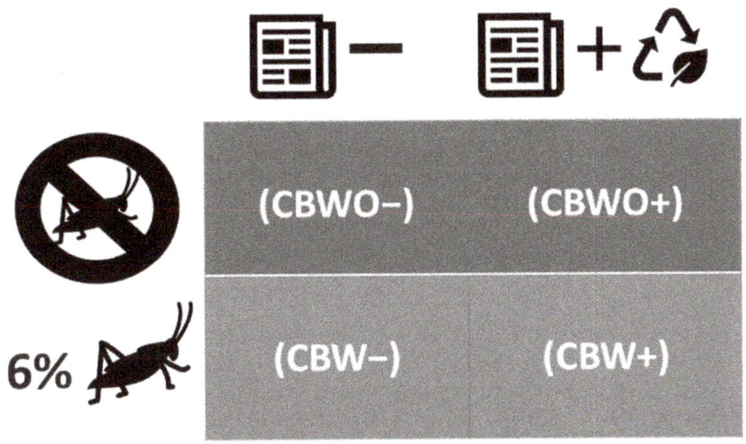

Figure 1. Factorial arrangement for the chocolate brownie (CB) treatments. ECP = Edible-cricket protein. WO: Without; W: With. ECP−: "No ECP added" disclosed information; ECP+: "Contains ECP + benefits" disclosed information. (1) CBWO− = CB without ECP (CBWO) presented under the ECP− disclosed information; (2) CBWO+ = CBWO presented under the ECP+ disclosed information; (3) CBW− = CB with 6% ECP (CBW) presented under the ECP− disclosed information; (4) CBW+ = CBW presented under the ECP+ disclosed information.

The evaluation consisted of (1) reporting elicited emotions before tasting (based on the sample's visual evaluation and the disclosed information) on a Check-all-that-apply (CATA) basis from a list of twenty-five emotion terms from the Essense25 profile emotion word list [27]; (2) rating expected (before-tasting) likings with a 9-point-hedonic scale (left-anchored dislike extremely and right-anchored like extremely); (3) reporting elicited emotions upon tasting on the CATA list mentioned above; (4) rating actual (after-tasting) likings with the previously mentioned 9-point-hedonic scale; and (5) indicating consumption intent (CI) and purchase intent (PI) if the sample were commercially available with a binomial scale (Yes or No).

2.4. Statistical Analysis

The sensory evaluation of CB treatments (resulting from the 2 × 2 factorial arrangement of formulation and disclosed information levels) followed a balanced and randomized block design (panelists as blocks). Statistical data analysis was conducted using the Statistical Analysis Software (SAS) version 9.4 (Cary, NC, USA), R software version 4.0.3 (RStudio, Inc., Boston, MA, USA), and the XLSTAT (Addinsoft, New York, NY, USA) statistical software version 2020 [41] with $\alpha = 0.05$ significance level. The effect of formulation (CBWO vs. CBW), disclosed information (ECP− vs. ECP+), demographics, tasting condition (before vs. after) and up to three-way interactions between gender (females vs. males), formulation, and disclosed information and between tasting condition, formulation, and disclosed information on overall liking (OL) was investigated with multi-way analysis of variance (ANOVA) in a mixed-effects model having panelists as a random effect and Tukey's HSD post hoc test. Check-all-that-apply (CATA) binary data from emotions (before- and after-tasting) were analyzed according to the procedures reported by Meyners et al. [42] and Ares et al. [43] segmented by tasting condition and gender. Global/individual Cochran Q tests determined the overall/individual effect of treatments within tasting condition and tasting condition within treatment in emotions distribution/each emotion term frequency distribution. Subsequently, all pairwise comparisons were conducted for treatment groups as well as tasting conditions following the Marascuilo and McSweeney procedure based on minimum required difference [44]. The proportion of discriminant emotions across genders within tasting conditions and across tasting conditions within genders were compared with two-population proportions Z-tests and two-tailed McNemar tests for correlated proportions, respectively. Emotions (segmented by tasting condition and gender), consumption intent (CI) and purchase intent (PI) were then input to a correspondence analysis based on Chi-square distances. For each tasting condition (before and after) and gender, the relationship between elicited emotions and product liking was unfolded through penalty-lift analysis of before-tasting and after-tasting OL to identify drivers/inhibitors of product liking. Overall liking mean impact was calculated as the mean OL difference from present vs. absent categories for each emotion with a 20% population threshold [38]. This difference was then standardized, and its significance ($p < 0.05$) was tested with a two-sample t-test. The random forest (RF) algorithm, an ensemble of decision trees, which are combined to predict a single outcome and modelled to provide diversity between the trees [45] was used to model PI prediction (using mtry = 32 features out of 68 in the random selection at each node of the $n = 1000$ decision trees) from formulation, disclosed information, demographic variables, sensory likings (before- and after-tasting), emotions (before- and after-tasting), and CI using full data as interest was on model performance. Because RF is an ensemble of several low-bias-high-variance components (decision trees), the RF variance is reduced, and its resulting discrimination is on average more accurate than its individual components. To increase the diversity among the decision trees, RF fits each tree on a random subset of the dataset, of the same length, selected with replacement (bootstrap replicate). In addition, diversity is increased during the growing of the trees as for each node, RF picks a small random subset of predictor attributes and uses only this subset to search for the best split of the data into their observed classes or numerical outcome. A noteworthy feature of RF is the overfitting control. Although RF can be composed of a large number of decision trees, the error rate for new samples converges to a limiting value when the number of trees goes to infinity [35]. The misclassification rate for RF was estimated using the out-of-bag observations and the classifier's performance was displayed on the Receiver Operating Characteristic (ROC) curve. Plots of variables relative importance from RF were obtained based on the mean decrease in accuracy and mean decrease in Gini index, which measures node impurity for classification trees.

3. Results and Discussion

3.1. Significance of Main Effects in Product Liking

The significance of the main effects and their interactions of interest (up to 3-way) on treatments' OL is summarized in the analysis of variance (ANOVA) shown in Table 2.

Tasting condition, formulation, and their 2-way interaction were significant ($p < 0.05$) for OL. Disregarding all other effects, OL was significantly ($p < 0.05$) lower after-tasting (6.30) than before-tasting (6.55) and was significantly ($p < 0.05$) lower for CBW than for CBWO (6.26 vs. 6.60, respectively). The levels of formulation (CBWO vs. CBW) influenced the way subjects rated their OL for treatments depending on the tasting condition (before vs. after tasting). Although the OL ratings were not significantly ($p = 0.08$) influenced by the levels of disclosed information (ECP− vs. ECP+), there was a significant ($p < 0.05$) interaction of disclosed information with tasting condition. On the other hand, gender levels (female vs. male) significantly ($p < 0.05$) interacted with the formulation effect causing differences in the OL ratings. Previous research indicated that males exhibited higher acceptability for edible insects than females [31,46] possibly because they had lower disgust sensitivity, experienced more curiosity, or associated novelty with edible insects more than females, which drove their willingness to try and ultimate acceptability of edible insects.

Table 2. ANOVA † table for the overall sensory acceptability of CB treatments ‡.

Effects	Overall Liking §	
	F Value	Pr > F
Gender	2.54	0.11
Age	1.43	0.22
Race	1.53	0.20
Education	1.32	0.27
High protein consumption	0.40	0.53
Previous edible insect	2.54	0.11
Tasting condition	17.95	<0.01
Formulation	35.00	<0.01
Disclosed information	3.07	0.08
Tasting condition * Formulation	15.58	<0.01
Tasting condition * Disclosed information	29.10	<0.01
Formulation * Disclosed information	2.84	0.09
Gender * Formulation	6.32	0.01
Gender * Disclosed information	0.35	0.55
Tasting condition * Formulation * Disclosed information	0.00	0.98
Gender * Formulation * Disclosed information	0.16	0.69

† ANOVA = Analysis of variance [2 genders (female and male), 6 age groups (18–22, 23–29, 30–39, 40–49, 50–59, ≥60 years old), 5 races (Asian, Black/African American, Latino, White/Caucasian, Other), 3 education levels (college, graduate/professional degree, high school or lower degree), 2 levels of high protein consumption (yes and no), 2 levels of previous edible insect (yes and no), 2 levels of tasting condition (before and after), 2 levels of formulation (CBWO and CBW), 2 levels of disclosed information (ECP− and ECP+). ‡ Treatments are described in Figure 1. § Overall liking data from $n = 210$ consumers were collected using a 9-point hedonic scale (1 = dislike extremely, 9 = like extremely) and analyzed by a mixed-effects model with panelists as a random effect. * Denotes interaction.

3.2. Effects of Formulation, Disclosed Information, and Gender on Expected and Actual Overall Liking

Figure 2 shows the treatments' OL least-square means in the before (expected) and after (actual) tasting conditions from the female and male groups. The CBWO expected OL was negatively affected ($p < 0.05$) by the ECP+ disclosed information in both genders, which could be attributed to food neophobia [47], disgust feeling [48], and other product-elicited mental associations with unpleasant variables [22]. Food neophobia is mainly related to unfamiliarity with novel foods while disgust is thought to be originated from mental associations with other disgusting variables, which makes it more complex to be understood and overcome or counterbalance. Both negative-product-elicited traits are considered the major limitation for the willingness to try edible insects in Western societies [21,49,50] although La Barbera et al. [22] found them uncorrelated and determined that "disgusting" feelings were more important than neophobia when predicting the willingness to eat insects. Although ECP+ disclosed information communicated environmental and nutritional benefits associated with anthropo-entomophagy, the negative feelings and expectations exerted a stronger effect than the environmental or nutritional consciousness and positive sensations. Possibly, sustainability and nutritional consciousness were not significant drivers for the expected OL of CB containing ECP [49]. On the other hand, the formulation had no significant effect ($p > 0.05$) on OL expectation regardless of the disclosed information. The perceived difference in appearance among formulations was not large enough to yield significant differences in liking expectations.

In the after-tasting condition, the female group rated a significantly higher ($p < 0.05$) OL for CBWO than for CBW for either disclosed information, but the male group rated similar ($p > 0.05$) OL across formulations for either disclosed information. The female group's mean OL (5.46) was significantly ($p < 0.05$) lower than that of the male group (6.26) only for CBW−. Possibly, the female group presented a lower taste rejection threshold than the male group for ECP, which suggests males are more likely to accept products containing ECP than females. Previous studies have found similar results claiming males had a lower aversion to consuming products containing edible insects than females [46,50–53]. However, other studies have suggested food neophobia [54], disgust [55], indirect (via disgust effect) implicit attitudes derived from implicit associations with edible insects [22], social and cultural norms [31], and perceived behavioral control [56] rather than gender as stronger determinants for the willingness to consume insects and actual-consumption behavior. Lower perceived behavioral control, higher measurements for neophobia and disgust, and more traditional food culture decrease the likelihood of edible insect consumption.

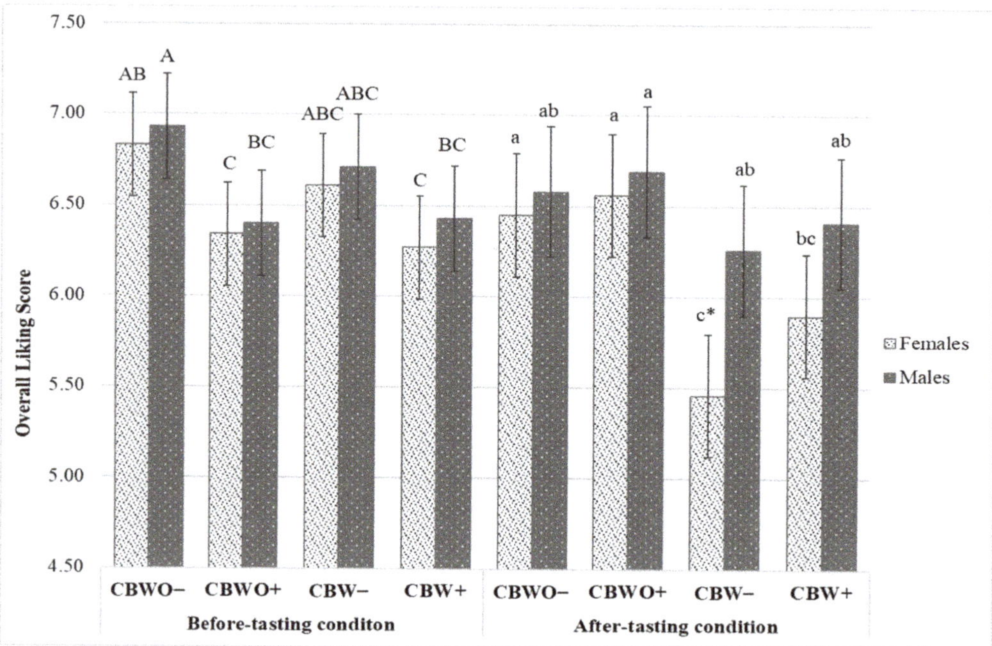

Figure 2. Treatments overall liking (OL) bar chart segmented by tasting condition (before vs. after tasting). Data are OL least square means and standard errors from n = 112 female and n = 98 male groups. Treatments are described in Figure 1. Different uppercase/lowercase letters indicate significantly ($p < 0.05$) different before tasting/after tasting OL scores (Tukey's means separation) across treatments and gender. * Denotes significantly ($p < 0.05$) lower after-tasting OL score than its corresponding before-tasting OL score (Tukey's means separation).

The disclosed information had no significant effect ($p > 0.05$) on actual OL ratings for either group (female and male) and either formulation (CBWO and CBW).

Other authors have also concluded that communicating environmental and health benefits of entomophagy is insufficient to alter the sensory acceptability of foods containing edible insects [25,46,57]. When consumers evaluate (taste/interact) products, their expectations for a given attribute or product's performance can be met (if actual performance after interacting with the product is as expected) or disconfirmed (negatively when intensity/liking expectations are higher than the actual perceptions/likings, or positively when they are higher than the intensity/liking expectations). When disconfirmation occurs, product acceptability can be: (1) aligned with expectations, (2) affected (positively or negatively) to a greater extent than if expectations had not been present, (3) negatively affected regardless of the direction of the disconfirmation, or (4) assimilated/contrasted with expectations depending on the perceived magnitude of the discrepancy [58]. Moreover, when sensory expectations are negatively disconfirmed, the probability of repeated purchase/consumption may decrease [57]. Comparing the before and after-tasting scenarios, CBW− had a significant ($p < 0.05$) negative liking disconfirmation among the female group, but the OL expectations for CBW+ were not significantly ($p > 0.05$) disconfirmed upon tasting (Figure 2). This suggests a positive effect of the disclosed information [24,59], which is possibly associated with the subjects' degree of environmental or nutritional consciousness [51]. The significant negative disconfirmation observed in the female group for CBW− could be explained by the deception caused by ECP− (appearance of CBW and claim of ECP absence made them believe they would taste and experience the characteristic sweet and chocolate flavors from regular brownies but instead they tasted ECP− added flavor). In the female group, the experienced discrepancies between the CBW− expected OL and the OL perceived after tasting were sufficiently large leading CBW− into a rejection region. In this region, an increase in the perceived real difference resulted in an under-rated actual OL when compared to a scenario without expectations. On the contrary, ECP+ disclosed information "prepared" the female group to experience the sensory characteristics of ECP (based on experience, beliefs, or mental associations) so no negative disconfirmation occurred for CBWO+ or CBW+. Rather, an assimilation effect was observed for CBWO+, in which the perceived OL after tasting was matched to the expected OL.

Overall, actual OL scores of at least 7 on a 9-point-hedonic scale are considered promising for regular food products [60] but given ECP represents a new concept for Westerners, the obtained actual OL for CBW+ (female group = 5.90 and male group = 6.41) represents an encouraging starting point for the incorporation of ECP into similar bakery products especially if targeting male consumers [28]. Moreover, the way information is conveyed can affect the consumers' perceptions and liking. In this study, the ECP−associated benefits were presented in the form of a statement accompanied by a picture of the ECP, but delivering the same information on the packaging or in informative sessions could improve the actual acceptability of CBW+ [56].

3.3. *Effects of Formulation and Disclosed Information on Emotional Profiles before and after Tasting*

3.3.1. Emotional Profiles of Males vs. Females before Tasting

The treatments' emotional profile based on self-reported applicable emotion terms from the Essense25 list [61] was evaluated separately for each gender and tasting condition. In the before-tasting condition, the female group exhibited a significantly ($p < 0.05$) higher proportion (17/25) of discriminant emotions than the male group (6/25). Other researchers have also reported higher emotional discrimination for food products among females when compared to males [62]. Table 3 shows the emotional profiles from the before-tasting condition exploring the observed differences between treatments separately for each gender.

For the female group, the ECP+ disclosed information led to a significant ($p < 0.05$) increase in the frequency of "adventurous," "interested," and "wild" regardless of the formulation while reducing the observed frequency of "bored" only for the CBW formulation. Similarly, the ECP+ disclosed information increased the frequency of the "adventurous" and "wild" emotions for both formulations among the male group and reduced the frequency of "bored" only for CBW. This pattern of emotional terms is common for individuals seeking pronounced sensations [63]. Sensation seeking is considered a powerful predictor of edible insect acceptability [49], exhibiting a strong positive correlation (0.30) with the acceptability of insect flour in foods [59]. Interest in the environment together with neophobia, familiarity, convenience, and affinity for meat are considered determinant variables for the readiness to adopt edible insects [64]. Neophobic subjects unconcerned with the environmental impact of food choices and with a high affinity for meat-based diets are less likely to adopt edible insects [51]. Still, presenting edible insects as invisible ingredients in familiar food products [19] with an appropriate sensory profile has been effective to improve their willingness to try [65].

Nevertheless, disclosing the presence of ECP and its benefits (ECP+) in CB also elicited unfavorable effects for both genders before-tasting emotional profiles. For the female group, a significant ($p < 0.05$) decrease in the proportion of "good," "happy," and "safe" positive emotion terms occurred for CBW+ when compared to CBW− while "worried" occurred more frequently for ECP+ disclosed information than for ECP− for either formulation. Similarly, for the male group, ECP+ significantly ($p < 0.05$) decreased the "calm" and "safe" terms and increased the choice frequency for "worried" for CBW. Additionally, ECP+ decreased the frequency of the "warm" term for the male group for CBWO when compared to ECP−. The observed negative effect of ECP+ triggering unsafety, mental discomfort, and lack of confidence in both genders agrees with other studies reporting "worry" and "concern" emotions from individuals regarding their safety (health risks) when eating foods containing edible insects [5]. These concerns arise mainly because of the limited availability of information about the process used to guarantee the innocuity and quality of the insect-derived ingredient [48,66] and its regulations [67] when incorporated into foods. However, this could be substantially improved if potential consumers are educated about the safety and regulations governing edible insect process throughout the added-value chain starting in farms until presented in a meal [68] and by repeated exposure to tasting events involving edible insects without any health-related adverse outcome [69].

Table 3. Emotional profile † elicited by treatments ‡ in the before-tasting condition.

Emotions	Females				Males			
	CBWO−	CBWO+	CBW−	CBW+	CBWO−	CBWO+	CBW−	CBW+
Active	13 ^{A}	9 ^{A}	10 ^{A}	8 ^{A}	10 ^{a}	15 ^{a}	7 ^{a}	11 ^{a}
Adventurous	11 ^{B}	***47*** ^{A}	10 ^{B}	***43*** ^{A}	7 ^{b}	30 ^{a}	11 ^{b}	***39*** ^{a}
Aggressive	3 ^{A}	4 ^{A}	4 ^{A}	4 ^{A}	2 ^{a}	1 ^{a}	1 ^{a}	2 ^{a}
Bored	8 ^{B}	1 ^{B}	20 ^{A}	5 ^{B}	8 ^{a,b}	4 ^{a,b}	11 ^{a}	3 ^{b}
Calm	31 ^{A}	15 ^{B}	25 ^{A,B}	14 ^{B}	27 ^{a}	19 ^{a,b}	28 ^{a}	15 ^{b}
Disgusted	0 ^{B}	8 ^{A}	1 ^{A,B}	8 ^{A}	2 ^{a}	5 ^{a}	4 ^{a}	5 ^{a}
Enthusiastic	***20*** ^{A}	***29*** ^{A}	***18*** ^{A}	***25*** ^{A}	14 ^{a}	19 ^{a}	16 ^{a}	21 ^{a}
Free	11 ^{A}	7 ^{A}	7 ^{A}	***8*** ^{A}	8 ^{a}	13 ^{a}	8 ^{a}	10 ^{a}
Good	47 ^{A,B}	40 ^{A,B}	***51*** ^{A}	34 ^{B}	50 ^{a}	43 ^{a,b}	41 ^{a,b}	35 ^{b}
Good natured	10 ^{A}	12 ^{A}	8 ^{A}	11 ^{A}	10 ^{a}	16 ^{a}	13 ^{a}	12 ^{a}
Guilty	4 ^{A}	4 ^{A}	5 ^{A}	6 ^{A}	5 ^{a}	3 ^{a}	5 ^{a}	3 ^{a}
Happy	32 ^{A,B}	20 ^{B,C}	34 ^{A}	13 ^{C}	27 ^{a}	17 ^{a}	20 ^{a}	24 ^{a}
Interested	***46*** ^{B}	***71*** ^{A}	***39*** ^{B}	***74*** ^{A}	***45*** ^{a}	***53*** ^{a}	***40*** ^{a}	***51*** ^{a}
Joyful	21 ^{A}	13 ^{A,B}	13 ^{A,B}	9 ^{B}	16 ^{a}	12 ^{a}	12 ^{a}	8 ^{a}
Loving	6 ^{A}	4 ^{A}	6 ^{A}	6 ^{A}	7 ^{a}	4 ^{a}	6 ^{a}	4 ^{a}
Mild	28 ^{A}	23 ^{A}	26 ^{A}	23 ^{A}	11 ^{a}	12 ^{a}	20 ^{a}	12 ^{a}
Nostalgic	7 ^{A,B}	4 ^{B}	11 ^{A}	4 ^{B}	6 ^{a}	4 ^{a}	5 ^{a}	2 ^{a}
Pleasant	27 ^{A}	13 ^{B}	***31*** ^{A}	8 ^{B}	21 ^{a}	15 ^{a}	27 ^{a}	20 ^{a}
Safe	29 ^{A,B}	16 ^{B,C}	***33*** ^{A}	11 ^{C}	18 ^{a,b}	17 ^{a,b}	23 ^{a}	9 ^{b}
Satisfied	23 ^{A}	11 ^{A,B}	21 ^{A,B}	10 ^{B}	15 ^{a}	10 ^{a}	13 ^{a}	10 ^{a}
Tame	11 ^{A}	2 ^{B}	9 ^{A,B}	5 ^{A,B}	6 ^{a}	5 ^{a}	7 ^{a}	4 ^{a}
Understanding	4 ^{A}	***13*** ^{A}	5 ^{A}	10 ^{A}	4 ^{a}	9 ^{a}	6 ^{a}	9 ^{a}
Warm	***17*** ^{A}	6 ^{B}	11 ^{A,B}	5 ^{B}	***14*** ^{a}	4 ^{b}	11 ^{a,b}	6 ^{a,b}
Wild	2 ^{B}	***12*** ^{A}	3 ^{B}	***12*** ^{A}	2 ^{c}	11 ^{a,b}	3 ^{b,c}	13 ^{a}
Worried	3 ^{C}	***13*** ^{A,B}	5 ^{B,C}	***15*** ^{A}	4 ^{a,b}	5 ^{a,b}	1 ^{b}	8 ^{a}

† Frequency of emotions in the before-tasting condition from $n = 112$ female and $n = 98$ male groups analyzed by two-sided Cochran's Q test with Marascuilo and McSweeney procedure (multiple-pairwise-comparisons-minimum-required difference) [44]. Different uppercase/lowercase letters within a row represent significant ($p < 0.05$) differences in the female/male group's emotion across treatments. Bolded and italicized frequency was significantly ($p < 0.05$) higher than its corresponding emotion in the after-tasting condition (Table 4). ‡ Treatments are described in Figure 1.

Table 4. Emotional profile † elicited by treatments ‡ in the after-tasting condition.

Emotions	Females				Males			
	CBWO−	CBWO+	CBW−	CBW+	CBWO−	CBWO+	CBW−	CBW+
Active	9 ^{A}	5 ^{A}	9 ^{A}	5 ^{A}	6 ^{a}	10 ^{a}	6 ^{a}	8 ^{a}
Adventurous	7 ^{B}	35 ^{A}	9 ^{B}	28 ^{A}	6 ^{b}	25 ^{a}	8 ^{b}	23 ^{a}
Aggressive	2 ^{A}	4 ^{A}	2 ^{A}	4 ^{A}	1 ^{a}	1 ^{a}	0 ^{a}	2 ^{a}
Bored	***18*** ^{A}	***6*** ^{B}	17 ^{A,B}	11 ^{A,B}	11 ^{a,b}	7 ^{a,b}	15 ^{a}	6 ^{b}
Calm	25 ^{A}	12 ^{B}	18 ^{A,B}	18 ^{A,B}	30 ^{a}	22 ^{a,b}	20 ^{a,b}	18 ^{b}
Disgusted	***4*** ^{A}	9 ^{A}	***13*** ^{A}	14 ^{A}	3 ^{a}	4 ^{a}	7 ^{a}	7 ^{a}
Enthusiastic	8 ^{A}	10 ^{A}	10 ^{A}	8 ^{A}	8 ^{a}	12 ^{a}	10 ^{a}	14 ^{a}
Free	6 ^{A,B}	4 ^{A,B}	8 ^{A}	1 ^{B}	10 ^{a}	11 ^{a}	8 ^{a}	10 ^{a}
Good	53 ^{A,B}	***56*** ^{A}	39 ^{B,C}	36 ^{C}	44 ^{a}	42 ^{a}	44 ^{a}	***50*** ^{a}
Good natured	6 ^{A}	11 ^{A}	9 ^{A}	6 ^{A}	15 ^{a}	20 ^{a}	10 ^{a}	16 ^{a}
Guilty	5 ^{A}	4 ^{A}	4 ^{A}	5 ^{A}	3 ^{a}	3 ^{a}	3 ^{a}	3 ^{a}
Happy	38 ^{A}	26 ^{A,B}	29 ^{A,B}	17 ^{B}	35 ^{a}	23 ^{a}	27 ^{a}	26 ^{a}
Interested	28 ^{B}	48 ^{A}	25 ^{B}	49 ^{A}	21 ^{a}	33 ^{a}	24 ^{a}	28 ^{a}
Joyful	19 ^{A}	18 ^{A}	13 ^{A}	9 ^{A}	14 ^{a}	15 ^{a}	16 ^{a}	***16*** ^{a}
Loving	6 ^{A}	8 ^{A}	8 ^{A}	3 ^{A}	6 ^{a}	5 ^{a}	8 ^{a}	5 ^{a}
Mild	23 ^{A}	28 ^{A}	30 ^{A}	23 ^{A}	***23*** ^{a}	14 ^{a}	20 ^{a}	17 ^{a}
Nostalgic	13 ^{A}	***10*** ^{A,B}	8 ^{A,B}	4 ^{B}	8 ^{a}	5 ^{a}	6 ^{a}	4 ^{a}
Pleasant	31 ^{A}	***29*** ^{A}	21 ^{A}	***24*** ^{A}	31 ^{a}	***39*** ^{a}	30 ^{a}	26 ^{a}
Safe	25 ^{A}	17 ^{A}	13 ^{A}	17 ^{A}	15 ^{a}	18 ^{a}	18 ^{a}	15 ^{a}
Satisfied	***37*** ^{A}	***39*** ^{A}	26 ^{A}	***39*** ^{A}	***36*** ^{a}	***41*** ^{a}	***31*** ^{a}	***30*** ^{a}
Tame	11 ^{A}	5 ^{A}	11 ^{A}	6 ^{A}	10 ^{a}	7 ^{a}	***14*** ^{a}	9 ^{a}
Understanding	4 ^{A,B}	5 ^{A,B}	1 ^{B}	10 ^{A}	5 ^{a}	9 ^{a}	5 ^{a}	10 ^{a}
Warm	9 ^{A}	6 ^{A}	7 ^{A}	9 ^{A}	8 ^{a}	8 ^{a}	10 ^{a}	6 ^{a}
Wild	4 ^{A}	5 ^{A}	4 ^{A}	2 ^{A}	2 ^{b}	11 ^{a}	2 ^{b}	11 ^{a}
Worried	1 ^{B}	6 ^{A,B}	7 ^{A,B}	8 ^{A}	1 ^{a}	3 ^{a}	2 ^{a}	3 ^{a}

† Frequency of emotions in the after-tasting condition from n = 112 female and n = 98 male groups analyzed by two-sided Cochran's Q test with Marascuilo and McSweeney procedure (multiple-pairwise-comparisons-minimum-required difference) [44]. Different uppercase/lowercase letters within a row represent significant ($p < 0.05$) differences in the female/male group's emotion across treatments. Bolded and italicized frequency was significantly ($p < 0.05$) higher than its corresponding emotion in the before-tasting condition (Table 3).
‡ Treatments are described in Figure 1.

In the female group, the "calm" and "tame" emotions were selected fewer times when CBWO was presented under ECP+ disclosed information; yet this effect is difficult to interpret as it could be both, positive and negative because it could reflect an "energetic" but also "nervous" or "anxious" short-term response or long-lasting state [70]. In fact, other researchers have categorized the "tame" emotion as an unclassified term [71,72]. Another adverse effect of the ECP+ disclosed information among the female group was the decreased frequency of the "pleasant" emotion's proportion for both formulations and increased frequency of the "disgust" term for CBWO when contrasting against ECP−. Disgust sensitivity has been identified as one of the major and most challenging constraints to entomophagy in the Western world [49], which is more frequent in young [73] females than in male consumers [74]. Overcoming disgust is key to improve the willingness to eat and/or buy insect foods because it is one of its most important predictors [75]. On the other hand, treatments' emotional profile in the before-tasting condition showed a minimal effect of formulation for either gender (Figure 3A,B).

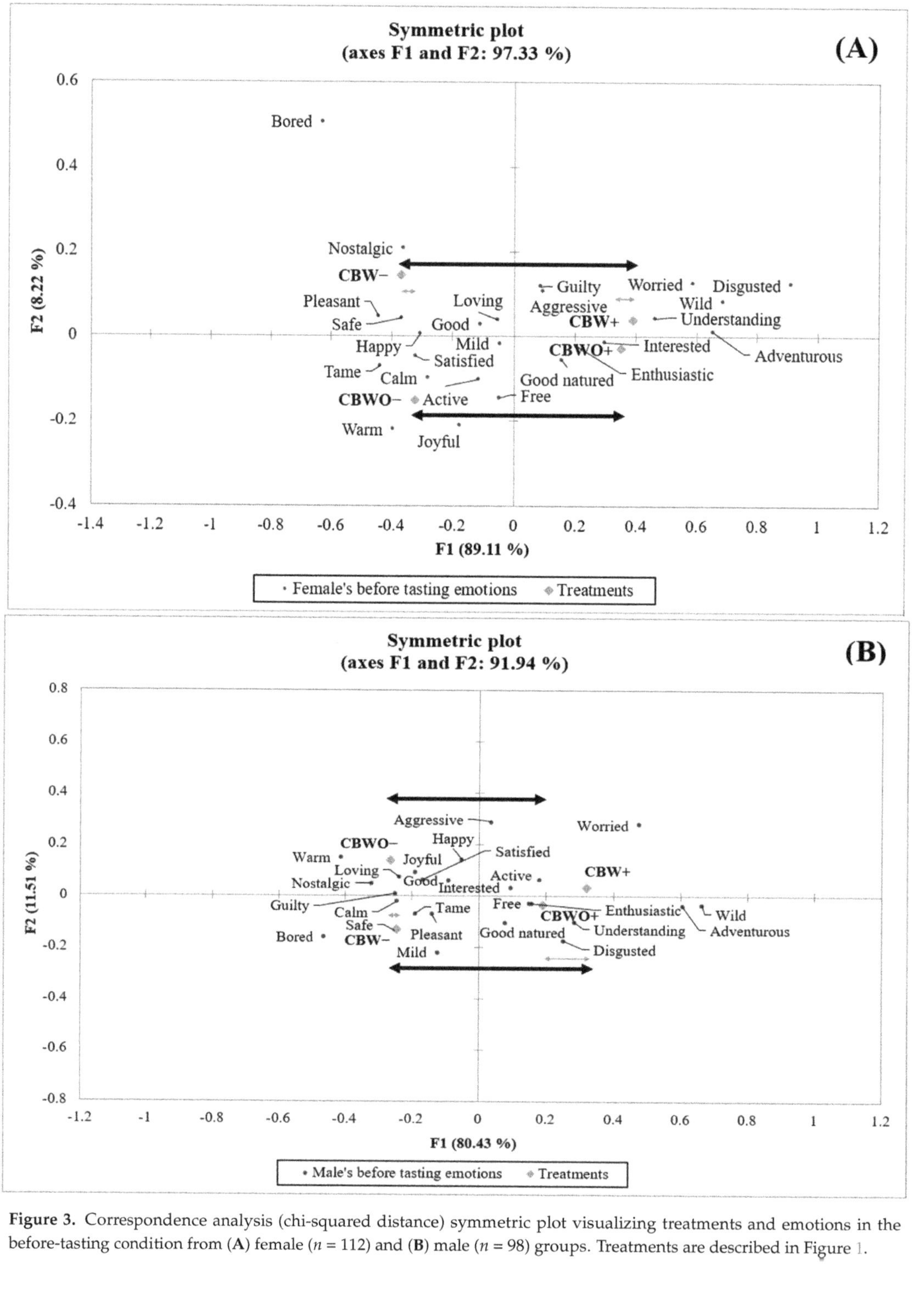

Figure 3. Correspondence analysis (chi-squared distance) symmetric plot visualizing treatments and emotions in the before-tasting condition from (**A**) female (n = 112) and (**B**) male (n = 98) groups. Treatments are described in Figure 1.

3.3.2. Emotional Profiles of Males vs. Females after Tasting

Contrary to the before-tasting condition, the proportion of discriminant after-tasting emotion terms for the female group (9/25) was not significantly different ($p > 0.05$) from that of the male group (4/25). Table 4 shows the effect of formulation and disclosed information on the treatments' emotional profile in the after-tasting condition by gender. For the female group, the "adventurous" and "interested" emotions were positively affected by the ECP+ disclosed information in both formulations, and the "bored" emotion was less frequent for CBWO+ than for CBWO−.

On the other hand, the male group was positively influenced by the ECP+ disclosed information for both formulations regarding the "adventurous" and "wild" emotions, which belong to the *active* dimension (which reflects characteristic emotions of an "energetic" state or mood elicited upon tasting foods and/or reading food names) [76] while the "bored" term was less frequent for CBW+ than for CBW−, which is generally considered a negative term with a high arousal state that commonly decreases food liking and intake [77]. These results suggest that an appropriate marketing campaign for ECP should lie in the context with novelty, adventure, and wild sensations [49,78].

The "understanding" emotion in the female group became more frequent for CBW when presented with the ECP+ disclosed information than when presented with the ECP− disclosed information. Although "understanding" emotion has been considered an unclassified term in some studies [72], others have placed it in the positive dimension or have found a significant positive correlation between "understanding" and product liking [79–81]. In this study, the female group possibly felt more understanding about the sensory profile of CBW+ (different flavor notes and texture characteristics compared to a regular brownie) because they were informed that ECP was present in the formulation. CBW− exhibited a lower proportion of the "understanding" emotion among the female group because of the disconfirmed sensory profile experienced for this treatment, which agrees with the observed behavior in the OL ratings previously discussed.

However, the female group's "free" emotion was negatively affected by the ECP+ disclosed information in the CBW formulation while the "calm" term significantly ($p < 0.05$) decreased in CBWO when presented under the ECP+ disclosed information compared to when presented under ECP−. Although the "worried" emotion was most frequent for CBW+ among the female group, it was not significantly ($p > 0.05$) different from CBWO+ or CBW−, evidencing an effect of the interaction between formulation and disclosed information. A formulation effect was observed among the female group only for the "good" emotion, which was significantly ($p < 0.05$) less frequent for CBW+ than for CBWO+. Still, among both groups (female and male), the disclosed information affected the treatments' emotional profile in the after-tasting condition more than the formulation (Figure 4A,B).

3.3.3. Differences in Emotional Profiles by Gender between Tasting Conditions

The female group exhibited significantly a higher ($p < 0.05$) proportion of discriminant emotion terms in the before tasting (17/25) condition than in the after tasting (9/25) condition, respectively, whereas the male group presented no significant ($p > 0.05$) differences in the proportion of discriminant emotions between tasting conditions (6/25 vs. 4/25 before- and after-tasting, respectively). This was expected as other researchers have reported a greater effect of informative claims on before-tasting elicited emotions [71].

Among the female group, the "adventurous" emotion significantly decreased upon tasting for both formulations (CBWO and CBW) when appearing with the ECP+ disclosed information, but for the male group, it decreased upon tasting only for CBW+. This could partially be explained by the need for optimization in CBW formulation; yet, since the effect was not observed for CBW−, it can also reflect bias triggered by the disclosed information or the need for a different/additional context of ECP+ emphasizing adventure, novelty, activeness, or a different product application closely related to "adventurous" feeling (e.g., energy drink, high-protein shakes, energy bars) [28,82].

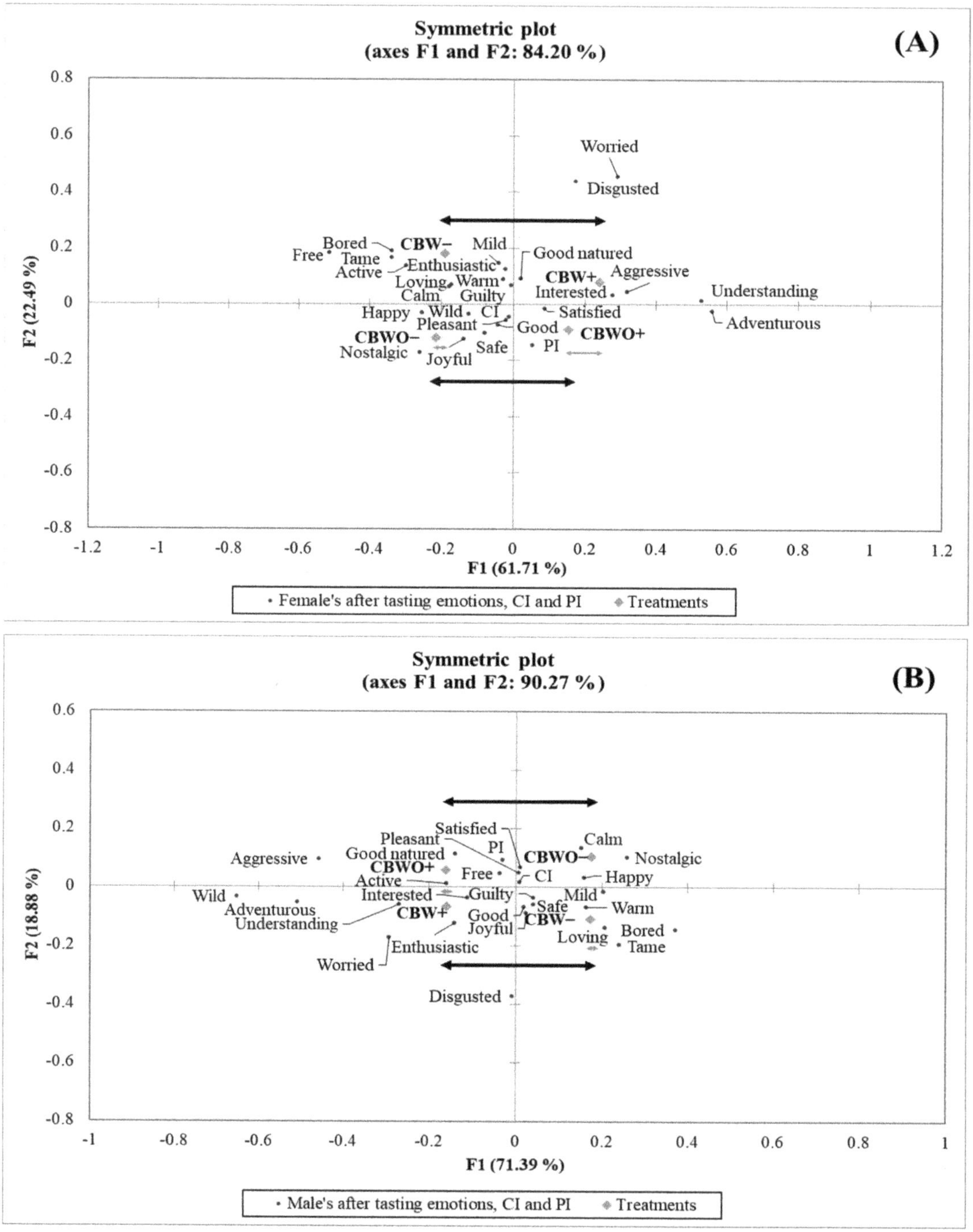

Figure 4. Correspondence analysis (chi-squared distance) symmetric plot visualizing emotions in the after-tasting condition, consumption intent (CI), purchase intent (PI), and treatments from (**A**) female (n = 112) and (**B**) male (n = 98) groups. Treatments are described in Figure 1.

Unexpected effects across tasting conditions were observed for both genders. The female group's "bored" emotion significantly increased for both disclosed information (ECP− and ECP+) upon tasting but only for CBWO formulation, suggesting a positive effect of the CBW formulation. The female group selected "disgust" emotion more frequently in the after-tasting condition than in the before-tasting condition for CBWO− and CBW− whereas the male group presented a similar proportion of "disgust" emotion for all treatments across the before- and after-tasting conditions. Generally, females are likely to experience the "disgust" emotion more than males due to a higher disgust sensitivity [74]. On the other hand, the female group exhibited a decrease in the "enthusiastic" emotion upon tasting for all treatments and decreased "free" frequency for CBW+. The "good" emotion occurred more frequently in the after-tasting condition than in the before-tasting condition for CBWO+ among the female group and for CBW+ among the male group. Other studies have reported a lower likelihood for acceptability and/or willingness to consume edible insects for females [46,52,83] than for males.

Both genders exhibited an overall negative response towards all treatments upon tasting, which was evidenced by a decreased frequency of the "interested" emotion after-tasting when compared to the before-tasting condition. This behavior was possibly driven by a generalized negative state upon tasting disconfirmation regarding flavor, texture, or aroma characteristics that may or may not have affected the treatments' likings but decreased their "interest" feeling. Alternatively, their curiosity regarding the sensory profile of samples or their identity was satisfied/deciphered upon tasting and their initial interest (before tasting) was mostly related to verifying their expectations. The "wild" and "worried" terms significantly decreased upon tasting only for the female group when presenting either formulation under ECP+ disclosed information. Schouteten et al. [84] reported that consumers elicited fewer negative emotions upon tasting insect-based burgers (insect ingredient was disclosed) when compared to the expected condition (ingredient was disclosed but no tasting took place), which supports our findings for the female group emotions of "worried" and "wild" towards CBWO+ and CBW+ upon tasting.

An overall positive effect of ECP+ for both genders was observed regarding "joyful" and "pleasant" positive-strong-valence emotions. The male group had an increased occurrence of the "joyful" and "mild" emotions upon tasting for CBW+ and CBWO−, respectively whereas in the female group, "pleasant" and "safe" emotions significantly decreased upon tasting for CBW− but, for CBWO+ and CBW+, the "pleasant" emotion increased significantly after tasting. The male group had a higher frequency of "pleasant" after tasting than before tasting only for CBWO+. Moreover, all treatments presented an increased frequency of "satisfied" emotion upon tasting for both genders (except for CBW− for the female group). King et al. [83] reported that males' acceptability for food products was associated with "satisfied" and "disgust" emotions whereas for females "joyful," "good," "happy," "pleasant," and "disgusted" were accentuated out of the 25 emotions associated with acceptability.

3.4. Relationship between Product-Evoked Emotions and Liking

3.4.1. Effect of Before-Tasting Emotional Profiles on Expected OL by Gender

Elicited emotions from the female and male groups in the before-tasting condition responsible for a significant ($p < 0.05$) effect on the expected OL of treatments are shown in Figure 5A,B, respectively. In the female group, the expected OL presented fewer and different significant emotion terms for either formulation when presented under the ECP+ disclosed information. Although ECP+ triggered a variety of emotions in both formulations, only a few of them significantly affected the expected OL [80]. Different formulations presented under the same disclosed information presented almost the same significant emotion terms. The emotions "happy," "good," "satisfied," "pleasant," and "safe," for CBWO−, and "happy," "safe," "good," and "pleasant" for CBW− positively affected the expected OL. Critical emotions for CBWO− and CBW− lie in the positive valence (pleasantness) dimension, which is strongly associated with product liking [80,84,85] and choice when involved with tasting [33] albeit "safe" is considered both, a positive and low activation/arousal emotion.

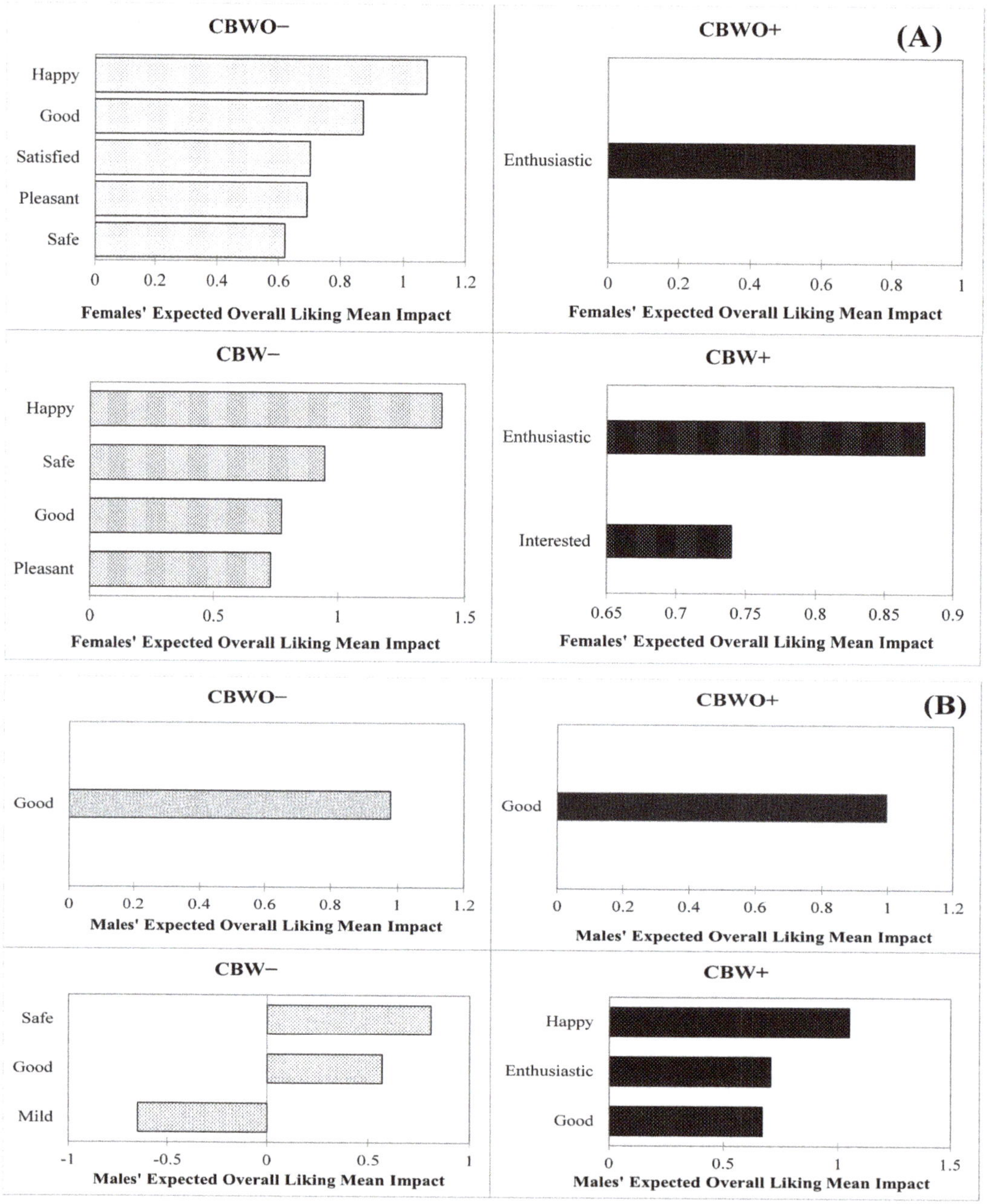

Figure 5. Treatments before-tasting overall liking (OL) mean impact (mean OL difference from present vs. absent categories for each emotion with a 20% population threshold size) vs. significant ($p < 0.05$, 2-sample t-test) emotions in the before-tasting condition (%) for (**A**) female ($n = 112$) and (**B**) male ($n = 98$) groups. Treatments are described in Figure 1.

On the other hand, when both formulations were presented under the ECP+ condition, the "enthusiastic" (for CBWO+), and the "enthusiastic" and "interested" (for CBW+) emotions positively affected the expected OL. These feelings belong to the "sensation seeking" [86] emotions lying on the high activation/arousal dimension. High activation/arousal emotions together with liking, and valence emotions have strong predictive power for product choice based on extrinsic cues [33], but on their own, they are associated with the motivation state of wanting rather than liking [85]. For example, when feeling hungry, subjects tend to experience arousal emotions that assist in the food search. Contrariwise, low levels of emotional arousal are closely related to less food consumption [82].

This suggests that the female group may have perceived differences in the appearance between formulations, which made the term "safe" a more critical attribute for the expected OL of CBW− than for CBWO− and "interested" for CBW+ than for CBWO+.

Among the male group, emotions affecting the expected OL differed across the disclosed information only for the CBW treatments. The "good," (positive-valence emotion) "safe," and "mild" (low activation/arousal emotions) significantly affected CBW− expected OL [33]. The "happy" emotion (positively associated with the pleasantness dimension) enhanced CBW+ expected OL the most followed by the "enthusiastic" (high activation/arousal emotion) and the "good" (positive emotion) terms. Differences in critical emotion terms across formulations presented under the same disclosed information were due to the extra emotion terms present for CBW− ("safe" and "mild") and CBW+ ("happy" and "enthusiastic"). The "good" positive emotion term positively affected the expected OL the most for CBWO for either disclosed information but was also critical for CBW− and CBW+.

3.4.2. Effect of After-Tasting Emotional Profiles on Actual OL by Gender

Figure 6A,B illustrate elicited emotions in the after-tasting condition from the female and male groups that significantly ($p < 0.05$) affected treatments' actual OL, respectively. Among the female group, the actual OL presented similar significant emotion terms across formulations for either disclosed information but for CBW treatments, fewer critical emotion terms affected the actual OL. When comparing across disclosed information, the "safe" and "mild" low activation/arousal emotions positively and negatively affected CBWO− and CBWO+ actual OL, respectively, while "happy" and "pleasant" positively affected CBW− and CBW+ actual OL for the female group, respectively. CBWO+ "mild" sensation reduced its actual OL possibly because female participants expected extravagant flavors or aroma from ECP, which were disconfirmed upon tasting. However, the disconfirmation experienced for CBW− did not elicit emotions that significantly inhibited its actual OL (considering a 20% selection threshold to evaluate significance). Although CBW− and CBW+ presented the lowest actual OL (5.46 and 5.90, respectively) within the female group (Figure 2) none of the after-tasting elicited emotions were significant inhibitors for it; on the contrary, the significant drivers for CBW− and CBW+ actual OL were all positive emotions in the valence continuum [87]. Product liking sometimes does not correlate well with emotions; products exhibiting low OL may elicit positive emotions and vice versa [80,83]. Nevertheless, liking and emotions together can better explain consumption behavior and food choices [76,88].

Among the male group, treatments presented the same drivers for actual OL except for CBW− ("interested" was not a significant OL driver), which belong to the positive valence dimension representing pleasantness and to the high activation/arousal dimension in the case of the "interested" emotion. The actual OL drivers for CBWO− and CBWO+ had the same order of importance whereas the order differed for CBW− and CBW+. These results further support the observed similarity in the male group's actual OL (Figure 2) across treatments given that they shared similar critical drivers for the actual OL. Gutjar et al. [80] stated that emotions are weakly correlated with product acceptability because they provide further information not explained by liking. Hence, positive-valence emotions associated with pleasantness are common drivers of liking whereas low or high activation/arousal emotions are not associated with OL. This represents an interesting orthogonal dimension to liking that should be further explored to better understand consumers' perceptions and behaviors [81].

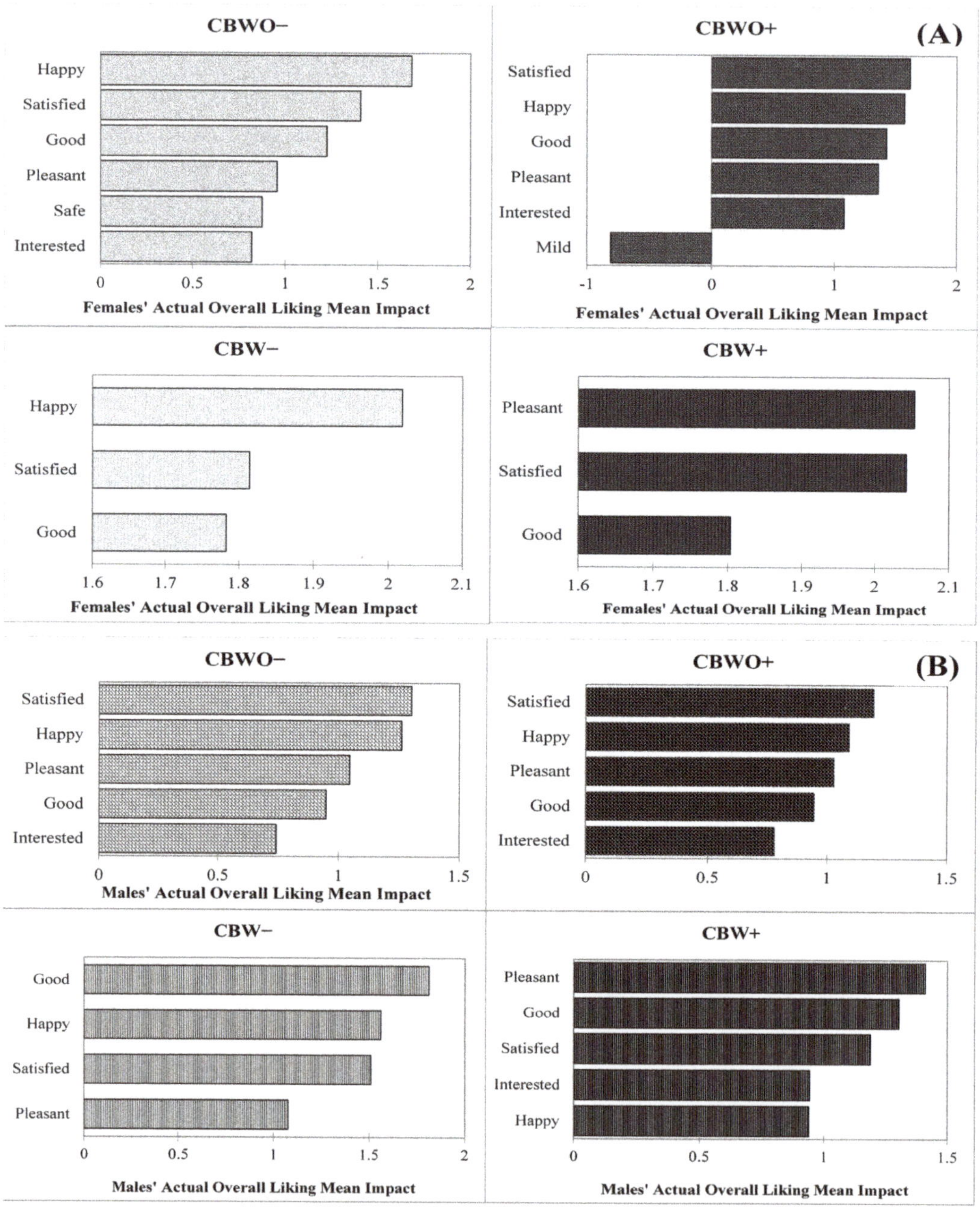

Figure 6. Treatments after-tasting overall liking (OL) mean impact (mean OL difference from present vs. absent categories for each emotion with a 20% population threshold size) vs. significant ($p < 0.05$, 2-sample t-test) emotions in the after-tasting condition (%) for (**A**) female ($n = 112$) and (**B**) male ($n = 98$) groups. Treatments are described in Figure 1.

3.5. Purchase Intent (PI) Predictive Importance of Socio-Demographic and Experimental Variables, Product-Evoked Emotions and Liking

The importance of measuring elicited emotions and their associations with product acceptability, consumption intent (CI), and PI has been emphasized because they provide information beyond liking about consumers' eating behaviors [77,78,81,89].

The performance of a random-forest PI prediction model using demographic variables (Table 1), likings, emotions, and experimental design variables as input is presented in Figure 7. The variables' importance derived from this model with an out-of-bag misclassification error rate of 14.64% is presented in Figure 8. Consumption intent [50], overall flavor liking, overall liking, texture liking [90], race [89], education level [91], and expected texture liking were among the top 10 most important variables for the correct prediction of PI as determined by mean decrease in classification accuracy and mean decrease in node impurity when the variable is permuted and split, respectively. "Satisfied" and "happy" after-tasting positive-valence emotions [33,80,84,85] and age [46,51] were critical for accurate PI prediction whereas expected and actual aroma liking, and appearance liking were critical PI predictors to obtain higher node purity. Although previous edible insect consumption [50,90], formulation [8,65], gender [8,50,52,61,79], disclosed information [8,28,57,83], after-tasting disgust [22,59] and worried were considered important for the PI prediction, the aforementioned variables were more critical to determine consumers' PI. Based on this model, the probability of purchase is higher for the consumer who is willing to consume the product upon tasting (CI = Yes), is Latino, has achieved or is pursuing a higher education degree, is satisfied and happy upon tasting, and is aged 18–29 years old. Additionally, the higher his/her liking ratings for actual overall flavor, OL, texture, aroma, appearance, and expected texture and aroma liking, the more likely the consumer purchases the product. These results suggest that marketing strategies should target consumers who match this ideal "profile," as they are more likely to purchase CB containing ECP. Furthermore, these results highlight the importance of sensory profile optimization for products containing ECP and appropriate benefits communication that evoke positive valence emotions known to improve overall acceptability and PI.

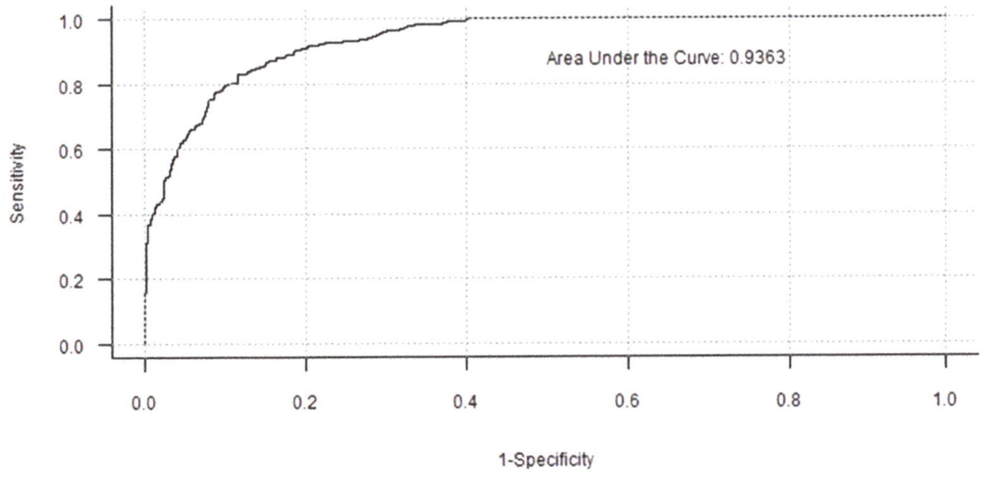

Figure 7. Receiver Operating Characteristic (ROC) curve illustrating the area under the curve (AUC) for the random forest classifier.

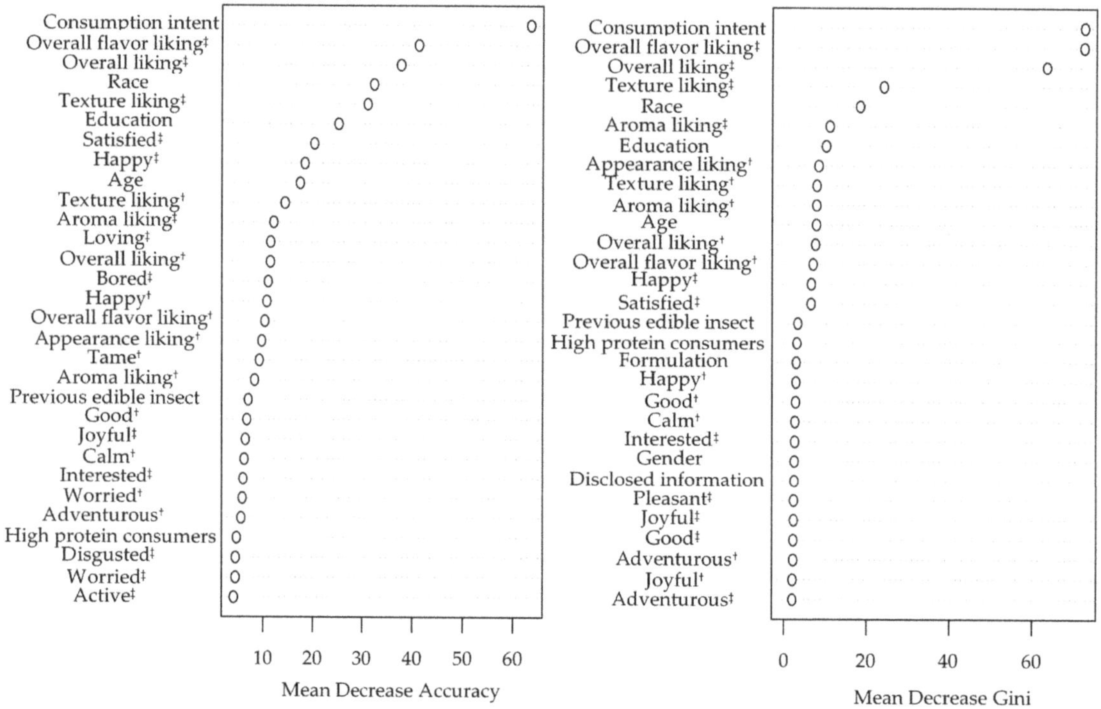

Figure 8. Random forest classifier variables importance plots for purchase intent (PI) prediction. † Before-tasting condition; ‡ after-tasting condition.

4. Study Limitations

The screening and recruitment of participants for this study did not follow a prespecified demographic criterion except for the minimum required age to participate in studies involving human subjects according to the guidelines of the Institutional Review Board of Louisiana State University Agricultural Center (IRB # HE 18-9 and IRB # HE 18-22). Hence, the participants' age and race distributions (Table 1) neither are equally balanced nor reflect the actual distribution of the US population; therefore, the findings from this study should be interpreted with caution and should not be generalized for the entire population. Likewise, the sample size in this study (n = 210) was not large enough to represent the entire US population; thus, a study with a much larger sample size is needed to confirm our findings. Lastly, because this study used only one product (chocolate brownies) and one concentration (6% *w/w*) of a commercial cricket protein powder, different results can be expected if different test samples and other concentrations or sources of cricket protein powder are to be used.

5. Conclusions and Future Studies

A better understanding of consumers' attitudes toward ECP and recommended approaches for incorporating edible insects into foods were achieved in this study. Actual OL was more affected by formulation than by disclosed information among the female group (showing higher acceptability for CBWO than for CBW) whereas the male group's actual OL was similar across all treatments. Yet, the female group presented significant negative disconfirmation upon tasting only for CBW−. Disclosed information had a greater effect than formulation on product-evoked emotions (before and after tasting) with "happy," "satisfied," "good," "pleasant," and "interested" being significant drivers for actual OL in both genders whereas "mild" inhibited actual OL among the female group. Consumption intent, race, education level, positive-valence after-tasting emotions, age, and sensory liking profile were top determinants for PI prediction. Because entomophagy is a new concept for the Western culture, information regarding the consumption of edible insects, including safety (potential biological, chemical, and physical hazards), environmental impact, and nutritional benefits may improve familiarity and alleviate aversion to entomophagy. Thus, our findings may guide future development of products incorporated with ECP for the Westerner diets. From our findings, we suggest that marketing strategies for ECP bakery applications target younger Latinos with higher education as they are more likely to purchase products containing ECP. According to our results, ECP acceptability can be improved through an appropriate food application and context for ECP whose formulation is optimized for sensory liking and emphasizing benefits from ECP consumption, which in turn evokes positive-valence emotions such as "happy" and "satisfied" that positively affect OL and PI. This relationship is important to the food industry to guide them in the development and marketing of foods containing edible insects, particularly for baked goods containing ECP. Product-elicited emotions (whose distribution in the before-tasting condition was independent of gender for CBW+ but associated with gender in the after-tasting condition for CBW+) add predictive power to solely liking ratings to understand consumers' PI behavior. This may guide the food industry in the development of "unique" products different from the ones existing in the market but with similar liking.

In the future, we recommend a consumer-based descriptive analysis to correlate the observed results with sensory descriptors and obtain additional insight as to what other sensory attribute may also affect product liking, consumers' emotions, and PI.

Author Contributions: Conceptualization: C.E.G., D.D.T., B.L., G.T. and W.P.; Performing the research work: C.E.G.; Data analysis and interpretation: C.E.G., D.D.T., B.L., G.T. and W.P.; Writing—original draft preparation: C.E.G.; Writing—review and editing: C.E.G., D.D.T., B.L., G.T. and W.P.; Supervision: W.P.; Project administration and resource/funding acquisition: W.P. All authors have read and agreed to the published version of the manuscript.

Funding: This work was partially supported by the USDA National Institute of Food and Agriculture Hatch project (Accession No. 1022270; Hatch Project No. LAB94473) and the LSU AgCenter.

Institutional Review Board Statement: The study was conducted according to the guidelines of the Declaration of Helsinki and approved by the Institutional Review Board of Louisiana State University Agricultural Center (IRB # HE 18-9 and IRB # HE 18-22).

Informed Consent Statement: Informed consent was obtained from all subjects involved in the study.

Data Availability Statement: The data that support the findings of this study are available from the corresponding author upon reasonable request.

Acknowledgments: We acknowledge the graduate students under the direct supervision of the corresponding author in the Sensory Services Lab, LSU AgCenter for assistance with the lab work.

Conflicts of Interest: All authors declare no conflict of interest. The funders had no role in the design of the study; in the collection, analyses, or interpretation of data; in the writing of the manuscript, or in the decision to publish the results.

References

1. Gurdian, C.; Reyes, V.; Kyereh, E.; Bonilla, F.; Galindo, C.; Chouljenko, A.; Solval, K.M.; Boeneke, C.; King, J.M.; Sathivel, S. Incorporating flaxseed (linum usitatissimum) oil into queso blanco at different stages of the cheese manufacturing process. *J. Food Process. Preserv.* **2017**, *41*, e13279. [CrossRef]
2. Gurdian, C.; Chouljenko, A.; Solval, K.M.; Boeneke, C.; King, J.M.; Sathivel, S. Application of edible films containing oregano (*Origanum vulgare*) essential oil on Queso Blanco cheese prepared with flaxseed (*Linum usitatissimum*) Oil. *J. Food Sci.* **2017**, *82*, 1395–1401. [CrossRef] [PubMed]
3. Ghosh, S.; Suri, D.; Uauy, R. Assessment of protein adequacy in developing countries: Quality matters. *Br. J. Nutr.* **2012**, *108*, S77–S87. [CrossRef] [PubMed]
4. Wu, G.; Fanzo, J.; Miller, D.D.; Pingali, P.; Post, M.; Steiner, J.L.; Thalacker-Mercer, A.E. Production and supply of high-quality food protein for human consumption: Sustainability, challenges, and innovations. *Ann. N. Y. Acad. Sci.* **2014**, *1321*, 1–19. [CrossRef] [PubMed]

5. Imathiu, S. Benefits and food safety concerns associated with consumption of edible insects. *NFS J.* **2020**, *18*, 1–11. [CrossRef]
6. Melgar-Lalanne, G.; Hernández-Álvarez, A.J.; Salinas-Castro, A. Edible insects processing: Traditional and innovative technologies. *Compr. Rev. Food Sci. Food Saf.* **2019**, *18*, 1166–1191. [CrossRef]
7. Montowska, M.; Kowalczewski, P.Ł.; Rybicka, I.; Fornal, E. Nutritional value, protein and peptide composition of edible cricket powders. *Food Chem.* **2019**, *289*, 130–138. [CrossRef]
8. Gurdian, C.E.; Torrico, D.D.; Li, B.; Tuuri, G.; Prinyawiwatkul, W. Effect of Informed Conditions on Sensory Expectations and Actual Perceptions: A Case of Chocolate Brownies Containing Edible-Cricket Protein. *Foods* **2021**, *10*, 1480. [CrossRef]
9. Murefu, T.; Macheka, L.; Musundire, R.; Manditsera, F. Safety of wild harvested and reared edible insects: A review. *Food Control* **2019**, *101*, 209–224. [CrossRef]
10. Delicato, C.; Schouteten, J.J.; Dewettinck, K.; Gellynck, X.; Tzompa-Sosa, D.A. Consumers' perception of bakery products with insect fat as partial butter replacement. *Food Qual. Prefer.* **2020**, *79*, 103755. [CrossRef]
11. Ardoin, R.; Romero, R.; Marx, B.; Prinyawiwatkul, W. Exploring New and Modified Rejection-Type Thresholds Using Cricket Snack Crackers. *Foods* **2020**, *9*, 1352. [CrossRef]
12. Ardoin, R.; Marx, B.D.; Boeneke, C.; Prinyawiwatkul, W. Effects of cricket powder on selected physical properties and US consumer perceptions of whole-wheat snack crackers. *Int. J. Food Sci. Technol.* **2021**, in press. [CrossRef]
13. Zielińska, E.; Pankiewicz, U.; Sujka, M. Nutritional, Physiochemical, and Biological Value of Muffins Enriched with Edible Insects Flour. *Antioxidants* **2021**, *10*, 1122. [CrossRef]
14. Biró, B.; Sipos, M.A.; Kovács, A.; Badak-Kerti, K.; Pásztor-Huszár, K.; Gere, A. Cricket-Enriched Oat Biscuit: Technological Analysis and Sensory Evaluation. *Foods* **2020**, *9*, 1561. [CrossRef] [PubMed]
15. Osimani, A.; Milanović, V.; Cardinali, F.; Roncolini, A.; Garofalo, C.; Clementi, F.; Pasquini, M.; Mozzon, M.; Foligni, R.; Raffaelli, N.; et al. Bread enriched with cricket powder (Acheta domesticus): A technological, microbiological and nutritional evaluation. *Innov. Food Sci. Emerg. Technol.* **2018**, *48*, 150–163. [CrossRef]
16. Adámek, M.; Adámková, A.; Mlček, J.; Borkovcová, M.; Bednářová, M. Acceptability and sensory evaluation of energy bars and protein bars enriched with edible insect. *Potr. Slovak J. Food Sci.* **2018**, *12*, 431–437. [CrossRef]
17. Hirunyophat, P.; Chalermchaiwat, P.; On-nom, N.; Prinyawiwatkul, W. Selected nutritional quality and physicochemical properties of silkworm pupae (frozen or powdered) from two species. *Int. J. Food Sci. Technol.* **2021**, *56*, 3578–3587. [CrossRef]
18. Duda, A.; Adamczak, J.; Chełmińska, P.; Juszkiewicz, J.; Kowalczewski, P. Quality and Nutritional/Textural Properties of Durum Wheat Pasta Enriched with Cricket Powder. *Foods* **2019**, *8*, 46. [CrossRef]
19. Megido, R.C.; Gierts, C.; Blecker, C.; Brostaux, Y.; Haubruge, É.; Alabi, T.; Francis, F. Consumer acceptance of insect-based alternative meat products in Western countries. *Food Qual. Prefer.* **2016**, *52*, 237–243. [CrossRef]
20. Smarzyński, K.; Sarbak, P.; Musiał, S.; Jeżowski, P.; Piątek, M.; Kowalczewski, P.Ł. Nutritional analysis and evaluation of the consumer acceptance of pork pâté enriched with cricket powder—Preliminary study. *Open Agric.* **2019**, *4*, 159–163. [CrossRef]
21. Ardoin, R.; Prinyawiwatkul, W. Consumer perceptions of insect consumption: A review of western research since 2015. *Int. J. Food Sci. Technol.* **2021**, in press. [CrossRef]
22. La Barbera, F.; Verneau, F.; Amato, M.; Grunert, K. Understanding Westerners' disgust for the eating of insects: The role of food neophobia and implicit associations. *Food Qual. Prefer.* **2018**, *64*, 120–125. [CrossRef]
23. Verneau, F.; La Barbera, F.; Kolle, S.; Amato, M.; Del Giudice, T.; Grunert, K. The effect of communication and implicit associations on consuming insects: An experiment in Denmark and Italy. *Appetite* **2016**, *106*, 30–36. [CrossRef]
24. Barsics, F.; Megido, R.C.; Brostaux, Y.; Barsics, C.; Blecker, C.; Haubruge, E.; Francis, F. Could new information influence attitudes to foods supplemented with edible insects? *Br. Food J.* **2017**, *119*, 2027–2039. [CrossRef]
25. Schäufele, I.; Albores, E.B.; Hamm, U. The role of species for the acceptance of edible insects: Evidence from a consumer survey. *Br. Food J.* **2019**, *121*, 2190–2204. [CrossRef]
26. Orsi, L.; Voege, L.L.; Stranieri, S. Eating edible insects as sustainable food? Exploring the determinants of consumer acceptance in Germany. *Food Res. Int.* **2019**, *125*, 108573. [CrossRef]
27. Sogari, G.; Bogueva, D.; Marinova, D. Australian consumers' response to insects as food. *Agriculture* **2019**, *9*, 108. [CrossRef]
28. Ardoin, R.; Prinyawiwatkul, W. Product appropriateness, willingness to try and perceived risks of foods containing insect protein powder: A survey of U.S. consumers. *Int. J. Food Sci. Technol.* **2020**, *55*, 3215–3226. [CrossRef]
29. Petersen, M.; Olson, O.; Rao, S. University Student Perspectives of Entomophagy: Positive Attitudes Lead to Observability and Education Opportunities. *J. Insect Sci.* **2020**, *20*, 30. [CrossRef]
30. Tan, H.S.G.; van den Berg, E.; Stieger, M. The influence of product preparation, familiarity and individual traits on the consumer acceptance of insects as food. *Food Qual. Prefer.* **2016**, *52*, 222–231. [CrossRef]
31. Cunha, L.M.; Ribeiro, J.C. Sensory and consumer perspectives on edible insects. In *Edible Insects in the Food Sector*; Springer: Cham, Switzerland, 2019; pp. 57–71.
32. Fischer, A.R.; Steenbekkers, L.B. All insects are equal, but some insects are more equal than others. *Br. Food J.* **2018**, *120*, 852–863. [CrossRef]
33. Gutjar, S.; Dalenberg, J.R.; de Graaf, C.; de Wijk, R.A.; Palascha, A.; Renken, R.J.; Jager, G. What reported food-evoked emotions may add: A model to predict consumer food choice. *Food Qual. Prefer.* **2015**, *45*, 140–148. [CrossRef]
34. Gurdian, C.E.; Torrico, D.D.; Li, B.; Prinyawiwatkul, W. Effect of serving plate types and color cues on liking and purchase intent of cheese-flavored tortilla chips. *Foods* **2021**, *10*, 886. [CrossRef]

35. Granitto, P.M.; Gasperi, F.; Biasioli, F.; Trainotti, E.; Furlanello, C. Modern data mining tools in descriptive sensory analysis: A case study with a Random forest approach. *Food Qual. Prefer.* **2007**, *18*, 681–689. [CrossRef]
36. Denuit, M.; Hainaut, D.; Trufin, J. Bagging Trees and Random Forests. In *Effective Statistical Learning Methods for Actuaries II*; Springer: Cham, Switzerland, 2020; pp. 107–130.
37. Mora, M.; Romeo-Arroyo, E.; Torán-Pereg, P.; Chaya, C.; Vázquez-Araújo, L. Sustainable and health claims vs sensory properties: Consumers' opinions and choices using a vegetable dip as example product. *Food Res. Int.* **2020**, 109521. [CrossRef] [PubMed]
38. Carabante, K.M.; Ardoin, R.; Scaglia, G.; Malekian, F.; Khachaturyan, M.; Janes, M.E.; Prinyawiwatkul, W. Consumer acceptance, emotional response, and purchase intent of rib-eye steaks from grass-fed steers, and effects of health benefit information on consumer perception. *J. Food Sci.* **2018**, *83*, 2560–2570. [CrossRef]
39. Sukkwai, S.; Chonpracha, P.; Kijroongrojana, K.; Prinyawiwatkul, W. Influences of a natural colourant on colour and salty taste perception, liking, emotion and purchase intent: A case of mayonnaise-based dipping sauces. *Int. J. Food Sci. Technol.* **2017**, *52*, 2256–2264. [CrossRef]
40. Samant, S.S.; Seo, H.-S. Quality perception and acceptability of chicken breast meat labeled with sustainability claims vary as a function of consumers' label-understanding level. *Food Qual. Prefer.* **2016**, *49*, 151–160. [CrossRef]
41. Addinsoft. XLSTAT Statistical and Data Analysis Solution. Available online: https://www.xlstat.com (accessed on 15 June 2021).
42. Meyners, M.; Castura, J.C.; Carr, B.T. Existing and new approaches for the analysis of CATA data. *Food Qual. Prefer.* **2013**, *30*, 309–319. [CrossRef]
43. Ares, G.; Dauber, C.; Fernández, E.; Giménez, A.; Varela, P. Penalty analysis based on CATA questions to identify drivers of liking and directions for product reformulation. *Food Qual. Prefer.* **2014**, *32*, 65–76. [CrossRef]
44. Sheskin, D.J. *Handbook of Parametric and Nonparametric Statistical Procedures*, 5th ed.; CRC Press: Boca Raton, FL, USA, 2020.
45. Breiman, L. Random forests. *Mach. Learn.* **2001**, *45*, 5–32. [CrossRef]
46. Hartmann, C.; Shi, J.; Giusto, A.; Siegrist, M. The psychology of eating insects: A cross-cultural comparison between Germany and China. *Food Qual. Prefer.* **2015**, *44*, 148–156. [CrossRef]
47. La Barbera, F.; Verneau, F.; Coppola, A. Entomophagy: A contribution to the understanding of consumer intention. *Calitatea* **2019**, *20*, 329–334.
48. Mishyna, M.; Chen, J.; Benjamin, O. Sensory attributes of edible insects and insect-based foods–Future outlooks for enhancing consumer appeal. *Trends Food Sci. Technol.* **2020**, *95*, 141–148. [CrossRef]
49. Lammers, P.; Ullmann, L.M.; Fiebelkorn, F. Acceptance of insects as food in Germany: Is it about sensation seeking, sustainability consciousness, or food disgust? *Food Qual. Prefer.* **2019**, *77*, 78–88. [CrossRef]
50. Sogari, G.; Menozzi, D.; Mora, C. The food neophobia scale and young adults' intention to eat insect products. *Int. J. Consum. Stud.* **2019**, *43*, 68–76. [CrossRef]
51. Verbeke, W. Profiling consumers who are ready to adopt insects as a meat substitute in a Western society. *Food Qual. Prefer.* **2015**, *39*, 147–155. [CrossRef]
52. Woolf, E.; Zhu, Y.; Emory, K.; Zhao, J.; Liu, C. Willingness to consume insect-containing foods: A survey in the United States. *Lwt* **2019**, *102*, 100–105. [CrossRef]
53. Schardong, I.S.; Freiberg, J.A.; Santana, N.A.; Richards, N.S.P.d.S. Brazilian consumers' perception of edible insects. *Cienc. Rural* **2019**, *49*, e20180960. [CrossRef]
54. Hartmann, C.; Siegrist, M. Becoming an insectivore: Results of an experiment. *Food Qual. Prefer.* **2016**, *51*, 118–122. [CrossRef]
55. Ammann, J.; Hartmann, C.; Siegrist, M. Does food disgust sensitivity influence eating behaviour? Experimental validation of the Food Disgust Scale. *Food Qual. Prefer.* **2018**, *68*, 411–414. [CrossRef]
56. Mancini, S.; Sogari, G.; Menozzi, D.; Nuvoloni, R.; Torracca, B.; Moruzzo, R.; Paci, G. Factors predicting the intention of eating an insect-based product. *Foods* **2019**, *8*, 270. [CrossRef]
57. Gross, S.; Waldrop, M.E.; Roosen, J. How does animal welfare taste? Combining sensory and choice experiments to evaluate willingness to pay for animal welfare pork. *Food Qual. Prefer.* **2021**, *87*, 104055. [CrossRef]
58. Anderson, R.E. Consumer Dissatisfaction: The Effect of Disconfirmed Expectancy on Perceived Product Performance. *J. Mark. Res.* **1973**, *10*, 38–44. [CrossRef]
59. Ruby, M.B.; Rozin, P.; Chan, C. Determinants of willingness to eat insects in the USA and India. *J. Insects Food Feed* **2015**, *1*, 215–225. [CrossRef]
60. Everitt, M. Consumer-targeted sensory quality. In *Global Issues in Food Science and Technology*; Elsevier: Amsterdam, The Netherlands, 2009; pp. 117–128.
61. Nestrud, M.A.; Meiselman, H.L.; King, S.C.; Lesher, L.L.; Cardello, A.V. Development of EsSense25, a shorter version of the EsSense Profile®. *Food Qual. Prefer.* **2016**, *48*, 107–117. [CrossRef]
62. Mora, M.; Urdaneta, E.; Chaya, C. Emotional response to wine: Sensory properties, age and gender as drivers of consumers' preferences. *Food Qual. Prefer.* **2018**, *66*, 19–28. [CrossRef]
63. Norbury, A.; Husain, M. Sensation-seeking: Dopaminergic modulation and risk for psychopathology. *Behav. Brain Res.* **2015**, *288*, 79–93. [CrossRef]
64. Gere, A.; Székely, G.; Kovács, S.; Kókai, Z.; Sipos, L. Readiness to adopt insects in Hungary: A case study. *Food Qual. Prefer.* **2017**, *59*, 81–86. [CrossRef]

65. Tan, H.S.G.; Verbaan, Y.T.; Stieger, M. How will better products improve the sensory-liking and willingness to buy insect-based foods? *Food Res. Int.* **2017**, *92*, 95–105. [CrossRef]
66. Dobermann, D.; Swift, J.; Field, L. Opportunities and hurdles of edible insects for food and feed. *Nutr. Bull.* **2017**, *42*, 293–308. [CrossRef]
67. Belluco, S.; Halloran, A.; Ricci, A. New protein sources and food legislation: The case of edible insects and EU law. *Food Secur.* **2017**, *9*, 803–814. [CrossRef]
68. Rumpold, B.A.; Langen, N. Consumer acceptance of edible insects in an organic waste-based bioeconomy. *Curr. Opin. Green Sustain. Chem.* **2020**, *23*, 80–84. [CrossRef]
69. Woolf, E.; Maya, C.; Yoon, J.; Shertukde, S.; Toia, T.; Zhao, J.; Zhu, Y.; Peter, P.; Liu, C. Information and taste interventions for improving consumer acceptance of edible insects: A pilot study. *J. Insects Food Feed* **2020**, *7*, 129–139. [CrossRef]
70. Köster, E.P.; Mojet, J. From mood to food and from food to mood: A psychological perspective on the measurement of food-related emotions in consumer research. *Food Res. Int.* **2015**, *76*, 180–191. [CrossRef]
71. Schouteten, J.J.; De Steur, H.; De Pelsmaeker, S.; Lagast, S.; De Bourdeaudhuij, I.; Gellynck, X. Impact of health labels on flavor perception and emotional profiling: A consumer study on cheese. *Nutrients* **2015**, *7*, 10251–10268. [CrossRef]
72. King, S.C.; Meiselman, H.L. Development of a method to measure consumer emotions associated with foods. *Food Qual. Prefer.* **2010**, *21*, 168–177. [CrossRef]
73. Sogari, G.; Menozzi, D.; Hartmann, C.; Mora, C. How to Measure Consumers Acceptance Towards Edible Insects?—A Scoping Review About Methodological Approaches. In *Edible Insects in the Food Sector: Methods, Current Applications and Perspectives*; Sogari, G., Mora, C., Menozzi, D., Eds.; Springer International Publishing: Cham, Switzerland, 2019; pp. 27–44.
74. Hamerman, E.J. Cooking and disgust sensitivity influence preference for attending insect-based food events. *Appetite* **2016**, *96*, 319–326. [CrossRef]
75. Berger, S.; Christandl, F.; Bitterlin, D.; Wyss, A.M. The social insectivore: Peer and expert influence affect consumer evaluations of insects as food. *Appetite* **2019**, *141*, 104338. [CrossRef]
76. Cardello, A.V.; Meiselman, H.L.; Schutz, H.G.; Craig, C.; Given, Z.; Lesher, L.L.; Eicher, S. Measuring emotional responses to foods and food names using questionnaires. *Food Qual. Prefer.* **2012**, *24*, 243–250. [CrossRef]
77. Jiang, Y.; King, J.M.; Prinyawiwatkul, W. A review of measurement and relationships between food, eating behavior and emotion. *Trends Food Sci. Technol.* **2014**, *36*, 15–28. [CrossRef]
78. Tuccillo, F.; Marino, M.G.; Torri, L. Italian consumers' attitudes towards entomophagy: Influence of human factors and properties of insects and insect-based food. *Food Res. Int.* **2020**, *137*, 109619. [CrossRef] [PubMed]
79. Ng, M.; Chaya, C.; Hort, J. Beyond liking: Comparing the measurement of emotional response using EsSense Profile and consumer defined check-all-that-apply methodologies. *Food Qual. Prefer.* **2013**, *28*, 193–205. [CrossRef]
80. Gutjar, S.; de Graaf, C.; Kooijman, V.; de Wijk, R.A.; Nys, A.; Ter Horst, G.J.; Jager, G. The role of emotions in food choice and liking. *Food Res. Int.* **2015**, *76*, 216–223. [CrossRef]
81. Spinelli, S.; Masi, C.; Dinnella, C.; Zoboli, G.P.; Monteleone, E. How does it make you feel? A new approach to measuring emotions in food product experience. *Food Qual. Prefer.* **2014**, *37*, 109–122. [CrossRef]
82. Kim, J.-Y.; Prescott, J.; Kim, K.-O. Emotional responses to sweet foods according to sweet liker status. *Food Qual. Prefer.* **2017**, *59*, 1–7. [CrossRef]
83. King, S.C.; Meiselman, H.L.; Carr, B.T. Measuring emotions associated with foods in consumer testing. *Food Qual. Prefer.* **2010**, *21*, 1114–1116. [CrossRef]
84. Schouteten, J.J.; De Steur, H.; De Pelsmaeker, S.; Lagast, S.; Juvinal, J.G.; De Bourdeaudhuij, I.; Verbeke, W.; Gellynck, X. Emotional and sensory profiling of insect-, plant-and meat-based burgers under blind, expected and informed conditions. *Food Qual. Prefer.* **2016**, *52*, 27–31. [CrossRef]
85. Berridge, K.C. 'Liking'and 'wanting'food rewards: Brain substrates and roles in eating disorders. *Physiol. Behav.* **2009**, *97*, 537–550. [CrossRef]
86. Kim, J.; Almanza, B.; Ghiselli, R.; Sydnor, S. The effect of sensation seeking and emotional brand attachment on consumers' intention to consume risky foods in restaurants. *J. Foodserv. Bus. Res.* **2017**, *20*, 336–349. [CrossRef]
87. Lagast, S.; Gellynck, X.; Schouteten, J.J.; De Herdt, V.; De Steur, H. Consumers' emotions elicited by food: A systematic review of explicit and implicit methods. *Trends Food Sci. Technol.* **2017**, *69*, 172–189. [CrossRef]
88. Dalenberg, J.R.; Gutjar, S.; Ter Horst, G.J.; De Graaf, K.; Renken, R.J.; Jager, G. Evoked emotions predict food choice. *PLoS ONE* **2014**, *9*, e115388. [CrossRef]
89. Kauppi, S.-M.; Pettersen, I.N.; Boks, C. Consumer acceptance of edible insects and design interventions as adoption strategy. *Int. J. Food Des.* **2019**, *4*, 39–62. [CrossRef]
90. Wendin, K.; Nyberg, M. Factors influencing consumer perception and acceptability of insect-based foods. *Curr. Opin. Food Sci.* **2021**, *40*, 67–71. [CrossRef]
91. Hunts, H.; Dunkel, F.; Thienes, M.; Carnegie, N. Gatekeepers in the food industry: Acceptability of edible insects. *J. Insects Food Feed* **2020**, *6*, 231–243. [CrossRef]

foods

MDPI

Article

Studying the Emotional Response to Insects Food Products

Michelangelo Serpico [1], Dominic Rovai [2], Kristine Wilke [3], Ruta Lesniauskas [4], Jeff Garza [3] and Amy Lammert [5,*]

1 ESCP Business School–Turin Campus, Corso Unione Sovietica 218 bis, 10134 Torino, Italy; michelangelo.serpico@edu.escp.eu
2 Department of Food, Bioprocessing and Nutrition Sciences, North Carolina State University, Raleigh, NC 27606, USA; drovai@ncsu.edu
3 Garza Consulting, Grand Rapids, MI 49501, USA; kristine.wilke@the-gc.com (K.W.); jeff.garza@the-gc.com (J.G.)
4 Garza Consulting, Evanston, IL 60201, USA; ruta@the-gc.com
5 Department of Food Science and Nutrition, California Polytechnic State University, San Luis Obispo, CA 93407, USA
* Correspondence: alammert@calpoly.edu

Abstract: Insects have been proposed as a sustainable food solution due to their environmental, nutritional, and socioeconomic value; however, in the western world, insects are viewed as disgusting. This research aimed to understand the acceptance of insect-based products in the US market by studying the emotional response to such. A survey of 826 consumers was conducted using (1) a modified version of the EsSense Profile® questionnaire to capture the emotional response to pictures of different kinds of foods, (2) images to evaluate the influence of the presence or absence of non-visible insects in food products, (3) information about the environmental value of insects, and (4) socioeconomic demographics. Disgust was found as a barrier to product acceptance. Insect food products were positively correlated with the emotions of interest, understanding, daring, adventurous, and worried, and negatively correlated with the emotions satisfied, good, pleasant, happy, calm, warm, nostalgic, and secure. The influence of sustainability-related information on the emotional response to such products is complex and should be carefully considered.

Keywords: EsSense Profile®; edible insects; entomophagy; emotions; sustainability

Citation: Serpico, M.; Rovai, D.; Wilke, K.; Lesniauskas, R.; Garza, J.; Lammert, A. Studying the Emotional Response to Insects Food Products. *Foods* **2021**, *10*, 2404. https://doi.org/10.3390/foods10102404

Academic Editor: Witoon Prinyawiwatkul

Received: 9 September 2021
Accepted: 6 October 2021
Published: 11 October 2021

1. Introduction

To date, food systems face the challenge of having to adapt to emerging social changes. Per capita demand for major commodities is expected to increase significantly by 2050, as the global population is predicted to grow to 9–10 billion people. In this regard, the FAO predicts a 60% increase in food demand and a 100% increase in meat demand [1]. Moreover, a recent Intergovernmental Panel on Climate Change (IPCC) report estimated the need for a 40 to 70% reduction of total greenhouse gas (GHG) emissions to maintain the average global temperature within a 2 °C increase above pre-industrial levels [2]. One of the most significant environmental challenges for food systems is reducing GHG emissions, as food systems contribute 20–30% of all man-made emissions. The most significant contributors are farming, agriculture, land-use change, and livestock, contributing around 14 to 18% of total emissions [3,4].

In this context, insects have been proposed by various authors as a potential sustainable food source, specifically as an alternative source to common animal protein sources [3,5–7]. Strong environmental benefits are associated with insect production in comparison to other types of livestock and include (1) specific emissions of greenhouse gases (GHGs) and ammonia during rearing are considerably lower; (2) their rearing requires fewer resources, including water and land; and (3) their efficiency of converting feed into protein is higher [8–11]. Despite their recognized value, barriers still exist to adopting insects as food, especially in the Western world [8,12]. Those barriers include production problems [3,8,13], legislative matters [5,8,14], and health-related issues [12,15]. However, one of the most important barriers to insect consumption is consumer acceptance.

Some psychological factors have been found to be related to insects acceptance, such as food neophobia, disgust perception, and sensation seeking [16–18]. Disgust is recognized as one of the main barriers [18], as in the western world, insects are often seen as vectors of disease [19].

On the other hand, positive reactions to insects as food have been associated with beliefs that insects are nutritious and positive for the environment [20]. It may then be feasible to influence these beliefs through education [21,22]. For example, Mancini et al. [23] found that the provision of information regarding the nutritional, gastronomical, and environmental value of insects may lower their rejection and decrease disgust reactions.

However, insects still fail to be accepted by western consumers [24,25]. This suggests that a more complex perspective on food choice has to be taken into account, as suggested by some researchers who suggest a transdisciplinary approach involving both biological and social sciences to understand the various and complex interactions between shaping the acceptance of insect food products [26]. Most researches have strongly focused on psychological factors (such as disgust sensitivity and food neophobia) and the rational aspect of decision making (as shown by the attention given to education), while some important aspects regarding food acceptance have been overlooked, such as the role of emotions and sociocultural factors [25,27].

The importance of emotions has been recognized by researchers for a long time, and emotions are important in the decision making processes, affecting what we think (as for example, influencing our perception of risk), how we think about it (pushing for a more heuristic or systematic thinking), and implicitly shaping our decisions through behavioral responses [28]. More specifically, emotions and food consumption are two strictly related concepts, as emotions are both a consequence of food consumption and a powerful influence on our eating behavior and decision-making processes [29]. For this reason, studies evaluating emotions can provide information beyond traditional hedonic and preference measurements [30] and, therefore, can help to better understand consumer behavior and choice [31].

In the case of entomophagy, the relevance of emotions has been suggested by Onwezen et al. [32], who highlighted that affective factors could help better explain the acceptance of insects products, as consumers may rely more on affective processing given the novelty of the insects as food. Furthermore, Gmuer, Guth, Hartmann, and Siegrist [33], who studied consumers' emotional responses to pictures of products containing insects, highlighted the importance of minimizing the negative emotional reaction (mainly related to disgust), and enhancing the positive experience and expectations of consumers. Le Goff and Delarue [34] evaluated facial expressions to study emotional responses to chips containing insects. They pointed out that consumers have strongly negative expectations of this kind of products, but the actual emotional response while consuming the product is significantly less negative, again suggesting the importance of building positive emotional expectations. Finally, Tan and House [27] emphasized that strong positive experiences toward entomophagy in Thailand, where it is traditionally consumed, suggest that familiarity and positive experiences seem to be an important factor in repeated consumption.

To increase consumers' acceptance of insects foods, Looy and Wood [25] advised that a strategy should consider both the emotional and the rational dimension, integrating the understanding of consumer emotions with the provision of information and education about insects' environmental values. This kind of strategy recognizes both the role of rationality (including the importance of values and goals) and emotions in shaping the formation of attitudes and in the translations of them into behavior [35]. However, some studies suggested that the emotional response to food products strongly depends on the product category considered [27]. Additionally, the practice of informing participants about the environmental value of insects may influence the emotional response [36].

Another proposed strategy to increase the acceptability of edible insects is the incorporation of powdered insects in familiar food products, as the incorporation of insects in familiar products may decrease neophobic reactions [37]. Additionally, research suggests that making insects non-visible may encourage consumption [27,33,38,39], possibly by minimizing the disgust related to the textural and optical attributes, which are the main elicitors of the aversive response [37,40]. Other researchers have also found that blindfolded participants tend to have difficulties in identifying insects by taste alone, further supporting this approach [41].

Incorporating insects in the non-visible form may help to remove culinary barriers related to the lack of knowledge regarding insect preparation [25,27,38]. Additionally, Tan and House [27] suggest the incorporation of processed insects in familiar food as a way to give importance to the fit of insects product in established eating practices, and therefore with the possibility of making insects product easier to be incorporated into existent diet patterns.

However, to the authors' knowledge, no study has considered the practice of informing participants about the environmental value of insects in relation to the emotional response to different food categories incorporating non-visible insects. The objective of this research is therefore to understand the acceptance of different potential food products containing non-visible insects in the US market by studying emotions associated with such products, the influence of information given about the environmental value of insects, and relating these measurements to the willingness to try the products shown.

2. Materials and Methods

2.1. Recruitment and Screening

All online questionnaires were developed and completed using Red Jade software (Martinez, CA, USA). The questionnaires were distributed via social media (Facebook), email, and in-person referrals. No subject compensation was provided for participation. Prior to distribution, 51 people completed the questionnaire to evaluate the time required for questionnaire completion, and their responses were not considered for the analysis. Only data from subjects who accepted the informed consent, were over 18 years old, completed the entire questionnaire, and resided in the US were evaluated. Additionally, the time it took to participants to take the questionnaire was evaluated, and participants who completed the questionnaire too quickly were assumed not to have answered the questions seriously and were discarded, a practice that is common in the academic field [33]. A total of 1765 subjects completed the survey, but only 826 respondents met these criteria.

2.2. Experimental Design

Each participant first received a single questionnaire that was composed of three independent questionnaires. Questionnaire 1 (Age Screener) was a screener for age. Questionnaire 2 (Pictures Evaluation) was a questionnaire containing pictures of different product categories followed by a question regarding their emotions associated with the product categories, expected hedonic acceptance, and expected willingness to try these products. Questionnaire 3 (Demographics) was general sociodemographic questions.

2.2.1. Age Screener

Participants provided information about their age (17 and under, 18–24, 25–34, 35–44, 45–54, 55–64, and 65 and over).

2.2.2. Picture Evaluation

Three different factors were evaluated to understand their effect on the emotional response of participants to the idea of food concepts that may contain insects: Product Category—different food product category images; Insect Presence—information about the presence or absence of non-visible insects in the products (as shown in Product Category images); and Sustainability Information—sustainability information regarding the environmental value of insects. The Product Category had five different levels, as five different pictures were shown representing the following product categories: (1) bread and pasta, (2) proteins supplemented products, (3) processed meat products, (4) sauces and dips, and (5) snacks. Figure 1 is an example of a Product Category image (all pictures were taken by one of the authors). The Insect Presence had two levels. The first option indicated the presence of non-visible insects in the product images shown, and the second option did not indicate the presence of insects shown in the images (please note that all images were the same, only the description changed). Sustainability Information also had two levels, as participants were given or not given sustainability information about the environmental value of insects. To decrease the amount of time spent on the questionnaire and decrease the cognitive fatigue of the participants, which is especially important when measuring emotional responses [42], while still ensuring the study of all main effects and the 2-way interactions, a D-optimal fractional factorial design with 17 treatments was created using Design Expert (Version 13, Stat-Ease, Inc., Minneapolis, MN, USA). Each participant, therefore, evaluated only 2 of the 17 treatments (Table 1). Treatment 6 and 17 (Table 1) are replicates to get an estimation of the true error in the design.

Figure 1. An example picture used to represent a Product Category. This image was used to represent the snack category. All pictures were taken by one of the authors.

The Sustainability Information options were presented first, if required, by presenting three infographics that showed the environmental benefits of insects (greenhouse gas emissions, land usage, feed, and water requirements, respectively) compared to beef, pork, and poultry. These infographics were obtained from littleherds.org and used with the company's permission. There was no information provided if the treatment did not require Sustainability Information.

The image for the selected Product Category in the treatment was presented next. Preceding the image was the following, only if the treatment was to include information about the presence of non-visible insects, sentence: "The following products contain INSECTS in a NON-VISIBLE FORM". Beneath each picture was a brief description of its content; for example, the picture related to 7Sn having the following sentence: "This product category includes SNACK FOODS such as: CHIPS, CRACKERS, PUFFED SNACKS, and others".

After each treatment image, the participant was asked to indicate their feelings with respect to products they typically consumed from the designated Product Category: "Considering this category of products, imagine eating a product that you frequently consume. Please select the words which describe how you FEEL RIGHT NOW. Select all that apply". If Insect Presence Information were given, then the following phrase was added to the end of the first sentence: "[. . .] made with NON-VISIBLE INSECTS".

Table 1. The 17 treatments evaluated in this experiment. The Product Category was indicated by the first one or two letters (C for bread and pasta, M for processed meat products, P for protein supplemented products, Sa for sauces, and Sn for snacks). Sustainability Information was indicated by the letter S only if the information was provided in the treatment. The letter i indicated the Insect Presence Information only if the information was to be given. For the treatment designated 1P, the subjects saw a picture of protein supplemented products (P), with no Sustainability Information or Insect Presence Information, but for the treatment designated 9SnSi, the subjects saw a picture of snacks with both Sustainability Information and Insect Presence Information.

	Factor 1	Factor 2	Factor 3
Treatment Code	Product Category	Insect Presence Information	Sustainability Information
1P	Proteins supplemented products	No	No
2CSi	Bread and pasta	Yes	Yes
3Sa	Sauces	No	No
4CS	Bread and pasta	No	Yes
5MSi	Processed meat products	Yes	Yes
6Ci	Bread and pasta	Yes	No
7Sn	Snacks	No	No
8Sni	Snacks	Yes	No
9SnSi	Snacks	Yes	Yes
10Mi	Processed meat products	Yes	No
11MS	Processed meat products	No	Yes
12SaS	Sauces	No	Yes
13Sai	Sauces	Yes	No
14Pi	Proteins supplemented products	Yes	No
15PS	Proteins supplemented products	No	Yes
16SnS	Snacks	No	Yes
17Ci	Bread and pasta	No	Yes

The list of emotions provided was from the EsSense25 questionnaire [42], which was chosen for the following reasons: (1) it is suitable for application with a large number of participants, as it requires little time in gathering response; (2) it is easy to apply, as it does not require any specific instrumentation; (3) it is a non-specific questionnaire, therefore, it is suited for comparing products from different categories; (4) it has been already applied in online studies, and its application in this context has been proven suitable; (5) the number of emotional terms has been considered high enough to capture differences between different product categories, but not too high as to create cognitive fatigue of participants [43]; and (6) it has been proven comparably reliable to the EsSense Profile®, which has been successfully applied in a variety of situations and is extensively applied in the industry [44].

The term daring was added, taken from the EsSense Profile® questionnaire [45], as it may be an important emotion when food products are made with insects. If the participants checked the term disgusted, another question was presented to them asking them to rate their feeling of disgust (Disgust Rating) on a 5-point scale (1 = Not Disgusted at All to 5 = Extremely Disgusted). Participants also indicated the expected overall liking of a product they frequently consumed considering the category shown in the image using a 9-point hedonic scale (1 = Dislike Extremely; 9 = Like Extremely) and their willingness to try using a 5-point scale (1 = Very Unwilling; 5 = Very Willing).

2.2.3. Socioeconomic Demographics

Finally, participants answered sociodemographic-related questions, gathering information about their (1) gender, (2) residence, (3) ethnic background, and (4) educational level.

2.3. Data Analysis

2.3.1. Screener and Socioeconomic Demographics

The age responses were recoded to three categories: 18–24 and 25–34 were recoded to Millennials (18–34), 35–44 and 45–54 were recoded to Generation X (35–54), and 55–64 and 65 and over were recoded to Baby Boomers (55 and over). The percentage of respondents (RStudio, Version 1.4.1717, RStudio, PBC, Boston, MA, USA) in each category was reported.

The gender (male; female; prefer not to answer), ethnic background (White or Caucasian; Hispanic or Latino; Black or African-American; Native American or American Indian; Asian or Pacific Islander; Other), education (Some High School; High School Graduate or Equivalent; Some College; Trade, Technical, or Vocational Training; Associate Degree; Bachelor's Degree; Master's Degree; Professional Degree; Doctorate Degree), and residence (San Luis Obispo County, CA; California; Other state in United States) were computed as a percentage of respondents (RStudio, Version 1.4.1717, RStudio, PBC, Boston, MA, USA) in each category for each socioeconomic demographic.

2.3.2. Picture Evaluation

The emotional responses, collected as binary responses, were rescaled to percentages, to get a better reading of the results. Afterward, a model (RStudio, Version 1.4.1717, RStudio, PBC, Boston, MA, USA) was used to calculate the estimated percentages and means in respect to all responses except the Disgust Rating. ANOVA (RStudio, Version 1.4.1717, RStudio, PBC, Boston, MA, USA) was used considering three factors: (1) the treatment received, (2) the order in which the participants received the treatment, (3) the participant. For the Disgust Rating, the unadjusted mean was calculated, since the disgust question was given only to the participants who checked the term disgusted in the CATA evaluation. The data was analyzed using a one-way ANOVA (RStudio, Version 1.4.1717, RStudio, PBC, Boston, MA, USA), considering the treatment as a factor.

A multivariate principal component analysis (PCA) was completed (RStudio, Version 1.4.1717, RStudio, PBC, Boston, MA, USA) on the emotional responses, the Disgust Rating, and the expected liking and willingness to try responses to: 1) identify key similarities and differences among the treatments, and 2) illustrate the relationships between the emotional responses, the expected liking, the Disgust Rating, and the expected willingness to try. The estimated percentages of the emotional responses were used to calculate the principal components, excluding the emotion terms with a range of less than 10% across the treatments (free, tame, and aggressive), as these were considered not to have an impact on the emotional response. These three terms, the expected liking, the expected willingness to try, and the Disgust Rating, were included as supplemental variables. Expected liking and expected willingness to try were included as supplemental variables as the focus was given to the emotional responses. The supplemental variables and the 17 treatments presented in Table 1 were used to build the biplot, using the factor loadings for the selected emotional responses, the supplemental variables, and the factor scores for the 17 treatments.

The emotional responses, the Disgust Rating, the expected liking and the expected willingness to try responses, and the effects of the three experimental design factors (Product Category, Insect Presence Information, and Sustainability Information) were analyzed using ANOVA (RStudio, Version 1.4.1717, RStudio, PBC, Boston, MA, USA). The full model included the main effects and each two-way interaction of the experimental design factors. The ANOVA model (Design Expert, Version 13, Stat-Ease, Inc., Minneapolis, MN, USA) was fit using backward elimination, and model terms were removed based on significance at the 90% confidence level. The coefficient of determination (R^2) was calculated with respect to each variable considered to measure the percentage of variance in the response explained by the design factors selected.

3. Results

3.1. Age Screener and Socioeconomic Demographics

The final study population (*n* = 826) was primarily female (87.89%), White or Caucasian (81.23%), had bachelor's or higher degrees (72.59%), and resided in California (66.83%). The full demographic breakout can be found in Table 2.

Table 2. The age and socioeconomic results from the participants in this study.

Sample Size	***n* = 826**
Gender	
Male	11.14%
Female	87.89%
Prefer not to answer	0.97%
Age	
Millennials (18–34)	19.49%
Generation X (35–54)	42.01%
Baby Boomers (55 and Over)	38.50%
Educational Level	
Some High School	0.00%
High School Graduate or Equivalent	2.78%
Some College	18.04%
Trade, Technical or Vocational	1.57%
Associate Degree	4.96%
Bachelor's Degree	37.65%
Master's Degree	24.94%
Professional Degree	4.24%
Doctorate Degree	5.81%
Residence	
San Luis Obispo County, CA	12.59%
California	54.24%
Other state in United States	33.17%
Ethnicity	
White or Caucasian	81.23%
Hispanic or Latino	6.30%
Black or African-American	0.36%
Native American or American Indian	0.00%
Asian or Pacific Islander	10.29%
Other	1.82%

3.2. Picture Evaluation

3.2.1. Principal Component Analysis

The emotional responses, Disgust Rating, expected liking and expected willingness to try responses were used to extract the Principal Components, and the first two dimensions have been considered (PC1 and PC2). These two dimensions explain 69% of the variability among the 17 treatments; more specifically, PC1 explains 60% of the total variability, and PC2 explains 9%. PC1 and PC2 (data not shown) have been used to build the biplot (Figure 2). PC1 has the highest percentage of variability explained and is the most important dimension to understand the results. Treatments that are more associated with satisfied, the Expected Liking, good, the Willingness to Try, calm, happy, secure, pleasant, warm, good-natured, nostalgic, joyful, free, loving, enthusiastic, and tame tend to be positively correlated, and negatively correlated with disgusted, interested, worried, adventurous, the Disgust Rating, daring, understanding and aggressive.

Figure 2. Biplot of the first two principal components, PC1 and PC2. The treatments are in blue, the CATA emotions from the EsSense Profile® are in green, and the scaled responses are in pink.

The biplot (Figure 1) separates treatments for which the Insect Presence Information was given from treatments for which it was not given. When the Insect Presence Information was given, the treatments were more associated with interested, understanding, daring, adventurous, disgusted, and worried and higher ratings for Disgust. When Insects Presence Information was not given, the Treatments were more associated with warm, joyful, nostalgic, enthusiastic, secure, happy, pleasant, satisfied, good, calm, good-natured, tame, active, and bored, and higher ratings on the Liking and Willingness to Try scales. Furthermore, treatments 15PS, 16SnS, and 1P are more associated with tame, active, and bored and less associated with interested and understanding.

3.2.2. Analysis of Variance

Results from the ANOVA for all responses are shown in Table 3. For each main factor and two-way interaction, p-values are presented (90% confidence level). Insect Presence Information, Product Category, and Sustainability Information are referred to as Insects, Product, and Sustainability in all figures and graphs. Each two-way interaction showed a significant effect for at least one response. The main effects and two-way interactions showing a significant effect (p-value < 0.10) were included in the regression model of each response. The regression models were used to build all the figures that will be presented below and to interpret the results.

Table 3. Results from the ANOVA for all responses. For each main factor and two-way interaction *p*-values are presented (90% confidence level). The R^2 coefficient is indicated as well. The notation "ns" indicate a non-significant effect ($p \leq 0.1$). Significant factors are designated as bold.

Response	R^2	Product	Information	Insect	Product-by-Information	Product-by-Insect	Information-by-Insect
Active	0.98	**<0.0001**	ns	**0.0004**	ns	**0.0002**	ns
Adventurous	0.82	ns	**0.0951**	**<0.0001**	ns	ns	ns
Aggressive	0.70	0.8784	0.5487	ns	**0.0627**	ns	ns
Bored	0.90	**0.0202**	ns	**0.0012**	ns	**0.0689**	ns
Calm	1.00	**0.0912**	0.9460	**0.0016**	**0.0501**	**0.0636**	ns
Daring	1.00	**0.0053**	0.1004	**0.0002**	**0.0116**	**0.0391**	ns
Disgusted	1.00	0.1044	0.6751	**<0.0001**	ns	**0.0334**	**0.0006**
Enthusiastic	0.37	ns	ns	**0.0091**	ns	ns	ns
Free	0.93	0.1470	0.2935	**0.0015**	**0.0695**	ns	ns
Good	0.88	ns	ns	**<0.0001**	ns	ns	ns
Good-natured	0.64	ns	ns	**0.0001**	ns	ns	ns
Guilty	0.99	**<0.0001**	ns	**<0.0001**	ns	**0.0001**	ns
Happy	0.98	0.1416	ns	**<0.0001**	ns	**0.0334**	ns
Interested	0.99	**0.0797**	**0.0848**	**0.0075**	**0.0531**	**0.0546**	ns
Joyful	0.94	**0.0170**	ns	**0.0002**	ns	**0.0196**	ns
Loving	0.48	ns	ns	**0.0019**	ns	ns	ns
Mild	0.00	ns	ns	ns	ns	ns	ns
Nostalgic	0.94	**0.0172**	ns	**0.0001**	ns	**0.0718**	ns
Pleasant	0.85	ns	**0.0351**	**<0.0001**	ns	ns	ns
Satisfied	0.93	ns	ns	**<0.0001**	ns	ns	ns
Secure	0.98	**0.0052**	ns	**<0.0001**	ns	**0.0020**	ns
Tame	0.25	ns	ns	**0.0418**	ns	ns	ns
Understanding	0.61	ns	0.0104	**0.0008**	ns	ns	ns
Warm	1.00	**0.0291**	0.1105	**0.0032**	**0.0560**	**0.0307**	ns
Wild	0.00	ns	ns	ns	ns	ns	ns
Worried	1.00	**0.0727**	0.1388	**0.0006**	**0.0181**	**0.0577**	ns
Disgust	0.85	ns	**0.0020**	**<0.0001**	ns	ns	**0.0010**
Liking	0.99	**0.0105**	0.2742	**<0.0001**	**0.0236**	ns	ns
Willing to Try	0.99	**0.0718**	0.7186	**<0.0001**	**0.0118**	ns	ns

The Insect Presence Information did not have a significant effect on the emotions mild, wild, and aggressive, but did have a significant effect, independently from the Sustainability Information and the Product Category (i.e., no significant two-way interactions), for enthusiastic, free, good, good-natured, loving, pleasant, satisfied and tame (Figure 3). The Insects Presence Information decreased in the proportion of respondents indicating the emotions for satisfied, good, and pleasant. Conversely, the Insect Presence Information increased the proportion of respondents indicating adventurous and understanding. A lower average Liking and Willingness to Try was observed when the Insect Presence Information was given: Liking decreased on average by about 3 points (from 7.1 to 4.1) on a 9-point hedonic scale ((1 = Dislike Extremely; 9 = Like Extremely); Willingness to try decrease on average by about 1.5 points (from 4.2 to 2.7) on a 5-point scale (1 = Very Unwilling; 5 = Very Willing).

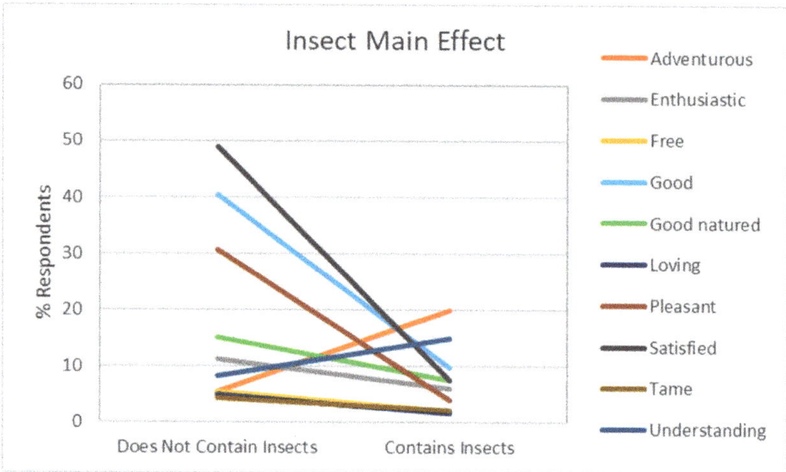

Figure 3. Effect of the Insect Presence Information on the emotional responses adventurous, enthusiastic, free, good, good-natured, loving, pleasant, satisfied, tame, and understanding. The y-axis indicates the predicted mean rating of Liking and Willingness to Try, and the x-axis is the presence or absence of Insect Information.

When evaluating the impact of Sustainability Information on products that did or did not contain insects, pleasant, adventurous, and understanding emotions were significant. For pleasant and adventurous, the impact of the Sustainability Information decreased for products that contained sustainability information: when Sustainability Information was given, the percentage of participants who checked the term "pleasant" decreased by around 7%, and the percentage who checked "adventurous" decreased by around 3.5%. The opposite was true for understanding, as the percentage of participants who checked this term increased by around 4.7%.

Liking and Willingness to Try responses had a significant interaction between the Product Category and the Sustainability Information (Figure 4). Results suggest a positive impact of the Sustainability Information on Bread and Pasta and a negative impact on Sauces.

The Insects Presence Information had a significant interaction effect with the Product Category for the following responses with three different patterns of interactions. The pattern for daring, disgusted, interested, worried (Figure 5) indicated an increased percentage of respondents checking the emotions, with a particularly strong effect for disgusted and worried, but had the opposite effect for interested in the Sauce category. The pattern for guilty, joyful, and active (Figure 5) tended to be more strongly related to the Product Category. For example, for guilty, a strong decrease in the percentage of respondents checking this emotion is observed for Carbohydrate and Snacks, a less strong decrease for Processed Meat Products, and a slight increase for Protein and sauces. The pattern for bored, calm, happy, nostalgic, secure and warm (Figure 6) tended to decrease the percentage of respondents indicating the emotions, with a particularly strong effect for happy and calm. The effect for bored, nostalgic, and warm was stronger for Snacks and Protein, Carbohydrate and Snacks, and Carbohydrate and Sauces, respectively. Other interesting patterns have been also observed. In particular, the significant interaction effects between the Product Category and Sustainability Information for daring, interested, worried, calm, and warm (Figure 7).

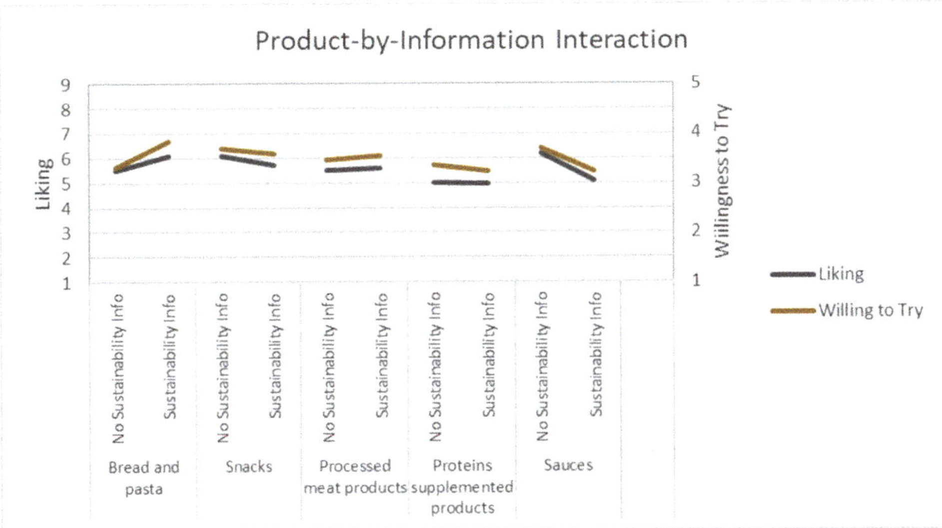

Figure 4. Effect of the interaction between the Sustainability Information and the Product Category on the expected liking (9-point hedonic scale, 1 = Dislike Extremely, 9 = Like Extremely) and expected willingness to try (5-point scale, 1 = Very Unwilling; 5 = Very Willing).

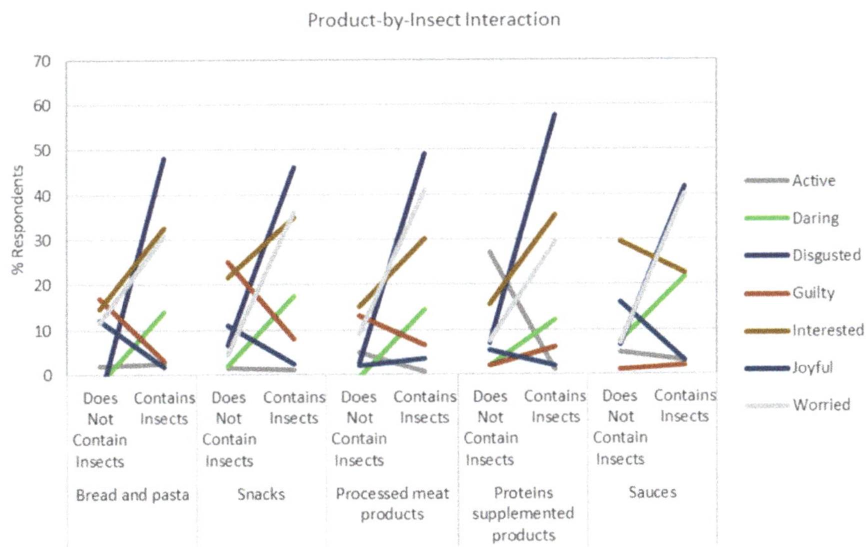

Figure 5. Effect of the interaction of the Insect Presence Information and the Product Category on Table 3. Effect of the interaction of the Insect Presence Information and the Product Category on the feelings bored, calm, happy, nostalgic, secure, and warm. The y-axis indicates the percentage of respondents checking the emotions, values that have been plotted using the regression model built from the ANOVA results, and the x-axis is the Insect Presence Information by each Product Category.

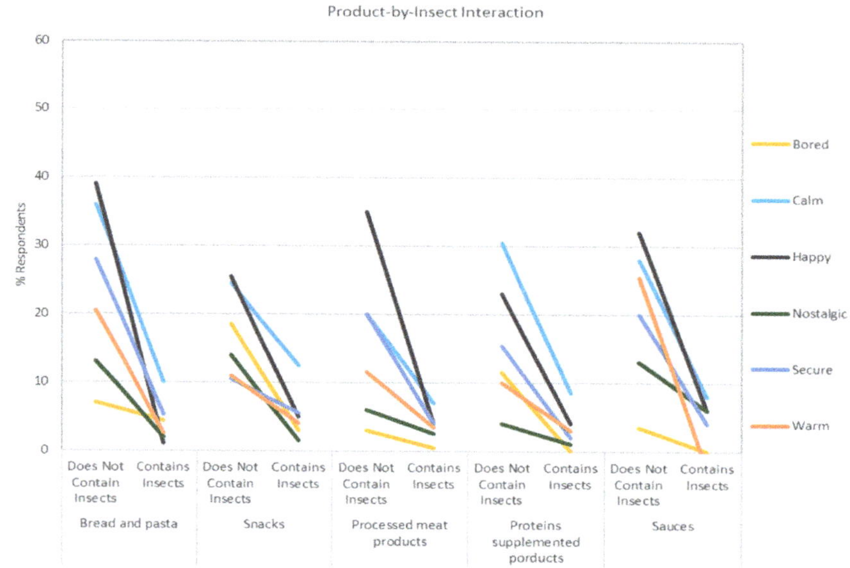

Figure 6. Effect of the interaction of the Insect Presence Information and the Product Category on the feelings bored, calm, happy, nostalgic, secure, and warm. The y-axis indicates the percentage of respondents checking the emotions, values which have been plotted using the regression model build from the ANOVA results, and the x-axis is the Insect Pres-ence Information by each Product Category.

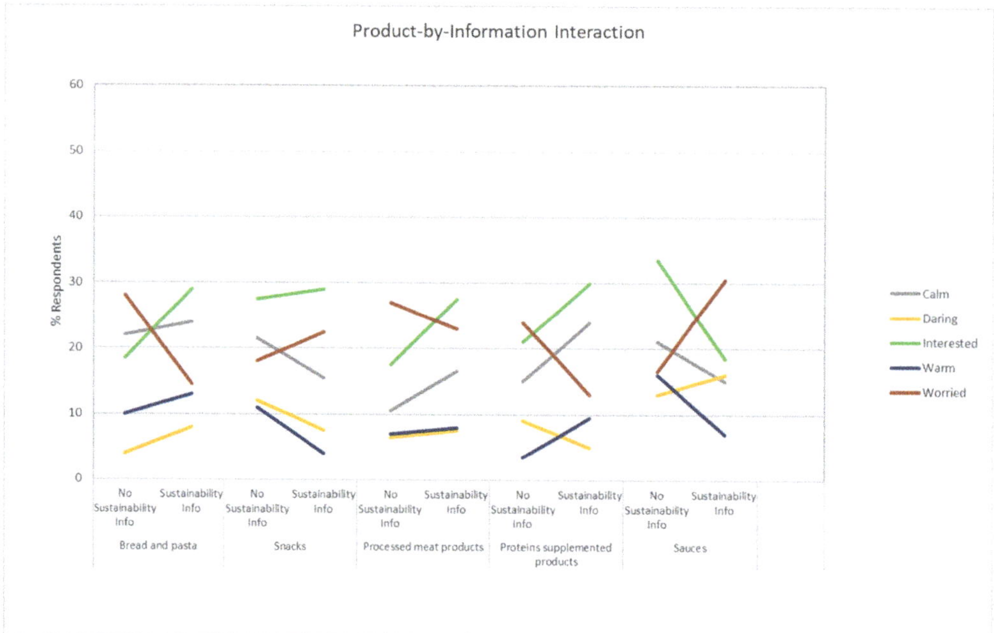

Figure 7. Effects of the interaction between the Sustainability information and the Product Category for the feelings calm, daring, interested, warm, and worried. The y-axis indicates the percentage of respondents checking the emotions, values that have been plotted using the regression model built from the ANOVA results, and the x-axis is the Insect Presence Information by each Product Category.

The Insects' Presence Information showed a significant interaction effect with the Sustainability Information in relation to both the disgusted and the Disgust Rating. For both responses, the Sustainability information showed an attenuation effect in respect to the Insects Presence Information, which tended to increase the percentage of participants checking the term disgusted and the mean for the Disgust Rating. When respondents were told that insects were present in the food products, the percentage of participants who checked the term "disgusted" increased by around 34% when Sustainability Information was given, and by 54% when no information was given, compared to the percentage of participants when insects were not present in the food products. Disgust Rating also increased by about 1.5 points on a 5-point scale (1 = Not Disgusted at All to 5 = Extremely Disgusted) when the Sustainability Information was not given, and of about 0.3 points when it was given.

3.2.3. Treatments Comparison

Treatment 2CSi was further studied to understand the potential of marketing of insects in carbohydrate-rich products such as bread and pasta, as the Product Category Bread and Pasta was associated with an acceptable occurrence of the disgust feeling when the Insect Presence Information was given (Figure 5), and high Willingness to Try and Liking when the Sustainability Information was given (see Figure 4).

First, treatment 2CSi was compared to the other treatments for which the Insect Presence Information was given, considering the estimated percentages of respondents for the emotions for which either the Product Category or the Sustainability Information had a significant effect (Figures 8 and 9). For Figures 8 and 9, only the emotions for which the incidence was more than 10% have been considered, as these were considered to have a decisive impact on the overall emotional response. From these figures, it may be observed that treatment 2CSi, in comparison with the other treatments for which the Insect Presence Information was given, showed good performances in respect to all emotions, including for disgusted, interested, and worried.

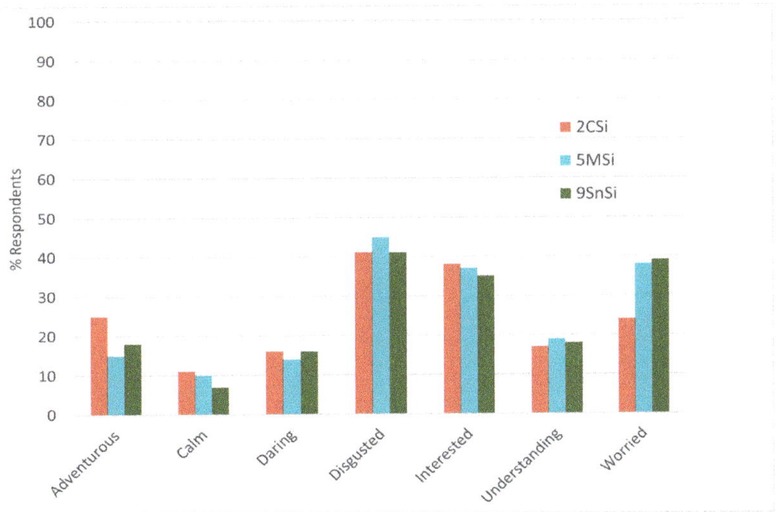

Figure 8. Estimated percentage of respondents considering the emotions for which either the Sustainability Information or the Product Category had a significant effect, and the treatments for which both the Sustainability Information and Insects Presence Information were given. Significance testing was not done across the treatments because the main focus of the study was to study the effects of the experimental design factors. Treatment abbreviations can be found in Table 1.

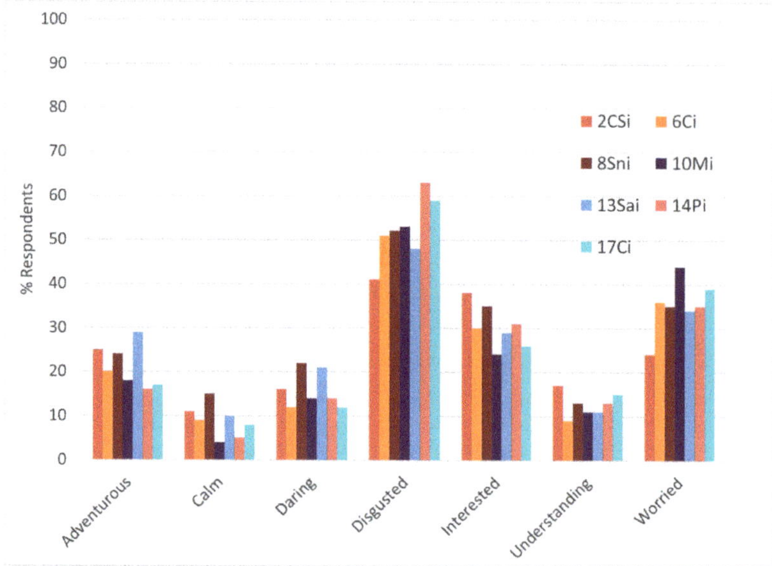

Figure 9. Estimated percentages of respondents considering the emotions for which either the Sustainability Information or the Product Category had a significant effect, and the treatments for which only the Insect Presence Information were given, and the treatment 2CSi. Treatment abbreviations can be found in Table 1.

Figure 10 presents the comparison between Treatment 2CSi and Treatment 4CS, i.e., the only treatment for which Bread and Pasta was used as Product Category and no Insect Presence Information was given; also here, only the emotions for which the incidence was more than 10% have been considered, as these were considered to have a decisive impact on the overall emotional response. From this figure, it can be observed that 2CSi performs better for adventurous, daring, guilty, interested, and understanding; on the other hand, 4CS performs better for calm, disgusted, good, good-natured, happy, joyful, nostalgic, pleasant, satisfied, secure, warm, and worried. Furthermore, 4CS had a higher Expected Liking and Willingness to Try. These results also suggest that if a product containing non-visible insects are marketed in a condition like 2CSi, acceptance of the product may be increased by improving the performance in respect to emotions such as calm, disgusted, good, good-natured, happy, joyful, nostalgic, pleasant, satisfied, secure, warm, or worried.

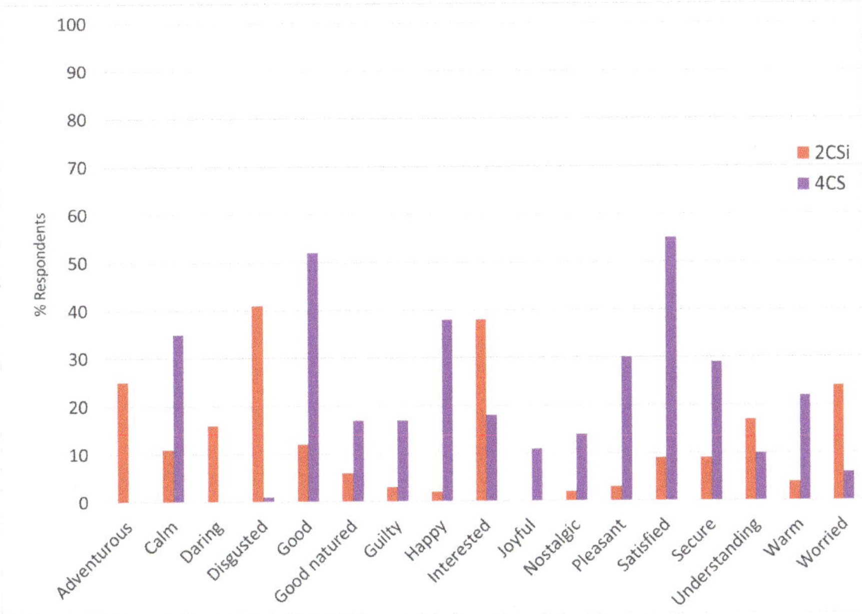

Figure 10. Estimated percentages of respondents considering all emotions and the Treatments 2CSi and 4CS. Treatment abbreviations can be found in Table 1.

4. Discussion

The importance of the emotional reaction to insect products has been recognized by various authors [25,33]; this study suggests that all products containing insects in a non-visible form are associated with a set of specific feelings which go beyond disgust. Such insect-based products have been found to be positively correlated with emotions such as interested, understanding, daring, adventurous, and worried, and strongly negatively correlated with feelings such as satisfied, good, pleasant, happy, calm, warm, nostalgic, and secure. Moreover, this study confirms the role of disgust, showing that feelings of disgust are strongly associated with insect products which aligns with previous work done by Le Goff and Delarue [34] with facial expressions and Gmuer, Guth, Hartmann, and Siegrist [33] who studied the emotional responses of consumers to pictures of products containing insects.

The fact that the Insect Presence Information increased the occurrence of feelings such as adventurous and daring, and decreased the occurrence of feelings such as bored and tame is consistent with other works that found sensation seeking as a factor for insect acceptance [16–19], as well as research that proposed that a viable target for insect products could be consumers seeking novel and unique food experiences and adventurous eating [24,46]. However, insect products were strongly negatively associated with expected liking and expected willingness to try in this study, which was more associated with the positive emotions (satisfied, good, happy) negatively correlated with insect products. These results also suggest the need for creating more positive emotional expectations and a reduction in the negative emotional responses, as suggested by previous research [27,33].

This research also indicates that presenting insects in the non-visible form is not enough to increase acceptability, even though some authors have proposed such a solution as a way to remove disgust-related reactions and increase acceptance [6,37]. This research indicates that negative emotions like disgust and worried are associated with products containing non-visible insects as well, and that the presence of insects in a non-visible form still significantly decreases the expected liking and willingness to try such food products.

Thus, additional solutions need to be found for marketing insect foods. Various authors proposed the use of education to provide information about the sustainability of insects to increase insect-based food acceptance [21,22]. Results from this study highlight some important issues about the use of such information regarding food products containing non-visible insects. Education on the sustainability value of insects had a slight positive effect on disgust responses towards insect products, which is not consistent with previous studies on willingness to consume burgers containing non-visible insects, which suggested no effect of sustainability information on disgust reactions [36]. This may be due to differences in consumer values when it comes to answering a survey, compared to consuming an actual product where sensory characteristics may influence consumer attitudes. In addition, the study demographics between the two studies differed significantly. However, the low mean scores for expected liking and willingness to try products for which both the Sustainability Information and the Insect Presence Information was given suggest that more than just educational information alone should be considered. Emotions related to insect-based foods are complex and should be considered. For example, the emotions adventurous and pleasant decreased when Sustainability Information was presented, while the opposite is true for the emotion understanding. Moreover, significant interactions have been found between the use of sustainability-related information and the product form regarding various emotions, including daring, calm, interested, warm and worried, with effects strongly varying according to the specific Product Category.

The use of information had a positive impact on the expected liking and willingness to try of products such as bread, pasta, and meat processed foods, and a negative impact on products such as sauces (Figure 4). The use of sustainable information should be carefully considered as it could be negative in certain contexts. For example, if adventurous eaters were to be targeted, the emotions such as daring and adventurous need to be increased, the use of sustainability-related information, if combined with the incorporation of non-visible insects in snacks and protein-supplemented products such as protein bars, may be counterproductive.

This research suggests that the choice of the product form has a strong impact on the emotional reactions of consumers to products containing non-visible insects. Interesting interactions were observed in relation to active, guilty, joyful, disgusted, interested, and secure. For sauces, the incorporation of non-visible insects provoked a lower occurrence of disgust feelings with respect to other categories; however, there was also observed a decrease in interested and an increase of guilty, contrarily to other categories, and a lower decrease in bored. Carbohydrate-rich products such as bread and pasta were associated with a lower increase of disgust feelings as well; additionally, it was observed a slight increase of active and a strong decrease of guilty when associated with non-visible insects. Protein-supplemented products such as protein bars showed a lower increase of disgust feelings; however, the impact was overall negative, including a strong decrease in the occurrence of active, secure, and joyful, and an increase in the occurrence of guilty. For meat processed foods, interesting interactions were observed; this was the only category for which the presence of non-visible insects was associated with an increase of the occurrence of joyful; moreover, it was also observed a strong decrease for the emotion guilty, contrarily to protein supplemented products and sauces. For snacks, a very strong decrease in guilty was observed, even though this Product Category was also associated with a strong increase in disgusted, and a strong decrease in joyful and secure.

Considering treatment 2CSi, such a solution, compared to all other treatments for which the Insect Presence Information was given, showed better overall performances in respect to the most occurring feelings, including adventurous, calm, daring, disgusted, interested, understanding, and worried. Furthermore, comparing treatment 2CSi to treatment for which Insect Presence Information was not given (treatment 4CS), treatment 2CSi performed better for some emotions, including adventurous, daring, guilty, interested, and understanding; however, in respect to treatment 2CSi, the comparison suggests that the expected liking and the performance in terms of various emotions (calm, disgusted, good,

good-natured, happy, joyful, nostalgic, pleasant, satisfied, secure, warm, and worried) need to be improved to successfully market products in such conditions.

Overall, such observations reinforce the need to carefully take into consideration the product form, highlighting the potential of incorporating non-visible insects in carbohydrate-rich products such as bread and pasta.

Limitations and Future Research

The results from this study are limited to the population taken characterized by a high proportion of white females, highly educated, and mainly living in California; more specifically, females represented 87.89% of the sample, white people represented 81.23% of the total participants, and 66.84% of the participants resided in California. Therefore, further development is needed to understand the application of these results for a broader population. These results should be compared with physical product testing, as this study is limited to the use of pictures in an online environment. Moreover, the influence of different types of insects should be considered, as participants were given a general indication about the presence of non-visible insects without any detail on the species. Additionally, other methods of measuring emotions should be taken into consideration as this research only used a self-report questionnaire. Finally, the study of the emotional responses should be compared to a segmentation analysis to explore the potential different emotional responses by individual segments and to confirm the existence of a segment of consumers attracted by the novelty of insects.

5. Conclusions

The emotional response to foods containing non-visible insects goes beyond disgust, as they are positively correlated with feelings such as interested, understanding, daring, adventurous, and worried, and negatively correlated with feelings such as satisfied, good, pleasant, happy, calm, warm, nostalgic, and secure. Effective solutions to improve acceptance of these products and increase positive emotional expectation is needed, as in the current context, an overall low expected liking and willingness to try characterizing these products. The use of sustainability-related information for marketing insects is complex and may not be beneficial or positive. Careful consideration should be taken with the choice of the product form, as the impact is complex and varies in combination with the use of sustainability information and the final product form. From this point of view, these results suggest that a good starting point could be the use of sustainability information in combination with the incorporation of non-visible insects in products such as bread and pasta (treatment 5CSi).

Author Contributions: Conceptualization, M.S., D.R. and A.L.; data curation, R.L. and J.G.; formal analysis, M.S., D.R., R.L. and J.G.; funding acquisition, D.R. and A.L.; investigation, M.S. and D.R.; methodology, M.S., D.R., R.L., K.W., J.G. and A.L.; project administration, M.S., D.R. and A.L.; resources, A.L.; software, R.L. and J.G.; supervision, M.S., D.R. and A.L.; validation, M.S., D.R. and A.L.; visualization, M.S. and D.R.; writing–original draft, M.S.; writing–review and editing, M.S., D.R. and A.L. All authors have read and agreed to the published version of the manuscript.

Funding: This work was supported by the National Institute of Food and Agriculture, National Needs Graduate Fellowship Program (Award No. 2017-38420-26767) and the California State University Agricultural Research Institute (Award No. 21-03-105).

Institutional Review Board Statement: The study was conducted in accordance with the Declaration of Helsinki and approved by the Cal Poly San Luis Obispo Institutional Review Board (2020-078-OL).

Informed Consent Statement: Informed consent was obtained from all subjects involved in the study.

Data Availability Statement: The data presented in this study are available on request from the corresponding author. The data are not publicly available due to privacy and ethical restrictions.

Conflicts of Interest: The authors declare no conflict of interest. The sponsors (National Institute of Food and Agriculture, National Needs Graduate Fellowship Program; California Agricultural Research Institute) had no role in the design, execution, interpretation, or writing of the study. Garza Consulting is a consumer, sensory, and statistical consulting company. Their role in this research was to provide guidance to the statistical design of the experiment, suggestions on the questionnaire, and data analysis. All three co-authors were not compensated, but were project collaborators.

References

1. Alexandratos, N.; Bruinsma, J. *World Agriculture towards 2030/2050: The 2012 Revision*; FAO: Rome, Italy, 2012.
2. IPCC (Intergovernmental Panel on Climate Change). *Climate Change 2014: Synthesis Report*; Contribution of Working Groups I, II and III to the Fifth Assessment Report of the Intergovernmental Panel on Climate Change; IPCC: Geneva, Switzerland, 2014.
3. Van Huis, A.; Van Itterbeeck, J.; Klunder, H.; Mertens, E.; Halloran, A.; Muir, G.; Vantomme, P. *Edible Insects: Future Prospects for Food and Feed Security*; FAO: Rome, Italy, 2013.
4. Vermeulen, S.; Campbell, B.; Ingram, J. Climate Change and Food Systems. *Annu. Rev. Environ. Resour.* **2021**, *37*, 195–222. [CrossRef]
5. Baiano, A. Edible insects: An overview on nutritional characteristics, safety, farming, production technologies, regulatory framework, and socio-economic and ethical implications. *Trends Food Sci. Technol.* **2020**, *100*, 35–50. [CrossRef]
6. Patel, S.; Suleria, H.; Rauf, A. Edible insects as innovative foods: Nutritional and functional assessments. *Trends Food Sci. Technol.* **2019**, *86*, 352–359. [CrossRef]
7. Meyer-Rochow, V.B. Can insects help to ease the problem of world food shortage? *Search* **1975**, *6*, 261–262.
8. Govorushko, S. Global status of insects as food and feed source: A review. *Trends Food Sci. Technol.* **2019**, *91*, 436–445. [CrossRef]
9. Nadeau, L.; Nadeau, I.; Franklin, F.; Dunkel, F. The potential for entomophagy to address undernutrition. *Ecol. Food Nutr.* **2015**, *54*, 200–208. [CrossRef] [PubMed]
10. Oonincx, D.; De Boer, I. Environmental impact of the production of mealworms as a protein source for humans—A life cycle assessment. *PLoS ONE* **2012**, *7*, e51145. [CrossRef] [PubMed]
11. Vandeweyer, D.; Crauwels, S.; Lievens, B.; Van Campenhout, L. Metagenetic analysis of the bacterial communities of edible insects from diverse production cycles at industrial rearing companies. *Int. J. Food Microbiol.* **2017**, *261*, 11–18. [CrossRef]
12. De Castro, R.; Ohara, A.; dos Santos Aguilar, J.; Domingues, M. Nutritional, functional and biological properties of insect proteins: Processes for obtaining, consumption and future challenges. *Trends Food Sci. Technol.* **2018**, *76*, 82–89. [CrossRef]
13. Yen, A. Insects as food and feed in the Asia Pacific region: Current perspectives and future directions. *J. Insects Food Feed* **2015**, *1*, 33–55. [CrossRef]
14. Belluco, S.; Halloran, A.; Ricci, A. New protein sources and food legislation: The case of edible insects and EU law. *Food Secur.* **2017**, *9*, 803–814. [CrossRef]
15. Belluco, S.; Losasso, C.; Maggioletti, M.; Alonzi, C.; Paoletti, M.; Ricci, A. Edible insects in a food safety and nutritional perspective: A critical review. *Compr. Rev. Food Sci. Food Saf.* **2013**, *12*, 296–313. [CrossRef]
16. Hamerman, E. Cooking and disgust sensitivity influence preference for attending insect-based food events. *Appetite* **2016**, *96*, 319–326. [CrossRef] [PubMed]
17. Ruby, M.; Rozin, P.; Chan, C. Determinants of willingness to eat insects in the USA and India. *J. Insects Food Feed* **2015**, *1*, 215–225. [CrossRef]
18. Ruby, M.; Rozin, P. Disgust, sushi consumption, and other predictors of acceptance of insects as food by Americans and Indians. *Food Qual. Prefer.* **2019**, *74*, 155–162. [CrossRef]
19. Tan, H.; Fischer, A.; Tinchan, P.; Stieger, M.; Steenbekkers, L.; van Trijp, H. Insects as food: Exploring cultural exposure and individual experience as determinants of acceptance. *Food Qual. Prefer.* **2015**, *42*, 78–89. [CrossRef]
20. Menozzi, D.; Sogari, G.; Veneziani, M.; Simoni, E.; Mora, C. Eating novel foods: An application of the Theory of Planned Behaviour to predict the consumption of an insect- based product. *Food Qual. Prefer.* **2017**, *59*, 27–34. [CrossRef]
21. Verneau, F.; La Barbera, F.; Kolle, S.; Amato, M.; Del Giudice, T.; Grunert, K. The effect of communication and implicit associations on consuming insects: An experiment in Denmark and Italy. *Appetite* **2016**, *106*, 30–36. [CrossRef]
22. Pambo, K.; Okello, J.; Mbeche, R.; Kinyuru, J.; Alemu, M. The role of product information on consumer sensory evaluation, expectations, experiences and emotions of cricket-flour-containing buns. *Food Res. Int.* **2018**, *106*, 532–541. [CrossRef]
23. Mancini, S.; Sogari, G.; Menozzi, G.; Nuvoloni, R.; Torracca, B.; Moruzzo, R.; Gisella, P. Factors Predicting the Intention of Eating an Insect-Based Product. *Foods* **2019**, *8*, 270. [CrossRef]
24. Baker, M.; Shin, J.; Kim, Y. Customer Acceptance, Barriers, and Preferences in the US. In *Edible Insects in Sustainable Food Systems*; Springer: Berlin/Heidelberg, Germany, 2018; pp. 387–399.
25. Looy, H.; Wood, J. Attitudes toward Invertebrates: Are Educational "Bug Banquets" Effective? *J. Environ. Educ.* **2006**, *37*, 37–48. [CrossRef]
26. Ghosh, S.; Jung, C.; Meyer-Rochow, V.B. What governs selection and acceptance of edible insect species? In *Edible Insects in Sustainable Food Systems*; Springer: Berlin/Heidelberg, Germany, 2018; pp. 331–351.
27. Tan, H.; House, J. Consumer acceptance of insects as food: Integrating psychological and socio-cultural perspectives. In *Edible Insects in Sustainable Food Systems*; Springer: Berlin/Heidelberg, Germany, 2018; pp. 375–386.

28. Lerner, J.; Li, Y.; Valdesolo, P.; Kassam, K. Emotion and decision making. *Annu. Rev. Psychol.* **2015**, *66*, 799–823. [CrossRef] [PubMed]
29. Macht, M. How emotions affect eating: A five-way model. *Appetite* **2008**, *50*, 1–11. [CrossRef] [PubMed]
30. Kenney, E.; Koushik, A. Recent developments in identifying and quantifying emotions during food consumption. *J. Sci. Food Agric.* **2016**, *96*, 3627–3630. [CrossRef]
31. King, S. Emotions elicited by foods. In *Emotion Measurement*; Woodhead Publishing: Sawston, UK, 2016; pp. 455–472.
32. Onwezen, M.C.; van den Puttelaar, J.; Verain, M.C.D.; Veldkamp, T. Consumer acceptance of insects as food and feed: The relevance of affective factors. *Food Qual. Prefer.* **2019**, *77*, 51–63. [CrossRef]
33. Gmuer, A.; Guth, J.; Hartmann, C.; Siegrist, M. Effects of the degree of processing of insect ingredients in snacks on expected emotional experiences and willingness to eat. *Food Qual. Prefer.* **2016**, *54*, 117–127. [CrossRef]
34. Le Goff, G.; Delarue, J. Non-verbal evaluation of acceptance of insect-based products using a simple and holistic analysis of facial expressions. *Food Qual. Prefer.* **2017**, *56*, 285–293. [CrossRef]
35. Bohner, G.; Dickel, N. Attitudes and attitude change. *Annu. Rev. Psychol.* **2011**, *62*, 391–417. [CrossRef]
36. Schouteten, J.J.; De Steur, H.; De Pelsmaeker, S.; Lagast, S.; Juvinal, J.G.; De Bourdeaudhuij, I.; Verbeke, W.; Gellynck, X. Emotional and sensory profiling of insect-, plant-and meat-based burgers under blind, expected and informed conditions. *Food Qual. Prefer.* **2016**, *52*, 27–31. [CrossRef]
37. Hartmann, C.; Shi, J.; Giusto, A.; Siegrist, M. The psychology of eating insects: A cross-cultural comparison between Germany and China. *Food Qual. Prefer.* **2015**, *44*, 148–156. [CrossRef]
38. Hartmann, C.; Siegrist, M. Insects as food: Perception and acceptance. Findings from current research. *Ernahr. Umsch.* **2017**, *64*, 44–50.
39. Castro, M.; Chambers, E. Consumer avoidance of insect containing foods: Primary emotions, perceptions and sensory characteristics driving consumers considerations. *Foods* **2019**, *8*, 351. [CrossRef]
40. Verbeke, W. Profiling consumers who are ready to adopt insects as a meat substitute in a Western society. *Food Qual. Prefer.* **2015**, *39*, 147–155. [CrossRef]
41. Meyer-Rochow, V.B.; Hakko, H. Can edible grasshoppers and silkworm pupae be tasted by humans when prevented to see and smell these insects? *J. Asia-Pac. Entomol.* **2018**, *21*, 616–619. [CrossRef]
42. King, S.; Meiselman, H.; Carr, B. Measuring emotions associated with foods: Important elements of questionnaire and test design. *Food Qual. Prefer.* **2013**, *28*, 8–16. [CrossRef]
43. Nestrud, M.; Meiselman, H.; King, S.; Lesher, L.; Cardello, A. Development of EsSense25, a shorter version of the EsSense Profile®. *Food Qual. Prefer.* **2016**, *48*, 107–117. [CrossRef]
44. Lagast, S.; Gellynck, X.; Schouteten, J.; De Herdt, V.; De Steur, H. Consumers' emotions elicited by food: A systematic review of explicit and implicit methods. *Trends Food Sci. Technol.* **2017**, *69*, 172–189. [CrossRef]
45. King, S.; Meiselman, H. Development of a method to measure consumer emotions associated with foods. *Food Qual. Prefer.* **2010**, *21*, 168–177. [CrossRef]
46. Rovai, D.; Michniuk, E.; Roseman, E.; Amin, S.; Lesniauskas, R.; Wilke, K.; Garza, J.; Lammert, A. Insects as a sustainable food ingredient: Identifying and classifying early adopters of edible insects based on eating behavior, familiarity, and hesitation. *J. Sens. Stud.* **2021**, e12681. [CrossRef]

foods

MDPI

Article

Exploring Text Mining for Recent Consumer and Sensory Studies about Alternative Proteins

Ziyang Chen [1], Cristhiam Gurdian [2], Chetan Sharma [1], Witoon Prinyawiwatkul [2] and Damir D. Torrico [1,*]

1 Centre of Excellence—Food for Future Consumers, Department of Wine, Food and Molecular Biosciences, Faculty of Agriculture and Life Sciences, Lincoln University, Lincoln 7647, New Zealand; Ziyang.Chen@lincolnuni.ac.nz (Z.C.); Chetan.Sharma@lincoln.ac.nz (C.S.)

2 Agricultural Center, School of Nutrition and Food Sciences, Louisiana State University, Baton Rouge, LA 70803, USA; cgurdi3@lsu.edu (C.G.); wprinya@lsu.edu (W.P.)

* Correspondence: Damir.Torrico@lincoln.ac.nz; Tel.: +64-3-423-0641

Abstract: Increased meat consumption has been associated with the overuse of fresh water, underground water contamination, land degradation, and negative animal welfare. To mitigate these problems, replacing animal meat products with alternatives such as plant-, insect-, algae-, or yeast-fermented-based proteins, and/or cultured meat, is a viable strategy. Nowadays, there is a vast amount of information regarding consumers' perceptions of alternative proteins in scientific outlets. Sorting and arranging this information can be time-consuming. To overcome this drawback, text mining and Natural Language Processing (NLP) are introduced as novel approaches to obtain sensory data and rapidly identify current consumer trends. In this study, the application of text mining and NLP in gathering information about alternative proteins was explored by analyzing key descriptive words and sentiments from n = 20 academic papers. From 2018 to 2021, insect- and plant-based proteins were the centers of alternative proteins research as these were the most popular topics in current studies. Pea has become the most common source for plant-based protein applications, while spirulina is the most popular algae-based protein. The emotional profile analysis showed that there was no significant association between emotions and protein categories. Our work showed that applying text mining and NLP could be useful to identify research trends in recent sensory studies. This technique can rapidly obtain and analyze a large amount of data, thus overcoming the time-consuming drawback of traditional sensory techniques.

Keywords: alternative proteins; text mining; natural language processing; sentiment analysis

Citation: Chen, Z.; Gurdian, C.; Sharma, C.; Prinyawiwatkul, W.; Torrico, D.D. Exploring Text Mining for Recent Consumer and Sensory Studies about Alternative Proteins. *Foods* **2021**, *10*, 2537. https://doi.org/10.3390/foods10112537

Academic Editor: Koushik Adhikari

Received: 9 September 2021
Accepted: 18 October 2021
Published: 21 October 2021

1. Introduction

Several environmental problems have been associated with the rapid increase in meat consumption and related industries. These problems include increased greenhouse gas emissions, nitrates leaching, land compaction, over-consumption of water, and antimicrobial resistance [1–4]. Thus, to meet the increasing demand for high-quality protein sources in a more environmentally friendly manner, replacing traditional meat with alternative proteins is a potential solution. Currently, there are five main approaches to alternative proteins including plant-based, insect-based, algae-related, fermented by yeast, and cultured meat (or in vitro meat) [5]. Many companies have started to explore the possibility of replacing animal meat-based products with these five types of alternative proteins [1]. To increase the likelihood of successfully commercializing novel products, sensory evaluation plays an important role in product development to optimize foods according to the feedback obtained from consumers [6].

As a key part of sensory science, the development of lexica through traditional approaches requires a large amount of effort, resources, time, and budget, which may sometimes raise barriers and hinder research and development [7]. Simultaneously, the increasing use of web-based platforms to gather information about consumers generates a massive amount of data (so-called big data), which could be of specific interest for fast-moving food companies to identify newer trends, niches, or advantages over competitors. In response to the aforementioned constraints and opportunities, many newer methods, especially those based on advanced computation and artificial intelligence, are paving the way for the development of rapid, efficient, and accurate techniques of data processing. One such technique is text mining, which helps evaluate big data to find meaningful relationships and assertions that would otherwise remain buried in the mass of textual content [8,9]. Analyses of words, sentences, paragraphs, or articles can offer hidden insights that might not be possible to obtain from questionnaires or surveys. Data that can be classified as text are obtained from different sources, including the internet, social media, and scientific reports. However, due to their characteristics and high freedom of word choices, the unprocessed texts tend to be harder to analyze and more time consuming [9,10]. The analyzed text matrix may lead up to thousands of words, and one word may have different meanings in different sentences. To structure this type of analysis, a text's basic workflow is followed by text segmentation (the process of dividing the main document into smaller parts that are called segments), sentence tokenization (the process of turning sentences into a string of characters called tokens), lemmatization (the process of clustering words and removing inflectional endings), and stemming (the process of removing the suffix from words, which reduces them to root words) [9,10].

All these analyses are barely possible to be finished through manual operation; thus, under this situation, an automatic approach (algorithms) shows significant advantages regarding the optimization of time. Recently, text mining and Natural Language Processing were introduced to help researchers obtain sensory data easier and faster from the internet instead of using repeated sensory tests [11–14]. This technique can obtain information from different sources (i.e., websites, journals, magazines with consumers' information), which creates a vast dataset of descriptive words. In general, the obtained lexica from these data mining techniques tend to be "consumer-based" in structure. However, this automation can decrease the time and money spent on research. In addition, it can read a significant amount of sensory data and reform that information into a structured and justified form that is suitable for further analyses. With this technique, sensory research can be conducted more efficiently at the early steps of product development.

For the past few decades, to save time and money in descriptive and consumer analyses, researchers have developed several types of rapid technique. However, all these methods have several shortcomings compared to traditional tests on various levels. The limitation of human processing data has been eliminated with the use of automated algorithms to analyze descriptive data. This research aimed to use text mining and Natural Language Processing to explore structures and meanings about alternative proteins based on the text data collected from scientific reports (n = 20 research papers). This research represents a prototype for text mining applications on identifying future food science trends and associations.

2. Materials and Methods

2.1. Selection of Papers

To obtain the data, one of the most important things is that it is accessible. All of these 20 papers (Table S1) were accessible for hypertext markup language (HTML) and in portable document format (.pdf), which means that they could be scraped by web crawler as well as .pdf text mining commands in R (Version 1.3.1093, Free Software Foundation, Boston, MA, USA) [15] after downloading. Thus, an alternative approach could be developed if the first scraping method did not work. To obtain meaningful insights into the current trends and consumer perception of alternative protein, the criteria for the selection of the scientific papers in this study considered only recently published articles (between 2018 and 2021). Papers' selection was based on the keywords "alternative protein", "plant-based", "insect-based", "algae-based", "yeast", and "cultured meat". Because they are recent studies, they can provide the latest information and trends of alternative proteins.

2.2. Processing of Papers and Texts

All the work was performed in the statistical computing language R (Version 1.3.1093) [15]. The packages applied in R were *rvest* and *xml2* (for web scraping), *pdftool* (for PDF document scraping), *tm* (for text mining), *SnowballC* (for text stemming), *RColorBrewer* (for coloring bar chat and word cloud), *syuzhet* (for emotion analysis and classification), *ggplot2* (for plotting charts) and *wordcloud* (for developing word cloud). Some results were exported as pictures by taking screenshots in portable network graphics (.png) document type in order to improve the pixel of the image.

2.3. Text Mining

2.3.1. Web Scraping

Although grabbing information from a website manually is feasible in some cases [14,16], applying a web crawler would be more advantageous because it saves time. In this case, all the data were collected from scientific reports, which were formatted in PDF. The below figure (Figure 1a) is an example of a simple web crawler performed on a single page (website) to illustrate the basic steps behind web scraping.

```
Codes:

‘

library(xml2)

library(rvest)

read_html("https://www.dataquest.io/blog/web-scraping-in-r-rvest/")

web=read_html("https://www.dataquest.io/blog/web-scraping-in-r-rvest/")

web %>% html_nodes("body span, p, ul, li") %>% html_text

’
```

(**a**)

```
[62] "Before we can start learning how to scrape a web page, we need to understand how a web page itself is structure
d."

[63] "From a user perspective, a web page has text, images and links all organized in a way that is aesthetically pleas
ing and easy to read. But the web page itself is written in specific coding languages that are then interpreted by our w
eb browsers. When we're web scraping, we'll need to deal with the actual contents of the web page itself: the code befor
e it's interpreted by the browser."

[64] "The main languages used to build web pages are called Hypertext Markup Language (HTML), Cascasing Style Sheets (C
SS) and Javascript. HTML gives a web page its actual structure and content. CSS gives a web page its style and look, inc
luding details like fonts and colors. Javascript gives a webpage functionality."

[65] "In this tutorial, we'll focus mostly on how to use R web scraping to read the HTML and CSS that make up a web pag
e."

[66] "Unlike R, HTML is not a programming language. Instead, it's called a markup language - it describes the content a
nd structure of a web page. HTML is organized using tags, which are surrounded by <> symbols. Different tags perform dif
ferent functions. Together, many tags will form and contain the content of a web page."

[67] "The simplest HTML document looks like this:"
```

(**b**)

Figure 1. Code (**a**) and a part of the text captured from the website (**b**) by the crawler.

The first step was to load the packages which supported the web scraping. In this case, *xml2* (R code) and *rvest* were loaded in the first and second lines, respectively. By applying the "*read_html()*" command and typing the URL into the brackets, this page's source file was captured in the third line. After this, the Cascading Style Sheets (CSS) information (in the .html document) was used to locate the text, which was needed to be scraped from the page. Normally, these elements of the website could be reached by opening the developing tool in the browser. Finally, by typing the CSS information into the brackets in the "*html_nodes()*" command, all of the text from this webpage was scraped and illustrated in the R console. An example of the scraped information is showed in Figure 1b.

2.3.2. PDF Scraping and Text Processing

Instead of websites, actual .pdf documents were used to scrape the information in this study. The .pdf document scrapping process was similar to the one used for web scraping. The codes applied in this study are shown in Supplementary File S1 and were written by Cristhiam Gurdian from Louisiana State University, USA. The first step was to download the academic articles that were suitable for the research topic. As detailed in Supplementary File S1, the codes required that the working directory was set to the folder containing the PDF files. After the directory was set, the codes were run for the Natural Language Processing (NLP) (Figure 2. text segmentation, sentence tokenization, lemmatization, and stemming). When this step was complete, the text matrix was ready to be analyzed. Word count and other data visualization techniques were produced by applying packages in the R program such as *syuzhet*, *ggplot2*, and word cloud. Additionally, these codes were used to count the keywords in the texts. A more detailed explanation of the procedure and specific codes used to analyze and process the data are shown in Supplementary File S1.

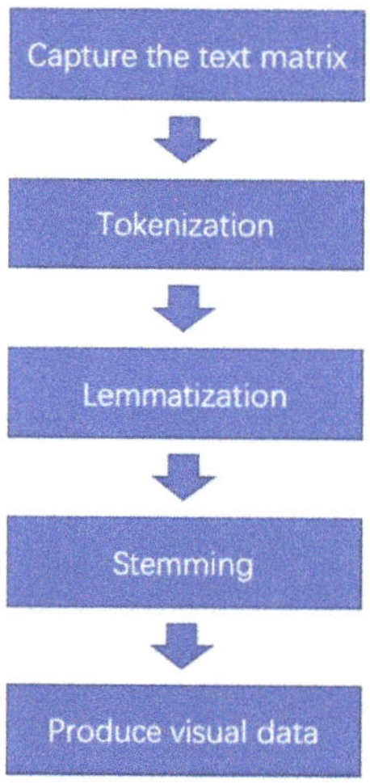

Figure 2. The basic workflow of Natural Language Processing.

2.3.3. Text Scraping and Natural Language Processing

To obtain more specific data regarding the sensory characteristics of alternative proteins, the objects of analysis in this study were the texts containing the findings from the selected academic papers. The introduction, materials and methods, conclusion, and references sections were excluded, and only the results and discussions parts were extracted for further analysis. The text from academic papers was copied and pasted into a text (.txt) document. There were $n = 20$ analyzed papers, and each result and discussion section of the papers was individually pasted into a new .txt document. After that, the vector containing all the .txt documents were combined to produce a .txt matrix, which was the main object of analysis in this study. Thus, 20 .txt documents that contain the texts from 20 academic papers and one .txt document named "Main Text Matrix" that contained all of the texts from the 20 .txt documents were generated. In total, 21 .txt documents were analyzed. The "Main Text Matrix" was produced to investigate the whole picture of these twenty academic papers regarding the sensory attributes of alternative proteins. All these documents were captured and processed using Natural Language Processing text segmentation, sentence tokenization, lemmatization, and stemming by running the respective codes (shown in Supplementary File S2) before producing any data visualization outputs.

The frequencies of each word occurring in the "Main Text Matrix" were counted and showed in a table and bar chart. In this manner, a preliminary relationship between words and alternative proteins was developed. Sentiment analysis and emotion classification was performed using a package called *syuzhet* (R code) [17]. The frequency of sentiments was counted and the proportion of each emotion in the matrix was illustrated in a bar chart. The emotion classification of the 20 .txt documents was run individually to obtain the proportion of emotional data in each paper. The types of alternative proteins mentioned in each article were also indicated; thus, the emotions associated with each type of alternative protein were explored. A word cloud was produced during the analysis to provide an intuitive image of the frequency of words in the matrix. Based on the word frequency results, the association between words was investigated. This process can show the vocabularies around the terms which were aimed at, as well as the strength of their relationship. More specific and reliable details regarding alternative proteins can be collected by following the word association data.

2.4. Statistical Analysis

To obtain the visual relationship between emotions and the types of alternative proteins, the correspondence analysis test was conducted using the XLSTAT software (Version 2018.1.1.62926, Addinsoft Inc., New York, NY, USA) in Excel with a $p < 0.05$ threshold for statistical significance.

3. Results and Discussion

The word frequency results from the "Main Text Matrix" are shown in Figure 3. The detailed word frequency data are shown in Table S2. A word cloud was generated to show the word frequency more intuitively (Figure 4). In the word cloud, the most frequent word appears in the center and the words with higher frequency appear with bigger font size, while the words with lower frequency appear with smaller font size. The proportion of each emotion in the text matrix is indicated in Figure 5. Partial results from the relevance analysis between keywords and other words are shown in Table 1. All the associations between words in the text mining analysis are shown in Supplementary File S3. The proportion of emotions in each paper (20 articles in total) were generated and are shown in Table 2. All the words shown in the tables, figures, and Supplementary Files were in their root form. For instance, "*consum*" would represent "*consumer*", "*consume*", "*consumes*", "*consuming*", "*consumed*", and "*consumption*". Thus, when the frequency of "*consum*" was 264 times, it meant that all the words related to this root appeared 264 times in total.

3.1. Word Frequency

A high frequency of "*meat*", "*protein*", "*product*", "*food*", and "*consum*" root words (531, 432, 404, 356, and 264, respectively) was observed.

The words related to a type of alternative protein were *insect* (179 times), *plantbas* (97 times), *pea* (82 times), *spirulina* (76 times), and *plant* (67 times) as indicated by the top 50 frequent words in the matrix (Table S2). Because the word roots *plantbas* and *plant* have a similar meaning, they were summarized together, representing the total words of plant-based alternative proteins. Thus, among the main categories of research focused on alternative proteins, the insect-based was the most common (179 times) followed by the plant-based (164 times). This suggests that insect- and plant-based proteins were the trendiest topics in alternative proteins scientific research. However, the number of academic papers used in this analysis limits these insights. The performance of text mining could be improved by increasing the size of the text matrix [14].

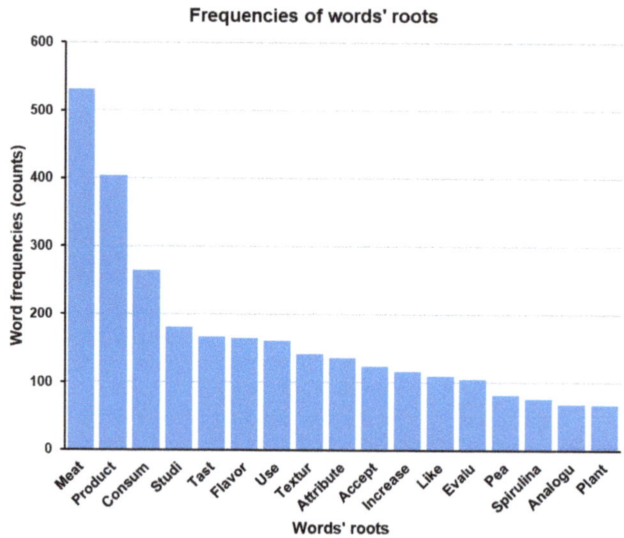

Figure 3. Example of a bar chart showing the frequency of keywords (roots) in the text matrix (n = 20 published papers).

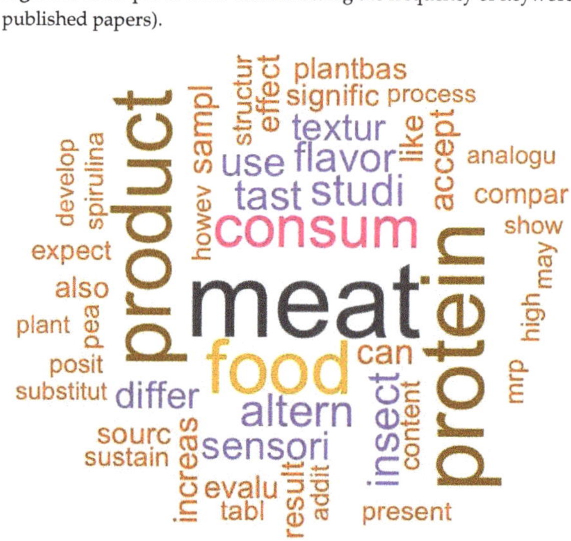

Figure 4. Word cloud obtained of the text matrix (n = 20 published papers).

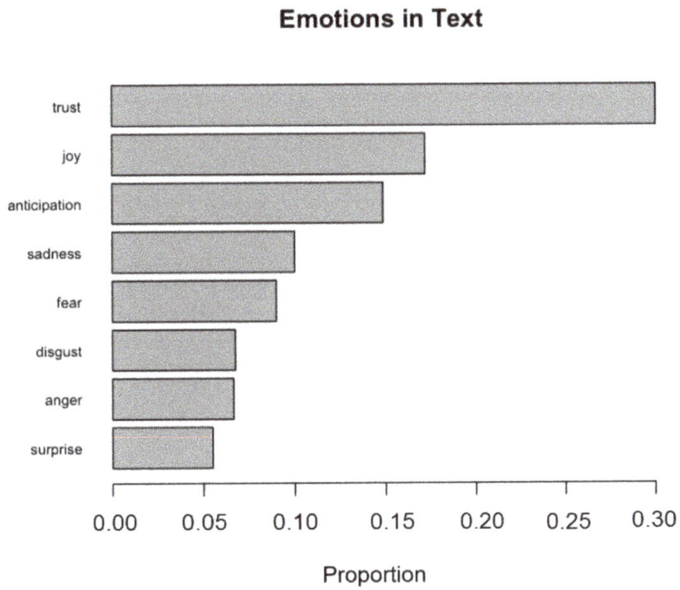

Figure 5. Proportions of emotion terms mentioned in the text matrix (n = 20 published papers).

There were no word-roots associated with cultured meat and/or yeast-fermented proteins that were shown in the top 50 frequency word list (Table S2), which means that they were not relevant in the findings of these 20 articles. The insect-based and plant-based proteins were the most frequent topics in this study, while algae-related proteins (76 times) were less explored, and the cultured meat and yeast-fermented proteins were the least explored alternatives in these studies. A potential explanation for this is that cultured meat and yeast-fermented proteins are in an early stage of development and not many findings have been cited in the current research papers analyzed in this study [6]. Interestingly, instead of the word *"soy"*, the word *"pea"* was the only plant-related word that appeared in the top 50 frequency of words list (Table S2). Based on this result, it can be concluded that researchers are shifting their attention to pea in terms of producing plant-based proteins in recent studies (from 2018 to 2021, which was the year range considered for the analyzed papers). Cosson et al. [18], García-Segovia et al. [19], Kaleda et al. [20], Martin et al. [21], Sha and Xiong [22], Stephan et al. [23], and Yuliarti et al. [24] investigated pea as a plant-based meat alternative protein while only two papers investigated soy as an alternative protein. Moreover, according to Cosson et al. [18], pea protein has become more popular in food products as a plant-based alternative protein due to the enhancement of the food systems' sustainability. The same approach could also be applied to the word root *"spirulina"*. When the applications of algae alternative protein were explored, *"spirulina"* was the most frequent word in research involving algae protein, which has been recently used in food products [25].

Thus, based on the word frequency in the text matrix, key points can be investigated for further analysis. For instance, in this study, it was found that researchers are mainly focused on exploring insect- and plant-based proteins as potential meat-protein alternatives while cultured meat and yeast-fermented proteins have been mentioned less frequently. Interestingly, pea has become the most common source for plant-based protein application in current research, while spirulina was the most popular algae-based alternative protein in the current studies.

The word roots, which may indicate the attributes of alternative proteins such as *"differ"*, *"accept"*, *"increas"* (the root of increase), *"like"*, and *"posit"* (the root of positive), are also illustrated in Table S2 with frequencies of 141, 124, 116, 110, and 73 times, respectively. To analyze these types of words, caution must be taken in the assumptions derived from the word roots' frequencies because they are counted regardless of their positive or negative implications in the article. Taking the word root *"differ"* as an example in the text, whether it is positive or not, it would be counted as a word root. The proportions of positive and negative differences in the text matrix were unknown. Thus, no assumption could be made regarding how positive or negative traditional meat products were in relationship to alternative proteins. The same rule applies for the word roots *"accept"*, *"increas"*, and *"like"* because they may represent not acceptable, not increased, and not liked, respectively. Although the antonym of these words could be written as unacceptable, decrease, and dislike, respectively, researchers tend to use their own descriptive words, making it possible that the same root words have been used in both positive and negative connotations. Hence, analyzing the relevance between keywords and other words can be used to support the frequencies of the words findings and improve the reliability of the assumptions made based on text mining.

3.2. Relevance between Different Words

The word frequency results can show important insights into the text matrix but neither positive nor negative statements can be inferred. In Table 1, partial results from the relevance analysis (proportion of association) between keywords and other words are shown. For the full results of the relevance analysis, Supplementary File S3 can be referred to. It is noticed that the word *"insect"* had high associations with the words *"willing"*, *"neophobia"*, *"cockroach"*, *"disgust"*, *"novel"*, and *"bit"*. The sensory profile and the acceptance of insect-based alternative protein are reflected by these words [26]. Moreover, the words *"willing"* and *"neophobia"* had similar coefficients (0.37–0.38). It could be assumed that insect neophobia can affect the willingness of trying insect-based alternative proteins [27]. Several articles supported this finding. De Koning et al. [28] found that food neophobia affected the willingness to consume insect protein and impacted plant-based proteins. Similarly, food neophobia caused a negative influence on the acceptability of entomophagy and the sensory appeal of insect-based products [29–31]. The words *"cockroach"*, *"disgust"*, and *"novel"* also showed high and similar relationships with the word *"insect"*. It can be concluded that *"cockroach"* was a trendy topic regarding insect-based alternative proteins because this word was mentioned in Chow et al. [30] and García-Segovia et al. [19] studies. In terms of the descriptive words, *"disgust"* and *"novel"* had the highest associations (0.33) with the word *"insect"*. Indeed, entomophagy is still considered a novel practice in Western cultures and "disgusting" was a commonly elicited emotion among participants when they were introduced to the concept of entomophagy [28,29]. Furthermore, insect-based bread has been considered disgusting by participants [19], and the disgust emotion has contributed to the rejection of entomophagy to a greater extent than food neophobia [27,30].

As mentioned above, an assumption based solely on the frequency of words such as *"accept"*, *"like"*, and *"expect"* is not a reliable approach. These three words were surrounded by negative words sharing a high association level: *"Don't"* (0.45) was related to *"accept"*; *"Negat"* (0.29), which was the word root of *"negative"*, was related to *"like"*, and *"disappoint"* (0.35) and *"reject"* (0.29) were associated with *"expect"*. These results suggest that alternative proteins had still not been accepted/liked/expected in the studies covered for this research (the negative words were not 100% related to the keywords). To better understand which type of alternative proteins have a negative effect on product acceptability, more data need to be considered. Firstly, plant-, insect-, and algae-based proteins were the text matrix's main objectives based on the result from the words' frequencies (Table S2). According to the relevance of insects with other words, it is expected that insect protein could negatively affect the product's acceptability [26]. Furthermore, in the relevance analysis of words, *"spirulinarel"* (the word associated to spirulina, an alga) was found to be related to *"like"*. Finally, there were no negative words shown in the relevance analysis of *"plantbas"*. Hence, it could be expected that insect-based proteins have lower acceptability among consumers, while the plant- and algae-based proteins have higher expected acceptability [32]. In this study, the information on cultured meat and yeast-fermented proteins was limited, which restricted the development of insights or possible inferences regarding its consumer acceptability.

Table 1. Relevance (proportion of association) between keywords and other words.

		1. Insect		
Willing (0.38)	Neophobia (0.37)	Cockroach (0.34)	Disgust (0.33)	Novel (0.33)
		2. Accept		
Don't (0.45)	Adult (0.31)	Barrier (0.29)	Elder (0.29)	
		3. Like		
Tomato (0.41)	Spirulinarel (0.30)	Negat (0.29)		
		4. Plantbase		
Health (0.26)	Insectsbas (0.26)	Asia (0.25)		
		5. Expect		
Disappoint (0.35)	Novel (0.30)	Reject (0.29)		
		6. Pea		
Lupin (0.54)	Mushroomi (0.51)	Dusti (0.49)	Green (0.45)	Nut (0.40)
Earthi (0.38)	Bitter (0.34)			

Indeed, a previous study found that insect-based products tend to have lower acceptability among consumers [27] when compared to other products formulated with alternative proteins, or when compared to control formulations without edible insects. In a different study, according to the sensory evaluation results from $n = 71$ participants, plant-based (soy) meat analogues were as acceptable as beef samples regarding visual appearance [1,33]. In another study, the sausage made from wheat and soy isolates was not significantly different compared to the traditional meat sausage in terms of texture liking [1,34]. Other research also showed that there were no significant differences between a plant-based (soy) meat patty and the all-beef patty regarding overall liking [35]. Furthermore, consumers prefer to adopt plant-based alternative proteins rather than insect-based proteins [28,36]. Usually, consumers refuse to eat insects because of food neophobia and feelings of disgust [28]. Although participants' acceptability might be improved occasionally through education about the nutritional and environmental benefits derived from insect protein consumption, at first glance, consuming insects was considered a disgusting and unadoptable practice most of the time [19,28–30,37].

In Table 1, a part of the developed descriptive lexicon for plant-based and pea alternative proteins is presented. The word *"plantbas"* was related to *"health"*, *"insectbas"*, and *"Asia"*. The consumers tended to agree that plant protein was healthier than meat protein, which has been previously documented by other scientific research [1,18,19,21,28]. Many plant-based protein products, such as tofu, were first introduced in Asia [1,28]. This might explain the observed high relevance between the plant-based protein word and the word *"Asia"*. According to Table S2, the plant- and insect-based proteins presented comparable high-frequency counts. Moreover, these words had a high relevance between them (Table 1). Because the plant-based protein high importance, and pea being the only type of plant that was shown in the top 50 words frequency list, the word *"pea"* was also analyzed as a keyword. According to Table 1, *"pea"* was in high relevance with *"mushroom"* (0.51) and *"lupin"* (0.54) because there were two articles in the text matrix that compared pea protein to mushroom and lupin [18,23]. Because of this, it was difficult to judge whether the following descriptive words were related to pea, mushroom, or lupin. However, in the article, pea was described as green, beany, fresh, and grassy, while lupin was evaluated as beany/green, mushroom/earthy, nutty, and other descriptors [18]. In other words, the dusty and earthy were not related to pea protein [18].

Based on the relevance analysis, it was also found that the acceptability of alternative proteins was related to age. The words *"adults"* (65–70 years old) (association level = 0.31) and *"elder"* (70 or above years old) (association level = 0.29) were highly associated with the word *"like"* in the context of alternative proteins [31]. Food neophobia may be mitigated as age increases [28–31]. All of these findings agree with the results from the word frequency and relevance analyses from the text mining approach, which were essential for making inferences about the alternative protein topic in this study.

3.3. Emotions Analysis

The outcome from the analysis of the emotions for the whole text matrix is illustrated in Figure 5. Overall, there was no significant association between the protein types and the emotions ($p = 0.41$) because the observed value for the Chi-square statistic was lower than the established critical value (Chi-square observed = 50.73, and Chi-square critical value = 66.34; Degree of Freedom = 49). Hence, the null hypothesis of independence between the rows and the columns of the contingency table (protein type × emotions) was not rejected. The results from the analysis of emotions present in each paper and the correspondence analysis symmetric plot are shown in Table 2 and Figure 6, respectively.

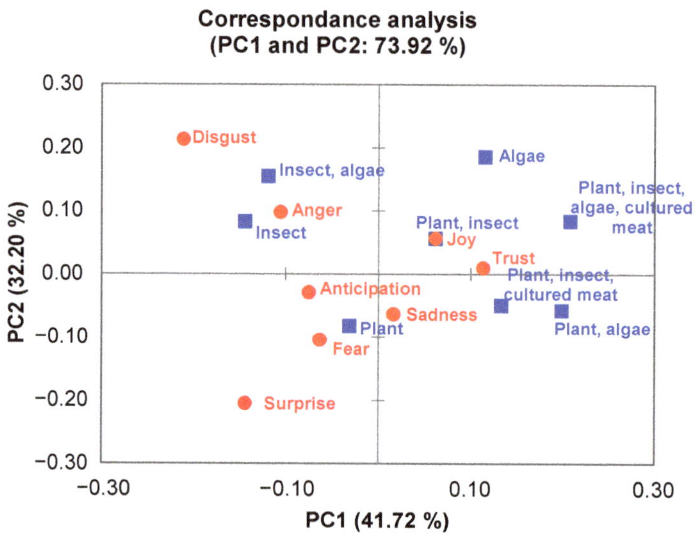

Figure 6. The symmetric plot of correspondence analysis.

"Trust" dominated the sentiments in the text matrix with an observed occurrence of 30% (Figure 5). However, "trust" cannot be classified as either positive or negative as this proportion is showing the cumulative occurrence for all contexts. "Joy" had the second-highest proportion in the text matrix, which might indicate that researchers or consumers felt optimistic for the future of the alternative proteins. In previous research, consumers have well accepted plant-based proteins due to their health benefits and relatively pleasant sensory attributes [1,18,19,21,28]. For the insect-based proteins, although cultural barriers affect their preference in the short term, it is expected that insect-based products will eventually become acceptable for the public because of the frequent exposure over time through advertisement, education, and marketing [19]. Algae-based protein, cultured meat, and yeast-fermented proteins had a competitive advantage over other alternative proteins because they require fewer resources to be produced, such as soil and freshwater [6,38,39]. However, these alternatives might require more expensive technological instruments to produce them. The negative emotions, including "sadness", "fear", and "disgust", represented 11, 9, and 6% of the total emotions, respectively (Figure 5). A more detailed profile of the proportions of emotions for each paper can be observed in Table 2. The negative emotions in the text matrix might be viewed as the potential drawbacks of alternative proteins. For example, the text containing food neophobia may reflect fear of the alternative protein sources regarding their effects on health, and the descriptive word "*disgusting*" may indicate mental associations with other disgusting elements that are evoked when exposed to alternative protein sources.

Table 2. Emotion analysis (percentage of representing the emotion) of each academic paper (details can be found in Table S1).

Paper No.	Alternative Protein Types	Percentage of Each Emotion in the Text (%)							
		Trust	Joy	Anticipation	Sadness	Fear	Disgust	Anger	Surprise
1	Plant	30	18	14	8	10	7	7	6
2	Plant, insect	34	17	16	9	7.5	7.5	9	0
3	Plant, insect	23	14	17	15	11	6	7	7
4	Plant	22	13.5	17	15	11	6.5	7.5	7.5
5	Plant, insect	40	21	13	6.5	5.5	4	5	5
6	Insect, algae	27.5	17.5	19	8	5	11	6	6
7	Plant, insect, cultured meat	32	20	14	11	10	4	4	5
8	Insect	23	16	14	11	11	9.5	11	4.5
9	Plant	35	17	29	4	4	3.5	0	7.5
10	Plant	28	14	16	16.5	13	1	4	7.5
11	Plant	19	14.5	13.5	12.5	13.5	8.5	7.5	11
12	Plant	34	12	26	14	4	4	6	0
13	Plant, algae	41	16	8.5	10.5	7	5	4	8
14	Plant	26	14	13	12	5.5	12.5	10	7
15	Algae	32	23	14	10	5	8	5	3
16	Plant, insect, algae, cultured meat	36	20	15	10	7	4	6	2
17	Plant	30	15	15.5	11.5	10	5	6.5	6.5
18	Insect	31	16.5	20	9.5	2.5	6	5	9.5
19	Plant	32	27.5	11.5	6	11.5	0	0	11.5
20	Insect	24	16	11	9	14.5	14	8	3.5

Figure 6 indicates the relationship between emotions and the categories of alternative proteins (based on the keywords of the n = 20 papers studied). In the symmetric plot, "*insect*" and "*insect, algae*" were separated from other categories as well as associated with "disgust" and "anger". Besides this, there was no other noticeable relationship shown. This result agrees with the Chi-square test of independence.

3.4. Comparison with Other Text Mining Works

In Bakhtin et al. [11] research, several documents were analyzed to achieve a better understanding of the core research topics and trends in agriculture and food production. The data were collected from several databases, including media, websites, and organizations' files. Based on text mining, the authors concluded that using fertilizers and chemical agents in farming were the major issues studied in food security and determined that embryo DNA, gene editing, and CRISPR/Cas9 were becoming the centers of genetic research instead of gene modification, which had been popular for years. They foresaw that edible insects, industrial meat production, and industrial food systems would become the focus of extensive research and remarked the higher relevance between food security and biological hazards, fungicides, and pesticides [11]. This research could be classified as a big data analysis with a robust approach to finding the underlying relationship between the terms embedded in a significant amount of text.

In another study, social media, websites, and databased papers were collected to study the underlying relationship between food safety, dietary pattern characterization, consumer opinion, product development, food knowledge discovery, and food supply chain management by text mining [40]. In total, n = 57 papers were analyzed in that study using a similar approach to the one used in our study. Their approach included word frequency, word association analysis, and sentiment analysis. Furthermore, these authors explored the application of other novel text mining techniques such as text classification, text clustering, and topic modeling [40]. However, a major drawback of their research was the absence of data visualization techniques, which provide a better appreciation of the relationship among the studied variables than the descriptive text. Compared to our research, although more papers were investigated in their study, two important text mining techniques, including the association and sentiment analyses, were used [40]. The result of the present study would tend to be more science-based compared to works using other databases; this can be an advantage when looking for reliable information in text mining [40]. This research showed the recent trends in scientific exploration regarding the use of alternative proteins. Nevertheless, a higher proportion of social media and internet data can be also beneficial for a consumer-based lexicon development.

4. Conclusions

In conclusion, this study analyzed $n = 20$ scientific reports about alternative proteins to explore the application of text mining in sensory research. According to the word frequency results, the insect- and plant-based alternative proteins were the centers of attention in recent research (2018–2021). Moreover, pea was the most studied plant source rather than soy among all plants. According to the results from the word association analysis, the insect-based protein was related to terms such as "*neophobia*", "*cockroach*", "*disgust*", and "*novel*", while plant-based protein was associated with "*health*" and "*Asia*". Furthermore, the insect-based protein contributed the most to the observed negative sentiments in the text matrix. Correspondence analysis showed that there was no evident association between the emotion terms and the alternative protein sources, although these associations may become significant by increasing the dataset or the emotion terms under analysis. Despite this, this research shows the implementation of a useful tool to obtain information rapidly on current trends in food science. Further research is recommended with larger datasets, which can include social media and websites.

Supplementary Materials: The following are available online at https://www.mdpi.com/2304-8158/10/11/2537/s1, Table S1: The list of scientific reports analyzed by the Natural Language Processing, Table S2: The frequency of words in the text matrix (top 50), Supplementary File S1: PDF document text mining codes and explanation produced by Cristhiam Gurdian, Supplementary File S2: Text (TXT) document mining codes obtained from https://www.red-gate.com/simple-talk/sql/bi/text-mining-and-sentiment-analysis-with-r/ (accessed on 1 September 2021), Supplementary File S3: The relevance (association levels) between keywords and other words.

Author Contributions: Conceptualization, D.D.T. and Z.C.; methodology, D.D.T., Z.C. and C.G.; formal analysis, Z.C.; investigation, Z.C.; data curation, Z.C.; writing—original draft preparation, Z.C.; writing—review and editing, Z.C., D.D.T., C.G, C.S. and W.P.; supervision, D.D.T. and C.S.; project administration, D.D.T.; funding acquisition, D.D.T. All authors have read and agreed to the published version of the manuscript.

Funding: This research was funded by Lincoln University, New Zealand, through the Centre of Excellence-Food for Future Consumers.

Data Availability Statement: The data presented in this study are available on request from the corresponding author.

Conflicts of Interest: The authors declare no conflict of interest. The funders had no role in the design of the study; in the collection, analyses, or interpretation of data; in the writing of the manuscript, or in the decision to publish the results.

References

1. Fiorentini, M.; Kinchla, A.J.; Nolden, A.A. Role of Sensory Evaluation in Consumer Acceptance of Plant-Based Meat Analogs and Meat Extenders: A Scoping Review. *Foods* **2020**, *9*, 1334. [CrossRef]
2. Foote, K.J.; Joy, M.K.; Death, R.G. New Zealand Dairy Farming: Milking our Environment for all Its Worth. *Environ. Manag.* **2015**, *56*, 709–720. [CrossRef]
3. Giacomelli, M.; Salata, C.; Martini, M.; Montesissa, C.; Piccirillo, A. Antimicrobial Resistance of Campylobacter jejuni and Campylobacter coli from Poultry in Italy. *Microb. Drug Resist.* **2014**, *20*, 181–188. [CrossRef] [PubMed]

4. Nawaz, M.F.; Bourrie, G.; Trolard, F. Soil Compaction Impact and Modelling. A Review. *Agron. Sustain. Dev.* **2013**, *33*, 291–309. [CrossRef]
5. Meat Alternatives and the Alternative Protein Market. Available online: https://www.fairr.org/article/alternative-protein-market/ (accessed on 14 January 2021).
6. Verbeke, W.; Sans, P.; Van Loo, E.J. Challenges and Prospects for Consumer Acceptance of Cultured Meat. *J. Integr. Agric.* **2015**, *14*, 285–294. [CrossRef]
7. Heymann, H.; King, E.S.; Hopfer, H. *Classical Descriptive Analysis*, 1st ed.; CRC Press: New York, NY, USA, 2014; pp. 9–40.
8. Lawless, H.T.; Heymann, H. *Sensory Evaluation of Food: Principles and Practices*, 1st ed.; Springer Science & Business Media: New York, NY, USA, 2010; pp. 227–252.
9. Spinelli, S.; Dinnella, C.; Masi, C.; Zoboli, G.P.; Prescott, J.; Monteleone, E. Investigating Preferred Coffee Consumption Contexts using Open-Ended Questions. *Food Qual. Prefer.* **2017**, *61*, 63–73. [CrossRef]
10. Delarue, J.; Lawlor, B. *Rapid Sensory Profiling Techniques: Applications in New Product Development and Consumer Research*, 1st ed.; Woodhead Publishing: Cambridge, UK, 2014; pp. 27–50.
11. Bakhtin, P.; Khabirova, E.; Kuzminov, I.; Thurner, T. The Future of Food Production–A Text-Mining Approach. *Technol. Anal. Strateg. Manag.* **2020**, *32*, 516–528. [CrossRef]
12. Bécue-Bertaut, M.; Álvarez-Esteban, R.; Pagès, J. Rating of Products through Scores and Free-Text Assertions: Comparing and Combining Both. *Food Qual. Prefer.* **2008**, *19*, 122–134. [CrossRef]
13. Zong, C.M.; Wang, H.B.; Du, X.J. The Application of Data Mining in the Food Manufacturing Industry. *Appl. Mech. Mater.* **2010**, *20*, 856–860. [CrossRef]
14. Hamilton, M.L.; Jacob, L. Fast and Automated Sensory Analysis: Using Natural Language Processing for Descriptive Lexicon Development. *Food Qual. Prefer.* **2020**, *83*, 103926. [CrossRef]
15. R: A Language and Environment for Statistical Computing. Available online: https://www.r-project.org (accessed on 29 January 2021).
16. Ickes, C.M.; Lee, S.Y.; Cadwallader, K.R. Novel Creation of a Rum Flavor Lexicon through the Use of Web-Based Material. *J. Food Sci.* **2017**, *82*, 1216–1223. [CrossRef]
17. Red-Gate. Text Mining and Sentiment Analysis. Available online: https://www.red-gate.com/simple-talk/sql/bi/text-mining-and-sentiment-analysis-with-r/ (accessed on 29 January 2021).
18. Cosson, A.; Delarue, J.; Mabille, A.C.; Druon, A.; Descamps, N.; Roturier, J.M.; Souchon, I.; Saint-Eve, A. Block Protocol for Conventional Profiling to Sensory Characterise Plant Protein Isolates. *Food Qual. Prefer.* **2020**, *83*, 103927. [CrossRef]
19. García-Segovia, P.; Igual, M.; Martínez-Monzó, J. Physicochemical Properties and Consumer Acceptance of Bread Enriched with Alternative Proteins. *Foods* **2020**, *9*, 933. [CrossRef]
20. Kaleda, A.; Talvistu, K.; Tamm, M.; Viirma, M.; Rosend, J.; Tanilas, K.; Kriisa, M.; Part, N.; Tammik, M.L. Impact of Fermentation and Phytase Treatment of Pea-Oat Protein Blend on Physicochemical, Sensory, and Nutritional Properties of Extruded Meat Analogs. *Foods* **2020**, *9*, 1059. [CrossRef]
21. Martin, C.; Lange, C.; Marette, S. Importance of Additional Information, as a Complement to Information Coming from Packaging, to Promote Meat Substitutes: A Case Study on a Sausage Based on Vegetable Proteins. *Food Qual. Prefer.* **2021**, *87*, 104058. [CrossRef]
22. Sha, L.; Xiong, Y.L. Plant Protein-Based Alternatives of Reconstructed Meat: Science, Technology, and Challenges. *Trends Food Sci. Technol.* **2020**, *102*, 51–61. [CrossRef]
23. Stephan, A.; Ahlborn, J.; Zajul, M.; Zorn, H. Edible Mushroom Mycelia of Pleurotus sapidus as Novel Protein Sources in a Vegan Boiled Sausage Analog System: Functionality and Sensory Tests in Comparison to Commercial Proteins and Meat Sausages. *Eur. Food Res. Technol.* **2018**, *244*, 913–924. [CrossRef]
24. Yuliarti, O.; Kovis, T.J.K.; Yi, N.J. Structuring the Meat Analogue by Using Plant-Based Derived Composites. *J. Food Eng.* **2021**, *288*, 110138. [CrossRef]
25. Grahl, S.; Strack, M.; Mensching, A.; Mörlein, D. Alternative Protein Sources in Western Diets: Food Product Development and Consumer Acceptance of Spirulina-Filled Pasta. *Food Qual. Prefer.* **2020**, *84*, 103933. [CrossRef]
26. Gurdian, C.E.; Torrico, D.D.; Li, B.; Tuuri, G.; Prinyawiwatkul, W. Effect of Disclosed Information on Product Liking, Emotional Profile and Purchase Intent: A Case of Chocolate Brownies Containing Edible-Cricket Protein. *Foods* **2021**, *10*, 1769. [CrossRef] [PubMed]
27. Gurdian, C.E.; Torrico, D.D.; Li, B.; Tuuri, G.; Prinyawiwatkul, W. Effect of Informed Conditions on Sensory Expectations and Actual Perceptions: A Case of Chocolate Brownies Containing Edible-Cricket Protein. *Foods* **2021**, *10*, 1480. [CrossRef] [PubMed]
28. De Koning, W.; Dean, D.; Vriesekoop, F.; Aguiar, L.K.; Anderson, M.; Mongondry, P.; Oppong-Gyamfi, M.; Urbano, B.; Luciano, C.A.G.; Jiang, B.; et al. Drivers and Inhibitors in the Acceptance of Meat Alternatives: The Case of Plant and Insect-Based Proteins. *Foods* **2020**, *9*, 1292. [CrossRef] [PubMed]
29. Ardoin, R.; Prinyawiwatkul, W. Product Appropriateness, Willingness to Try and Perceived Risks of Foods Containing Insect Protein Powder: A Survey of US Consumers. *Int. J. Food Sci. Technol.* **2020**, *55*, 3215–3226. [CrossRef]
30. Chow, C.Y.; Riantiningtyas, R.R.; Sørensen, H.; Frøst, M.B. School Children Cooking and Eating Insects as Part of a Teaching Program–Effects of Cooking, Insect Type, Tasting Order and Food Neophobia on Hedonic Response. *Food Qual. Prefer.* **2021**, *87*, 104027. [CrossRef]

31. Grasso, A.C.; Hung, Y.; Olthof, M.R.; Verbeke, W.; Brouwer, I.A. Older Consumers' Readiness to Accept Alternative, More Sustainable Protein Sources in the European Union. *Nutrients* **2019**, *11*, 1904. [CrossRef]
32. Schouteten, J.J.; De Steur, H.; De Pelsmaeker, S.; Lagast, S.; Juvinal, J.G.; De Bourdeaudhuij, I.; Verbeke, W.; Gellynck, X. Emotional and Sensory Profiling of Insect-, Plant-and Meat-Based Burgers under Blind, Expected and Informed Conditions. *Food Qual. Prefer.* **2016**, *52*, 27–31. [CrossRef]
33. Gómez, I.; Ibañez, F.C.; Beriain, M.J. Physicochemical and Sensory Properties of Sous Vide Meat and Meat Analog Products Marinated and Cooked at Different Temperature-Time Combinations. *Int. J. Food Prop.* **2019**, *22*, 1693–1708. [CrossRef]
34. Kamani, M.H.; Meera, M.S.; Bhaskar, N.; Modi, V.K. Partial and Total Replacement of Meat by Plant-Based Proteins in Chicken Sausage: Evaluation of Mechanical, Physico-Chemical and Sensory Characteristics. *J. Food Sci. Technol.* **2019**, *56*, 2660–2669. [CrossRef]
35. Wong, K.M.; Corradini, M.G.; Autio, W.; Kinchla, A.J. Sodium Reduction Strategies through Use of Meat Extenders (White Button Mushrooms vs. Textured Soy) in Beef Patties. *Food Sci. Nutr.* **2019**, *7*, 506–518. [CrossRef]
36. Gómez-Luciano, C.A.; de Aguiar, L.K.; Vriesekoop, F.; Urbano, B. Consumers' Willingness to Purchase Three Alternatives to Meat Proteins in the United Kingdom, Spain, Brazil and the Dominican Republic. *Food Qual. Prefer.* **2019**, *78*, 103–732. [CrossRef]
37. Tucker, C.A. The Significance of Sensory Appeal for Reduced Meat Consumption. *Appetite* **2014**, *81*, 168–179. [CrossRef]
38. Rodríguez-Limas, W.A.; Tannenbaum, V.; Tyo, K.E. Blocking Endocytotic Mechanisms to Improve Heterologous Protein Titers in Saccharomyces cerevisiae. *Biotechnol. Bioeng.* **2015**, *112*, 376–385. [CrossRef] [PubMed]
39. Van Krimpen, M.M.; Bikker, P.; Van der Meer, I.M.; Van der Peet-Schwering, C.M.C.; Vereijken, J.M. Cultivation, Processing and Nutritional Aspects for Pigs and Poultry of European Protein Sources as Alternatives for Imported Soybean Products. *Foods* **2013**, *9*, 1292.
40. Tao, D.; Yang, P.; Feng, H. Utilisation of Text Mining as a Big Data Analysis Tool for Food Science and Nutrition. *Compr. Rev. Food Sci. Food Saf.* **2020**, *19*, 875–894. [CrossRef] [PubMed]

foods

MDPI

Article

Understanding the Key Factors of Older Adults' Continuance Intention in Congregate Meal Halls

Wang-Chin Tsai [1] and Xuqi Chen [2,*]

[1] Department of Creative Design, National Yunlin University of Science & Technology, Yunlin 640, Taiwan; wangwang@yuntech.edu.tw
[2] Department of Agricultural and Resource Economics, University of Tennessee, Knoxville, TN 37996, USA
* Correspondence: xchen88@utk.edu

Abstract: Eating congregate/community meals with friends promotes a balanced and healthy diet among older adults. It is helpful for postponing aging, preventing chronic diseases, and improving their quality of life. However, little research has examined the continuance intention for older adults with the congregate meal program in Taiwan. This study established a model for key factors of older adults' continuance intention dining at senior meal halls, and hypotheses to explain them, and subsequently designed questionnaires and scales. By analyzing the longitudinal data collected from 416 individuals using survey questionnaires, we found that the perceived service quality is the main factor that affects the perceived satisfaction, and the perceived satisfaction of the older adults plays an important role in this survey. It showed that if the older adults are satisfied with the service quality provided by the senior meal halls, which will accordingly affect the post-use trust, they will show a positive continuance intention to participate in the senior meal halls. We also found that the older adults have positive views on the planning and service contents of the existing senior meal halls. Together, these results illustrate the process and provide comprehensive insights and evidence to create a better user experience and improve the satisfaction of the congregate meal for older adults.

Keywords: elderly; congregate meal; trust; continued intention

Citation: Tsai, W.-C.; Chen, X. Understanding the Key Factors of Older Adults' Continuance Intention in Congregate Meal Halls. *Foods* **2021**, *10*, 2638. https://doi.org/10.3390/foods10112638

Academic Editor: Witoon Prinyawiwatkul

Received: 23 September 2021
Accepted: 28 October 2021
Published: 30 October 2021

1. Introduction

As defined by the World Health Organization (WHO), a society with 20% of the total population aged over 65 years old is classified as a super-aged society [1]. According to the statistics of the Department of Household Registration, Ministry of the Interior, Taiwan is expected to enter the super-aged society in 2025, meaning that 20% of the population will be over 65 years old [2] (Figure 1), which would imply a severe aging population structure in Taiwan. Retired older adults may lose their goals in life, and may feel insignificant [3]. Additionally, when older adults lose their goals in life, consequent brain and physiology degeneration may occur, which can lead to serious social problems, such as dementia accelerated by loneliness [4], suicide due to depression [5], and even dying alone. Since these issues involve enormous consumption of medical and social resources, the current situation presents a challenge for the successful aging of older adults [6].

In previous decades, there has been a growing emphasis on "aging in place" globally [7], hoping to call more attention to the health, residential environment, psychology, diet, and interpersonal relationships of older adults [8]. Social Eating Programs for older adults have mentioned the nutritional risks and social isolation of older adults living in communities. According to Fulkerson [9], using Meals on Wheels, Eating with Friends, and Congregate/Community Meals can promote a balanced and healthy diet among older adults, which is helpful for postponing aging, preventing chronic diseases, and improving their quality of life [10]. Meanwhile, these programs can also alleviate problems, such as inconvenience in preparing meals, loneliness and depression, and a lack of activities. The Ministry of the Interior (Taiwan) has listed congregate/community meals in Welfare Law concerning older adults since 1997. For communities near rural areas, the establishment of congregate/community meals and senior meal halls is essential, due to the severe aging population structure and migration of young and middle-aged populations [11]. In addition, although senior meal halls have achieved initial success, there are still a considerable number of older adults who feel insecure about local aging and senior meal halls, which even lead to distrust, rejection, aversion [12], and reluctance in dining out. Taking Yunlin County in this study as an example, where the seniors account for the second-highest percentage of the population of Taiwan's 22 cities and counties, the county government has subsidized the establishment of several senior meal halls, where nearly 8000 people eat every day. This fact indicates that the older adults in Yunlin County have a certain demand for catering services. Therefore, it is necessary to look for experiences from the older adults who have dined at senior congregate meal halls, understand their pre-use trust, post-use trust, satisfaction, and continuance intention, through which we can better understand their experience, feelings, and mental processes when experiencing these meal halls (Figure 2). Our results can contribute to their specification by providing information about the reasons that drive the factors of this model.

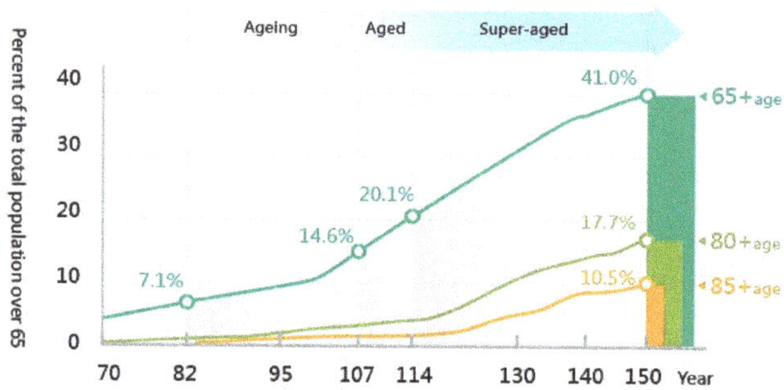

Figure 1. Aging process of Taiwan [2].

Figure 2. Older adults eat in the senior meal hall. (The image was taken by the author).

This study established a model for key factors of older adults' continuance intention dining at senior meal halls, and hypotheses to explain them. From a questionnaire formed from scales used in previous studies, data was gathered to enable structural equation modeling to be conducted to test these hypotheses. The findings allow suggestions to be put forth in order to shape aging policy to guide practitioners, and to support government decision-making.

2. Theoretical Framework and Research Hypotheses

2.1. Trust

As one of the factors explaining consumers' attitude, satisfaction, commitment, or behavioral intention, trust plays a significant role in uncertain and risky circumstances [13,14]. Generally, trust is a mental state interpreted as positive results of "perceived probabilities", "confidence", or "expectations" [15]. Trust is a dynamic concept [16] that changes, develops, and accumulates through repeated behavior [17]. The trust level of consumers will change through use or experience, and there is a feedback cycle of "trust-action-study-trust", which may repeatedly run [18]. Therefore, trust can be classified as pre-use trust and post-use trust, by which to describe differences of consumers' trust in products or services through potential influencing factors, and understand the value and influence of products or services on consumers [19]. While the social participation of older adults and the environment complement each other, trust is the basis of social participation [20]. Therefore, to improve older adults' participation level in the senior meal halls, and understand its influence on older adults, it is necessary to understand the difference between pre-use trust and post-use trust. Older adults' welfare and quality of life could be focused on two aspects: food, and service. In this study, pre-use trust represented the trust level before the older adults participated in community senior meal halls, while post-use trust represented the trust level after the older adults had participated in the community senior meal halls. Pre-use trust is the prerequisite and influencing factor of post-use trust [16]. That is, in addition to the relevant influencing factors of the topic, post-use trust can also be affected by pre-use trust. Therefore, the hypothesis is as follows:

Hypothesis 1 (H1). *There is a significant positive correlation between pre-use trust and post-use trust in older adults' participation in senior meal halls.*

2.2. Perceived Risk (PR) and Perceived Value (PV)

Initially belonging to psychology, perceived risk refers to the expected negative effects that consumers feel when they buy specific products [21]. Due to factors such as retirement, the decline of physical functions, and social and economic status, older adults are likely to have social-psychological pressure, leading to a higher risk of depression [22]. In comparison with others, older adults are more likely to feel insecure. Although senior meal halls have been implemented for many years in Taiwan, many people still do not know much about them. Therefore, older adults may have concerns about issues such as food security [23] and economic cost [24]. The community service of senior meal halls mainly includes two aspects of "food" and "service". "Food" belongs to the physical health dimension, which also includes the economic dimension, while "service" belongs to the psychological dimension [25]. In this study, perceived risk mainly explored the practical problems that older adults might face before or when participating in senior meal halls. Meanwhile, trust is one of the determinants to reduce perceived risk and uncertainty [26]. The higher the trust of older adults in senior meal halls, the lower the expected perceived risks are when they participate in community senior meal halls. Therefore, the hypothesis is as follows:

Hypothesis 2 (H2). *There is a significant negative correlation between pre-use trust and perceived risk in older adults' participation in senior meal halls.*

Value is the reflection of a capital cost optimization related to purchasing or production, which creates high-performing (value in use) and attractive (noble value) products or services for consumers [27]. On the other hand, perceived value is the value of products or services as perceived by consumers, which can be seen as the trade-off between perceived benefits and perceived costs [28]. Consumers will repeatedly patronize companies or enterprises with higher perceived value [29]. Although there is usually a negative correlation between perceived value and perceived risk [30], the perceived risk of this study focused more on reality, while the perceived value focused more on psychology. Therefore, no discussion will be made concerning the two factors. Mentally speaking, with certain accessibility, senior meal halls have lower participation requirements and offer opportunities to interact with others, which is one of the positive ways in which to achieve successful aging [31]. Therefore, taking senior meal halls as a platform and medium promotes older adults maintaining their physical and mental health, while older adults can encourage and support each other, and chat and interact to expand their social network [32]. Furthermore, the improvement of trust can also promote the improvement of perceived value [33]. Thus, the higher the trust in senior meal halls among older adults, the higher the probability of a positive perception of the community senior meal halls. Therefore, the hypothesis is as follows:

Hypothesis 3 (H3). *There is a significant positive correlation between pre-use trust and perceived value in older adults' participation in senior meal halls.*

2.3. Perceived Service Quality (PQ)

Perceived service quality is the consumers' overall evaluation of the perceived service, and whether it meets their expectations [34]. The evaluation standard of the service quality is dynamic, which varies according to different service items [35], and is usually classified as interaction quality, physical service environment, and outcome quality [36]. Consumers' previous experience can be an effective reference for perception and service quality analysis [37]. In other words, consumers must have experienced relevant services before having the perceived service quality, which will affect subsequent perception, such as satisfaction or brand loyalty [38], and further decisions about whether to use it again or not. Meanwhile, in the same way that perception is generated after using or experiencing a product or service, the perceived service quality will be affected by the previous factors of pre-use trust, perceived risk, and perceived value. In this study, perceived service quality represented the perceived service quality of the community senior meal halls participated by older adults. Therefore, the hypotheses are as follows:

Hypothesis 4 (H4). *There is a significant negative correlation between perceived risk and perceived service quality in older adults' participation in senior meal halls.*

Hypothesis 5 (H5). *There is a significant positive correlation between perceived value and perceived service quality in older adults' participation in senior meal halls.*

Hypothesis 6 (H6). *There is a significant positive correlation between pre-use trust and perceived service quality in older adults' participation in senior meal halls.*

2.4. Perceived Satisfaction (PS)

Perceived satisfaction is seen as an evaluation by comparing expectations and actual experience of products or services [39]. The service and experience evaluation results are considered to be the main predictors of behavioral intention [40]. Perceived service quality and satisfaction can be converted, as both are consumers' evaluation factors or opinions on particular products or services [34]. The difference between the two factors is that the judgment criteria of perceived service quality are more realistic and concrete, while satisfaction depends more on personal tendency and emotion [41]. Therefore, older adults' satisfaction with senior meal halls represents the relative relationship before and after the participation in community senior meal halls, which reflects the internal feeling of the older adults involved. Therefore, the hypothesis is as follows:

Hypothesis 7 (H7). *There is a significant positive correlation between perceived service quality and perceived satisfaction in older adults' participation in senior meal halls.*

2.5. Continuance Intentiion (CI)

Continuance intention, one of the means to measure whether products or services are successful [42], refers to the behavior shown after using or experiencing [43]. To promote the senior meal halls and make them widely accepted, in addition to the initial participation, continuance intention or behavior of the older adults is important [44]. As mentioned above, satisfaction is one of the main predictors [40,45] that affect the post-use trust [46] and continuance intention through trust [47]. In this study, continuance intention represented that, after participating in community senior meal halls, the older adults' perceived probability of continuance use in the future from a subjective perspective. Therefore, the hypotheses are as follows:

Hypothesis 8 (H8). *There is a significant positive correlation between perceived satisfaction and post-use trust in older adults' participation in senior meal halls.*

Hypothesis 9 (H9). *There is a significant positive correlation between perceived satisfaction and continuance intention in older adults' participation in senior meal halls.*

Hypothesis 10 (H10). *There is a significant positive correlation between post-use trust and continuance intention in older adults' participation in senior meal halls.*

2.6. Proposed Theoretical Model

Based on the discussions above, ten research hypotheses and the following model (e.g., Figure 3) were proposed. In particular, the model includes seven dimensions, namely pre-use trust, perceived risk, perceived value, perceived service quality, perceived satisfaction, post-use trust, and continuance intention.

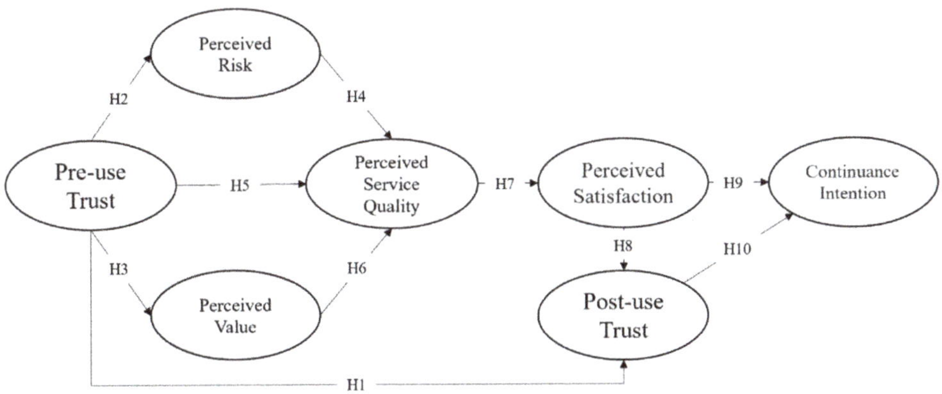

Figure 3. A model for continuance intention in congregate meal halls.

2.7. Definition and Measure of Variables

Based on the research topic and related literature, this questionnaire for this study was designed by selecting items from previous studies. The variables, operationalized definitions, and source of the items are shown in Table 1.

Table 1. Definition of variable operability and reference scales.

Research Variable	Operability Definition	Reference Scale
Pre-use trust	Trust level before individuals participate in community senior meal halls	[16,48,49]
Perceived risk	Expected perceived risk probability of individuals who participate in community senior meal halls	[50]
Perceived value	Perceived positive results probability of individuals participate in community senior meal halls	[51]
Perceived service quality	Perceived service quality level of individuals who participate in community senior meal halls	[52]
Perceived satisfaction	Relative relationship between experience before and after individuals participate in community senior meal halls	[19,48,53]
Post-use trust	Trust level after individuals participate in community senior meal halls	[16,48,49]
Continuance intention	Perceived probability of continuance of use in the future from a subjective perspective by individuals who participate in community senior meal halls	[54]

3. Data and Methodology

3.1. Sample and Data Collection

Taking older adults who have participated in senior meal halls as subjects, from March to May 2021, the questionnaires of this study were distributed to older adults in 88 senior meal halls in Yunlin County in written form. Ethical approval for this study was obtained from the National Cheng Kung University Human Research Ethics Committee (110-417-2). All participants originally gave informed consent to participate in this survey, including consent for their data to be used for research purposes. There were 20 to 30 older adults in each hall on average, and five to six people were randomly selected. In addition to demographic variables, the Likert 7-point scale was used for the items, ranging from 1 point (Strongly disagree) to 7 points (Strongly agree). Finally, 440 samples were collected. After eliminating invalid samples (logic error or too many similar options), there were 416 valid questionnaires. The estimated parameters of this research questionnaire include 38 questions, while all 416 questionnaires met Jackson's standard, where the ratio of estimated parameters to sample number should be higher than 1:10 [55]. Therefore, the following data analysis is based on these valid questionnaires.

Overall, among these respondents, 46.9% were male, and 53.1% were female. Over half of the participants(60.6%) had received an elementary school education. 67.1% were lived with family members, whereas 32.9% were lived independently. Most participants reported living in rural areas (62.5%). Approximately 38% of older adults reported that their main mode of transportation was using a private vehicle (motorcycle or car), 26.2% by bicycle, and 35.8% by foot. Slightly more than half of the respondents (51%) reported preparing food materials on their own, and 34% older adults received food prepared by family members, and others by neighborhood, friends, and caregivers (15%). Furthermore, older adult respondents reported "occasionally (48.1%)" for their weekly cooking frequency, followed by "frequently (21.1%)", and "always (14.9%)".

3.2. Reliability Analysis

Reliability tests and item analyses were carried out based on the results of the pre-test questionnaire to make the research results more accurate. The unstable items were deleted to ensure the reliability and avoid discrimination of the scales. As indicated in Table 2, the Cronbach's α value of all dimensions is greater than 0.7, meaning that all the dimensions have high reliability. Furthermore, the Cronbach's α value of each dimension will be lower than the current results when any of the items is deleted, suggesting that no item should be deleted. In general, the scales have high reliability for further analysis.

Table 2. Reliability and items analysis of the dimension items.

Dimension	Question	CITC	Cronbach's Alpha Value after Deleting the Item	Cronbach's Alpha Value
Pre-use trust	PT1	0.801	0.867	0.900
	PT2	0.731	0.883	
	PT3	0.735	0.882	
	PT4	0.720	0.885	
	PT5	0.772	0.874	
perceived risk	PR1	0.850	0.920	0.937
	PR2	0.848	0.920	
	PR3	0.820	0.925	
	PR4	0.775	0.933	
	PR5	0.871	0.916	
Perceived value	PV1	0.739	0.871	0.893
	PV2	0.675	0.879	
	PV3	0.692	0.877	
	PV4	0.631	0.884	
	PV5	0.653	0.881	
	PV6	0.698	0.876	
	PV7	0.747	0.870	
Perceived service quality	PQ1	0.790	0.850	0.892
	PQ2	0.749	0.865	
	PQ3	0.749	0.866	
	PQ4	0.760	0.862	
Perceived satisfaction	PS1	0.787	0.877	0.904
	PS2	0.741	0.887	
	PS3	0.734	0.888	
	PS4	0.740	0.887	
	PS5	0.795	0.875	
Post-use trust	PO1	0.769	0.893	0.911
	PO2	0.752	0.895	
	PO3	0.687	0.904	
	PO4	0.741	0.896	
	PO5	0.751	0.895	
	PO6	0.810	0.887	
Continuance intention	CI1	0.761	0.915	0.925
	CI2	0.779	0.912	
	CI3	0.697	0.923	
	CI4	0.778	0.913	
	CI5	0.836	0.905	
	CI6	0.857	0.902	

Note: CITC:corrected item-total correlation.

3.3. Descriptive Analysis and Normal Distribution Test

The hypothesis model of this study includes seven dimensions, namely pre-use trust, perceived risk, perceived value, perceived service quality, perceived satisfaction, post-use trust, and continuance intention. Moreover, the 38 items, along with the mean and standard deviation of the samples are shown in Table 3. At the same time, Table 4 shows the normal distribution of the sample data. The absolute value of skewness of all components is less than 3, while the absolute value of kurtosis is less than 7. Thus, it is considered that the data in this study conform with the normal distribution [56].

Table 3. Descriptive analysis and normal distribution test of dimension items.

	Number	Average Value	SD	Skewness	Kurtosis
Pre-use trust	416	5.10	1.13	−1.39	2.31
Perceived risk	416	4.20	1.52	−0.47	−1.04
Perceived value	416	4.72	1.10	−0.86	0.55
Perceived service quality	416	5.24	1.16	−0.77	0.46
Perceived satisfaction	416	5.09	1.17	−0.59	−0.67
Post-use trust	416	4.90	1.16	−0.61	−0.27
Continuance intention	416	5.12	1.24	−1.08	0.77

Note: SD: Standard Deviation

Table 4. Measurement model.

Dimension	Item	Unstd.	S.E.	Unstd./S.E.	*p*-Value	Std.	CR	AVE
PT	PT1	1.000	-	-	-	0.867	0.900	0.644
	PT2	0.839	0.045	18.834	0.000	0.774		
	PT3	0.826	0.044	18.926	0.000	0.777		
	PT4	0.826	0.045	18.397	0.000	0.762		
	PT5	0.892	0.043	20.909	0.000	0.828		
PR	PR1	1.000	-	-	-	0.890	0.938	0.752
	PR2	0.920	0.035	26.180	0.000	0.883		
	PR3	0.902	0.037	24.196	0.000	0.850		
	PR4	0.822	0.038	21.710	0.000	0.804		
	PR5	0.961	0.035	27.687	0.000	0.905		
PV	PV1	1.000	-	-	-	0.793	0.894	0.546
	PV2	0.888	0.057	15.462	0.000	0.721		
	PV3	0.920	0.057	16.039	0.000	0.743		
	PV4	0.899	0.063	14.340	0.000	0.677		
	PV5	0.870	0.059	14.726	0.000	0.693		
	PV6	0.921	0.057	16.105	0.000	0.746		
	PV7	0.981	0.057	17.295	0.000	0.790		
PQ	PQ1	1.000	-	-	-	0.861	0.892	0.674
	PQ2	0.914	0.046	19.765	0.000	0.811		
	PQ3	0.859	0.045	19.254	0.000	0.797		
	PQ4	0.931	0.047	19.904	0.000	0.815		
PS	PS1	1.000	-	-	-	0.834	0.904	0.655
	PS2	0.896	0.048	18.557	0.000	0.791		
	PS3	0.890	0.049	18.263	0.000	0.782		
	PS4	0.909	0.049	18.363	0.000	0.785		
	PS5	1.000	0.049	20.559	0.000	0.849		
PO	PO1	1.000	-	-	-	0.819	0.911	0.632
	PO2	0.966	0.052	18.568	0.000	0.796		
	PO3	0.813	0.051	16.007	0.000	0.713		
	PO4	0.905	0.050	18.223	0.000	0.785		
	PO5	0.936	0.052	18.140	0.000	0.783		
	PO6	0.997	0.048	20.956	0.000	0.867		
CI	CI1	1.000	-	-	-	0.805	0.926	0.678
	CI2	0.958	0.049	19.540	0.000	0.830		
	CI3	0.785	0.048	16.325	0.000	0.726		
	CI4	0.931	0.049	18.843	0.000	0.809		
	CI5	1.009	0.049	20.800	0.000	0.867		
	CI6	1.066	0.049	21.702	0.000	0.893		

Note: Unstd.: Unstandardized factor loadings, S.E.: Standard Error, Std.: Standardized factor loadings, CR: Composite Reliability, AVE: Average Variance Extracted.

3.4. Confirmatory Factor Analysis

3.4.1. Convergent Validity

AMOS v22.0 software was adopted to analyze the structural equation model in this study. Through a large amount of research, AMOS was proven to be a reliable structural equation model software. In addition, the data analysis includes two stages, according to the study of Anderson and Gerbing [57]. The first stage is the measurement model, where the maximum likelihood estimation method is used to estimate parameters, including factor loading, reliability, convergent validity, and discriminant validity. Congruent with the studies on convergence validity by Hair et al. [58], Nunnally and Bernstein [59] and Fornell and Larcker [60], and the study on loading of standardized factors by Chin [61] and Hooper et al. [62], the standardized factor loading in this study is higher than 0.6, the reliability of study dimension composition is higher than 0.7, and the average variance extraction (AVE) is higher than 0.5. These results indicate that the dimensions have good convergence validity [58]. The above numbers are listed in Table 4.

For discriminant validity, according to the study of Fornell and Larcker [60], if the square root of the AVE of each dimension is greater than the correlation coefficient between dimensions, it indicates that the model has discriminant validity. Meanwhile, this study showed that the values of all diagonal lines are greater than those outside the diagonal lines, indicating that every dimension in this study has a good discriminant validity (e.g., Table 5).

Table 5. Discriminant validity for the measurement model.

	PT	**PR**	**PV**	**PQ**	**PS**	**PO**	**CI**
PT	0.802						
PR	−0.151	0.867					
PV	0.452	−0.228	0.738				
PQ	0.315	−0.262	0.220	0.820			
PS	0.303	−0.173	0.431	0.290	0.809		
PO	0.301	−0.160	0.319	0.500	0.242	0.794	
CI	0.316	−0.176	0.254	0.398	0.164	0.287	0.823

Note: The items on the diagonal in bold represent the square roots of the Average Variance Extracted(AVE); off-diagonal elements are the correlation estimates.

3.4.2. Model Fit Test

According to Jackson et al. [63], Kline [56], Schumacker [64], and Hu and Bentler [65], several indicators (MLχ^2, DF, χ^2/DF, RMSEA, SRMR, TLI, CFI, NFI, GFI, PGFI, PNFI, and IFI) should be selected to evaluate the structural model fit. Based on the research hypothesis and models, as shown in Table 6, most of the standard model fit evaluation indicators meet the recommended fit's independent and combination rules. Hence, the structural model has a good fit.

3.5. Path Analysis

The path analysis results of Table 7 indicate that Pre-use Trust(PT) significantly influences Post-use Trust(PO)(b = 0.296, $p < 0.001$), Perceived Risk(PR) (b = −0.226, $p = 0.001$), and PV (b = 0.469, $p < 0.001$). Perceived Risk (PR) (b = −0.165, $p < 0.001$) and Pre-use Trust(PT) (b = 0.306 $p < 0.001$) significantly influence Perceived Service Quality(PQ). Perceived Service Quality (PQ) (b = 0.360, $p < 0.001$) significantly influences Perceived Satisfaction (PS). Perceived Satisfaction (PS) (b = 0.175, $p = 0.001$) significantly influences Post-use Trust (PO). Perceived Satisfaction (PS) (b = 0.127, $p = 0.021$) and Post-use Trust(PO) (b = 0.307, $p < 0.001$) significantly influence Continuance Intention (CI).

Table 6. Evaluation results.

Indicators	Norm	Results	Judgment
$ML\chi^2$	The small the better	1349.508	
DF	The large the better	655.000	
χ^2/DF	$1 < \chi^2/DF < 5$	2.060	Yes
RMSEA	<0.08	0.051	Yes
SRMR	<0.08	0.036	Yes
TLI (NNFI)	>0.9	0.931	Yes
CFI	>0.9	0.936	Yes
NFI	>0.9	0.883	No
GFI	>0.8	0.858	Yes
PGFI	>0.5	0.759	Yes
PNFI	>0.5	0.823	Yes
IFI	>0.9	0.936	Yes

Note: $ML\chi^2$: ML chi-square, DF: Degrees of Freedom, χ^2/DF: normed Chi-square, RMSEA: Root Mean Square Error Approximation, SRMR: Standardized Root Mean Square Residual, TLI: Tucker-Lewis Index, CFI: Comparative Fit Index, NFI: Normative Fit Index, GFI: Goodness of Fit index, PGFI: Parsimony Goodness of Fit Index, PNFI: Parsimony Normed Fit Index, IFI: Incremental Fit Index.

Table 7. Regression coefficient.

Hypothesis	DV	IV	Unstd	S.E.	Unstd./S.E.	*p*-Value	Std.	Results
H1	PO	PT	0.296	0.051	5.823	0.000	0.308	Yes
H2	PR	PT	−0.226	0.070	−3.251	0.001	−0.171	Yes
H3	PV	PT	0.469	0.050	9.451	0.000	0.508	Yes
H4	PQ	PR	−0.165	0.037	−4.413	0.000	−0.222	Yes
H5	PQ	PV	0.063	0.064	0.992	0.321	0.060	No
H6	PQ	PT	0.306	0.061	5.052	0.000	0.312	Yes
H7	PS	PQ	0.360	0.053	6.738	0.000	0.361	Yes
H8	PO	PS	0.175	0.051	3.426	0.001	0.177	Yes
H9	CI	PS	0.127	0.055	2.306	0.021	0.121	Yes
H10	CI	PO	0.307	0.057	5.369	0.000	0.289	Yes

Note: DV: Dependent Variable, IV: Independent Variable

3.6. Hypothesis Explanation

This study used the structural equation model to determine the influence factors of continuance intention on older adults' participation in senior meal halls.

Table 7 shows the normalization coefficient of the Structural equation modeling (SEM) model in this study. The higher coefficient implies that the independent variable plays a significant role in the dependent variable. Meanwhile, Figure 4 shows the influence between variables in the structural model.

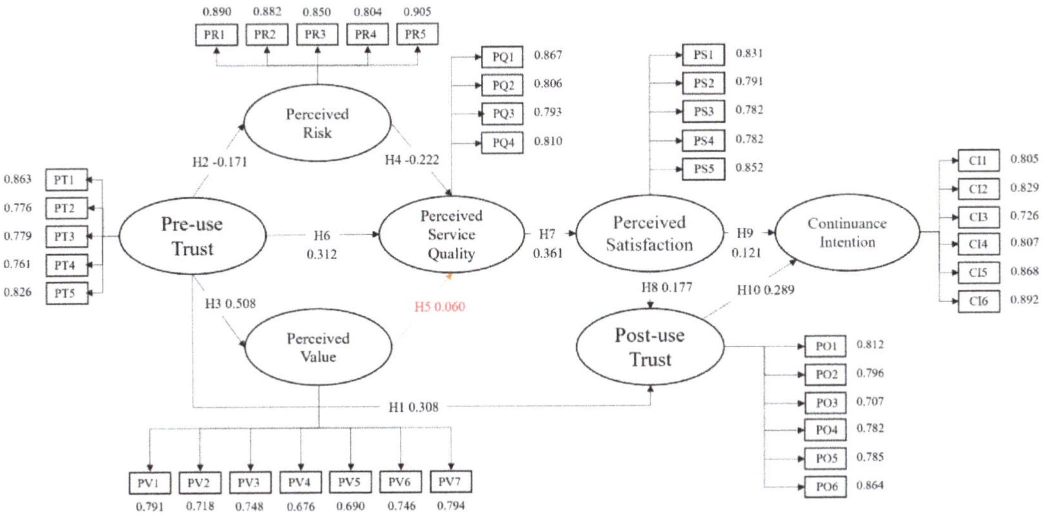

Figure 4. Research structure pattern diagram.

4. Discussion

This study identified critical factors affecting older adults' continuous participation in senior meal halls through structural equation modeling. Our study results in several important findings.

First, this result provides evidence that older adults' pre-use trust has a strong positive effect on perceived value. The hypothesis from extended valence framework along with the hypothesized path from pre-use trust to usage in the perceived value stage of the model is fully supported. Further, the path coefficient is the highest, suggesting that the more positive pre-use trust the older adults hold, the more value they can feel from the senior meal halls. Through relevant government policy advocacies, shared peer and community recommendations, trial and guide promotion meetings, such information on the space equipment, food concepts, and functions of senior meal halls, older adults have already known the content of attending the meal program. It has generated public reputation on its effectiveness, and constructed the participants' pre-use trust in senior meal halls. Moreover, after their actual participation in senior meal halls, the older adults feel motivated to enter the senior meal halls, and participate in the well-designed food promotion activities and catering services. In addition, they can get to know, talk, and share life trivia with other older adults and enthusiastic volunteers to achieve the effects on physical and mental health. Consequently, the trust level of senior meal hall participants will positively correlate with the perceived value.

Among the survey samples, 67% of older adults live with their family members. In addition to their own efforts, they need the assistance of family members or caregivers to prepare their meals. Although 84% of the older adults have cooking habits, they easily fail to have a balanced diet [9], because of their children's eating out, a small number of diners, and an unstable meal preparation. Therefore, it is necessary for older adults to eat in a community society. Older adults of the same age may relieve their psychological pressure, and find a way to communicate with others through the interaction of eating together [66]. In addition, the correct dietary concept could be communicated in the process through the transmission and exchange of nutritional information. These findings are consistent with the results of previous studies [67,68]. Furthermore, the older adults feel at ease to eat at a fixed meal time, feeling that it can reduce the burden and trouble of their children in the meal preparation [69], and that they need not rely on relatives and friends to live. Ultimately, they feel the value of participation in senior meal halls.

Second, the results indicate pre-use trust has a strong negative effect on perceived risk. Perceived risks existed due to the limitations of individual economic ability, food preparation ability, mobility, and physical and mental degradation to older adults' participation in senior meal halls [70,71]. The results of this study suggest that more than 60% of the older adults get to and from the senior meal halls by cars and motorcycles, while 40% walk, which may lead to traffic risks [72,73]. In addition, due to the recent improvement of dietary quality and concept, the concern for food security has also increased. Moreover, with the help of nutritionists in planning the menu, there is no need to eat leftovers several times, or excessively pickled food as in the past, as an imbalanced diet may lead to imbalanced nutrition [74]. Therefore, the risk factors such as cost standard, taste, nutritional balance, and food security are relatively reduced, proving a significant negative correlation between the degree of pre-use trust and the perceived risk. Moreover, the higher the pre-use trust is, the lower the degree of perceived risk older adults hold.

Third, we find that there is a significant negative correlation between the abovementioned older adults' perceived risk and the senior consumers' perceived service quality. In addition, the older adults' pre-use trust in senior meal halls and their perceived service quality has a significant positive correlation. This represents the concept of pre-use trust fully established by senior meal halls through government policies and community word-of-mouth. Currently, senior meal halls in Taiwan also encourage older adults not only to be passive recipients of care services, but also to become service providers (Figure 5). Therefore, many volunteers are involved in various service projects of the senior meal halls. The image they have established also promotes the perceived service quality of the older adults' participation [75].

Figure 5. Older adults engage as volunteers for food service. (The image was taken by the author).

However, there is no significant correlation between the perceived value and the perceived service quality in older adults' participation in senior meal halls. This result is different from the proposed hypothesis (H5). The reason why the perceived value does not influence perceived service quality may be that the perceived value of older adults belongs to the spiritual and psychological level of services, such as atmosphere [76], interaction mode, and diet education. Meanwhile, the perceived service quality corresponds with clear service contents (taste, space construction, charging price, and others). Older adults may pay more attention to the actual perceived service quality, while less attention is paid to the influence of psychological value.

In terms of the overall perceived service quality, the current senior meal halls, in combination with the Long-term Care Plan of Taiwan 2.0, shall provide nutritional assessment and interview and nutritional consultation for dietary guidance, alongside providing rich and varied dishes according to the needs and physical conditions of the older adults. As for management and control of service quality, a standard menu for cooking dishes should be established to ensure stable color, flavor, and taste in each dish. In daily quality control management, the service quality of internal staff should be emphasized, and workflow should be formulated to avoid unstable quality.

An irregular satisfaction survey of the meals should be conducted to serve as a basis for adjusting and improving dishes. In addition, services, such as simple blood pressure measurement, prevention of dementia, and nutrition lectures could be added. At the same time, the circular economy model is applied to increase the participation opportunities of older adults of the same age, by which to improve their participation in the perceived satisfaction of senior meal halls. The overall evaluation of the above specific service items indicates the experience and feelings of the older adults, making them feel that perceived service quality is essential for the perceived satisfaction of the senior meal halls. As one of the core services, perceived service quality is an important reference indicator for the overall high satisfaction of the senior meal halls.

Fourth, we find that there is a significant positive correlation among perceived satisfaction, post-use trust, and continuance intention of older adults participating in senior meal halls, which shows that the perceived satisfaction influences trust attitude and continuance intention. Furthermore, the higher the perceived satisfaction, the greater the influence on the post-use trust and continuance intention among older adults. Through meal planning and food security, cleaning, interpersonal interaction, price, space layout and arrangement, and other service factors such as meal serving, the sense of community is established, where post-use trust is consequently generated.

Finally, there is a significant positive correlation between post-use trust and continuance intention on older adults' participation in senior meal halls. This suggests that the post-use trust of this study has an influence on the continuance intention behavior. Use trust will become one of the factors in considering whether older adults continue to participate in the senior meal hall plan. The higher the trust, the more concerned they will be about hall equipment, space cooperation, and dining atmosphere. Consequently, a higher continuance intention of participation is generated [65]. In contrast, continuous participation in senior meal halls gradually changes the diet and lifestyle of older adults and promotes successful aging.

5. Conclusions and Suggestions

5.1. Conclusions

With the rapid development of the aging society, the older adults' catering services and lifestyles are constantly changing. In Taiwan, by eating together in the form of senior meal halls run by communities and relying on stable government planning and mechanism, it is expected that the dietary needs of the older adults could be met, and successful and healthy aging could be achieved. Previous studies on the senior meal halls for older adults are mostly qualitative analyses, lacking empirical data. Based on various important theories, the contribution of this study is to take pre-use trust as the first level of influence source of post-use trust, perceived risk, and perceived value, and establish a theoretical framework to influence the initial attitude of older adults participating in senior meal halls. The structural equation model was used to determine the main factors affecting the perceived service quality, perceived satisfaction, and continuance intention. Moreover, through an analysis of the related influence of this study, most of the dimensions have a direct or indirect influence on the continuance intention of participating in senior meal halls. The model of influencing factors on continuance intention of older adults' participation in senior meal halls established in this study is an acceptable model, which suggests that this model has a certain effect on explaining the intention of the older adults' participation in senior meal halls. Consequently, this can also provide an important reference for future policy and industrial development.

In addition, the results also show that the perceived service quality is the main factor that predicts the perceived satisfaction, while the perceived satisfaction of the older adults plays an important role in this survey. It shows that if the older adults are satisfied with the service quality (such as food flavor, space environment, atmosphere, among other factors) provided by the senior meal halls, which will accordingly affect the post-use trust, they will show a positive continuance intention to participate in the senior meal halls.

Through this survey, we believe that our study provided comprehensive insights and evidence into older adult's continuance intention process in congregate meal halls. The survey also indicated that, in addition to their concern about the food provided, the older adults also pay attention to how they can avail of health education publication services, nutrition consulting services, welfare information, or recommendation services through senior meal halls [77]. Therefore, it is essential to determine how to develop other regular activities to increase interaction frequency, social interaction, and willingness to continue participating [78]. This may include increasing the participation of older adults by recommendation, front-end tasks such as picking vegetables, washing vegetables, or menu planning, and follow-up tasks such as serving dishes, placing dishes, tidying up, or space arrangement, so that older adults can have a sense of achievement in social participation and labor efforts. At the same time, efforts should be made to improve the mobility limitations of older adults due to aging or inconvenient transportation. All of these will be indispensable factors for older adults to continue participating in senior meal halls.

5.2. Research Limitations and Future Research Suggestions

Some limitations of this study may indicate future research directions.

Firstly, this study explored the mutual influencing factors of the older adults' trust attitude, perceived value, and continuance intention of participating in senior meal halls. Thus, researchers can continue to explore multiple dietary plans with different meal-sharing patterns and service scales in the future. Secondly, due to the limitation of time and resources, this study only collected questionnaires from senior meal halls run by public sectors in central Taiwan. Although a highly aged county in Taiwan was taken as a reference value, individual differences still exist in different counties and cities. Therefore, different opinions may be held on the topic of this study. It is suggested that future research explores senior meal halls run by different counties, cities, or organizations to provide a reference for the government, communities, and relevant industries.

Author Contributions: Conceptualization, W.-C.T.; Data curation, W.-C.T.; Formal analysis, W.-C.T.; Supervision, W.-C.T. and X.C.; Writing—original draft, W.-C.T.; Writing—review and editing, W.-C.T. and X.C. All authors have read and agreed to the published version of the manuscript.

Funding: This research was funded by the Taiwan Ministry of Science and Technology (grant number: MOST 110-2410-H-224 -017).

Institutional Review Board Statement: The study was conducted according to the guidelines of the Declaration of Helsinki, and approved by National Cheng Kung University Human Research Ethics Committee under number (110-417-2).

Informed Consent Statement: Informed consent was obtained from all subjects involved in the study. All respondents gave their informed consent for inclusion of their answers before they participated in the study. Respondent details have been collected in an anonymous way with no personally identifiable information. Written informed consent has been obtained from the patient(s) to publish this paper.

Data Availability Statement: The data presented in this study are available on request from the first author.

Acknowledgments: I am grateful to Guan-Ting Lu for his work in data collection and the grant of Ministry of Science and Technology. I also thank the anonymous reviewers who provided valuable comments on the manuscript.

Conflicts of Interest: The authors declare no conflict of interest.

References

1. WHO. *Ageing, Active: A Policy Framework*; WHO: Geneva, Switzerland, 2002.
2. National Development Council, Population Projection 2020-2070, Taiwan. Available online: https://www.ndc.gov.tw/ (accessed on 26 October 2021).
3. Han, J.; Richardson, V.E. The relationships among perceived discrimination, self-perceptions of aging, and depressive symptoms: A longitudinal examination of age discrimination. *Aging Ment. Health* **2014**, *19*, 747–755. [CrossRef] [PubMed]
4. Carandang, R.R.; Shibanuma, A.; Kiriya, J.; Asis, E.; Chavez, D.C.; Meana, M.; Murayama, H.; Jimba, M. Determinants of depressive symptoms in Filipino senior citizens of the community-based ENGAGE study. *Arch. Gerontol. Geriatr.* **2019**, *82*, 186–191. [CrossRef] [PubMed]
5. Holt-Lunstad, J.; Smith, T.B.; Baker, M.; Harris, T.; Stephenson, D. Loneliness and Social Isolation as Risk Factors for Mortality. *Perspect. Psychol. Sci.* **2015**, *10*, 227–237. [CrossRef]
6. Santini, Z.I.; Jose, P.E.; Cornwell, E.Y.; Koyanagi, A.; Nielsen, L.; Hinrichsen, C.; Meilstrup, C.; Madsen, K.R.; Koushede, V. Social disconnectedness, perceived isolation, and symptoms of depression and anxiety among older Americans (NSHAP): A longitudinal mediation analysis. *Lancet Public Health* **2020**, *5*, e62–e70. [CrossRef]
7. Buffel, T.; Handler, S. *Age-Friendly Cities and Communities: A Global Perspective*; Policy Press: Bristol, UK, 2018.
8. Lewis, C.; Buffel, T. Aging in place and the places of aging: A longitudinal study. *J. Aging Stud.* **2020**, *54*, 100870. [CrossRef]
9. Fulkerson, J.A.; Larson, N.; Horning, M.; Neumark-Sztainer, D. A Review of Associations Between Family or Shared Meal Frequency and Dietary and Weight Status Outcomes Across the Lifespan. *J. Nutr. Educ. Behav.* **2014**, *46*, 2–19. [CrossRef] [PubMed]
10. Divert, C.; Laghmaoui, R.; Crema, C.; Issanchou, S.; Van Wymelbeke, V.; Sulmont-Rossé, C. Improving meal context in nursing homes. Impact of four strategies on food intake and meal pleasure. *Appetite* **2015**, *84*, 139–147. [CrossRef] [PubMed]
11. Welfare, M.O.H.A. Ten-Year Long-Term Care Program. Available online: https://1966.gov.tw/LTC/cp-3635-42393-201.html (accessed on 26 October 2021).
12. Phillips, J.; Walford, N.; Hockey, A. How do unfamiliar environments convey meaning to older people? Urban dimensions of placelessness and attachment. *Int. J. Ageing Later Life* **2012**, *6*, 73–102. [CrossRef]
13. Hartmann, M.; Klink, J.; Simons, J. Cause related marketing in the German retail sector: Exploring the role of consumers' trust. *Food Policy* **2015**, *52*, 108–114. [CrossRef]
14. Sengupta, A.S.; Balaji, M.; Krishnan, B.C. How customers cope with service failure? A study of brand reputation and customer satisfaction. *J. Bus. Res.* **2014**, *68*, 665–674. [CrossRef]
15. Şahin, A.; Kitapçi, H.; Zehir, C. Creating Commitment, Trust and Satisfaction for a Brand: What is the Role of Switching Costs in Mobile Phone Market? *Procedia Soc. Behav. Sci.* **2013**, *99*, 496–502. [CrossRef]
16. Lee, J.-N.; Choi, B. Effects of initial and ongoing trust in IT outsourcing: A bilateral perspective. *Inf. Manag.* **2011**, *48*, 96–105. [CrossRef]
17. Gefen, D.; Karahanna, E.; Straub, D. Inexperience and experience with online stores: The importance of tam and trust. *IEEE Trans. Eng. Manag.* **2003**, *50*, 307–321. [CrossRef]
18. Urban, G.L.; Amyx, C.; Lorenzon, A. Online Trust: State of the Art, New Frontiers, and Research Potential. *J. Interact. Mark.* **2009**, *23*, 179–190. [CrossRef]
19. Lin, J.; Wang, B.; Wang, N.; Lu, Y. Understanding the evolution of consumer trust in mobile commerce: A longitudinal study. *Inf. Technol. Manag.* **2013**, *15*, 37–49. [CrossRef]
20. Veenstra, G.; Luginaah, I.; Wakefield, S.; Birch, S.; Eyles, J.; Elliott, S. Who you know, where you live: Social capital, neighbourhood and health. *Soc. Sci. Med.* **2005**, *60*, 2799–2818. [CrossRef]
21. Dunn, M.G.; Murphy, P.E.; Skelly, G.U. Research note: The influence of perceived risk on brand preference for supermarket products. *J. Retail.* **1986**, *62*, 204–216.
22. Koster, A.; Bosma, H.; Kempen, G.I.; Penninx, B.W.; Beekman, A.T.; Deeg, D.J.; van Eijk, J.T. Socioeconomic differences in incident depression in older adults: The role of psychosocial factors, physical health status, and behavioral factors. *J. Psychosom. Res.* **2006**, *61*, 619–627. [CrossRef]
23. Bergmans, R.S.; Sadler, R.C.; Wolfson, J.A.; Jones, A.D.; Kruger, D. Moderation of the Association Between Individual Food Security and Poor Mental Health by the Local Food Environment Among Adult Residents of Flint, Michigan. *Health Equity* **2019**, *3*, 264–274. [CrossRef]
24. Murayama, Y.; Yamazaki, S.; Yamaguchi, J.; Hasebe, M.; Fujiwara, Y. Chronic stressors, stress coping and depressive tendencies among older adults. *Geriatr. Gerontol. Int.* **2020**, *20*, 297–303. [CrossRef]
25. Huang, J.-P. *Discussion on the Satisfaction and Depression Emotion of the Elderly in the Community and the Related Factors*; National Taipei University of Nursing and Health Sciences: Taipei, Taiwan, 2018.
26. Kim, S.S.; Malhotra, N.K. A Longitudinal Model of Continued IS Use: An Integrative View of Four Mechanisms Underlying Postadoption Phenomena. *Manag. Sci.* **2005**, *51*, 741–755. [CrossRef]
27. Eggert, A.; Ulaga, W.; Frow, P.; Payne, A. Conceptualizing and communicating value in business markets: From value in exchange to value in use. *Ind. Mark. Manag.* **2018**, *69*, 80–90. [CrossRef]
28. Lovelock, C. *Services Marketing People, Technology, Strategy, 5/e*; Pearson Education: London, UK, 2008.

29. Ryu, K.; Lee, H.-R.; Kim, W.G. The influence of the quality of the physical environment, food, and service on restaurant image, customer perceived value, customer satisfaction, and behavioral intentions. *Int. J. Contemp. Hosp. Manag.* **2012**, *24*, 200–223. [CrossRef]
30. Ho, S.S.; Leong, A.D.; Looi, J.; Chen, L.; Pang, N.; Tandoc, E. Science Literacy or Value Predisposition? A Meta-Analysis of Factors Predicting Public Perceptions of Benefits, Risks, and Acceptance of Nuclear Energy. *Environ. Commun.* **2018**, *13*, 457–471. [CrossRef]
31. Cheng, M.-H.; Zheng, M.-C. Arrangements of the congregate meal site based on elderly interactive behavior. In Proceedings of the 2017 International Conference on Applied System Innovation (ICASI), Sapporo, Japan, 13–17 May 2017; pp. 154–157. [CrossRef]
32. Cassel, J. Psychosocial Processes and "Stress": Theoretical Formulation. *Int. J. Health Serv.* **1974**, *4*, 471–482. [CrossRef]
33. Kim, D.J.; Ferrin, D.L.; Rao, H.R. Trust and Satisfaction, Two Stepping Stones for Successful E-Commerce Relationships: A Longitudinal Exploration. *Inf. Syst. Res.* **2009**, *20*, 237–257. [CrossRef]
34. Chen, C.-F. Investigating structural relationships between service quality, perceived value, satisfaction, and behavioral intentions for air passengers: Evidence from Taiwan. *Transp. Res. Part A Policy Pr.* **2008**, *42*, 709–717. [CrossRef]
35. Chen, L.; Li, Y.-Q.; Liu, C.-H. How airline service quality determines the quantity of repurchase intention—Mediate and moderate effects of brand quality and perceived value. *J. Air Transp. Manag.* **2018**, *75*, 185–197. [CrossRef]
36. Brady, M.K.; Cronin, J.J. Some New Thoughts on Conceptualizing Perceived Service Quality: A Hierarchical Approach. *J. Mark.* **2001**, *65*, 34–49. [CrossRef]
37. Ju, Y.; Back, K.-J.; Choi, Y.; Lee, J.-S. Exploring Airbnb service quality attributes and their asymmetric effects on customer satisfaction. *Int. J. Hosp. Manag.* **2018**, *77*, 342–352. [CrossRef]
38. Cheng, C.-C.; Chang, Y.-Y.; Chen, C.-T. Construction of a service quality scale for the online food delivery industry. *Int. J. Hosp. Manag.* **2021**, *95*, 102938. [CrossRef]
39. Oliver, R.L. A Cognitive Model of the Antecedents and Consequences of Satisfaction Decisions. *J. Mark. Res.* **1980**, *17*, 460–469. [CrossRef]
40. Gabbott, M.; Hogg, G. An empirical investigation of the impact of non-verbal communication on service evaluation. *Eur. J. Mark.* **2000**, *34*, 384–398. [CrossRef]
41. Oliver, R.L. *Satisfaction: A Behavioral Perspective on the Consumer: A Behavioral Perspective on the Consumer*; Routledge: Oxford, UK, 2014.
42. Teo, T.S.H.; Srivastava, S.C.; Jiang, L. Trust and Electronic Government Success: An Empirical Study. *J. Manag. Inf. Syst.* **2008**, *25*, 99–132. [CrossRef]
43. Montazemi, A.R.; Qahri-Saremi, H. Factors affecting adoption of online banking: A meta-analytic structural equation modeling study. *Inf. Manag.* **2015**, *52*, 210–226. [CrossRef]
44. Bhattacherjee, A. Understanding Information Systems Continuance: An Expectation-Confirmation Model. *MIS Q.* **2001**, *25*, 351–370. [CrossRef]
45. Dash, G.; Kiefer, K.; Paul, J. Marketing-to-Millennials: Marketing 4.0, customer satisfaction and purchase intention. *J. Bus. Res.* **2020**, *122*, 608–620. [CrossRef]
46. Yeh, Y.S.; Li, Y. Building trust in m-commerce: Contributions from quality and satisfaction. *Online Inf. Rev.* **2009**, *33*, 1066–1086. [CrossRef]
47. Jiang, Y.; Lau, A.K. Roles of consumer trust and risks on continuance intention in the sharing economy: An empirical investigation. *Electron. Commer. Res. Appl.* **2021**, *47*, 101050. [CrossRef]
48. Fernandes, T.; Calamote, A. Unfairness in consumer services: Outcomes of differential treatment of new and existing clients. *J. Retail. Consum. Serv.* **2016**, *28*, 36–44. [CrossRef]
49. Babu, M.M.; Dey, B.L. Appraisal of the potential of the growth of e-retailer based on impact of perceived value on trust for online purchases. *Strat. Chang.* **2018**, *27*, 477–487. [CrossRef]
50. Im, I.; Kim, Y.; Han, H.-J. The effects of perceived risk and technology type on users' acceptance of technologies. *Inf. Manag.* **2008**, *45*, 1–9. [CrossRef]
51. Hsiao, K.-L.; Chen, C.-C. What drives smartwatch purchase intention? Perspectives from hardware, software, design, and value. *Telemat. Inform.* **2018**, *35*, 103–113. [CrossRef]
52. Sok, P.; O'Cass, A. Achieving service quality through service innovation exploration—exploitation: The critical role of employee empowerment and slack resources. *J. Serv. Mark.* **2015**, *29*, 137–149. [CrossRef]
53. Kim, S.; Lee, J.-S. Is satisfaction enough to ensure reciprocity with upscale restaurants? The role of gratitude relative to satisfaction. *Int. J. Hosp. Manag.* **2013**, *33*, 118–128. [CrossRef]
54. Chen, S.C.; Chen, H.H. The empirical study of customer satisfaction and continued behavioural intention towards self-service banking: Technology readiness as an antecedent. *Int. J. Electron. Financ.* **2009**, *3*, 64. [CrossRef]
55. Jackson, D.L. Revisiting Sample Size and Number of Parameter Estimates: Some Support for the N:q Hypothesis. *Struct. Equ. Model. A Multidiscip. J.* **2003**, *10*, 128–141. [CrossRef]
56. Kline, R.B. Response to Leslie Hayduk's Review of Principles and Practice of Structural Equation Modeling, 4th Edition. *Can. Stud. Popul.* **2018**, *45*, 188–195. [CrossRef]

57. Anderson, J.C.; Gerbing, D.W. Structural equation modeling in practice: A review and recommended two-step approach. *Psychol. Bull.* **1988**, *103*, 411–423. [CrossRef]
58. Hair, J.F.; Black, W.C.; Babin, B.J.; Anderson, R.E.; Tatham, R.L. *Multivariate Data Analysis*; Prentice Hall: Upper Saddle River, NJ, USA, 1998; Volume 5.
59. Nunnally, J.; Jum, N.; Bernstein, I.H.; Bernstein, I. *Psychometric Theory*; McGraw-Hill Companies: New York, NY, USA, 1994.
60. Fornell, C.; Larcker, D.F. Evaluating Structural Equation Models with Unobservable Variables and Measurement Error. *J. Mark. Res.* **1981**, *18*, 39–50. [CrossRef]
61. Chin, W.W. Commentary: Issues and opinion on structural equation modeling. *Commentary* **1998**, *22*, 7–16.
62. Hooper, D.; Coughlan, J.; Mullen, M.R. Structural equation modelling: Guidelines for determining model fit. *Electron. J. Bus. Res. Methods* **2008**, *6*, 53–60.
63. Jackson, D.L.; Gillaspy, J.A., Jr.; Purc-Stephenson, R. Reporting practices in confirmatory factor analysis: An overview and some recommendations. *Psychol. Methods* **2009**, *14*, 6. [CrossRef]
64. Whittaker, T.A. *A Beginner's Guide to Structural Equation Modeling*; Taylor & Francis: Oxford, UK, 2011.
65. Hu, L.; Bentler, P.M. Cutoff criteria for fit indexes in covariance structure analysis: Conventional criteria versus new alternatives. *Struct. Equ. Model. A Multidiscip. J.* **1999**, *6*, 1–55. [CrossRef]
66. Sen, B.; Goldfarb, S.; Tarver, W. Family structure and risk behaviors: The role of the family meal in assessing likelihood of adolescent risk behaviors. *Psychol. Res. Behav. Manag.* **2014**, *7*, 53–66. [CrossRef]
67. Björnwall, A.; Sydner, Y.M.; Koochek, A.; Neuman, N. Eating Alone or Together among Community-Living Older People—A Scoping Review. *Int. J. Environ. Res. Public Health* **2021**, *18*, 3495. [CrossRef] [PubMed]
68. Winter, J.; McNaughton, S.; Nowson, C. Older adults' attitudes to food and nutrition: A qualitative study. *J Aging Re* **2016**, *5*, 114–119. [CrossRef]
69. Ishikawa, M.; Takemi, Y.; Yokoyama, T.; Kusama, K.; Fukuda, Y.; Nakaya, T.; Nozue, M.; Yoshiike, N.; Yoshiba, K.; Hayashi, F.; et al. "Eating together" is associated with food behaviors and demographic factors of older Japanese people who live alone. *J. Nutr. Health Aging* **2016**, *21*, 662–672. [CrossRef]
70. Rempe, H.M.; Sproesser, G.; Gingrich, A.; Spiegel, A.; Skurk, T.; Brandl, B.; Hauner, H.; Renner, B.; Volkert, D.; Sieber, C.C.; et al. Measuring eating motives in older adults with and without functional impairments with The Eating Motivation Survey (TEMS). *Appetite* **2019**, *137*, 1–20. [CrossRef]
71. Host, A.; McMahon, A.-T.; Walton, K.; Charlton, K. Factors Influencing Food Choice for Independently Living Older People—A Systematic Literature Review. *J. Nutr. Gerontol. Geriatr.* **2016**, *35*, 67–94. [CrossRef]
72. Deeming, C. Food and Nutrition Security at Risk in Later Life: Evidence from the United Kingdom Expenditure & Food Survey. *J. Soc. Policy* **2010**, *40*, 471–492. [CrossRef]
73. Chung, W.T.; Gallo, W.T.; Giunta, N.; Canavan, M.E.; Parikh, N.S.; Fahs, M.C. Linking Neighborhood Characteristics to Food Insecurity in Older Adults: The Role of Perceived Safety, Social Cohesion, and Walkability. *J. Hered.* **2011**, *89*, 407–418. [CrossRef]
74. Kim, C.-O. Food choice patterns among frail older adults: The associations between social network, food choice values, and diet quality. *Appetite* **2016**, *96*, 116–121. [CrossRef]
75. Liu, Y.; Duan, Y.; Xu, L. Volunteer service and positive attitudes toward aging among Chinese older adults: The mediating role of health. *Soc. Sci. Med.* **2020**, *265*, 113535. [CrossRef]
76. Stroebele-Benschop, N.; Depa, J.; De Castro, J.M. Environmental Strategies to Promote Food Intake in Older Adults: A Narrative Review. *J. Nutr. Gerontol. Geriatr.* **2016**, *35*, 95–112. [CrossRef] [PubMed]
77. Sulmont-Rossé, C.; Symoneaux, R.; Feyen, V.; Maitre, I. Improving Food Sensory Quality With and For Elderly Consumers. *Methods Consum. Res.* **2018**, *2*, 355–372. [CrossRef]
78. Thomas, N.; Emond, R. Living alone but eating together: Exploring lunch clubs as a dining out experience. *Appetite* **2017**, *119*, 34–40. [CrossRef] [PubMed]

foods

MDPI

Article

Emoji for Food and Beverage Research: Pleasure, Arousal and Dominance Meanings and Appropriateness for Use

Sara R. Jaeger [1], David Jin [1], Grace S. Ryan [1] and Joachim J. Schouteten [2,*]

[1] The New Zealand Institute for Plant and Food Research Limited, Mt Albert Research Centre, Private Bag 92169, Auckland 1142, New Zealand; sara.jaeger@plantandfood.co.nz (S.R.J.); david.jin@plantandfood.co.nz (D.J.); grace.ryan@plantandfood.co.nz (G.S.R.)
[2] Department of Agricultural Economics, Ghent University, Coupure Links 653, 9000 Gent, Belgium
* Correspondence: joachim.schouteten@ugent.be

Abstract: Emoji have been argued to have considerable potential for emotion research but are struggling with uptake in part because knowledge about their meaning is lacking. The present research included 24 emoji (14 facial, 10 non-facial) which were characterized using the PAD model (Pleasure-Arousal-Dominance) of human affect by 165 consumers in New Zealand and 861 consumers in the UK. The results from the two countries were remarkably similar and contributed further evidence that emoji are suitable for cross-cultural research. While significant differences between the emoji were established for each of the PAD dimensions, the mean scores differed most on the Pleasure dimension (positive to negative), then on the Arousal dimension (activated to deactivated), and lastly on the Dominance dimension (dominance to submissive). The research also directly measured the perceived appropriateness of the 24 emoji for use with foods and beverages. The emoji face savoring food, clapping hands and party popper were in the top-5 for the highest appropriateness in food and beverage context for both studies, as was a strong negative expression linked to rejection (Study 1: face vomiting; Study 2: nauseated face). On the other hand, zzz and oncoming fist were considered as the least appropriate to be used in a food and beverage context in both studies. Again, the results from the UK and NZ were in good agreement and identified similar groups of emoji as most and least suitable for food-related consumer research.

Keywords: emoji; emotions; pleasure; arousal; dominance; PAD model; consumer research

Citation: Jaeger, S.R.; Jin, D.; Ryan, G.S.; Schouteten, J.J. Emoji for Food and Beverage Research: Pleasure, Arousal and Dominance Meanings and Appropriateness for Use. *Foods* **2021**, *10*, 2880. https://doi.org/10.3390/foods10112880

Academic Editor: Witoon Prinyawiwatkul

Received: 27 October 2021
Accepted: 16 November 2021
Published: 22 November 2021

1. Introduction

1.1. Emoji in Food-Related Consumer Research

As a means to gain additional insights about consumers' product experiences, measurement of emotional associations to foods and beverages (F&B) has increased in the past decade [1,2]. Questionnaires are primarily used, and regarded as the default measurement approach [3]. They can be implemented using emotion words or emoji, and we focus here on the latter which are less studied than emotion words [4] but have high relevance in consumer research as graphical representations that convey emotions and feelings through their resemblance with real objects [5].

Emoji are now widely used on social platforms and in digital communications have become well known by many in the general public [6]. This has the advantage that participants in consumer research are likely to be used to expressing emotions and feelings through emoji which, relative to emotion words, positively contributes to ecological validity [4,7]. As emoji are non-verbal, they enable consumers to express emotions that otherwise cannot be expressed with words or that would be expressed differently [8]. The potential of emoji in consumer research has been illustrated by several studies showing that they can discriminate between F&B products (e.g., [9–11]) and increase food choice prediction [12]. However, it has been pointed out that knowledge is lacking about which emoji are appropriate to be used in the context of F&Bs and eating/drinking consumption situations more broadly [4,13]. This fits with the notion that textual context could reduce ambiguity in interpretation of some emoji [14]. Therefore, there is a need to establish suitability for individual emoji in food-related consumer research.

1.2. Dimensional Meanings of Emoji: Pleasure, Arousal and Dominance

Despite empirical evidence pointing to good similarity between the meanings and interpretations that research participants associate to different emoji [15,16], there is nonetheless limited knowledge on emoji meanings in the context of food-related research, and this has been suggested as a major barrier for uptake [4,13]. Furthermore, one needs to consider that emoji meanings can be established by obtaining more information about the verbal meaning of an emoji (e.g., by a CATA question or assessment of the use of the emoji in a certain food context) but also by assessing the meanings of the emoji related to the main dimensions of emotions. To bring a new aspect to this pursuit, we focus in this paper on emoji meaning on such underpinning dimensions of emotions and human affect [17] rather than profiling by individual emotion words.

Dimensions of emotions have not been an area of focus in food-related consumer research [18] and emphasis has de facto been on the valence dimension (positive to negative), where even emotion words with negative valence were underrepresented in early works (e.g., [19]). In more recent times, the arousal dimension (activated to deactivated) has been given greater attention, for example by the development of a circumplex-inspired emotion word questionnaire that purposefully spans the valence × arousal space [20]. Further, Self-Assessment Manikins (SAM: [21]) have been used to measure establish meanings for the dimensions of valance and arousal in the USA [22] and China [23].

However, the dominance dimension (dominance to submissive) remains largely neglected still. This dimension can be seen as a "power" dimension which is related to the extent that a person is in control of a situation and the people in that situation [24]. The dominance dimension can distinguish between similar arousal levels. For example, fear and anger are both high in arousal but fear is low in dominance while anger is high in dominance [8]. Data from a projective mapping exercise using emoji with pre-adolescents suggest that the dominance dimension might also be important for the meaning attached to emoji [8]. The authors hypothesized that, compared to adults, higher relevance of the dominance dimension among pre-adolescents might be due to a focus on situational control (thus dominance) rather than psychological activation.

When considering the different dimensions of human affect, the PAD model (Pleasure, Arousal and Dominance: [25]) has relevance since it incorporates the three aforementioned dimensions and has been widely used to examine human affect and emotion [17]. The PAD model has been validated as appropriate for studying consumer emotions and consumer behavior in different consumption contexts, including food-related settings [26,27], and is therefore appropriate for research into emoji meaning.

1.3. Research Aims and Contributions

Building on the research summarized above, and the recent direct recommendations of Jaeger, Vidal and Ares [4] and Schouteten and Meiselman [13], the present research had two objectives: (1) contribute further knowledge regarding emoji meanings, and (2) establish the extent to which consumers find certain emoji suitable in the F&B context.

We extend previous research by using the PAD scale to measure three dimensions of meaning and by including some different emoji to those previously considered. We also obtain data in a country where emoji meanings regarding the main emotional dimensions (valence, arousal, and power) have not previously been established: New Zealand (NZ) and the United Kingdom (UK).

Finally, we directly measure the perceived appropriateness of individual emoji in the context of F&Bs and associated consumption occasions. It should be noted that the goal was to explore the application of the PAD meanings of the selected emoji in two countries, rather than directly comparing the PAD meanings in these two countries.

2. Materials and Methods

2.1. Participants

Study 1 was conducted in NZ at a central location testing facility in Auckland. Adult participants (n = 180) (50% female, 20–65 years old) were recruited by a professional company. Study 2 took place in the UK (England, Scotland, Wales, and Northern Ireland) with participants (n = 1047) (50% female, 18–65 years old) who had self-registered with an ISO-accredited web panel provider. The Supplementary Material has full participant details.

Both studies were covered by a general approval for sensory and consumer research from the Human Ethics Committee at the New Zealand Institute for Plant and Food Research (PFR). Participants were assured that their responses would remain confidential. As compensation, participants in Study 1 received cash while those in Study 2 earned reward points and promotional offers.

2.2. Brief Overview of Empirical Approach

There were three parts to the empirical approach that participants took part in. The first was to complete the 18 PAD scales (i.e., emoji meaning) for several emoji. Then participants completed questions relating to Objective 2. These captured emoji specific use and suitability for F&B occasions. Finally, there were questions about emoji use and attitudes, as well as demographic and socio-economic questions. The procedures are described in full below, starting with the selection of emoji to be included in the research.

2.3. Emoji Selection and Evaluation

2.3.1. Emoji

A candidate set of emoji were proposed by two authors and revised through discussion with other authors and colleagues. The number of emoji included in the research (n = 24) was determined by constraints of Study 1 and the target was to have each emoji evaluated by ~40 participants.

The same set of 24 emoji was used in both studies (Figure 1). Their selection process was guided by several criteria, foremost seeking to span each of the PAD dimensions, through a combination of emoji previously used in food-related research [8,22], those known to be popular on social media in situations relating to eating and drinking [28] and emoji not previously considered in this research domain (e.g., 🎉, 🧘, 💥). Facial emoji were used primarily, but where a greater range in one or more of the PAD dimensions were expected by the inclusion of non-facial emoji, these were selected (42%). Inclusion of non-facial emotion was warranted since some of the most-used emoji on Twitter are non-facial (https://emojitracker.com/, (accessed on 1 April 2021)), and some have obvious links to emotional affect (e.g., red heart (❤️), thumbs up sign (👍) and thumbs down sign (👎)) [29]. Moreover, some previous studies excluded non-facial emoji [22], so less is known about their meaning.

2.3.2. PAD Scale

The PAD scale from Mehrabian and Russell [25], which is a well-established and valid measurement tool valid for the scientific study of human emotions [17], was used to measure emoji meaning (Obj. 1).

Specifically, each emoji was evaluated on the 18 semantic differentials (Table 1), which comprised six items for each of the three dimensions P (Pleasure), A (Arousal), and D (Dominance) (Obj. 1). These were implemented accordingly using 9-pt scales. There was one minor change: for P6, based on Detandt, Leys [30], the right anchor (9) was retained as 'bored' while the left anchor (1) was changed from 'relaxed' to 'amused'.

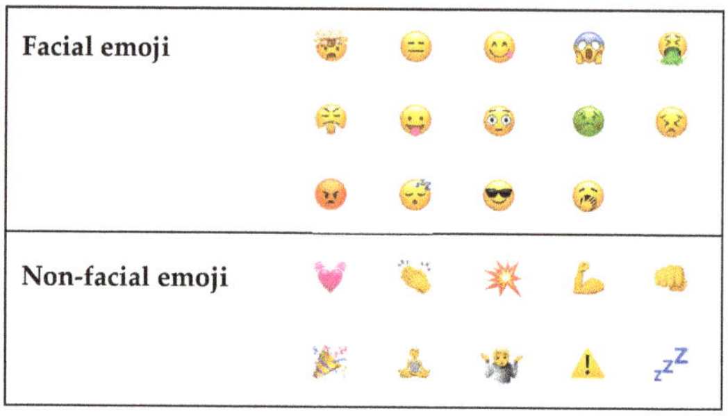

Figure 1. Emoji included in the research.

Table 1. The 18 semantic differentials in the pleasure-arousal-dominance (PAD) scale used in Study 1 and Study 2, with end-point anchors shown.

Dimension	PAD Code	Left Anchor (1)	Right Anchor (9)
Pleasure (Pleasure-Displeasure)	P1	Happy	Unhappy
	P2	Pleased	Annoyed
	P3	Satisfied	Unsatisfied
	P4	Contented	Melancholic
	P5	Hopeful	Despairing
	P6	Amused *	Bored
Arousal (Arousal-Nonarousal)	A1	Stimulated	Relaxed
	A2	Excited	Calm
	A3	Frenzied	Sluggish
	A4	Jittery	Dull
	A5	Wide-awake	Sleepy
	A6	Aroused	Unaroused
Dominance (Dominance-Submissiveness)	D1	Controlling	Controlled
	D2	Influential	Influenced
	D3	In control	Cared for
	D4	Important	Awed
	D5	Dominant	Submissive
	D6	Autonomous	Guided

Note. * Changed scale anchor relative to Mehrabian and Russell (1974), in accordance with Detandt, Leys [30].

In keeping with Mehrabian and Russell [25], *Pleasure* was regarded as a feeling along a continuum ranging from unhappiness to extreme happiness (i.e., valence). The *Arousal* dimension was interpreted as a mental activity that can be described along a single dimension ranging from sleep to excitement. Lastly, *Dominance* was depicted as being related to feelings of control along a continuum from submissive to dominance.

2.3.3. Emoji Specific Use and Suitability for F&B Occasions

In addition to consumers' evaluation of the emoji on the PAD scales, the emoji were also assessed in relation to use characteristics specific to F&Bs and/or eating/drinking situations (Obj. 2). Three questions were used, where the first asked participants to indicate frequency of use of the focal emoji when sending messages, emails, etc. A 5-pt response scale was used with anchors: 5 = 'always or almost every time (>90%)', 4 = 'frequently (more often than not, 65–90%)', 3 = 'sometimes (not all the time, but neither rarely, 35–65%)', 2 = 'infrequently (generally not, but on occasion, 10–35%)' and '1 = never or very infrequently (<10%)'. This question was only included in study 1. Next, participants were asked to indicate whether or not they could imagine a food/beverage or eating situation where using this emoji would be appropriate. The final question pertained to past use and participants responded yes, no, not sure, to the question of whether they recalled having used the focal emoji for a food, beverage or eating situation.

2.3.4. General Emoji Use and Attitudes

General emoji use was measured with the question: "In general, how often do you use emoji when sending messages, emails etc.?" which had the same five answer categories as used above for the emoji-specific use question.

A set of eight statements capturing general attitudes to emoji were scaled on fully labeled 7-pt Likert statements with endpoint anchors 1 = 'disagree strongly' and 7 = 'agree strongly'. The statements were: 'Emoji are more appropriate in private than professional communications', 'I generally use a small set of emoji over and over again', 'Using emoji helps me to express my moods/emotions', 'My favorite emoji are "face emoji" (e.g., , , )', 'I like to use many different emoji, of all kinds', 'I consider myself to be emoji savvy and literate', 'Emoji are fun to use and receive', and 'In computer-mediated communications, emoji help me to better express myself'.

2.4. Data Collection Procedures

Emoji were presented sequentially in both studies, following a Williams design (Study 1) or in randomised order (Study 2). Pen and paper ballots were used in Study 1, with the emoji shown at the top of the page in colour (4.2 × 4.2 cm) and the 18 PAD scales below. These were followed by the questions pertaining to emoji specific uses and perceptions. Each participant evaluated six emoji, and once completed a short ballot followed comprising questions relating to the general use of emoji. The procedure was very similar in Study 2, with the exception that each participant evaluated three emoji and each of the 18 PAD scales were presented horizontally and shown individually (Part 2 of Supplementary Material). Part 3 of Supplementary Material has details on the allocation of emoji to participants. The presentation order of PAD scales varied across respondents, in accordance with Ares, Reis et al. (2015) in Study 1 and randomly in Study 2.

Study 1 took part in a central location facility and participants were seated in standard sensory testing booths. Pen and paper ballots were used in Study 1 as the CLT location does not have access to computerized data collection infrastructure. Participants in Study 2 completed the survey in a location of their choosing using a desktop/laptop computer or tablet. In both studies, the data were obtained as part of surveys/research sessions that included tasks other than those described here (not considered further due to lack of relevance). Demographic and socio-economic information was obtained as the final part of the survey/session. Participants typically took 10–15 min to complete the study.

2.5. Data Analysis

The data for each study were analysed separately, and all analyses were performed in XLSTAT v.2020.5.1 [31] using a 5% significance level.

The first step was to exclude participants who stated that they never used emoji or only did so infrequently (i.e., <10% of communications). It was regarded as non-sensical to include such participants to address the aims of the research. The effective sample sizes were: Study (1) $n = 165$, with each emoji being evaluated by 40 to 43 participants, and Study (2) $n = 861$ with each emoji being evaluated by 99 to 116 participants.

Objective 1: The 18 PAD variables (Study 1 and Study 2) were input to a one-way analysis of variance (ANOVA) using emoji as a fixed factor and participants as a random factor. Tukey's HSD was used for post hoc tests. Principal components analysis (PCA) was performed using the correlation matrix based on means as input. RV coefficients were calculated to determine the similarity between the PCA matrixes (3 dim.) of the emoji and PAD variables for the NZ and UK study. One-way ANOVA was performed for each of the three dimensions (P, A and D) and supplemented with post hoc tests as above, after confirming that Cronbach alpha values exceeded the typical 0.7 threshold (α for, respectively, Pleasure, arousal and dominance was 0.95, 0.90 and 0.79, in Study 1 and 0.93, 0.88 and 0.76 in Study 2).

Objective 2: Emoji use responses (Study 1) were also analysed using a one-way ANOVA, while citation frequencies were calculated for responses relating to eating and drinking. Correlation analyses were used to determine the association between emoji-specific use frequencies and perceived appropriateness for use in situations relating to F&Bs or eating and drinking.

3. Results

The results for Study 1 and Study 2 were very similar and are jointly presented, progressing from Objective 1 (dimensional emoji meanings) to Objective 2 (emoji appropriateness and use in the context of F&B consumption).

3.1. Emoji Characterisation on PAD Variables

For the main insights relating to Objective 1, the 18 PAD variables were averaged to their three respective dimensions. Significant differences ($p < 0.001$) between the 24 emoji were established (Parts 4 to 6 of Supplementary Material contains detailed results by study incl. Tukey HSD post hoc tests). Figure 2 showed that the span in mean scores across the 24 emoji was larger in Study 1 than Study 2, although it should be noted that the same patterns were observed in both studies. For example, the top-5 ranked emoji for pleasure and displeasure were identical (respectively, party popper (🎉), face savouring food (😋), smiling face with sunglasses (😎), face with tongue (😛) and pouting face (😡), face vomiting (🤮), persevering face (😣), nauseated face (🤢), face with steam from nose (😤). For the deactivated pole of the *Arousal* dimension, the top-5 ranked emoji were also identical (yawning face (🥱), sleeping face (😴), zzz (💤), person in lotus position (🧘) and expressionless face (😑)), as were three of the top-5 ranked emoji for the activated pole of *Arousal* (party popper (🎉), collision (💥), face with tongue (😛)). For *Dominance*, agreement between the two studies was also very high: persevering face (😣), nauseated face (🤢), flushed face (😳), yawning face (🥱) and sleeping face (😴) for the most submissive emoji, and collision (💥), flexed biceps (💪), oncoming fist (👊) and party popper (🎉) as high in dominance. On this dimension, the largest discrepancy between the two studies was seen for smiling face with sunglasses (😎). Finally, Figure 2 drew attention to differences in meaning across the 24 emoji, with some being perceived as similar with regard to *Pleasure*, yet different on *Arousal* (e.g., face with steam from nose (😤) and yawning face (🥱)), or how emoji perceived similar with regard to *Dominance* would be associated with the different poles of the *Arousal* dimension (e.g., face screaming in fear (😱) and sleeping face (😴)).

To further support the exploration of emoji meaning according to the PAD model, PCA performed on the 18 PAD variables resulted in three-factor solutions being retained for both studies, accounting for >90% of total variance with ~61%, ~23% and ~8% accounted for by PC1, PC2 and PC3, respectively. Fitting with the univariate results, the spaces spanned by PC1 and PC2 were highly similar and captured variation in each of the *Pleasure*, *Arousal* and *Dominance* dimensions. The continuum from pleasure to displeasure is located on an axis spanning from the 2nd quadrant (top left) to the 4th quadrant (bottom right) (Figure 3, top) and separating smiling face emoji (face with tongue (😛), face savouring food (😋) and smiling face with sunglasses (😎)) from those perceived negatively, for example, face vomiting (🤮), pouting face (😡) and face with steam from nose (😤) (Figure 3, bottom). The *Arousal* dimension was located approximately perpendicular to the *Pleasure* dimension and spanned from the 3rd quadrant (lower right) to the 1st quadrant (upper left) separating emoji perceived as high in arousal (e.g., oncoming fist (👊), flexed biceps (💪) and collision (💥)) from those perceived as low in arousal (e.g., yawning face (🥱), zzz (💤) and sleeping face (😴)). The *Dominance* dimension ran along PC1 (Figure 3, top) separating high dominance emoji such as oncoming fist (👊) and flexed biceps (💪) from low dominance (i.e., submissive) emoji-like flushed face (😳) and persevering face (😣)

(Figure 3, bottom). Part 7 of Supplementary Materials contains the spaces for variables and observations spanned by PC2 and PC3, where the separation of emoji along the continuum from arousal to nonarousal was clearly seen. The RV coefficients calculated on three dimensions exceeded 0.95 [32] and confirmed the high degree of similarity in Study 1 and Study 2 results: emoji (0.956) and PAD variables (0.974).

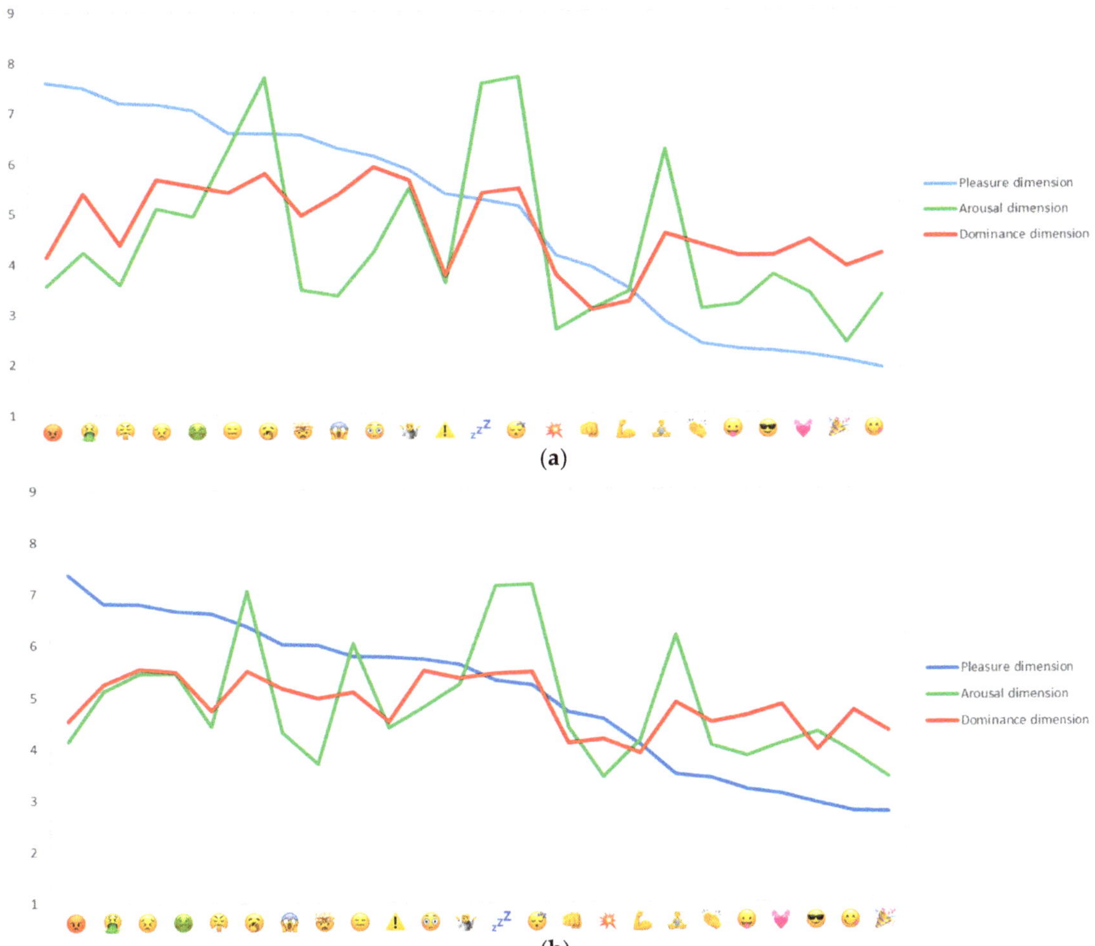

Figure 2. Emoji characterization by PAD dimensions for: (**a**) Study 1 (NZ) and (**b**) Study 2 (UK). *Pleasure*—low anchor (1) is associated with pleasure and high anchor (9) is associated with displeasure; *Arousal*—low anchor (1) is associated with arousal and high anchor (9) is associated with nonarousal (A); *Dominance*—low anchor (1) is associated with dominance and high anchor (9) is associated with submissiveness. Means are sorted by value for *Pleasure* within studies.

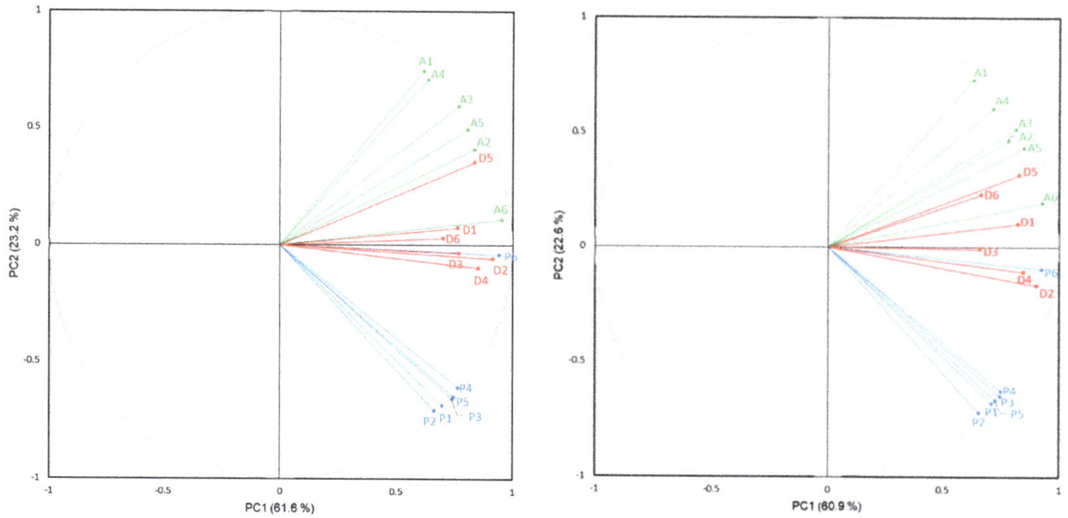

(a)

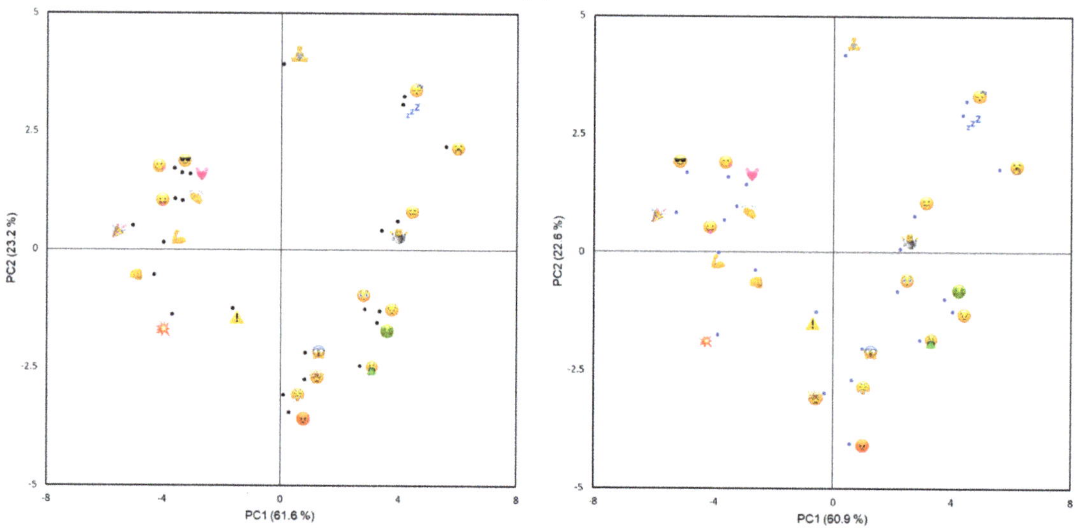

(b)

Figure 3. Two-dimensional spaces following Principal Components Analysis for PAD ratings of 24 emoji in Study 1 (NZ, left) and Study 2 (UK, right). (**a**) Variables plot spanned by PC1 and PC2, and (**b**) Observations plot spanned by PC1 and PC2. For variables plot, colors are used to visually highlight the PAD dimensions, with blue for P1 to P6, green for A1 to A6 and red for D1 to D6.

3.2. Emoji Appropriateness and Use

The results linked to Objective 2 are shown in Table 2. The data for frequency of emoji use (Study 1 only) identified a grouping of five emoji that were most frequently used (face with tongue (😛), face savouring food (😋), party popper (🎉), beating heart (💓) and smiling face with sunglasses (😎)), and significantly more than a group of eight emoji (persevering face (😣), exploding head (🤯), warning (⚠️), collision (💥), flexed biceps (💪), yawning face (🥱) and person in lotus position (🧘)). The average frequency of use for emoji in the former group was between the scale anchors 'sometimes (not all the time, but neither rarely, 35–65%)' and 'infrequently (generally not, but on occasion, 10–35%)' while the least used emoji, on average, were more likely to be used 'never or very infrequently (<10%)'. Correlation analysis on the values in Table 2 showed that the emoji-specific use frequencies were statistically associated with perceived appropriateness for use in situations relating to F&Bs or eating and drinking, and as one increased so did the other (r = 0.50 and 0.56, respectively; $p < 0.015$).

Table 2. Emoji use characteristics among participants in the two studies. In Study 1 (NZ) each emoji was evaluated by 40 to 43 participants and in Study 2 (UK) each emoji was evaluated by 99 to 116 participants.

Emoji Name and Image		Study 1 (NZ)			Study 2 (UK)	
		Avg. Frequency of Using This Emoji *	Emoji Use in F&B or Eating Situation Is Appropriate (%)	Have Used Emoji in F&B or Eating Situation (%)	Emoji Use in F&B or Eating Situation Is Appropriate (%)	Have Used Emoji in F&B or Eating Situation (%)
Beating heart	💓	2.7 ^{ef}	69	38	56	18
Clapping hands	👏	2.3 ^{bcdef}	88	33	56	22
Collision	💥	1.6 ^{abc}	77	19	42	16
Exploding head	🤯	1.7 ^{abc}	43	7	43	18
Expressionless face	😑	2.1 ^{abcdef}	60	20	33	13
Face savouring food	😋	2.8 ^{ef}	86	49	75	39
Face screaming in fear	😱	2.3 ^{bcdef}	63	35	44	20
Face vomiting	🤮	1.8 ^{abcd}	72	49	40	22
Face with steam from nose	😤	1.7 ^{abc}	43	10	32	11
Face with tongue	😛	2.9 ^{f}	71	29	55	25
Flexed biceps	💪	1.8 ^{abc}	65	15	39	12
Flushed face	😳	2.4 ^{cdef}	63	35	34	18
Nauseated face	🤢	2.1 ^{abcdef}	69	38	60	34
Oncoming fist	👊	2.2 ^{abcdef}	38	20	25	14
Party popper	🎉	2.7 ^{ef}	78	43	56	26
Persevering face	😣	1.8 ^{abc}	50	8	31	15
Person in lotus position	🧘	1.4 ^{a}	49	14	25	7
Person shrugging	🤷	2.2 ^{bcdef}	52	21	33	15
Pouting face	😡	2.0 ^{abcde}	48	30	39	12
Sleeping face	😴	2.4 ^{cdef}	50	15	27	9
Smiling face with sunglasses	😎	2.7 ^{def}	73	18	50	25
Warning	⚠️	1.6 ^{ab}	50	10	25	11
Yawning face	🥱	1.8 ^{abc}	48	13	28	12
Zzz	💤	2.1 ^{abcdef}	36	7	25	10

Notes. F&B = food and beverage. * Stated emoji use frequency measured on a 5-pt scale with the following anchors: 5 = 'Always or almost every time (>90%)', 4 = 'Frequently (more often than not, 65–90%)', 3 = 'Sometimes (not all the time, but neither rarely, 35–65%)', 2 = 'Infrequently (generally not, but on occasion, 10–35%)' and 1 = 'Never or very infrequently (<10%)'.*) Post hoc test performed using Tukey's HSD and emoji that share a letter are not significantly different at the 5% level. This set of data was only collected in Study 1.

Regarding perceived suitability of use of the 24 emoji for F&Bs or situations relating to eating and drinking, the two studies were quite similar (r = 0.82, $p < 0.001$) and for past emoji use in food-related situations the two sets of responses were also positively correlated (r = 0.73, $p < 0.001$). Face savouring food (😋), clapping hands (👏) and party popper (🎉) were in the top-5 for both studies, as was a strong negative expression linked to rejection (Study 1: face vomiting (🤮); Study 2: nauseated face (🤢)). The same two emoji were regarded as least appropriate in both studies: zzz (💤) and oncoming fist (👊).

There appeared to be a systematic difference between the two studies where frequency of selection of emoji as appropriate was higher among participants in Study 1 than in Study 2. Based on 50% citation frequency as the boundary value for "appropriateness", the emoji collision (💥), flexed biceps (💪) and person in lotus position (🧘) would be regarded as appropriate in Study 1 but not Study 2 (100%, 82% and 82% vs. 42%, 39% and 25%). Among the 24 emoji included in the research, only four did not exceed the 50% criterion in Study 1 (exploding head (🤯), oncoming fist (👊), yawning face (🥱) and zzz (💤)); while only seven emoji exceeded the 50% criterion in Study 2 (face savoring food (😋), nauseated face (🤢), clapping hands (👏), party popper (🎉), beating heart (💓), face with tongue (😛) and smiling face with sunglasses (😎)). This systematic difference between the two studies also extended to the responses about the previous use of the focal emoji for F&Bs or situations relating to eating and drinking. In Study 2, stated use linked to F&Bs or eating and drinking situations was above 33% for two emoji only face savouring food (😋) and nauseated face (🤢). In Study 1, 10 of the 24 emoji exceeded this criterion.

In both studies, participants' general attitudes to emoji were positive (Table 3), with two exceptions: "I consider myself emoji literate" and "I like to use many different emoji, of all kinds." For these statements, the average values were closer to the scale anchor 'neither agree nor disagree' and the standard deviations were larger. The latter pointed to heterogeneity, which histograms of response distributions confirmed. In fact, an exploratory hierarchical cluster analysis (Euclidean distance, Ward's method) based on these attitudinal statements established two consumer segments (Part 8 of Supplementary Material) where participants in one of the segments agreed much more strongly that emoji were best suited to private communications, and they described themselves as less emoji savvy and less likely to use many different emoji. The two clusters did, however, not differ with regard to the agreement on the statements that emoji were fun to use and a means to express moods/emotions and oneself more generally.

Table 3. Average score * (M) and standard deviation (SD) for general emoji attitudes among participants. Shown for Study 1 (NZ: n = 165 and Study 2 (UK: n = 861).

Statement	Study 1 (NZ)		Study 2 (UK)	
	M	**SD**	**M**	**SD**
Emoji are fun to use and receive	5.7	1.2	5.3	1.3
Emoji are more appropriate in private than professional communications	6.0	1.3	5.8	1.3
I consider myself to be emoji savvy and literate	4.5	1.7	4.3	1.7
I generally use a small set of emoji over and over again	6.0	1.0	5.4	1.3
I like to use many different emoji, of all kinds	4.3	1.8	4.4	1.7
In computer-mediated communications, emoji help me to better express myself	5.3	1.4	4.8	1.5
My favourite emoji are "face emoji" (e.g., 😊,😂,😠)	5.8	1.3	5.3	1.3
Using emoji helps me to express my moods/emotions	5.9	1.2	5.1	1.4

Notes. * 1 = Disagree Strongly; 2 = Disagree Moderately; 3 = Disagree Slightly; 4 = Neither agree nor disagree; 5 = Agree Slightly; 6 = Agree Moderately; 7 = Agree Strongly.

4. Discussion

4.1. Emoji Meanings

A lack of knowledge about their meaning makes in-depth interpretation of results from food-related consumer research that involves emoji more challenging. Against this background, the present research established emoji meanings for 24 emoji on the *Pleasure, Arousal* and *Dominance* (PAD) dimensions, and characterisation of individual emoji against the 18 pairs of semantic differentials and three dimensions was the primary contribution of the research (Figures 2 and 3; Parts 4 to 7 of Supplementary Materials). The finding that differences in the meaning of the 24 emoji were greatest on the *Pleasure* dimension followed by the *Arousal* dimension (Figure 2, Parts 5 and 6 of Supplementary Material) fit with past empirical research [23,33], and the focus in emotion research on these two dimensions as the key determinants of human affect [17]. It provided information to aid researchers in interpreting consumers' emoji use, be it in questionnaires that capture associations to F&Bs or on social media where emoji use is prevalent both in relation to situations involving eating and drinking and more generally [6,28,34].

In relation to the concern expressed by Jaeger, Jin et al. [35] that the emoji typically used in food-related consumer research may be less able to span the two-dimensional valence × arousal space compared to emotion words, the findings from the present research show that it is possible to identify emoji that span this space (e.g., face vomiting (🤮), face savouring food (😋), sleeping face (😴) and collision (💥)). Moreover, the positions of emoji in the space spanned by the first two components following PCA of the 19 PAD variables (Figure 3) were in line with prior research suggesting that emoji can span the valence × arousal configuration [9]. Their ability to do so is directly linked to their selection, and in this regard, emoji do not differ from emotion words. For example, the EsSense Profile™ [36] is dominated by words with positive valence and limited variation in arousal and as a result it poorly spans the full valence × arousal space [37].

The larger span in average scores on the *Pleasure* and *Arousal* dimensions than on the *Dominance* dimension meant that the 24 emoji differed least on the dominance to submissive continuum. While possibly an artefact of emoji selection, this result seemed to align well with the *Dominance* being given less consideration in emotion research [17] and imply that *Dominance* supplements emoji meanings rather than being a critical aspect hereof. The average scores on this dimension fitted expectations with oncoming fist (👊) and flexed biceps (💪) as representatives of high dominance, and flushed face (😳) and person shrugging (🤷) as representatives of submission.

As recommended by Jaeger, Vidal and Ares [4], the research included several emoji that until now have not been used in the context of F&B research, and many of these were non-facial (e.g., 💓, 💪, 🧘). Early research with emoji tended to focus on facial emoji because facial expressions are critical to understanding essential human emotions in face-to-face communication [38]. However, the finding that certain non-facial emoji contributed nuance in the expression of pleasure/positive valence (Figure 3, Part 5 of Supplementary Material) was interesting and suggested that they may contribute to a more complete representation of the core dimensions of human affect [22]. Prior research has found out that certain non-facial emoji are frequently used by consumers to describe their emotions in a food consumption context and have a high discriminative ability [10,39]. We recommend that non-facial be considered for inclusion in research, especially where it concerns F&B products or situations that vary in degree of positive valence. The results from the general emoji attitudes (Table 3) support this in the sense that there did not appear to be a strong preference for facial emoji.

Use of the PAD scale to establish emoji meanings was successful and there was good agreement between the average PAD scores of this study and the scores for valence and arousal obtained by Jaeger, Roigard et al. [22] using Self-Assessment Manikins (SAM: Bradley and Lang [21]). Since each PAD dimension is an average of six semantic differential, this measurement approach requires more effort than the SAMs, but it may be worthwhile considering that the manikins are not intuitive to everybody [40]. The use of the three different PAD dimensions and 7-point scale has the advantage that it offers a more nuanced perspective compared to traditional sentiment analysis which generally only provides a three-way classification of positive, neutral and negative sentiments [41]. A limitation of PAD and SAM approaches, however, is that they do not provide detailed semantic meanings. For example, persevering face (😣) and nauseated face (🤢), had about the same average scores on the PAD dimensions (Part 6 of Supplementary Material), with a high degree of displeasure, medium arousal, and high dominance. Yet, they mean different things, and this was not adequately captured. Therefore, emoji meanings on PAD dimensions should not be regarded as stand-alone as they are not comprehensive. However, as shown by Jaeger, Roigard et al. [22] combining SAM and textual descriptions to explore the meanings of facial emoji is useful, and neither is complete without the other.

Cross-cultural comparison was not a primary research aim, but the similarity in results obtained from adult participants in NZ and UK was noteworthy nonetheless and contributed new empirical evidence to underpin the viewpoint that the interpretation of emoji is very similar across different cultures [4]. The results also aligned with another recent study where similar emoji meanings were found among adult consumers from the USA and China [23] and further supported the suitability of emoji for cross-cultural research.

4.2. Emoji Appropriateness for Use in Research Linked to F&Bs and Eating/Drinking

In addition to establishing emoji meanings on the PAD dimensions, the present research considered the appropriateness of the 24 emoji for use with F&Bs and for situations involving eating and drinking (Obj. 2). It appeared that the results from the two studies differed considerably since a cut-off value of 50% citation frequency for appropriateness meant that 20 out of 24 emoji were considered as appropriate in Study 1 (NZ) while only 7 out of 24 emoji were regarded as appropriate in an F&B context by the participants in Study 2 (UK). While this at a first glance suggested that cultural differences could largely influence perceived emoji appropriateness, a more likely explanation for the differences in citation frequency for appropriateness of use was that data collection in NZ took place in a CLT setting while data collection in the UK was via an online survey. It could be that the participants at a CLT are more motivated or involved (e.g., they already make an effort to come to the CLT) while participants at an online survey might be more driven by incentives. This suggestion takes into consideration the strong positive correlation between the NZ and UK data for appropriateness ($r = 0.82$) and the fact that question format has a big impact on citation frequency. In a methodological study that compared five question formats for emoji research [9], Ares and Jaeger showed that forced yes/no questions resulted in a higher citation for individual emoji compared to CATA questions but not greater sample discrimination. Assuming this difference has to do with greater effort and attention required for yes/no questions, it could explain why responses from the NZ participants in a CLT setting was characterised by higher citation frequencies but a similar relative ordering of the most and least suitable emoji. Therefore, it may not be appropriate to apply a fixed cut-off value as the main criterion for whether an emoji is regarded as suitable or not. Moreover, the value used here 50% seems conservative considering the recent work by Sick et al. [8] who examined the appropriateness of 92 facial emoji using different eating contexts (e.g., breakfast, snack, birthday) and deemed that emoji which were selected by at least 20% of the participants in at least one eating context qualified as appropriate.

What this further suggests is that decisions about emoji selection should be study specific. This would contribute flexibility to address situations and topics that consumers may not have readily considered when asked to rate appropriateness. It is possible to imagine product-focused research into beverages that help people get a good night's sleep where 🥱 and 😴 would be appropriate, development of spicy-hot cooking sauces where 🤯 would be appropriate or that 🧘 could have relevance in describing tranquil situations such as sitting in the back garden with a glass of wine enjoying the sunset. Other examples include the consumption of insects or gene-edited foods, where certain emoji become relevant even though they would not be regarded as appropriate in the context of typical eating situations.

Participants' general attitudes to emoji were positive, but there was evidence of segmentation that could have influenced the perceived appropriateness of emoji use in the F&B context. It seemed that this segmentation was not country specific but differentiated between participants who found emoji less suitable for professional use and relied on a smaller set of emoji in computer-mediated communications. Thus, despite the widespread adoption and popularity of emoji [42,43], some people are more positive and frequent users. A study by Jaeger, Xia et al. [15] found that frequency of emoji use did not influence research participants ability to use emoji questionnaires to characterise F&B stimuli and it could be that despite differences in emoji attitude and personal use, research participants are similarly able to use emoji for expressing their emotional and conceptual associations to F&Bs and situations involving eating and drinking.

4.3. Limitations and Suggestions for Future Research

Limitations and suggestions for future research pertain to both objectives of the present research. The NZ study was carried out as a CLT while the data was collected online in the UK study due to budget restrictions. As such, data of less participants were collected during the NZ study so one needs to consider the NZ results as more preliminary data. Further, given the different testing's conditions, one need to be careful about a direct comparison of the NZ and UK results. With regard to emoji meaning (Obj. 1), the PAD model [25] can be seen as an alternative to SAM [21] for gaining insights about emoji meanings, although both should be combined with semantic information (e.g., through open questions as done in Jaeger, Roigard et al. [22]) to obtain a full understanding of the emoji meanings. We expect that a global study combining such information sources across many emoji would enhance their uptake in F&B research, especially if also linked to the appropriateness of emoji use in the F&B context. With regard to the latter, an interesting extension would be to examine the similarity between the perceived appropriateness indicated by consumers and their actual use of emoji in real life, hereby linking this work to past research characterising emoji use in the F&B context, including tweets [28] and restaurant reviews [44]. More generally, future research should examine the interplay between the meanings of emoji and their appropriateness in the F&B context. This is necessary considering that context is crucial for the meanings attached to emoji.

The perception that emoji are only suitable for research with younger persons has been suggested as a barrier to uptake of emoji [4]; and while beyond the scope of this study to examine potential age effects regarding emoji meaning and appropriateness, this is a relevant topic for future research. A recent study found that emoji might be differently understood depending on age and gender, showing, for example, that respondents aged 30 years and above tended to interpret emoji more literally while younger users interpreted them in more conventionalized ways [45]. Emoji also tend to be more frequently used by women than men [42], but more in private communication than public communication [46]. This study only included data from people who used emoji at least infrequently so it is relevant that future research includes comparisons between participants who differ in emoji usage (e.g., non-users, vs. regular users vs. superusers).

A better understanding of the meaning and appropriateness of emoji in the F&B context also has relevance and value for product development and marketing. For example, the inclusion of emoji in advertisements leads to higher purchase intentions and positive affect [47], and an analysis of the four largest Spanish beer companies revealed how emoji were used as a differentiating element for brand positioning [48]. Against this backdrop, it would be interesting to examine if emoji can be used in the same way as emotional/conceptual profiling to strengthen the product experience when the emotional message by the product is consonant with the brand [49]. In line with this, further research could examine the sensory drivers of product emotions which has been established using emotion words [50] but not emoji.

5. Conclusions

Emoji meanings and appropriateness for use in F&B research were investigated in the present research. A total of 24 emoji were considered, which included several non-facial emoji that had not been previously investigated. The research also extended past studies by including consumers from NZ and UK and hereby extending the geographical reach of empirical research into emoji meanings. The PAD model *Pleasure, Arousal, Dominance* was used to establish emoji meanings and this dimensional approach was successful in characterizing and differentiating between the included emoji. The biggest differences in meaning were for *Pleasure*, followed by *Arousal*, and based on the included emoji it was possible to identify a subset that spanned the valence $\times$ arousal space well. *Dominance* was, on average, the least important of the PAD dimensions for characterizing emoji meaning. The results from NZ and UK were remarkably similar with regard to emoji meanings, in support of past claims that emoji have high suitability for cross-cultural research. Regarding the appropriateness of emoji for use in F&B research, similarity in NZ and UK results remained quite high. Values for perceived appropriateness were established, but we leave it to individual researchers to decide which threshold they use as the cut-off regarding whether a certain emoji is suitable for a particular research activity.

Supplementary Materials: The following are available online at https://www.mdpi.com/article/10.3390/foods10112880/s1, (1) Summary of participant characteristics; (2) Exemplar ballots for PAD scales; (3) Emoji allocation to participants; (4) Means for 18 PAD variables on 24 emoji with Tukey HSD results; (5) Line plots of means for 18 PAD variables on 24 emoji; (6) Means for 3 PAD dimensions on 24 emoji; (7) Spaces for PAD variables spanned by PC2 and PC3 following PCA; (8) Cluster analysis on general emoji attitude data.

Author Contributions: S.R.J. and J.J.S. conceptualization, methodology, formal analysis, writing—original draft, writing—review and editing; D.J. investigation, data curation, visualization; G.S.R. investigation. All authors have read and agreed to the published version of the manuscript.

Funding: Financial support for authors S.R.J., G.S.R. and D.J. was received from two sources: (1) The New Zealand Ministry for Business, Innovation and Employment, and (2) The New Zealand Institute for Plant and Food Research Limited.

Institutional Review Board Statement: The study was conducted according to the guidelines of the Declaration of Helsinki. All studies were covered by a general approval for sensory and consumer research from the Human Ethics Committee at The New Zealand Institute for Plant and Food Research Limited. Participants were assured that their responses would remain confidential. Reward points and promotional offers were offered as compensation.

Informed Consent Statement: Informed consent was obtained from all subjects involved in the study.

Data Availability Statement: The data presented in this study are available on request from the corresponding author.

Acknowledgments: Staff at the Sensory and Consumer Science Team at PFR are thanked for help with data collection, especially Sok Leeang Chheang and Christina M. Roigard.

Conflicts of Interest: The authors declare no conflict of interest.

References

1. Lagast, S.; Gellynck, X.; Schouteten, J.J.; De Herdt, V.; De Steur, H. Consumers' emotions elicited by food: A systematic review of explicit and implicit methods. *Trends Food Sci. Technol.* **2017**, *69*, 172–189. [CrossRef]
2. Meiselman, H.L. The (gradual) development of emotion measurement for food. *Curr. Opin. Food Sci.* **2021**, *40*, 187–191. [CrossRef]
3. Cardello, A.V.; Jaeger, S.R. Questionnaires should be the default method in food-related emotion research. *Food Qual. Prefer.* **2021**, *92*, 104180. [CrossRef]
4. Jaeger, S.R.; Vidal, L.; Ares, G. Should emoji replace emotion words in questionnaire-based food-related consumer research? *Food Qual. Prefer.* **2021**, *92*, 104121. [CrossRef]
5. Riordan, M.A. Emojis as tools for emotion work: Communicating affect in text messages. *J. Lang. Soc. Psychol.* **2017**, *36*, 549–567. [CrossRef]
6. Bai, Q.; Dan, Q.; Mu, Z.; Yang, M. A systematic review of emoji: Current research and future perspectives. *Front. Psychol.* **2019**, *10*, 2221. [CrossRef]
7. Jaeger, S.R.; Cardello, A.V.; Schutz, H.G. Emotion questionnaires: A consumer-centric perspective. *Food Qual. Prefer.* **2013**, *30*, 229–241. [CrossRef]
8. Sick, J.; Monteleone, E.; Pierguidi, L.; Ares, G.; Spinelli, S. The Meaning of Emoji to Describe Food Experiences in Pre-Adolescents. *Foods* **2020**, *9*, 1307. [CrossRef]

9. Ares, G.; Jaeger, S.R. A comparison of five methodological variants of emoji questionnaires for measuring product elicited emotional associations: An application with seafood among Chinese consumers. *Food Res. Int.* **2017**, *99*, 216–228. [CrossRef]
10. Jaeger, S.R.; Lee, S.M.; Kim, K.O.; Chheang, S.L.; Jin, D.; Ares, G. Measurement of product emotions using emoji surveys: Case studies with tasted foods and beverages. *Food Qual. Prefer.* **2017**, *62*, 46–59. [CrossRef]
11. Gallo, K.E.; Swaney-Stueve, M.; Chambers, D.H. A focus group approach to understanding food-related emotions with children using words and emojis. *J. Sens. Stud.* **2017**, *32*, e12264. [CrossRef]
12. Schouteten, J.J.; Verwaeren, J.; Lagast, S.; Gellynck, X.; De Steur, H. Emoji as a tool for measuring children's emotions when tasting food. *Food Qual. Prefer.* **2018**, *68*, 322–331. [CrossRef]
13. Schouteten, J.J.; Meiselman, H.L. The potential of using emoji instead of words when measuring emotions associated with food. *Food Qual. Prefer.* **2021**, *92*, 104182. [CrossRef]
14. Miller, H.; Kluver, D.; Thebault-Spieker, J.; Terveen, L.; Hecht, B. Understanding emoji ambiguity in context: The role of text in emoji-related miscommunication. In Proceedings of the International AAAI Conference on Web and Social Media, Montreal, QC, Canada, 15–18 May 2017.
15. Jaeger, S.R.; Xia, Y.; Lee, P.Y.; Hunter, D.C.; Beresford, M.K.; Ares, G. Emoji questionnaires can be used with a range of population segments: Findings relating to age, gender and frequency of emoji/emoticon use. *Food Qual. Prefer.* **2018**, *68*, 397–410. [CrossRef]
16. Barbieri, F.; Kruszewski, G.; Ronzano, F.; Saggion, H. How cosmopolitan are emojis? Exploring emojis usage and meaning over different languages with distributional semantics. In Proceedings of the 24th ACM international conference on Multimedia, Amsterdam, The Netherlands, 15–19 October 2016.
17. Bakker, I.; Kruszewski, G.; Ronzano, F.; Saggion, H. Pleasure, arousal, dominance: Mehrabian and Russell revisited. *Curr. Psychol.* **2014**, *33*, 405–421. [CrossRef]
18. Prescott, J. Some considerations in the measurement of emotions in sensory and consumer research. *Food Qual. Prefer.* **2017**, *62*, 360–368. [CrossRef]
19. King, S.C.; Meiselman, H.L. Development of a method to measure consumer emotions associated with foods. *Food Qual. Prefer.* **2010**, *21*, 168–177. [CrossRef]
20. Jaeger, S.R.; Roigard, C.M.; Jin, D.; Xia, Y.; Zhong, F.; Hedderley, D.I. A single-response emotion word questionnaire for measuring product-related emotional associations inspired by a circumplex model of core affect: Method characterisation with an applied focus. *Food Qual. Prefer.* **2020**, *83*, 103805. [CrossRef]
21. Bradley, M.M.; Lang, P.J. Measuring emotion: The self-assessment manikin and the semantic differential. *J. Behav. Ther. Exp. Psychiatry* **1994**, *25*, 49–59. [CrossRef]
22. Jaeger, S.R.; Roigard, C.M.; Jin, D.; Vidal, L.; Ares, G. Valence, arousal and sentiment meanings of 33 facial emoji: Insights for the use of emoji in consumer research. *Food Res. Int.* **2019**, *119*, 895–907. [CrossRef]
23. Jaeger, S.R.; Vidal, L.; Ares, G. Consumer product emotion research using emoji. In *Emotion Measurement*; Elsevier: Amsterdam, The Netherlands, 2021; pp. 535–558.
24. Scherer, K.R.; Shuman, V.; Fontaine, J.; Soriano Salinas, C. The GRID meets the Wheel: Assessing emotional feeling via self-report. In *Components of Emotional Meaning: A Sourcebook*; Fontaine, J.R.J., Scherer, K.R., Soriano, C., Eds.; Oxford University Press: Oxford, UK, 2013; pp. 281–298.
25. Mehrabian, A.; Russell, J.A. *An Approach to Environmental Psychology*; The MIT Press: Cambridge, MA, USA, 1974.
26. Jang, H.-W.; Lee, S.-B. Applying effective sensory marketing to sustainable coffee shop business management. *Sustainability* **2019**, *11*, 6430. [CrossRef]
27. Jang, S.S.; Liu, Y.; Namkung, Y. Effects of authentic atmospherics in ethnic restaurants: Investigating Chinese restaurants. *Int. J. Contemp. Hosp. Manag.* **2011**, *23*, 662–680. [CrossRef]
28. Vidal, L.; Ares, G.; Jaeger, S.R. Use of emoticon and emoji in tweets for food-related emotional expression. *Food Qual. Prefer.* **2016**, *49*, 119–128. [CrossRef]
29. Jaeger, S.R.; Ares, G. Dominant meanings of facial emoji: Insights from Chinese consumers and comparison with meanings from internet resources. *Food Qual. Prefer.* **2017**, *62* (Suppl. C), 275–283. [CrossRef]
30. Detandt, S.; Leys, C.; Bazan, A. A French translation of the pleasure arousal dominance (PAD) semantic differential scale for the measure of affect and drive. *Psychol. Belg.* **2017**, *57*, 17. [CrossRef]
31. Addinsoft. *XLSTAT Statistical and Data Analysis Solution*; Addinsoft: Long Island, NY, USA, 2020.
32. Blancher, G.; Clavier, B.; Egoroff, C.; Duineveld, K.; Parcon, J. A method to investigate the stability of a sorting map. *Food Qual. Prefer.* **2012**, *23*, 36–43. [CrossRef]
33. Jaeger, S.R.; Roigard, C.M.; Chheang, S.L. The valence× arousal circumplex-inspired emotion questionnaire (CEQ): Effect of response format and question layout. *Food Qual. Prefer.* **2021**, *90*, 104172. [CrossRef]
34. Tang, Y.; Hew, K.F. Emoticon, emoji, and sticker use in computer-mediated communication: A review of theories and research findings. *Int. J. Commun.* **2019**, *13*, 27.
35. Jaeger, S.R.; Jin, D.; Hunter, D.C.; Roigard, C.M.; Hedderley, D.I. Multi-response approaches in product-focused investigations: Methodological variations across three case studies. *Food Res. Int.* **2020**, *132*, 109113. [CrossRef]
36. King, S.C.; Meiselman, H.L.; Carr, B.T. Measuring emotions associated with foods in consumer testing. *Food Qual. Prefer.* **2010**, *21*, 1114–1116. [CrossRef]

37. Jaeger, S.R.; Spinelli, S.; Ares, G.; Monteleone, E. Linking product-elicited emotional associations and sensory perceptions through a circumplex model based on valence and arousal: Five consumer studies. *Food Res. Int.* **2018**, *109*, 626–640. [CrossRef]
38. Cherbonnier, A.; Michinov, N. The recognition of emotions beyond facial expressions: Comparing emoticons specifically designed to convey basic emotions with other modes of expression. *Comput. Hum. Behav.* **2021**, *118*, 106689. [CrossRef]
39. Schouteten, J.J.; Verwaeren, J.; Gellynck, X.; Almli, V.L. Comparing a standardized to a product-specific emoji list for evaluating food products by children. *Food Qual. Prefer.* **2019**, *72*, 86–97. [CrossRef]
40. Silva, L.G.Z.D.; Guimarães, P.D.; de Souza Gomes, L.O.; de Almeida Neris, V.P. A comparative study of users' subjective feeling collection instruments. In Proceedings of the 19th Brazilian Symposium on Human Factors in Computing Systems, Diamantina, Brazil, 26–30 October 2020.
41. Miller, H.J.; Thebault-Spieker, J.; Chang, S.; Johnson, I.; Terveen, L.; Hecht, B. "Blissfully Happy" or "Ready toFight": Varying Interpretations of Emoji. In Proceedings of the Tenth international AAAI Conference on Web and Social Media, Cologne, Germany, 17–20 May 2016.
42. Prada, M.; Rodrigues, D.L.; Garrido, M.V.; Lopes, D.; Cavalheiro, B.; Gaspar, R. Motives, frequency and attitudes toward emoji and emoticon use. *Telemat. Inform.* **2018**, *35*, 1925–1934. [CrossRef]
43. Ljubešić, N.; Fišer, D. A global analysis of emoji usage. In Proceedings of the 10th Web as Corpus Workshop, Berlin, Germany, 12 August 2016.
44. Chik, A.; Vásquez, C. A comparative multimodal analysis of restaurant reviews from two geographical contexts. *Vis. Commun.* **2017**, *16*, 3–26. [CrossRef]
45. Herring, S.C.; Dainas, A.R. Gender and age influences on interpretation of emoji functions. *ACM Trans. Soc. Comput.* **2020**, *3*, 1–26. [CrossRef]
46. Chen, Z.; Lu, X.; Ai, W.; Li, H.; Mei, Q.; Liu, X. Through a gender lens: Learning usage patterns of emojis from large-scale Android users. In Proceedings of the 2018 World Wide Web Conference, Lyon, France, 23–27 April 2018.
47. Das, G.; Wiener, H.J.; Kareklas, I. To emoji or not to emoji? Examining the influence of emoji on consumer reactions to advertising. *J. Bus. Res.* **2019**, *96*, 147–156. [CrossRef]
48. Casado-Molina, A.M.; Rojas-de Gracia, M.M.; Alarcón-Urbistondo, P.; Romero-Charneco, M. Exploring the opportunities of the emojis in brand communication: The case of the beer industry. *Int. J. Bus. Commun.* **2019**, 2329488419832964. [CrossRef]
49. Thomson, D.M.H.; Crocker, C.; Marketo, C.G. Linking sensory characteristics to emotions: An example using dark chocolate. *Food Qual. Prefer.* **2010**, *21*, 1117–1125. [CrossRef]
50. Spinelli, S.; Jaeger, S.R. What do we know about the sensory drivers of emotions in foods and beverages? *Curr. Opin. Food Sci.* **2019**, *27*, 82–89. [CrossRef]

foods

Article

Sensory Features Introduced by Brewery Spent Grain with Impact on Consumers' Motivations and Emotions for Fibre-Enriched Products

Ana Curutchet [1], Maite Serantes [1], Carolina Pontet [1], Fatima Prisco [1], Patricia Arcia [1,2,*], Gabriel Barg [3], Juan Andres Menendez [3] and Amparo Tárrega [4]

1 Engineering Department, Catholic University of Uruguay, Montevideo 11600, Uruguay; ana.curutchet@ucu.edu.uy (A.C.); maite.serantes@correo.ucu.edu.uy (M.S.); caropontet96@gmail.com (C.P.); fatimapriscof@gmail.com (F.P.)
2 Latitud-LATU Foundation, Av. Italia 6201, Montevideo 11500, Uruguay
3 Neuroscience and Learning Department, Catholic University of Uruguay, Montevideo 11600, Uruguay; gbarg@ucu.edu.uy (G.B.); JUAN.MENENDEZ@ucu.edu.uy (J.A.M.)
4 Instituto de Agroquímica y Tecnología de Alimentos, CSIC, Avda. Agustín Escardino 7, 46980 Paterna, Spain; atarrega@iata.csic.es
* Correspondence: parcia@latitud.org.uy

Abstract: The aim of this work was both to formulate three different fibre-enriched products by the addition of Brewery Spent Grain (BSG), and to evaluate the impact of this fibre enrichment on sensory quality, acceptability, and purchase intention under blind conditions. BSG was incorporated into bread, pasta, and chocolate milk at levels of 8.3%, 2.8%, and 0.35% (*w*/*w*), respectively. The fibre-enriched products and their regular counterparts were evaluated together by consumers through a CATA questionnaire, the EsSense 25 Profile, an overall acceptability rating, and a purchase intention ranking. Although fibre-enriched bread and chocolate milk ranked lower in overall acceptability compared with their counterparts, no significant difference was found for fibre-enriched pasta ($p > 0.05$). Purchase intention did not differ significantly for both bread and pasta ($p > 0.05$), yet the reasons for purchasing them differed significantly ($p < 0.05$). Consumers recognised the fibre enrichment in these two products and, therefore, were willing to partially compromise on sensory attributes. The fibre-enriched chocolate milk, nonetheless, scored significantly ($p < 0.05$) lower in purchase intention than the control. This work demonstrates that the effect of BSG addition is product-specific, and that fibre perception makes consumers feel more confident.

Keywords: BSG; brewery; by-product; acceptability; valorisation; sensory

Citation: Curutchet, A.; Serantes, M.; Pontet, C.; Prisco, F.; Arcia, P.; Barg, G.; Menendez, J.A.; Tárrega, A. Sensory Features Introduced by Brewery Spent Grain with Impact on Consumers' Motivations and Emotions for Fibre-Enriched Products. *Foods* **2022**, *11*, 36. https://doi.org/10.3390/foods11010036

Academic Editor: Witoon Prinyawiwatkul

Received: 17 November 2021
Accepted: 20 December 2021
Published: 24 December 2021

Publisher's Note: MDPI stays neutral with regard to jurisdictional claims in published maps and institutional affiliations.

1. Introduction

In recent years, extensive research has been undertaken to identify ingredients and components that display some clear physiological activity, namely improving intestinal transit [1,2] and the immunological system [3], or reducing cholesterol [4], blood glucose [5], arterial hypertension [6], and obesity [7].

The consumption of dietary fibre has potential benefits for human health, such as preventing chronic non-communicable diseases (NCDs) including obesity, diabetes, and cardiovascular disease [8,9]. Insoluble fibre prevents constipation, while soluble fibre is positively associated with decreased absorption of cholesterol and glucose at the intestinal level [10]. Indeed, a high dietary fibre intake has been correlated with a lower glycaemic response in diabetic patients [11].

Despite the evidence for the benefits of fibre intake, a worldwide fibre deficiency is observed in both developed and underdeveloped countries. A review by [12] revealed the daily average dietary fibre intake in adults for ten countries (including the USA, UK, and Spain), none of which reached the recommended amount. Although this recommendation varies depending on age and lifestyle, on average, it should be greater than 25 g/day according to the World Health Organization (WHO).

In order to meet the demand of dietary fibre intake, fibre enrichment of food products has been widely proposed as a way to increase its consumption. This is especially appealing since NCDs are the leading causes of death in current times [13]. Among them, cardiovascular diseases are the most common NCD globally, leading to an estimated 17.8 million deaths in 2017 [14]. Furthermore, according to research published by the World Health Organization [13], the prevalence of obesity tripled between 1975 and 2016.

Brewery Spent Grain (BSG) is a by-product obtained during the brewing process of malted barley, with an overall composition in dry matter of fibre (30–50%), protein (19–30%), lipids (10%), and ash (2–5%). The high amount of fibre as well as the high protein content in BSG justifies its use as a functional ingredient in the preparation of food for human consumption, and has resulted in research in the inclusion of BSG as a raw material. To mention some of them, [15] studied the valorisation of BSG through fibre-enriched egg pasta, reaching an optimised formulation of 6.2% of BSG. It has also been studied together with mushrooms, as an animal protein replacer and fibre-enricher of smoked sausages containing 3–6% of BSG [16]. Moreover, [17] obtained promising results for a snack chip containing 40% BSG inclusion.

When formulating products with fibre, the fibre will likely impart a strong flavour, coarse texture, and dry mouthfeel [18]; therefore, consumers are sometimes reluctant to accept the reduction in sensory quality [19]. This is especially important if the product is mainly consumed for hedonic pleasure [20] as in the cases reported by [21,22], in which cakes and cookies, respectively, were enriched with fibre by adding fruit pomace and resulted in a reduction in overall acceptability and purchase intention. However, owing to the multiple health benefits derived from fibre enrichment, they can also be valued and consumed for their utilitarian nutritional value [20]. For instance, fibre enrichment did not have any perceivable effect on the overall acceptability of fermented pork sausages [23] and noodles [24]. According to [20], consumers' responses may vary between hedonic and utilitarian products; therefore, two utilitarian products (bread and pasta) and one hedonic product (chocolate milk) were selected to study the effect of BSG enrichment.

The motivations driving food choice and tolerance vary from one food product to another. A hedonic scale is a simple method of measuring the overall liking scores, and it is considered the most informative assessment by consumers. However, consumer perception of a product goes beyond overall liking, also including extrinsic elements such as perceived benefits, quality, and wellness of the associated food. Additionally, cognitive–affective factors including perceived emotions, expectations, and health attitudes influence food choice [25,26]. Those factors produce a top-down modulation of food perception and attention that can modify consumer behaviour [27]. These extrinsic elements can be measured using advanced approaches, such as the EsSense Profile and check-all-that-apply (CATA) as conscious methods [28].

This work aimed to evaluate the use of BSG as fibre enrichment in three different product categories, namely bread, pasta, and chocolate milk, and the impact on consumer motivations and emotions that the fibre enrichment using BSG caused.

2. Materials and Methods

2.1. Brewers Spent Grains Flour Obtention

BSG (a mixture of *Hordeum vulgare* and *Zea mays)* was obtained from the Uruguayan brewing company Fábricas Nacionales de Cerveza S.A. BSG flour was produced by first drying the BSG, the same day it was generated, in a convection oven at 55 °C for 72 h followed by grounding it in a laboratory mill (Retsch ZM 200). The fraction that passed through a 1 mm mesh screen was used for the bread and pasta, while a 0.25 mm mesh screen was used for the chocolate milk. BSG flour was packed in polyethylene bags and kept at room temperature until use. For BSG composition analysis, the fraction that passed through a 1 mm mesh screen was used, except for fibre analysis for which a 0.5 mm mesh was used.

2.2. BSG Composition

Proximate analyses were performed on the BSG flour according to methods proposed by the Association of Official Analytical Collaboration [29]. Protein content was determined by Kjeldahl method and performed according to method 981.10 of the AOAC International with a conversion factor of 6.25. Fat content was determined by the soxtec method, as described in AOAC 2003.5. Quantification of moisture and ash content was performed under AOAC 925.09 and AOAC 923.02, respectively. Total dietary fibre was measured according to AOAC 985.29. Carbohydrates were calculated by difference (i.e., all other nutrients were summed and subtracted from the total weight of the sample). Sodium was estimated based on the values obtained by [30].

2.3. Product Formulation

Formulations for fibre-enriched bread, pasta and chocolate milk were developed. The amount of BSG flour used in each product was adjusted to reach a total fibre content of 2.5 g of dietary fibre per serving size, thus allowing the products to bear the "source of fibre" claim according to Uruguayan law [31]. This resulted in a content of dietary fibre of 2.5%, 5.0%, and 1.25% for pasta, bread and chocolate milk, respectively, which were used as the lower limit for BSG addition in the fibre-enriched products. These products also comply with the Codex Alimentarius [32] source of fibre claim, in particularly with the condition which calls for a minimum of 10% of the daily reference value per serving. BSG-enriched samples were fibre-enriched bread, fibre-enriched pasta and fibre-enriched chocolate milk, with BSG flour concentrations of 8.3%, 2.8% and 0.35%, respectively. Non-added-BSG formulations were also developed, by replacing the BSG with flour (in the case of bread and pasta) or milk (for the chocolate milk) in the fibre-enriched formulations, and used as control. These are the regular bread, regular pasta and regular chocolate milk. In Table S1, the nutritional facts of each product are presented.

2.3.1. Bread

Bread was formulated according to [30], and its formulation is detailed in Table 1. All ingredients were bought locally. The dough was formed in a KitchenAid professional Bowl-Lift Stand Mixer 5KSM7990X using a spiral dough hook, 6.9 L bowl and a low-speed setting. For the fibre-enriched bread, dry ingredients were first mixed with water for 15 min. Oil and BSG flour (<1 mm) were then added in turns and mixed for 5 min after each addition. The regular bread dough was made likewise, except for the final mixing, which was omitted. Once the doughs were formed, they were left to bulk ferment at 35 °C and 60% relative humidity (RH) for 30 min. Doughs were shaped in 27.5 × 15.0 × 6.5 cm^3 (L × W × H) loaf pans and left to further ferment for 50 min under the same conditions. The proofed breads were baked at 190 °C for 30 min in a convection oven. After baking, they were cooled for 20 min, packed in sealed polyethylene bags, and stored refrigerated until sensory evaluation took place 24 h later.

Table 1. Formulations of fibre-enriched bread and regular bread.

Ingredients	Fibre-Enriched Bread (%)	Regular Bread (%)
Water	33.37	33.37
Wheat flour	45.00	53.30
BSG flour (<1 mm)	8.30	-
Sunflower oil	4.86	4.86
Granulated sugar	3.60	3.60
Skim milk powder	3.20	3.20
Salt	1.20	1.20
Instant yeast	0.47	0.47

2.3.2. Pasta

The ingredients for the pasta were: wheat flour, BSG flour (<1 mm), water and eggs. The formulation was adapted from the pasta maker machine recipe book (Table 2). All ingredients were bought at a local supermarket. Fresh pasta was made with a pasta maker machine (HR2355/08, Philips, Amsterdam, The Netherlands), using the standard program and following its instructions. Spaghetti were cut at a length of approximately 30 cm and left at room temperature until the moment of cooking. The pasta was always consumed the same day that it was made.

Table 2. Formulations of fibre-enriched pasta and regular pasta.

Ingredients	Fibre-Enriched Pasta (%)	Regular Pasta (%)
Water	11.97	11.97
Wheat flour	70.06	72.86
BSG flour (<1 mm)	2.80	-
Egg	15.97	15.97

2.3.3. Chocolate Milk

The regular chocolate milk was formulated according to a locally commercial chocolate milk, to which BSG flour was then blended in to create the fibre-enriched version. Chocolate milk was made in a mixer, by mixing semi-skimmed milk (1.5% fat content), BSG flour (<250 µm), cocoa powder (12% fat content), granulated sugar, carrageenan type A, carboxymethyl cellulose (CMC) and Givaudan chocolate flavouring for 2 min. Formulations are detailed in Table 3. All ingredients were bought locally.

Table 3. Formulations of the fibre-enriched chocolate milk and regular chocolate milk.

Ingredients	Fibre-Enriched Chocolate Milk (%)	Regular Chocolate Milk (%)
Semi-skim milk	89.47	89.82
Granulated sugar	7.00	7.00
Cacao powder	3.00	3.00
BSG flour (<250 µm)	0.35	-
Carrageenan Type A	0.075	0.075
CMC	0.075	0.075
Chocolate flavouring	0.03	0.03

2.4. Consumers' Evaluation Sessions

To study the acceptability of products, one session was held for each product, under blind conditions and on different days. The number of consumers that participated was 110, 119 and 111 in bread, pasta and chocolate milk product sessions, respectively. Participants were randomly recruited on the university campus and public places, being regular consumers of each product between June and September of 2021. The panellists, male and female, ranged between 18 and 70 years old. For the sensory evaluation, the participants were provided with two samples—one of the regular product and one of the fibre-enriched product. These samples were given without any information related to their composition or health effects. Water was provided to consumers as a palate cleanser between samples.

2.4.1. Bread

Breads were prepared the previous day and refrigerated in plastic bags to avoid moisture loss. On the day of evaluation, breads were removed from the refrigerator and left to warm up to room temperature. Half slices of fibre-enriched bread and regular bread (approximately 15 g) were served in biodegradable plastic plates coded with a three-digit number.

2.4.2. Pasta

Fresh pasta was made on the day of evaluation. BSG-added and regular pasta were cooked in abundant salted (1 tbl. spoon) water for 15 min, drained and seasoned with extra virgin olive oil (1 tbl. spoon). Approximately 15 g of cooked spaghetti was served in white polystyrene cups coded with a three-digit number. Samples were maintained warm until evaluation.

2.4.3. Chocolate Milk

Chocolate milk samples were prepared, refrigerated and evaluated in less than 24 h after being produced. Samples were served at 5 °C in plastic cups coded with a three-digit number.

2.5. Sensory Evaluation Questionnaire

Three-digit-coded samples of both products were randomly presented to regular consumers of each product. Participants completed a computer-based questionnaire. The questionnaire consisted of two sections. Section one measured consumers' perception of the product and section two determined the emotions provoked by the product in consumers.

In section one, participants were asked to rate the overall acceptability using a nine-point hedonic scale (1—extremely disliked to 9—extremely liked). Subsequently, consumers completed a CATA product-specific questionnaire to describe the sample. CATA attributes were determined through preliminary brainstorming sessions with consumers who tasted similar products. The attributes were randomly presented to participants to avoid order effects. Finally, purchase intention was measured by a five-point scale from 1—would definitely not buy to 5—would definitely buy, followed by a CATA questionnaire to uncover the reason behind their decision (Table 6).

In section two, the EsSense25 questionnaire was used to collect emotional response data based on the methodology suggested by [33]. The twenty-five emotions were first translated into Spanish by the authors. As Spanish and Portuguese are closely related due to their Latin roots, Spanish translations were then compared to the Portuguese terms suggested by [34] and then back-translated to English. Emotions in the EsSense 25 profile are classified into positive, negative and unclassified subgroups according to their nature [35]. Consumers were presented with a five-point intensity scale (1—Nothing, 2—Slightly, 3—Mildly, 4—Very and 5—Extremely) and asked to select the most accurate intensity for every one of the twenty-five emotions. The emotions were randomly presented to participants to avoid order effects. See Questionnaire S1 for the full questionnaire design.

2.6. Data Analysis

Acceptability, purchase intention and EsSense 25 results obtained were subjected to paired samples Student's *t*-test analysis with a significance level of 0.05. Raw CATA data were converted into binary data based on frequency of citation (1—cited, 0—non-cited). Nonparametric Cochran's Q test was then performed on the binary CATA data to detect significant differences ($p < 0.05$) among fibre-enriched and regular samples for each attribute. Based on recent research published by [35], EsSense25 results were subjected to paired samples Student's *t*-test analyses at a significance level of 0.05 to investigate difference in the intensity of emotions between fibre-enriched products and their counterparts. All statistical analysis was performed using XLSTAT software (2019 version, Addinsoft Inc., New York, NY, USA).

3. Results and Discussion

3.1. BSG Flour Composition

The BSG flour composition is shown in Table 4, including protein, fat, carbohydrates, dietary fibre, ash, sodium and moisture contents. Results show a notable feature: the total dietary fibre content is higher than the one obtained by [30], which was 44.61 ± 0.29%. General composition is in the reported range [36]. The significant amount of this nutrient allows the formulation of products with the addition of BSG as a functional ingredient, and for them to bear the "source of fibre" claim.

Table 4. BSG flour composition per 100 g.

Component	Mean Value ± SD
Proteins (g)	24.9 ± 0.02
Lipids (g)	6.28 ± 0.16
Carbohydrates (g)	16.24 *
Total dietary fibre (g)	45.47 ± 0.34
Ash (g)	2.81 ± 0.25
Sodium (mg)	240
Moisture (g)	4.3 ± 0.4

* Carbohydrates were determined by difference.

3.2. *Hedonic Analysis and Emotional Profile*

3.2.1. Bread

The acceptability differed significantly for regular and fibre-enriched bread (Table 5). Similar results were obtained by [30,37], where the BSG addition to bread dropped consumers overall acceptability. On the other hand, for purchase intention, no significant difference was found. Consumers may have identified the health benefits of fibre-enriched bread based on its appearance, even though no information was given [38]. Awareness of health benefits may have caused consumers to compromise on the product's sensory features. Indeed, for utilitarian products such as bread, sensory quality is not the main driver of consumer choice, as opposed to hedonic products [20].

Table 5. Acceptability and purchase intention for bread, pasta and chocolate milk; regular and fibre-enriched product.

Product	Acceptability *	Purchase Intention **
Bread		
Regular bread	7.10 ± 1.48 [a]	3.93 ± 1.00 [a]
Fibre-enriched bread	6.44 ± 2.06 [b]	3.75 ± 1.26 [a]
Pasta		
Regular pasta	6.87 ± 1.45 [a]	4.14 ± 0.93 [a]
Fibre-enriched pasta	6.91 ± 1.91 [a]	4.00 ± 1.20 [a]
Chocolate milk		
Regular chocolate milk	6.47 ± 1.86 [a]	3.55 ± 1.26 [a]
Fibre-enriched chocolate milk	5.68 ± 2.13 [b]	3.03 ± 1.34 [b]

Acceptability and Purchase intention expressed as mean ± SD. For each category, product scores not sharing letters are significantly different ($p < 0.05$) according to Student's *t*-test. * Evaluated in a structured nine-point scale. ** Evaluated in a structured five-point scale.

The most frequently mentioned reasons for consumers choosing to buy the fibre-enriched bread (67%) were that it was healthy (64%), tasty (81%), to include fibre in their diet (57%) and to avoid weight gain (19%) (Table 6). Except for being tasty, and because their kids would like it, all the other reasons were more frequently selected for fibre-enriched bread than for regular bread ($p < 0.05$). Despite the high frequency of choice of the tasty attribute for fibre-enriched bread, it was more commonly selected as a reason for buying the regular bread ($p < 0.05$), which accounted for a vast 98% of the consumers willing to buy the latter.

Table 6. Reasons for buying and not buying the regular and fibre-enriched samples of bread, pasta and chocolate milk.

	Frequency of Citation (%)					
	Bread		Pasta		Chocolate Milk	
	Regular	Fibre-enriched	Regular	Fibre-enriched	Regular	Fibre-enriched
Reasons to buy						
Because it is healthy	**21.3**	**63.5**	**23.8**	**58.9**	13.4	20.5
Because it is tasty	**97.5**	**81.1**	**94.1**	**86.7**	**97.0**	**95.5**
To avoid weight gain	**1.3**	**18.9**	3.0	11.1	0.0	0.0
Because it is high in calories	2.5	1.4	4.0	4.4	4.5	6.8
Because it is good for my family	10.0	20.3	9.9	13.3	4.5	9.1
Because I want to include fibre in my diet	**11.3**	**56.8**	**14.9**	**36.7**	11.9	15.9
Because my kids would like it	**16.3**	**4.1**	12.9	7.8	9.0	15.9
Reasons not to buy						
Because it does not seem healthy	30.0	8.3	**35.3**	**3.7**	11.4	13.4
Because it is distasteful	**33.3**	**61.1**	**29.4**	**66.7**	**52.3**	**56.7**
Because it is not the one I always buy	10.0	22.2	**17.6**	**48.1**	**18.2**	**26.9**
Because I do not need fibre	3.3	0.0	5.9	0.0	4.5	4.5
To prevent weight gain	20.0	19.4	23.5	14.8	15.9	14.9
Because I think it is bad for my family	0.0	0.0	17.6	3.7	2.3	1.5
Because it seems expensive	**0.0**	**11.1**	**0.0**	**18.5**	6.8	1.5
Because I don't want to include fibre in my diet	6.7	5.6	17.6	3.7	2.3	3.0
Because my family would not like it	**6.7**	**25.0**	23.5	18.5	4.5	10.4

Values in bold indicate statistically significant differences ($p < 0.05$) between samples.

Despite what [20] found, for both fibre-enriched and regular bread, the main reason for not buying them was related to their taste, where fibre-enriched bread ranked higher than regular bread ($p < 0.05$): 61% for fibre-enriched compared to 33% for regular bread (Table 6). Sensory properties have already been reported to be the main reason for not consuming a fibre-enriched product [39]. Unlike for regular bread, price resulted in a relatively common reason for not buying the fibre-enriched one ($p < 0.05$). This is supported by [40], which states expensiveness as a misperception consumers have towards fibre-enriched products. Furthermore, the reason related to consumers' families not liking this type of product may be due to the low popularity of fibre-enriched products in children and adolescents [41,42].

With respect to the CATA questionnaire, 12 out of the 21 attributes detailed showed a significant difference in terms of frequency of mention (Figure 1). The attributes most used to describe fibre-enriched bread were fibrous, natural, soft, compact, fluffy, and dry. For regular bread, soft, fluffy, natural, and moist were the attributes most frequently selected. Despite many of these attributes being recurrent in both bread samples, repeated frequent attributes, except for natural, differed significantly on the frequency of mention. Fibre-enriched breads have already been reported to have a decreased specific volume compared to non-enriched [37,43]. Additionally, consumers may have described the fibre-enriched bread as dry since fibre has a high-water holding capacity [44] and free water may seem scarce in this type of bread.

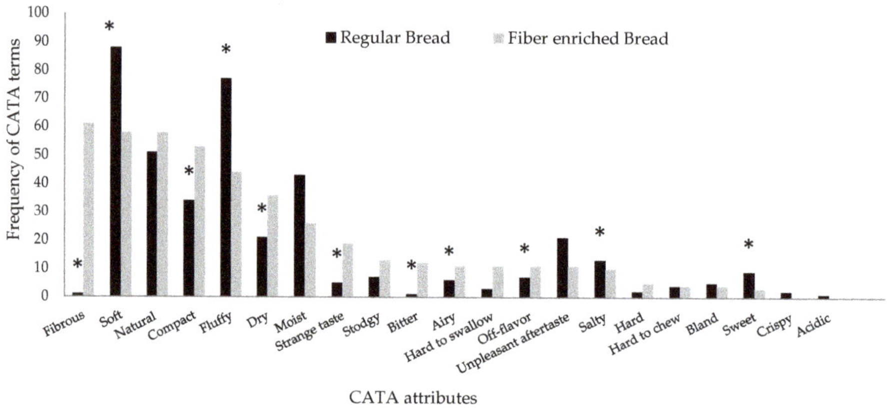

Figure 1. Frequency of CATA terms for fibre-enriched bread and regular bread. * indicates significant difference at $p < 0.05$.

Five of the twenty-five emotional terms showed significant difference between bread samples (Table 7). Two were positive emotions—warm ($p = 0.000$) and active ($p = 0.027$)—with participants being more active after testing the regular bread but less warm. Additionally, tame ($p = 0.001$) and guilty ($p = 0.005$) showed a significant difference. Participants felt more guilty and less tame after testing the regular bread. Finally, respondents felt significantly more bored ($p = 0.000$) when consuming the fibre-enriched bread.

Table 7. Effect of product on consumers' emotional response for bread, pasta and chocolate milk.

Mean Scores	Bread			Pasta			Chocolate Milk		
	Regular	Fibre-Enriched	*p* Value	Regular	Fibre-Enriched	*p* Value	Regular	Fibre-Enriched	*p* Value
Active (+)	**3.02**	**2.66**	**0.027**	**2.85**	**2.50**	**0.021**	2.67	2.64	0.872
Adventurous (+)	2.44	2.56	0.458	2.60	2.86	0.109	2.39	2.31	0.632
Aggressive	1.92	1.86	0.693	2.33	2.30	0.875	1.62	1.62	0.956
Bored (−)	**1.66**	**2.22**	**0.0001**	**1.54**	**2.70**	**<0.0001**	2.18	1.87	0.052
Calm (+)	2.57	2.81	0.123	**1.93**	**2.86**	**<0.0001**	**2.41**	**2.81**	**0.016**
Disgusted (−)	1.69	1.56	0.353	**2.37**	**1.67**	**<0.0001**	1.80	1.87	0.742
Enthusiastic (+)	2.98	2.97	0.947	2.99	3.08	0.512	2.71	2.70	0.931
Free (+)	2.8	2.59	0.227	**2.69**	**2.10**	**0.0003**	2.58	2.73	0.415
Good (+)	2.97	2.99	0.914	2.59	2.49	0.549	2.79	3.01	0.178
Good natured (+)	2.97	2.45	0.129	**2.92**	**2.32**	**0.0001**	2.50	2.67	0.267
Guilty	**1.93**	**1.52**	**0.005**	**2.28**	**1.58**	**<0.0001**	1.45	1.67	0.083
Happy (+)	3.07	2.85	0.187	**2.94**	**2.53**	**0.011**	2.71	3.01	0.070
Interested (+)	2.93	3.14	0.170	2.95	3.05	0.464	2.64	2.66	0.889
Joyful (+)	2.84	3.03	0.246	**2.54**	**2.99**	**0.003**	2.95	2.78	0.283
Loving (+)	2.45	2.46	0.931	**2.15**	**2.52**	**0.015**	2.44	2.49	0.744
Mild	2.18	2.16	0.885	2.42	2.27	0.339	**1.65**	**2.08**	**0.001**
Nostalgic (+)	2.36	2.19	0.298	2.67	2.72	0.782	2.37	2.36	0.988
Pleasant (+)	2.66	2.45	0.170	**2.48**	**2.13**	**0.029**	2.99	2.71	0.071
Satisfied (+)	2.88	2.71	0.310	2.53	2.24	0.079	2.97	2.92	0.716
Secure (+)	2.9	2.89	0.976	3.10	3.00	0.478	2.58	2.80	0.211
Tame	**1.6**	**2.1**	**0.001**	**1.69**	**2.55**	**<0.0001**	**2.02**	**1.57**	**0.002**
Understanding (+)	3.02	2.89	0.360	3.04	2.94	0.499	2.87	2.80	0.669
Warm (+)	**2.29**	**2.88**	**0.0001**	**2.25**	**3.05**	**<0.0001**	2.78	2.49	0.095
Wild	2.25	2.15	0.567	**2.64**	**2.30**	**0.036**	1.97	2.06	0.583
Worried (-)	2.17	2.29	0.437	**2.67**	**2.67**	**0.979**	2.33	2.22	0.503

Classification of terms described by EsSense profile in positive (+), negative (−) or unclassified. Emotions in bold are significant difference between regular and fibre-enriched version.

3.2.2. Pasta

For pasta, no significant differences were found in either acceptability or purchase intention between the regular and the fibre-enriched product (Table 6). This first result is in accordance with [39], who did not find a significant difference between the liking of traditional pasta and the liking of 10% wheat-bran-added pasta. The fact that consumers were equally willing to buy regular and fibre-enriched pasta is understandable based on the previous result on liking. This was an interesting finding, as only a 15.8% of the total grains consumed on a given day are whole grains, according to the U.S. Department of Health and Human Services [45]. It may thus seem that respondents were not against fibre-enriched pasta, but unaware of it as a replacement to regular pasta [39,46]. Indeed, [46] reported that their focus group participants chose whole grain breads more regularly than whole grain pasta, for which they usually opted for the refined version.

In this case, 76% of consumers were willing to buy the fibre-enriched pasta. The most frequently given reasons for buying it were: because it is tasty (86.7%), healthy (58.9%) and because they want to include fibre in their diets (3.7%) (Table 6). The main reason for buying regular pasta was that it is tasty (94.1%), while all the other reasons were negligible in terms of frequency of mention (<20%). Significant differences ($p < 0.05$) were detected in the frequency of mention of healthy and to include fibre in their diet between the two products.

On the other hand, the main reasons for not buying fibre-enriched pasta were because it is distasteful (66.7%) and because is not the one they always buy (48.1%) (Table 6). The first one is once again related to fibre imparting a strong flavour, coarse texture and dry mouthfeel, as argued for bread samples [18]. In addition, the last reason can be associated with food neophobia, which is defined as the reluctance to eat new foods [47] and has already been reported to be a barrier to novel foods consumption [48].

The CATA questionnaire results for pasta showed that 11 out of the 20 terms significantly differed in their frequency of choice between samples (Figure 2). Participants associated the fibre-enriched pasta with the attributes: natural, fibrous, granulated, coarse, dark, gritty and strange taste. On the other hand, the terms soft, uniform, homogeneous, and flexible were associated with regular pasta. Moreover, both pastas were frequently described as tasty with no significant difference between them.

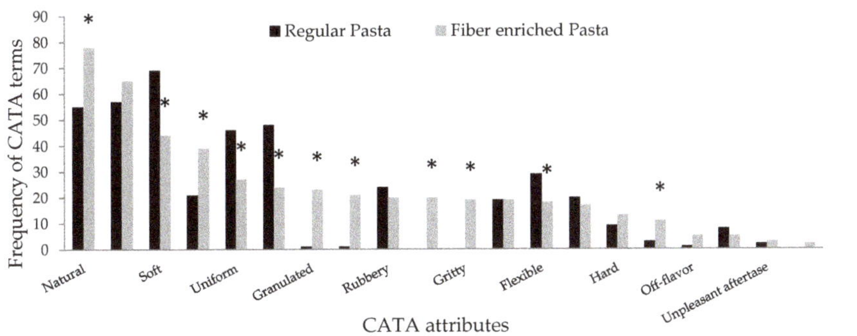

Figure 2. Frequency of CATA terms for fibre-enriched pasta and regular pasta. * indicates significant difference at $p < 0.05$.

Numerous texture attributes differed significantly in frequency of mention between both samples, suggesting that addition of BSG to pasta produces textural and flavour changes. This is not surprising, as this is similar to results obtained by [49], who studied whole barley incorporation into regular pasta for a total final dietary fibre concentration of 2.6% (w/w). Their results showed that both firmness and adhesiveness decreased with the amount of barley being added, whilst taste and overall liking ranked significantly lower for barley-added pasta when compared to the control sample. This is apparent by the results of CATA attributes—fibre-enriched pasta was described as more fibrous, granulated, coarse and gritty and less soft, uniform, homogenous and flexible compared to the regular sample. Certainly, the addition of fibre interferes with gluten network and prevents the cohesiveness of the dough. As a result, starch granules are more easily lost during cooking and the quality of the final product decreases [49]. It should be noted that a possible solution to these negative effects may be the addition of xanthan gum and vital gluten.

In [50], equal texture and flavour characteristics for both the 5% (w/w) β-glucans-added pasta and control sample were obtained, by adding 5% xanthan gum and vital gluten.

Moreover, the colour difference detected between samples is due to BSG flour being darker than wheat flour, as already observed by [51] in BSG-enriched pasta. A decrease in yellowness and increase in brownness and redness was observed for all 5%, 10% and 20% (w/w) BSG addition levels in dried pasta. This may have decreased its acceptability as the yellow colour is highly valued by consumers of durum wheat pasta [51].

Fourteen of the twenty-five emotional terms presented significant differences in their rating when comparing fibre-enriched and regular pasta (Table 7), indicating that the two products were perceived in very different ways. Two negative emotional terms—bored ($p < 0.0001$) and disgusted ($p < 0.0001$)—presented significant differences. While the first term was significantly more elicited for the BSG-added sample, the opposite occurred for the latter. Other emotions that presented significant difference included: guilty ($p < 0.0001$) and wild ($p = 0.036$), which were less frequently applied to fibre-enriched pasta when compared to regular pasta, and tame ($p < 0.0001$), which was rated higher for the fibre-enriched version.

3.2.3. Chocolate Milk

Significant differences were found in both the acceptability and purchase intention of chocolate milk products (Table 6). These differences were probably rooted in the fact that fibre inclusion introduces changes in the texture and flavour of the product, as evidenced by the results of CATA attributes and reasons for not purchasing the product. Among consumers willing to buy fibre-enriched chocolate milk (39.6%), most of them (95.5%) stated that taste was the main driver for purchasing (Table 6). However, this was also the main cited reason for buying regular chocolate milk, scoring a significantly higher frequency of mention compared to the first one ($p < 0.05$). This difference in frequency of choice is mostly due to the higher percentage of consumers buying the regular sample (60.4%).

In addition, 39.6% of consumers were willing to buy the fibre-enriched chocolate milk. The main reason given against buying it (60.4% of consumers) was due to its taste, this being more frequently selected than for the regular product ($p < 0.05$) (Table 6). This attribute was also chosen for bread and pasta as main reasons for not buying the products. In [52], it was reported that the incorporation of BSG into baked snacks altered the flavour profile of the snacks, in addition to their taste and overall acceptability. Researchers associated this unpleasant odour with compounds deriving from fermentation and the Maillard reaction, including 3-methyl-butanal, 2,3 butanedione and 2-methy-butanal, which were present in high quantities in both BSG flour and the BSG snacks.

For chocolate milk, 7 of the 20 terms presented in the CATA questionnaire demonstrated significant differences ($p < 0.05$) (Figure 3). Both chocolate milks were frequently associated with the attributes smooth, strong chocolate flavour and natural. However, the attributes gritty, artificial, coarse, aftertaste, strange taste, off-flavour and greasy distinguished the fibre-enriched chocolate milk from the regular one ($p < 0.05$). Those attributes may explain the significant lower acceptability obtained for the BSG-added product, affecting both product texture and flavour.

The acceptability of fibre-enriched milk has not been thoroughly studied in the literature. However, its addition to other fluid dairy products, such as yoghurt, has been investigated. Particularly, [53] evaluated the sensory characteristics and consumer acceptance of fibre-enriched yoghurts, reporting a significantly lower acceptability compared to the control sample, caused primarily by a gritty or sandy texture. This conclusion is in accordance with the results obtained in this study, where fibre-enriched chocolate milk was significantly less accepted than the regular one and the gritty attribute was selected by 30% of participants for the BSG product (12% for the regular chocolate milk). In [54], the perceived oral grittiness of multiparticulate formulations was studied, and it positively correlated with the amount and size of particles being dispersed in the media. The product matrix has also been reported to greatly influence the particle size detection threshold, concretely particles being dispersed in liquid matrices are easier to detect than the ones in solid matrices [55]. Although the BSG particle size used for fibre-enriched chocolate milk was smaller than for fibre-enriched bread and pasta, it was still considerably large (>200 μm) compared to detection thresholds for cellulose particles (1.5% w/w) in viscous low-fat quark and semi-solid high-fat processed cheese, which are estimated to be 52 μm and 82 μm, respectively [55]. In addition, the product matrix being liquid is likely to have facilitated the perception of grittiness compared to semi-solid or solid foods. For all the above, unpleasant roughness and grittiness was still perceived, albeit BSG concentration being considerably low (0.35% w/w).

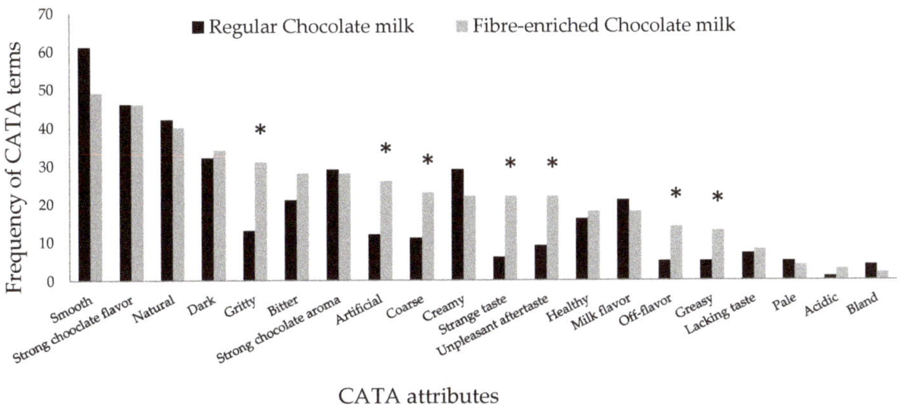

Figure 3. Frequency of CATA terms for fibre-enriched chocolate milk and regular chocolate milk. * indicates significant difference at $p < 0.05$.

Furthermore, although bitterness was relatively more frequently used to describe the fibre-enriched chocolate milk, its frequency of mention did not differ significantly between samples ($p < 0.05$). In [53], it was reported that bitterness had a negative influence in the flavour quality scores of 30 g/kg fibre-enriched yoghurts, which were significantly lower than for the 15 g/kg counterparts. This result suggests that fibre addition may have increased the bitterness of the fibre-enriched chocolate milk, yet its concentration was not high enough to exert a significant effect.

Just as taste scored significantly higher in terms of frequency of mention among the reasons for not buying fibre-enriched chocolate milk compared to the regular one ($p < 0.05$), significant differences ($p < 0.05$) were also found in frequency of mention of the following CATA attributes describing taste: unpleasant aftertaste, off-flavour and strange taste. Again, this shows the unpleasant and negative effects of incorporating BSG into foods [52]. Nonetheless, a decrease in sensorial properties (particularly aroma and flavour) has also been reported after incorporating cauliflower by-product extracts into apple juice [56], becoming apparent it is one of the greatest struggles of by-products' valorisation.

In the case of chocolate milk, three emotional terms showed significant difference in their mean scores between samples (Table 7): calm ($p = 0.016$), mild ($p = 0.001$) and tame ($p = 0.002$). Participants felt significantly milder after consuming the fibre-enriched chocolate milk, yet they felt tamer after testing the regular chocolate milk. A positive emotion also presented significant differences in its rating—calm with $p = 0.016$—this emotion being more frequently elicited for the fibre-enriched chocolate milk.

3.3. Comparison of the Fibre Impact on Different Category of Products

For different food product categories, consumers hold specific sensory and hedonic expectations which subsequently affect their food perception and acceptance [57]. In fact, previous experiences, product characteristics and context shape expectations. Sensory attributes of the product must meet consumer expectations for the acceptability to be high. In this work, consumers responded differently to the fibre enrichment of bread, pasta and chocolate milk. Interestingly, consumer responses towards bread and pasta were remarkably similar. In neither products consumer showed a statistically difference in purchase intention between the regular and fibre-enriched version. Amidst the reasons for buying the fibre-enriched bread and pasta, consumers cited healthy and to include fibre in their diets. These results confirm the ability of respondents to successfully recognise fibre-enriched products and, thus, infer their health benefits. Knowledge about dietary fibre and its health effects has already been reported to be substantial among the general population [58]. In [59], it was reported that information on fibre displayed on labels was well-understood by consumers and gave a positive preference for high-fibre products.

In the case of chocolate milk, consumers could not recognise the addition of fibre to the product, and instead, perceived a strong and unexpected flavour, coming from BSG's strong aromatic compounds, enhanced due to the product being liquid. Although some respondents mention the nutty flavour characteristic of the fibre-enriched version, most did not, as shown by the low acceptability and purchase intention rating. Lack of awareness of its nutritional benefits may have prevented the "halo" effect, which is defined as a positive influence on the perception of a product due to an unrelated impression from another attribute [60,61]. Indeed, communication of a product's health benefits can be a powerful marketing tool for the food sector due to its potential "halo" effect. For instance, [21] reported an increase in liking for fibre-enriched apple pomace cake when the claim source of fibre was displayed, showing the impact, or "halo" effect, of health claims on hedonic liking. Thus, although fibre-enriched chocolate milk was not liked as much as the control, including the corresponding nutritional claims may bridge the gap between them. Pasta is the product for which emotional response differed the most, despite being the only one with no significant difference in either acceptability or purchase intention. Differences in emotional response between regular and BSG-enriched samples were evoked due to differences in their sensory features, as no information was provided. Five emotions were repeated in at least two categories and show significant differences between the fibre-enriched and the regular one: tame, calm, warm, bored, guilty and active.

Only tame had the same behaviour in the three food categories; certainly, fibre enrichment made consumers feel more tame. However, although consumers felt more warm, bored, active and less guilty when tasting fibre-enriched pasta and bread, the opposite occurred for fibre-enriched chocolate milk. This seems to indicate that the knowledge or perception of fibre enrichment provokes a distinct set of emotions. In [62], the emotions evoked by beer consumption in blind and informed conditions were studied, finding that the label information provided to participants increased their acceptance, induced some positive emotions and increased their willingness to pay (WTP) for pure malt beers.

Emotions reported by consumers confirmed a top-down modulation produced by previous experiences related to fibre content. For both bread and pasta, sensory qualities derived from fibre addition were perceived as healthy. A top-down modulation is observed in the high score that the statement "because it is healthy" obtained as a motivation to purchase these products (Table 6). This is also seen in the emotional response. Fibre-enriched bread and pasta evoked emotions associated with health (warm, bored, tame, active), while regular bread and pasta induced guilt, an emotion associated with highly caloric non-healthy food intake, as was described by [63,64]. This pattern was not present in the chocolate milk, as chocolate is strongly associated with hedonic expectations and the fibre inclusion modified the standard characteristics of the product.

4. Conclusions

In this work, the effect of BSG addition to different food product categories was analysed by sensory evaluations of both control and fibre-enriched samples. The amount of BSG flour added to the fibre-enriched products was enough for them to bear the claim Source of fibre according to Codex Alimentarius. The addition of fibre led to some changes in acceptability and buying intention as well as in product perception. Results showed that BSG enrichment had a significant effect on the sensory properties of all three products ($p < 0.05$), affecting both texture and flavour. Although the fibre-enriched bread and fibre-enriched chocolate milk ranked lower in overall acceptability compared with their counterparts, no significant difference was found between fibre-enriched pasta and regular pasta ($p > 0.05$).

Interestingly, purchase intention did not differ significantly for either bread or pasta ($p > 0.05$), yet the reasons for purchasing them differed significantly ($p < 0.05$). It seems consumers were aware of fibre-enrichment in these two, and were willing to partially compromise on sensory attributes. Fibre-enriched chocolate milk, nonetheless, scored significantly lower in purchase intention than the regular product, probably due to the mouthfeel perception being affected by BSG particles and lack of awareness of it being a functional product.

Overall, the findings demonstrated that the effect of BSG addition is product-specific; while fibre-enriched pasta and fibre-enriched bread were approved by consumers, results for fibre-enriched chocolate milk suggested the opposite. Ambivalence was seen in the emotions generated by the different fibre enriched products, depending on whether the fibre was perceived or not. When consumers perceived the fibre enrichment in the product, they approved consumption, feeling in general more confident.

Finally, it should be noted that this study used convenience sampling of consumers and only three products were investigated. Further research considering the demographic characteristics of the consumers and including other products should be conducted to determine the full effect of fibre enrichment with BSG in consumers' response. Future research should also focus on the effect of information on the emotions provoked by fibre-enriched products, and their effects on attention, to draw more in-depth conclusions.

Supplementary Materials: The following are available online at https://www.mdpi.com/article/10.3390/foods11010036/s1, Table S1: Nutritional facts of fibre-enriched and regular bread, pasta and chocolate milk expressed in 100 g of product, Questionnaire S1: Sensory evaluation of pasta.

Author Contributions: Conceptualization, A.C., P.A., G.B., J.A.M. and A.T.; methodology, A.C., G.B., J.A.M. and P.A.; formal analysis, A.C., and P.A.; investigation, M.S., F.P. and C.P.; resources, A.C., P.A.; data curation, M.S., F.P., C.P.; writing—original draft preparation, M.S., F.P., C.P. and A.C.; writing—review and editing, P.A. and A.T.; visualization, A.C., P.A. and A.T.; supervision, A.C., P.A. and A.T.; project administration, A.C.; funding acquisition, A.C. and G.B. All authors have read and agreed to the published version of the manuscript.

Funding: This research received financial of MCIN/AEI/10.13039/501100011033—Proyect PID2019-107723RB-C21.

Informed Consent Statement: Informed consent was obtained from all subjects involved in the study.

Data Availability Statement: The datasets generated for this study are available on request to the corresponding author.

Acknowledgments: Authors are grateful to FNC S.A. (Montevideo, Uruguay) for providing the BSG.

Conflicts of Interest: The authors declare no conflict of interest.

References

1. Ranok, A.; Dissamal, P.; Kupradit, C.; Khongla, C.; Musika, S.; Mangkalanan, S. Physicochemical properties and antioxidant activity of gluten-free riceberry-cheese cracker under simulated gastrointestinal transit. *J. Food Sci. Technol.* **2021**, *58*, 2825–2833. [CrossRef] [PubMed]
2. Frame, L.A.; Costa, E.; Jackson, S.A. Current explorations of nutrition and the gut microbiome: A comprehensive evaluation of the review literature. *Nutr. Rev.* **2020**, *78*, 798–812. [CrossRef] [PubMed]
3. Reyes-Becerril, M.; Gijón, D.; Angulo, M.; Vázquez-Martínez, J.; López, M.G.; Junco, E.; Armenta, J.; Guerra, K.; Angulo, C. Composition, antioxidant capacity, intestinal, and immunobiological effects of oregano (Lippia palmeri Watts) in goats: Preliminary in vitro and in vivo studies. *Trop. Anim. Health Prod.* **2021**, *53*, 101. [CrossRef] [PubMed]

4. Fukami, H.; Higa, Y.; Hisano, T.; Asano, K.; Hirata, T.; Nishibe, S. A review of red yeast rice, a traditional fermented food in japan and east asia: Its characteristic ingredients and application in the maintenance and improvement of health in lipid metabolism and the circulatory system. *Molecules* **2021**, *26*, 1619. [CrossRef]
5. Niu, B.; Feng, S.; Xuan, S.; Shao, P. Moisture and caking resistant Tremella fuciformis polysaccharides microcapsules with hypoglycemic activity. *Food Res. Int.* **2021**, *146*, 110420. [CrossRef]
6. Krittanawong, C.; Isath, A.; Hahn, J.; Wang, Z.; Fogg, S.E.; Bandyopadhyay, D.; Jneid, H.; Virani, S.S.; Tang, W.H.W. Mushroom Consumption and Cardiovascular Health: A Systematic Review. *Am. J. Med.* **2021**, *134*, 637–642.e2. [CrossRef]
7. Zhu, M.; Zhou, F.; Ouyang, J.; Wang, Q.; Li, Y.; Wu, J.; Huang, J.; Liu, Z. Combined use of epigallocatechin-3-gallate (EGCG) and caffeine in low doses exhibits marked anti-obesity synergy through regulation of gut microbiota and bile acid metabolism. *Food Funct.* **2021**, *12*, 4105–4116. [CrossRef]
8. Lattimer, J.M.; Haub, M.D. Effects of Dietary Fiber and Its Components on Metabolic Health. *Nutrients* **2010**, *2*, 1266–1289. [CrossRef]
9. Selani, M.M.; Brazaca, S.G.C.; Dos Santos Dias, C.T.; Ratnayake, W.S.; Flores, R.A.; Bianchini, A. Characterisation and potential application of pineapple pomace in an extruded product for fibre enhancement. *Food Chem.* **2014**, *163*, 23–30. [CrossRef]
10. Chau, C.-F.; Chen, C.-H.; Lin, C.-Y. Insoluble fiber-rich fractions derived from Averrhoa carambola: Hypoglycemic effects determined by in vitro methods. *LWT-Food Sci. Technol.* **2004**, *37*, 331–335. [CrossRef]
11. Stephen, A.M.; Champ, M.M.-J.; Cloran, S.J.; Fleith, M.; van Lieshout, L.; Mejborn, H.; Burley, V.J. Dietary fibre in Europe: Current state of knowledge on definitions, sources, recommendations, intakes and relationships to health. *Nutr. Res. Rev.* **2017**, *30*, 149–190. [CrossRef] [PubMed]
12. Prasadi V. P., N.; Joye, I.J. Dietary Fibre from Whole Grains and Their Benefits on Metabolic Health. *Nutrients* **2020**, *12*, 3045. [CrossRef] [PubMed]
13. World Health Organization. *Obesity and Overweight*; World Health Organization: Geneva, Switzerland, 2020.
14. Kaptoge, S.; Pennells, L.; De Bacquer, D.; Cooney, M.T.; Kavousi, M.; Stevens, G.; Riley, L.M.; Savin, S.; Khan, T.; Altay, S.; et al. World Health Organization cardiovascular disease risk charts: Revised models to estimate risk in 21 global regions. *Lancet Glob. Health* **2019**, *7*, e1332–e1345. [CrossRef]
15. Cappa, C.; Alamprese, C. Brewer's spent grain valorization in fiber-enriched fresh egg pasta production: Modelling and optimization study. *LWT* **2017**, *82*, 464–470. [CrossRef]
16. Nagy, M.; Semeniuc, C.A.; Socaci, S.A.; Pop, C.R.; Rotar, A.M.; Sălăgean, C.D.; Tofană, M. Utilization of brewer's spent grain and mushrooms in fortification of smoked sausages. *Food Sci. Technol.* **2017**, *37*, 315–320. [CrossRef]
17. Garrett, R.; Bellmer, D.; McGlynn, W.; Rayas-Duarte, P. Development of new chip products from brewer's spent grain. *J. Food Qual.* **2021**, *2021*, 5521746. [CrossRef]
18. Mudgil, D.; Barak, S.; Khatkar, B.S. Cookie texture, spread ratio and sensory acceptability of cookies as a function of soluble dietary fiber, baking time and different water levels. *LWT-Food Sci. Technol.* **2017**, *80*, 537–542. [CrossRef]
19. Grigor, J.M.; Brennan, C.S.; Hutchings, S.C.; Rowlands, D.S. The sensory acceptance of fibre-enriched cereal foods: A meta-analysis. *Int. J. Food Sci. Technol.* **2016**, *51*, 3–13. [CrossRef]
20. Maehle, N.; Iversen, N.; Hem, L.; Otnes, C. Exploring consumer preferences for hedonic and utilitarian food attributes. *Br. Food J.* **2015**, *117*, 3039–3063. [CrossRef]
21. Curutchet, A.; Trias, J.; Tárrega, A.; Arcia, P. Consumer Response to Cake with Apple Pomace as a Sustainable Source of Fibre. *Foods* **2021**, *10*, 499. [CrossRef]
22. Curutchet, A.; Cozzano, S.; Tárrega, A.; Arcia, P. Blueberry pomace as a source of antioxidant fibre in cookies: Consumer's expectations and critical attributes for developing a new product. *Food Sci. Technol. Int.* **2019**, *25*, 642–648. [CrossRef] [PubMed]
23. Ham, Y.-K.; Hwang, K.-E.; Kim, H.-W.; Song, D.-H.; Kim, Y.-J.; Choi, Y.-S.; Kim, C.-J. Effects of fat replacement with a mixture of collagen and dietary fibre on small calibre fermented sausages. *Int. J. Food Sci. Technol.* **2016**, *51*, 96–104. [CrossRef]
24. Karthiayani, A.; Ganga, M.; Kumar, C. Fiber Enrichment in Noodles using Banana Pseudo-stem (Musa paradaisiaca). *Asian J. Dairy Food Res.* **2021**, *40*, 106–111. [CrossRef]
25. Leng, G.; Adan, R.A.H.; Belot, M.; Brunstrom, J.; De Graaf, K.; Dickson, S.; Hare, T.; Maier, S.; Menzies, J.; Preißl, H.; et al. The determinants of food choice. *Proc. Nutr. Soc.* **2017**, *76*, 316–327. [CrossRef]
26. Macht, M. How emotions affect eating: A five-way model. *Appetite* **2008**, *50*, 1–11. [CrossRef] [PubMed]
27. Wegman, J.; van Loon, I.; Smeets, P.A.M.; Cools, R.; Aarts, E. Top-down expectation effects of food labels on motivation. *Neuroimage* **2018**, *173*, 13–24. [CrossRef]
28. Ng, M.; Chaya, C.; Hort, J. Beyond liking: Comparing the measurement of emotional response using EsSense Profile and consumer defined check-all-that-apply methodologies. *Food Qual. Prefer.* **2013**, *28*, 193–205. [CrossRef]
29. Horwitz, W. *Official Methods of Analysis of AOAC International. Volume I, Agricultural Chemicals, Contaminants, Drugs/Edited by William Horwitz*; AOAC International: Gaithersburg, MD, USA, 2010.
30. Arcia, P.; Curutchet, A.; Cozzano, S.; Santiago, R. Bagazo de cervecería como ingrediente en el desarrollo de panificados. Impacto del rotulado en la intención de compra y aceptabilidad. *Innotec* **2018**, *16*, 40–46. [CrossRef]
31. Mercosur/Gmc/Res N°01/12. Reglam Ento Tecnico Mercosur Sobre Informacion Nutricional Complementaria (Declaraciones De Propiedades Nutricionalesj; Volume 148. Available online: http://www.puntofocal.gov.ar/notific_otros_miembros/pry33a1_t.pdf (accessed on 10 November 2021).

32. FAO; WHO. Codex guideline for use of nutrition and health claims-CAC/GL 23-1997. *Codex Aliment. Comm.* **2004**, *23*, 1–8.
33. Nestrud, M.A.; Meiselman, H.L.; King, S.C.; Lesher, L.L.; Cardello, A.V. Development of EsSense25, a shorter version of the EsSense Profile®. *Food Qual. Prefer.* **2016**, *48*, 107–117. [CrossRef]
34. Rocha, Y.J.P.; Lapa-Guimarães, J.; de Noronha, R.L.F.; Trindade, M.A. Evaluation of consumers' perception regarding frankfurter sausages with different healthiness attributes. *J. Sens. Stud.* **2018**, *33*, e12468. [CrossRef]
35. Yang, Q.; Shen, Y.; Foster, T.; Hort, J. Measuring consumer emotional response and acceptance to sustainable food products. *Food Res. Int.* **2020**, *131*, 108992. [CrossRef] [PubMed]
36. Lynch, K.M.; Steffen, E.J.; Arendt, E.K. Brewers' spent grain: A review with an emphasis on food and health. *J. Inst. Brew.* **2016**, *122*, 553–568. [CrossRef]
37. Amoriello, T.; Mellara, F.; Galli, V.; Amoriello, M.; Ciccoritti, R. Technological properties and consumer acceptability of bakery products enriched with brewers⇔ spent grains. *Foods* **2020**, *9*, 1492. [CrossRef] [PubMed]
38. Combest, S.; Warren, C. Perceptions of college students in consuming whole grain foods made with Brewers' Spent Grain. *Food Sci. Nutr.* **2019**, *7*, 225–237. [CrossRef]
39. Laureati, M.; Conte, A.; Padalino, L.; Del Nobile, M.A.; Pagliarini, E. Effect of Fiber Information on Consumer's Expectation and Liking of Wheat Bran Enriched Pasta. *J. Sens. Stud.* **2016**, *31*, 348–359. [CrossRef]
40. Quagliani, D.; Felt-Gunderson, P. Closing America's Fiber Intake Gap: Communication Strategies From a Food and Fiber Summit. *Am. J. Lifestyle Med.* **2017**, *11*, 80–85. [CrossRef]
41. Meynier, A.; Chanson-Rollé, A.; Riou, E. Main factors influencing whole grain consumption in children and adults—A narrative review. *Nutrients* **2020**, *12*, 2217. [CrossRef]
42. Kamar, M.; Evans, C.; Hugh-Jones, S. Factors influencing adolescent whole grain intake: A theory-based qualitative study. *Appetite* **2016**, *101*, 125–133. [CrossRef]
43. Olubunmi, I.P.; Babatunde, K.S.; Bolanle, O.O.; Seyioba, S.O.; Taiwo, L.-T.; Olukayode, O.A.; Nwankego, E.G. Quality Evaluation of Fibre- Enriched Bread. *Int. J. Nutr. Food Sci.* **2015**, *4*, 503. [CrossRef]
44. Zhou, Y.; Dhital, S.; Zhao, C.; Ye, F.; Chen, J.; Zhao, G. Dietary fiber-gluten protein interaction in wheat flour dough: Analysis, consequences and proposed mechanisms. *Food Hydrocoll.* **2021**, *111*, 10623. [CrossRef]
45. Ahluwalia, N.; Herrick, K.A.; Terry, A.L.; Hughes, J.P. Contribution of Whole Grains to Total Grains Intake Among Adults Aged 20 and Over: United States, 2013–2016. *NCHS Data Brief.* **2019**, *7*, 1–8.
46. Barrett, E.M.; Foster, S.I.; Beck, E.J. Whole grain and high-fibre grain foods: How do knowledge, perceptions and attitudes affect food choice? *Appetite* **2020**, *149*, 104630. [CrossRef] [PubMed]
47. Domjan, M. *Introduction to Food Neophobia: Historical and Conceptual Foundations*; Woodhead Publishing: Cambridge, UK, 2018.
48. Tuorila, H.; Hartmann, C. Consumer responses to novel and unfamiliar foods. *Curr. Opin. Food Sci.* **2020**, *33*, 1–8. [CrossRef]
49. Makhlouf, S.; Jones, S.; Ye, S.-H.; Sancho-Madriz, M.; Burns-Whitmore, B.; Li, Y.O. Effect of selected dietary fibre sources and addition levels on physical and cooking quality attributes of fibre-enhanced pasta. *Food Qual. Saf.* **2019**, *3*, 117–127. [CrossRef]
50. Krawecka, A.; Sobota, A.; Sykut-Domanska, E. Physicochemical, sensory, and cooking qualities of pasta enriched with oat β-glucans, xanthan gum, and vital gluten. *Foods* **2020**, *9*, 1412. [CrossRef]
51. Nocente, F.; Taddei, F.; Galassi, E.; Gazza, L. Upcycling of brewers' spent grain by production of dry pasta with higher nutritional potential. *LWT* **2019**, *114*, 108421. [CrossRef]
52. Ktenioudaki, A.; Crofton, E.; Scannell, A.G.M.; Hannon, J.A.; Kilcawley, K.N.; Gallagher, E. Sensory properties and aromatic composition of baked snacks containing brewer's spent grain. *J. Cereal Sci.* **2013**, *57*, 384–390. [CrossRef]
53. Tomic, N.; Dojnov, B.; Miocinovic, J.; Tomasevic, I.; Smigic, N.; Djekic, I.; Vujcic, Z. Enrichment of yoghurt with insoluble dietary fiber from triticale–A sensory perspective. *LWT-Food Sci. Technol.* **2017**, *80*, 59–66. [CrossRef]
54. Lopez, F.L.; Bowles, A.; Gul, M.O.; Clapham, D.; Ernest, T.B.; Tuleu, C. Effect of formulation variables on oral grittiness and preferences of multiparticulate formulations in adult volunteers. *Eur. J. Pharm. Sci.* **2016**, *92*, 156–162. [CrossRef]
55. Santagiuliana, M.; Marigómez, I.S.; Broers, L.; Hayes, J.E.; Piqueras-Fiszman, B.; Scholten, E.; Stieger, M. Exploring variability in detection thresholds of microparticles through participant characteristics. *Food Funct.* **2019**, *10*, 5386–5397. [CrossRef]
56. Amofa-Diatuo, T.; Anang, D.M.; Barba, F.J.; Tiwari, B.K. Development of new apple beverages rich in isothiocyanates by using extracts obtained from ultrasound-treated cauliflower by-products: Evaluation of physical properties and consumer acceptance. *J. Food Compos. Anal.* **2017**, *61*, 73–81. [CrossRef]
57. Cardello, A.V. Food quality: Relativity, context and consumer expectations. *Food Qual. Prefer.* **1995**, *6*, 163–170. [CrossRef]
58. Yalçln, E.; Kösemeci, C.; Correia, P.; Karademir, E.; Ferreira, M.; Florencą, S.G.; Guiné, R.P.F. Evaluation of consumer habits and knowledge about dietary fibre and fibre-rich products in Turkish population. *Open Agric.* **2020**, *5*, 375–385. [CrossRef]
59. Baixauli, R.; Salvador, A.; Hough, G.; Fiszman, S.M. How information about fibre (traditional and resistant starch) influences consumer acceptance of muffins. *Food Qual. Prefer.* **2008**, *19*, 628–635. [CrossRef]
60. Li, T.; Dando, R. Impact of common food labels on consumer liking in vanilla yogurt. *Foods* **2019**, *8*, 584. [CrossRef]
61. Biondi, B.; Camanzi, L. Nutrition, hedonic or environmental? The effect of front-of-pack messages on consumers' perception and purchase intention of a novel food product with multiple attributes. *Food Res. Int.* **2020**, *130*, 108962. [CrossRef] [PubMed]
62. Zanetta, L.D.; Umebara, M.T.C.; Costa, J.P.; Takeda, D.K.; Da Cunha, D.T. Hedonic, emotional and willingness-to-pay response to beers of a different type in Brazil. *Br. Food J.* **2021**, *123*, 87–107. [CrossRef]

63. Hur, J.; Jang, S. Anticipated guilt and pleasure in a healthy food consumption context. *Int. J. Hosp. Manag.* **2015**, *48*, 113–123. [CrossRef]
64. Wansink, B.; Chandon, P. Can "low-fat" nutrition labels lead to obesity? *J. Mark. Res.* **2006**, *43*, 605–617. [CrossRef]

foods

MDPI

Article

Sensory Drivers of Consumer Acceptance, Purchase Intent and Emotions toward Brewed Black Coffee

Ammaraporn Pinsuwan [1], Suntaree Suwonsichon [1,*], Penkwan Chompreeda [1] and Witoon Prinyawiwatkul [2]

[1] Kasetsart University Sensory and Consumer Research Center (KUSCR), Department of Product Development, Faculty of Agro-Industry, Kasetsart University, Bangkok 10900, Thailand; ammaraporn.pi@gmail.com (A.P.); fagipkc@ku.ac.th (P.C.)

[2] Agricultural Center, School of Nutrition and Food Sciences, Louisiana State University, Baton Rouge, LA 70803, USA; wprinya@lsu.edu

* Correspondence: suntaree.su@ku.ac.th or fagisrsu@ku.ac.th; Tel.: +66-2-562-5017

Abstract: The link between coffee aroma/flavor and elicited emotions remains underexplored. This research identified key sensory characteristics of brewed black coffee that affected acceptance, purchase intent and emotions for Thai consumers. Eight Arabica coffee samples were evaluated by eight trained descriptive panelists for intensities of 26 sensory attributes and by 100 brewed black coffee users for acceptance, purchase intent and emotions. Results showed that the samples exhibited a wide range of sensory characteristics, and large differences were mainly described by the attributes *coffee identity (coffee ID), roasted, bitter taste, balance/blended* and *fullness*. Differences also existed among the samples for overall liking, purchase intent and most emotion terms. Partial least square regression analysis revealed that liking, purchase intent and positive emotions, such as *active, alert, awake, energetic, enthusiastic, feel good, happy, jump start, impressed, pleased, refreshed* and *vigorous* were driven by *coffee ID, roasted, ashy, pipe tobacco, bitter taste, rubber, overall sweet, balanced/blended, fullness* and *longevity*. Contrarily, *sour aromatic, sour taste, fruity, woody, musty/earthy, musty/dusty and molasses* decreased liking, purchase intent and positive emotions, and stimulated negative emotions, such as *disappointed, grouchy* and *unfulfilled*. This information could be useful for creating or modifying the sensory profile of brewed black coffee to increase consumer acceptance.

Keywords: coffee; emotions; descriptive analysis; consumer perception; liking

Citation: Pinsuwan, A.; Suwonsichon, S.; Chompreeda, P.; Prinyawiwatkul, W. Sensory Drivers of Consumer Acceptance, Purchase Intent and Emotions toward Brewed Black Coffee. *Foods* **2022**, *11*, 180. https://doi.org/10.3390/foods11020180

Academic Editor: Edgar Chambers IV

Received: 9 December 2021
Accepted: 7 January 2022
Published: 11 January 2022

1. Introduction

Coffee is one of the most widely consumed beverages in the world. In 2021, revenue in the coffee segment amounts to around $409 billion, and the global coffee market is anticipated to grow annually by 6.9% during 2021–2025 [1]. Coffee-drinking is driven by different motivations, such as the psychological stimulation induced by caffeine and the sensory enjoyment provided by coffee aroma and flavor [2]. Descriptive sensory analysis was used in several studies as an effective evaluation method to determine the effects of various factors (e.g., coffee origin, roast level, brewing method, serving temperature) on coffee aroma and flavor [3–7]. Chambers et al. [8] developed a sensory lexicon containing a set of 110 well-defined and referenced attributes, which could be used to describe the aroma and flavor of a wide range of brewed coffee samples.

Recently, the emotional experiences elicited by foods and beverages have increasingly gained interest from researchers, because they are known to affect consumer acceptance and consumption, apart from product sensory characteristics [9]. The EsSense Profile™ (ESP) with 39 emotion terms developed by King and Meiselman [10] has been widely used to measure emotional responses toward food products [11]. As different foods evoke different emotions [12], emotion scales were developed for specific food products, such as dark chocolate [13], chocolate and hazelnut spreads [14], blackcurrant squashes [15], wine [16], beer [17], ice cream [18], muffin [19], and entomophagy [20]. For coffee, an emotion lexicon called the Coffee-drinking Experience (CDE) was developed by Bhumiratana et al. [21]. The CDE consists of 44 emotion terms that describe coffee-drinking experiences in two dimensions, a valence dimension (unpleasant-pleasant) and an activation dimension (low-high). With the CDE, more specialized emotions that described *alert* and *focused* mental states (e.g., *in control, motivated, productive, clear minded*) were captured. Later, Kanjanakorn and Lee [22] used the ESP and CDE to study emotions elicited by coffee-drinking, and results of the two scales were compared. They found that emotions evoked by coffee-drinking were captured as *energetic, satisfied* and *pleased* by both scales at similar ratings. However, additional significant changes in *boosted, jolted* and *jump start* were detected when the CDE was used. van Zyl and Meiselman [23] suggested that a language-specific emotion lexicon was necessary because studies in different countries revealed that emotions expressed by words were influenced by culture and context [14,24,25]. Thus, coffee emotion lexicons were developed in different languages, such as Chinese, Korean [26] and Thai [27] using the CDE and/or the ESP as baselines.

While emotions toward food products have been studied extensively, the link between emotion responses and product sensory characteristics remains underexplored. Research in this area is limited only to a few product categories, such as yogurt [28,29], chocolate/dark chocolate [13,30–32], cashew nut, processed tomato [31,32], potato chips [29,32], cheese, snack bars, and fruit [29]. Based on the results of previous research, the association between emotions and sensory characteristics was found to be specific to a product category, that is, the association found in one product category could not be extended to other product categories [11]. For example, *sweet* in cashew nut was found to be positively linked with *affectionate* and *aggressive*, and negatively with *bored*, while *sweet* in chocolate was found to positively link with *guilty*, and negatively with *interested* and *satisfied*. In canned tomato, *sweet* was associated with *curiosity, surprise* and *enthusiasm* [32]. To our knowledge, the work done by Bhumiratana et al. [33] was the only research that studied the link between emotions and sensory characteristics of brewed coffee using data obtained from consumer testing and sensory descriptive analysis. In their study, six coffee samples commercially available in the US market were evaluated by a trained coffee panel to rate the intensities of 12 sensory attributes. The same set of samples were used to elicit emotions in 94 US coffee drinkers. Results revealed key sensory characteristics of brewed coffee that stimulated positive and negative emotions for the US consumers.

The current research was aimed to study further on the sensory characteristics of brewed coffee that drove acceptance, purchase intent and emotion responses for Thai consumers, another group of coffee users. Recently, coffee-drinking has become a new cultural practice among Thai people and the coffee market in Thailand is expected to grow at an annual rate of 12.35% during 2021–2025 [34]. Thailand is also one of the top coffee producers, ranked third among the coffee-producing countries in Asia, following Vietnam and Indonesia [35]. As black coffee-drinking is becoming more popular [36], all samples in this study were evaluated in the form of brewed black coffee without the addition of other ingredients, such as sugar, milk and coffee creamer, and all of the participants for consumer test were brewed black coffee users.

2. Materials and Methods

2.1. Coffee Samples

Eight commercial Arabica roasted coffee bean samples were used in this study (Table 1). The samples were selected through a two-step screening process. Firstly, 87 coffee bean samples commercially available in Thailand were reviewed by researchers, and 15 samples were chosen to represent a diversity in roast levels (light, medium and dark) and manufacturing countries (Thailand, Lao People's Democratic Republic, Ireland, Italy and USA). Secondly, brewed black coffee drinks prepared from those 15 samples were tasted by eight trained panelists, and eight samples that represented a wide range of sensory characteristics were selected for the current study.

Table 1. A list of eight Arabica roasted coffee bean samples evaluated in the study.

Brand	Product Description	Manufacturing Country	Roast Level
Bon Café	Café Rama	Chonburi, Thailand	Light
Café Direct	Kilimanjaro	Dublin, Ireland	Light
Dao coffee	Arabica premium	Champasak, Lao People's Democratic Republic	Medium
My Choice	Arabica blend	Samutsakhon, Thailand	Medium
Royal Project	Royal project coffee	Chiangmai, Thailand	Medium
The Coffee Bean Roasting	Roasted and whole bean coffee	Pathumthani, Thailand	Medium
Illy	Espresso	Trieste, Italy	Dark
Starbucks	Espresso roast	Seattle, WA, USA	Dark

2.2. Sample Preparation

Each of the coffee samples was prepared by grinding the roasted beans to a medium grind using a coffee bean grinder (DēLonghi, model KG 89, Treviso, Italy); this was done no more than 15 min before brewing to preserve the volatile compounds and freshness of the coffee samples [5]. A drip coffee maker was used to brew the coffee samples in this study. According to Sanchez and Chambers [5], dripping was recommended as the most appropriate brewing method for use in sensory testing of coffee over cupping, espresso and filtered infusion methods because the dripping method provided better differentiation among coffee samples in terms of aroma, flavor and aftertaste attributes. Purified (reverse osmosis, carbon filtered) water (1000 mL) at room temperature was added to a drip coffee maker machine (Bosch, model TKA3A031, Munich, Germany). Ground coffee (55 g) was put into a coffee filter (1 × 4-inch, Bon Café, Bangkok, Thailand) and then into the coffee-maker machine. The ratio of ground coffee to water used in this study was in accordance with the International Standard for the preparation of coffee samples for use in sensory analysis [37], which suggested a ratio in a 5–9 g range per 100 mL of water, and was the same ratio that Di Donfrancesco et al. [4] used. Each brewed coffee sample was transferred into a stainless steel vacuum jug (Zebra, Bangkok, Thailand) to maintain its temperature until served. The serving temperature was in a range of 60–65 °C, which was similar to that used by Di Donfrancesco et al. [4].

2.3. Sample Evaluation

All samples were subjected to a descriptive analysis and consumer test. All testing was conducted at the sensory evaluation facilities of the Department of Product Development at Kasetsart University, Bangkok, Thailand. The rooms had appropriate lighting, were air-conditioned (about 25 °C), and did not have extraneous odors. The testing room was separated from the sample preparation room.

2.3.1. Descriptive Analysis

A panel consisting of eight trained panelists (all females, age range 38–56 years) from the Kasetsart University Sensory and Consumer Research (KUSCR) Center participated in the test. The panelists had 120 h of training in descriptive analysis and a minimum of 2000 h of testing experiences with various food products and beverages, including brewed coffee. The number of panelists used in this study fell within the range of 8–12, as suggested by Heymann et al. [38] for descriptive analysis. Previously, Maximo-Gacula and Rutenbeck [39] and Drake [40] indicated that a sensory panel with 6 to 14 trained panelists was adequate for descriptive analysis.

Prior to testing, 3 days of orientation (6 h each day) were held, during which panelists identified terms for describing flavor characteristics of all eight black coffee samples. To assist in identifying attributes, terminology developed by Sanchez and Chambers [5] and Chambers et al. [8] for brewed coffee was provided as an initial list. Panelists tasted all samples, discussed possible terms, and compiled a final consensus list of attributes for testing. Panelists were allowed to add new sensory terms detected in the samples which were not in the initial list. Subsequently, they discussed the definition, references and reference intensities of each attribute. During the orientation, panelists also practiced scoring of each attribute on a 15 cm line scale, with 0 meaning none and 15 meaning extremely high. Scores were monitored to ensure that variations among panelists were low (standard deviation ± 1) prior to product testing [41].

Product testing was completed in five 3 h sessions, with three to four samples being tested during each session. Two replicates were evaluated for each sample, and the evaluation was completed by replication. A randomized complete block design was used to determine the serving order within each replication. Panelists received 80 mL of each coffee sample in a pre-heated white porcelain cup labelled with a three-digit random code and covered with a lid. After tasting the sample, panelists rated the intensity of all attributes on 15 cm line scales (0 = none and 15 = extremely high) before moving onto the next sample. References were provided during evaluations to anchor values on the scale. Unsalted crackers (Jacob's Original Cream Cracker, Kraft Foods Malaysia, Petaling Jaya, Malaysia), sliced apples and reverse osmosis deionized water were provided for panelists to cleanse their palates between samples. The panelists had at least a 15 min break after each sample evaluation.

2.3.2. Consumer Test

One hundred Thai consumers (62 females and 38 males, age range 18–64 years) participated in the test. They were recruited based on their weekly brewed black coffee consumption of at least three times a week. Of the 100 participants, 53 drank brewed black coffee 3–5 times a week, while 32 did once daily, and 15 did more than once daily. Bhumiratana et al. [21] classified participants who drank coffee 1–2 times a week, 3–5 times a week, and at least once daily as 'light', 'medium', and 'heavy' users, respectively. Thus, according to these authors, the consumer participants of this study consisted of 53 medium users and 47 heavy users of brewed black coffee.

Each consumer evaluated all eight black coffee samples in two sessions, four samples per session with a 5 min break between samples and 20 min break between sessions. Samples were served one at a time in a randomized and balanced order across consumers. After receiving 40 mL of each coffee sample in a paper cup, consumers were instructed to taste the sample and rate their overall liking on a 9-point hedonic scale (1 = dislike extremely, 9 = like extremely) and purchase intent on a 2-point category scale (purchase/not purchase). The consumers also checked all appropriate terms listed on a check-all that-apply (CATA) question to describe their emotions as evoked by each coffee sample. The CATA question included 20 emotion terms (Table 2) selected from our previous study [27], in which a consumer-defined emotion lexicon for coffee-drinking was developed and further refined by Thai coffee drinkers. Of the 20 terms, 15 terms were similar to those listed in the EsSense Profile^{TM} (ESP) [10], WellSense Profile^{TM} [42] and/or Coffee-drinking Experience (CDE) Profile [21], while five terms were newly generated by Thai coffee drinkers. Terms were presented in a random order among consumers in accordance with other studies [15,26] to minimize the visual processing effect on participants [43]. Unsalted crackers (Jacob's Original Cream Cracker, Kraft Foods Malaysia, Petaling Jaya, Malaysia) and purified bottled water (Nestlé Pure Life®, Nestlé Thai, Ayuthaya, Thailand) were provided for consumers to cleanse their palates between samples.

2.4. *Data Analysis*

Multivariate analysis of variance (MANOVA) and Duncan's multiple range test (DMRT) were performed to determine differences among eight brewed coffee samples based on attribute intensities, liking scores, and percentages of purchase intent at a 95% confidence level ($p \leq 0.05$). Principal component analysis (PCA), with varimax rotation, was performed to uncover relationships between significant attributes and samples. Cochran's Q test was used to determine significant differences among the coffee samples based on the frequency counts of each emotion term included in the CATA questions at a 95% confidence level ($p \leq 0.05$) [44]. When significant differences were found, the Marascuilo and McSweeney procedure [44] was used for multiple pairwise comparisons. Correspondence analysis (CA) was then performed to determine relationships between significant emotion terms and samples. Partial least square regression (PLSR) analysis was also performed to determine sensory drivers of liking, purchase intent, and emotions using sensory attribute intensities as x variables and liking scores, purchase intent (%) and emotions (proportion of frequency counts) as y variables. Insignificant sensory attributes and emotion terms were excluded from the data set prior to PLSR analysis. The statistical software used for MANOVA and DMRT was IBM SPSS Statistics version 28.0 (Thaisoftup Co., Ltd., Bangkok, Thailand) and that used for PCA, Cochran's Q test, CA and PLSR was XLSTAT statistical software version 19.6 (Addinsoft, New York, NY, USA).

Table 2. A list of 20 emotion terms with Thai translation evaluated in the study.

Emotion Terms (Thai Translation) *			
Active (กระฉับกระเฉง) [a,b,c]	Energetic (มีแรง/มีพลัง) [a,b,c]	Happy (มีความสุข) [a,b]	Refreshed (สดชื่น) [b]
Alert (ตื่นตัว) [b]	Enthusiastic (กระตือรือร้น) [a]	Impressed (ประทับใจ)	Relaxed (ผ่อนคลาย) [b,c]
Awake (ตื่น) [c]	Feel good (รู้สึกดี)	Joyful (สบายใจ) [a,b,c]	Unfulfilled (ไม่เติมเต็ม) [b]
Bored (เบื่อ/เซ็ง) [a,c]	Good mood (อารมณ์ดี)	Jump start (พร้อมทำงาน) [c]	Vigorous (กระปรี้กระเปร่า)
Disappointed (ผิดหวัง) [c]	Grouchy (หงุดหงิด) [c]	Pleased (พึงพอใจ) [a,c]	Wistful (โหยหา)

* Emotion terms were selected from a previous study of Pinsuwan et al. [27] in which a consumer-defined emotion lexicon for coffee-drinking was developed and further refined by Thai coffee drinkers. Some of these terms were similar to those listed in other existing emotion profiles. [a] Terms that were also present in the EsSense Profile™ (ESP) [10]. [b] Terms that were also present in the WellSense Profile™ [42]. [c] Terms that were also present in the Coffee-drinking Experience (CDE) Profile [21].

Although sample description was given (Table 1), sample identification was not revealed when reporting the results to avoid conflicts of interest. The three-digit random codes were used to denote the samples.

3. Results and Discussion

3.1. Sensory Characteristics

Table 3 shows the final list of 26 attributes detected in the coffee samples along with their definitions, references and reference intensities for sensory evaluation. Most of the terms were consistent with those in the coffee lexicons developed by Chambers et al. [8] and Sanchez and Chambers [5] (n = 22 and 2, respectively), while two terms (*bitter aromatic* and *creosote/tar*) were newly added to the list. The definitions of the pre-existing attributes were consistent with those of previous studies; however, references for most of these attributes (n = 17) were modified because the ones used in the previous studies were not available in Thailand. Newly developed references were selected through a long discussion process among all panelists to fit with the attribute definitions [41]. Chambers et al. [8] introduced the concept of 'living' lexicons, that is, lexicons must be allowed to grow, change, or adapt as new samples are tested or new understandings of the attribute dimensions and new references arise. Thus, the current research expanded the pre-existing lexicons for brewed coffee by providing new sensory terms and references. A lexicon with well-defined and referenced descriptors helps to facilitate accurate and precise communication among the sensory panelists [45].

Mean intensities of sensory attributes for eight brewed black coffee samples are shown in Table 4. Most attributes were present in all samples, but at varying intensities. The attributes *fruity* and *rubber* were present only in a few samples (samples 562 and 187 for *fruity* and sample 712 for *rubber*), and thus described the unique sensory characteristics of the samples. MANOVA results revealed that almost all attributes were significantly different ($p \leq 0.05$) among samples, except *nutty*. Large differences across the samples were mainly described by *coffee ID*, *roasted*, *bitter taste*, *balance/blended* and *fullness*.

Table 3. Sensory attributes, definitions and references for brewed black coffee evaluation.

Attribute	Definition	References and Intensities
Coffee identity (Coffee ID) [a]	A distinctly roasted brown, slightly bitter aromatic characteristic of brewed coffee. Additional descriptors may/may not include woody, oily, acidic, and full bodied, and these notes may occur at varying intensities	Bon Aroma gold instant coffee [1] = 3.5 Giovanni Caffé American roasted & ground [2] = 7.0 Suzuki Coffee Arabica special blend [2] = 8.5
Roasted [b]	Dark brown impression characteristic of products cooked to a high temperature by dry heat. It does not include bitter or burnt notes	Medium roasted peanuts [3] = 6.5 Dark roasted peanuts [4] = 9.5 Over roasted peanut [5] = 15.0
Burnt [b]	The dark brown carbon impression of an over-cooked or over-roasted product that can be sharp, bitter and sour	Over roasted peanuts [5] = 7.5
Acrid [c]	The sharp, pungent, bitter, acidic aromatics associated with products that are excessively roasted or browned	Dark roasted peanuts [4] = 3.0 Over roasted peanuts [5] = 8.5
Bitter aromatic [d]	The perception of a bitter aromatic of coffee	Dark roasted peanuts [4] = 3.0 Over roasted peanuts [5] = 7.5
Smoky [c]	An acute pungent aromatic that is a product of combustion of wood, leaves or non-natural product	Blue diamond smoked almonds = 2.0
Ashy [c]	Dry, dusty, dirty or smoky aromatics associated with the residual of burnt products	Ash of Marlboro red cigarettes = 4.0 (only for smelling)
Woody [c]	The sweet, brown, musty, dark aromatics associated with a bark of a tree	Heritage premium shelled walnuts = 4.0 Popsicle sticks = 7.5 (only for smelling)
Molasses [b]	Dark caramelized top notes which may include slightly sharp, acrid and sulfur notes associated with molasses	Grandma's molasses = 6.5
Overall sweet [c]	The perception of a combination of sweet taste and aromatics	Post shredded Wheat = 1.5 Kellogg's Special K Breakfast Cereal = 3.0
Nutty [c]	A combination of slightly sweet, brown, woody, oily, musty, astringent, and bitter aromatics commonly associated with nuts, seeds, beans, and grains	Dr. Green wheat germ = 7.5
Dark chocolate [c]	A high-intensity blend of cocoa and cocoa butter that may include dark roast, spicy, burnt, musty notes which include increased astringency and bitterness	Lindt Excellence dark chocolate bar (85% cocoa) = 11.0
Fruity [c]	A sweet, floral aromatic blend of a variety of ripe fruits	Ceres Hanepoot white grape juice = 2.5
Sour aromatic [b]	An aromatic associated with the impression of a sour product	Bush's pinto beans = 2.0
Pipe tobacco [c]	The brown, sweet, slightly pungent, fruity, floral, spicy aromatics associated with cured tobacco	Captain Black pipe tobacco (gold) = 8.5 (only for smelling)
Musty/dusty [c]	The aromatics associated with dry closed air spaces such as attics and closets. May be dry, musty, papery, dry soil or grain	Dr. Green wheat germ = 3.5
Musty/earthy [c]	Somewhat sweet, heavy aromatics associated with damp black soil	Fresh beetroot cube = 7.0
Creosote/tar [d]	A pungent chemical aromatic associated with unrefined crude oil products	Tar = 12.0 (only for smelling)
Rubber [c]	A dark heavy slightly sharp and pungent aromatic associated with rubber	Mahakit Rubber bands = 5.0 (only for smelling)

Table 3. Cont.

Attribute	Definition	References and Intensities
Bitter taste [b]	The fundamental taste factor associated with a caffeine solution	0.5 g/L caffeine solution = 6.5 0.6 g/L caffeine solution = 8.5 0.7 g/L caffeine solution = 10.0 1.0 g/L caffeine solution = 12.0
Sour taste [b]	The fundamental taste factor associated with a citric acid solution	0.15 g/L citric acid solution = 1.5 0.5 g/L citric acid solution = 3.0
Astringent [e]	A drying, puckering or tingling sensation on the surface and/or edge of the tongue and mouth	0.5 g/L Alum Solution = 2.5
Overall impact [c]	The maximum overall sensory impression during the whole tasting time	Suzuki Coffee Arabica special blend [2] = 7.5 Giovanni Caffé American roasted & ground [2] = 9.0 Bon Aroma gold instant coffee [1] = 12.0
Balance/ Blended [c]	The melding of individual sensory notes such that the products present a unified overall sensory experience as opposed to spikes or individual notes	Bon Aroma gold instant coffee [1] = 3.0 Giovanni Caffé American roasted & ground [2] =6.0 Suzuki Coffee Arabica special blend [2] = 10.0
Longevity [c]	The time that the full integrated sensory experience sustains itself in the mouth and after swallowing	Suzuki Coffee Arabica special blend [2] = 7.5 Giovanni Caffé American roasted & ground [2] = 9.0 Bon Aroma gold instant coffee [1] = 12.0
Fullness [c]	The foundation of flavor notes that give substance to the product. The perception of robust flavor that is rounded with body	Bon Aroma gold instant coffee [1] = 5.0 Giovanni Caffé American roasted & ground [2] = 7.5 Suzuki Coffee Arabica special blend [2] = 10.0

[a] Attribute from a study of Sanchez and Chambers [5] with modification of reference samples. [b] Attribute from a study of Chambers et al. [8]. [c] Attribute from a study of Chambers et al. [8] with modification of reference samples. [d] Newly added attribute in the current study. [e] Attribute from a study of Sanchez and Chambers [5]. [1] Prepared by dissolving instant coffee (30 g) in hot (74 °C) water (1420 mL). [2] Prepared by brewing ground coffee (30 g) with water (1420 mL) using a drip coffee maker machine. [3,4,5] Prepared by roasting raw peanut in an oven at 218 °C for 12, 17 and 22 min, respectively.

Results from the PCA map (Figure 1) indicated that significant sensory attributes formed two key dimensions that explained 72.9% of the total variability (41.6% and 31.3%, respectively). For principal component (PC) 1, heavily loaded attributes with absolute loading values greater than 0.6 included *burnt, acrid, bitter aromatic, smoky, ashy, musty/dusty, musty/earthy, creosote/tar, rubber, bitter taste, overall impact* and *longevity* in the positive dimension and *overall sweet* and *dark chocolate* in the negative dimension. For PC2, the heavily loaded attributes in the positive dimension were *coffee ID, roasted, balance/blended* and *fullness*, while those in the negative dimension were *woody, fruity, sour aromatic* and *sour taste*. Spreading of the samples over the PCA map indicated that the selected samples represented a wide range of sensory characteristics as intended.

Sample 712, located on the far-right of the map (Figure 1), was differentiated from the rest of the samples by having the highest ($p \leq 0.05$) intensities in *burnt, acrid, ashy, creosote/tar, bitter taste* and *overall impact* (Table 4). Additionally, it tended to be rated higher in *bitter aroma, smoky, pipe tobacco, musty/dusty, musty/earthy, longevity, coffee ID, roasted, balance/blended* and *fullness*, but lower in *overall sweet, woody, sour aromatic* and *sour taste* than other samples. Sample 712 was the only sample with a *rubber* note, even though such a flavor was detected at very low intensity. It was also the only sample with no detectable *molasses* flavor.

Table 4. Mean intensity scores of sensory attributes for eight black coffee samples.

Attributes	Coffee Samples (Roast Level *)								F-Values
	562 (l)	187 (l)	139 (m)	311 (m)	849 (m)	968 (m)	745 (d)	712 (d)	
Coffee ID	9.53 ± 0.31^{a}	6.68 ± 0.42^{d}	9.59 ± 0.04^{a}	8.89 ± 0.09^{b}	8.72 ± 0.22^{b}	8.41 ± 0.22^{c}	9.31 ± 0.05^{a}	9.58 ± 0.42^{a}	94.90
Roasted	9.91 ± 0.04^{b}	8.72 ± 0.13^{c}	10.69 ± 0.35^{a}	9.74 ± 0.46^{b}	9.56 ± 0.27^{b}	9.09 ± 0.13^{c}	9.99 ± 0.19^{b}	11.06 ± 0.35^{a}	24.62
Burnt	$5.36 \pm 0.02^{b,c}$	4.91 ± 0.31^{d}	5.42 ± 0.29^{b}	4.77 ± 0.11^{d}	4.83 ± 0.17^{d}	$5.05 \pm 0.07^{c,d}$	5.57 ± 0.13^{b}	5.94 ± 0.09^{a}	12.89
Acrid	$4.99 \pm 0.07^{b,c}$	$5.02 \pm 0.33^{b,c}$	5.28 ± 0.40^{b}	$4.50 \pm 0.09^{d,e}$	$4.72 \pm 0.13^{c,d}$	4.23 ± 0.33^{e}	5.34 ± 0.13^{b}	5.76 ± 0.02^{a}	12.38
Bitter aromatic	$5.28 \pm 0.27^{b,c}$	$5.02 \pm 0.20^{c,d}$	$5.39 \pm 0.34^{b,c}$	4.81 ± 0.09^{d}	4.76 ± 0.07^{d}	4.38 ± 0.09^{e}	$5.59 \pm 0.04^{a,b}$	5.89 ± 0.19^{a}	15.10
Smoke	3.56 ± 0.04^{b}	3.62 ± 0.17^{b}	3.94 ± 0.26^{a}	$3.34 \pm 0.13^{b,c}$	3.14 ± 0.19^{c}	3.16 ± 0.04^{c}	3.53 ± 0.08^{b}	4.06 ± 0.18^{a}	9.31
Ashy	$1.94 \pm 0.14^{c,d}$	$1.67 \pm 0.26^{d,e}$	2.56 ± 0.09^{b}	$1.96 \pm 0.03^{c,d}$	$1.92 \pm 0.06^{c,d}$	1.61 ± 0.10^{e}	2.00 ± 0.09^{c}	2.94 ± 0.00^{a}	20.51
Woody	1.48 ± 0.20^{a}	1.53 ± 0.31^{a}	1.19 ± 0.18^{b}	1.10 ± 0.05^{b}	1.50 ± 0.00^{a}	1.49 ± 0.12^{a}	1.49 ± 0.24^{a}	1.20 ± 0.07^{b}	3.72
Molasses	1.16 ± 0.04^{a}	0.99 ± 0.11^{a}	0.97 ± 0.04^{a}	1.01 ± 0.11^{a}	1.14 ± 0.02^{a}	0.53 ± 0.04^{b}	1.21 ± 0.34^{a}	0.00 ± 0.00^{c}	27.42
Overall sweet	1.78 ± 0.17^{a}	$1.44 \pm 0.18^{b,c}$	$1.51 \pm 0.05^{b,c}$	$1.67 \pm 0.06^{a,b}$	1.79 ± 0.13^{a}	1.78 ± 0.19^{a}	$1.60 \pm 0.04^{a,b}$	1.29 ± 0.02^{c}	4.83
Nutty ^{ns}	1.34 ± 0.17	1.08 ± 0.24	1.29 ± 0.04	1.22 ± 0.04	1.38 ± 0.09	1.35 ± 0.18	1.39 ± 0.11	1.24 ± 0.07	1.74
Dark chocolate	2.81 ± 0.08^{c}	2.48 ± 0.47^{d}	3.06 ± 0.27^{b}	2.39 ± 0.11^{d}	2.38 ± 0.35^{d}	1.94 ± 0.09^{e}	3.59 ± 0.04^{a}	3.14 ± 0.06^{b}	36.29
Fruity	0.79 ± 0.03^{b}	1.05 ± 0.12^{a}	0.00 ± 0.00^{c}	0.00 ± 0.00^{c}	0.00 ± 0.00^{c}	0.00 ± 0.00^{c}	0.00 ± 0.00^{c}	0.00 ± 0.00^{c}	294.30
Sour aromatic	2.78 ± 0.49^{b}	3.00 ± 0.09^{a}	1.66 ± 0.12^{e}	1.43 ± 0.19^{f}	$2.17 \pm 0.24^{c,d}$	2.26 ± 0.28^{c}	2.04 ± 0.18^{d}	1.43 ± 0.17^{f}	79.43
Pipe tobacco	$1.28 \pm 0.22^{a,b,c}$	$1.04 \pm 0.15^{c,d}$	$1.41 \pm 0.13^{a,b}$	$1.16 \pm 0.05^{b,c}$	$1.03 \pm 0.01^{c,d}$	$1.03 \pm 0.22^{c,d}$	0.86 ± 0.55^{d}	1.44 ± 0.09^{a}	5.53
Musty/dusty	$1.63 \pm 0.00^{c,d}$	$1.93 \pm 0.16^{a,b,c}$	$1.60 \pm 0.21^{d,f}$	$1.66 \pm 0.35^{b,c,d}$	$1.81 \pm 0.09^{a,b,c,d}$	1.33 ± 0.29^{f}	2.01 ± 0.02^{a}	$1.95 \pm 0.07^{a,b}$	5.48
Musty/earthy	$1.34 \pm 0.04^{a,b,c}$	$1.29 \pm 0.09^{a,b,c}$	1.48 ± 0.12^{a}	1.09 ± 0.04^{c}	$1.19 \pm 0.18^{b,c}$	1.11 ± 0.19^{c}	$1.46 \pm 0.11^{a,b}$	$1.33 \pm 0.03^{a,b,c}$	2.76
Creosote/tar	1.36 ± 0.06^{b}	0.99 ± 0.07^{d}	1.36 ± 0.07^{b}	0.94 ± 0.04^{d}	$1.04 \pm 0.15^{c,d}$	0.97 ± 0.04^{d}	$1.26 \pm 0.06^{b,c}$	1.75 ± 0.32^{a}	11.75
Rubber	0.00 ± 0.00^{b}	0.00 ± 0.00^{b}	0.00 ± 0.00^{b}	0.00 ± 0.00^{b}	0.00 ± 0.00^{b}	0.00 ± 0.00^{b}	0.00 ± 0.00^{b}	1.33 ± 0.02^{a}	470.97
Bitter taste	10.22 ± 0.04^{c}	9.38 ± 0.18^{e}	10.79 ± 0.03^{b}	10.18 ± 0.16^{c}	9.76 ± 0.13^{d}	9.44 ± 0.27^{e}	11.04 ± 0.50^{b}	11.59 ± 0.10^{a}	52.43
Sour taste	1.98 ± 0.29^{b}	2.33 ± 0.34^{a}	1.10 ± 0.02^{e}	1.13 ± 0.06^{e}	$1.61 \pm 0.30^{c,d}$	$1.79 \pm 0.11^{b,c}$	1.51 ± 0.24^{d}	0.98 ± 0.15^{e}	30.37
Astringent	1.47 ± 0.06^{a}	1.35 ± 0.04^{a}	1.39 ± 0.06^{a}	$1.31 \pm 0.09^{a,b}$	1.39 ± 0.00^{a}	1.34 ± 0.27^{a}	1.16 ± 0.01^{b}	1.41 ± 0.04^{a}	2.54
Overall impact	10.38 ± 0.09^{b}	10.34 ± 0.39^{b}	$10.19 \pm 0.34^{b,c}$	$9.93 \pm 0.25^{c,d}$	9.80 ± 0.02^{d}	9.68 ± 0.07^{d}	$9.95 \pm 0.02^{c,d}$	11.09 ± 0.31^{a}	15.48
Balance/Blended	$9.00 \pm 0.44^{a,b}$	5.50 ± 0.27^{e}	$8.77 \pm 0.24^{b,c}$	9.28 ± 0.40^{a}	8.50 ± 0.35^{c}	8.03 ± 0.13^{d}	$8.66 \pm 0.04^{b,c}$	$9.06 \pm 0.22^{a,b}$	68.23
Longevity	$10.16 \pm 0.04^{b,c}$	$9.88 \pm 0.44^{c,d}$	$10.44 \pm 0.01^{a,b}$	$10.05 \pm 0.11^{c,d}$	$9.69 \pm 0.19^{d,e}$	9.46 ± 0.06^{e}	$9.78 \pm 0.04^{c,d,e}$	10.70 ± 0.19^{a}	9.61
Fullness	8.70 ± 0.02^{d}	7.71 ± 0.47^{e}	9.75 ± 0.18^{a}	$9.24 \pm 0.34^{b,c}$	8.73 ± 0.03^{d}	8.64 ± 0.02^{d}	9.06 ± 0.18^{c}	$9.55 \pm 0.28^{a,b}$	33.54

Scores were the average values of two replications, where each replication was evaluated by eight panelists. * Roast level: l = light, m = medium and d = dark. ^{a,b,c,d,e,f} When the F-test was significant (the critical F value = 2.21 at $\alpha = 0.05$, df = 7, 49, a two-tailed test, the mean intensities in the same row with different superscripts are significantly different ($p \leq 0.05$), based on Duncan's multiple range test. ^{ns} Mean intensities in the same row are not significantly different ($p > 0.05$).

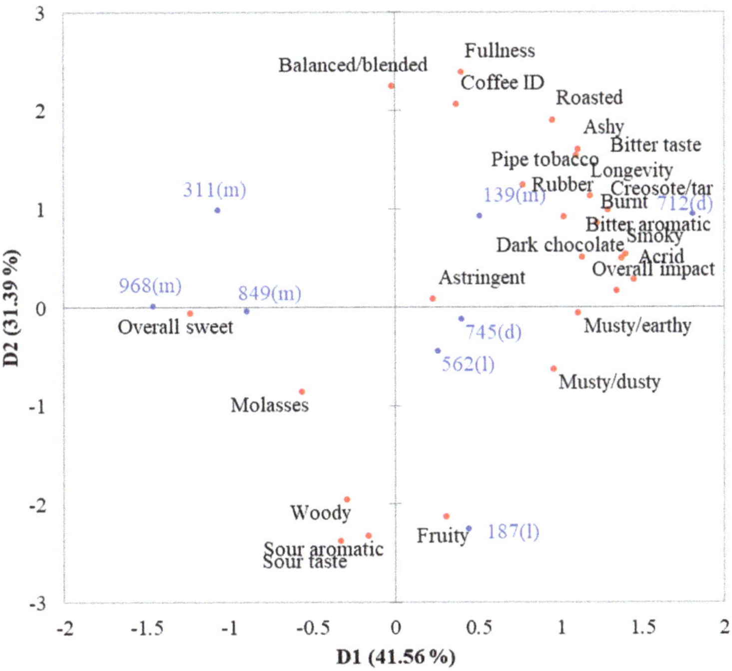

Figure 1. Principal component analysis (PCA) map of the sensory profiles of eight brewed black coffee. ● = sensory attributes, ● = samples with roast level in parenthesis (l = light, m = medium and d = dark).

Samples 139, 562 and 745 shared some common characteristics, as shown by a closer grouping on the PCA map (Figure 1). These samples were rated as high as ($p > 0.05$) sample 712 in *coffee ID* (Table 4). In addition, their intensities in *burnt, bitter aromatic, smoky, ashy, creosote/tar* and *bitter taste* were second to those of sample 712 and were higher ($p \leq 0.05$) than those of other samples. However, differences still existed among samples 139, 562 and 745. Sample 139 was different from the other two samples by having higher ($p \leq 0.05$) *smoky, ashy, roasted* and *fullness* intensities, but lower ($p \leq 0.05$) *woody, sour aromatic* and *sour taste* intensities. Sample 562 was rated higher ($p \leq 0.05$) in *fruity, sour aromatic* and *sour taste,* but lower ($p \leq 0.05$) in *bitter taste, dark chocolate* and *fullness* than the others, while sample 745 was higher ($p \leq 0.05$) in *musty/dusty* and *dark chocolate,* but lower ($p \leq 0.05$) in *pipe tobacco* and *astringent* than the others.

Samples 311, 849 and 968 were located closer to each other on the left of the PCA map (Figure 1) so they shared some common characteristics. These samples tended to be rated lower in *burnt, acrid, bitter aromatic, smoky, musty/dusty, musty/earthy, creosote/tar, bitter taste, overall impact* and *longevity* than the rest of the samples (Table 4). Differences among these three samples existed, especially between samples 311 and 968. Of the three samples, *woody* and *sour aromatic* were the lowest ($p \leq 0.05$) for sample 311, while *coffee ID, roasted, bitter aromatic, ashy, molasses dark chocolate, musty/dusty, bitter taste* and *balance/blended* were the lowest ($p \leq 0.05$) for sample 968.

Sample 187, located at the bottom of the map (Figure 1), was differentiated from the rest of the samples by having the highest ($p \leq 0.05$) intensities in *fruity, sour aromatic* and *sour taste,* but the lowest ($p \leq 0.05$) intensities in *coffee ID, balance/blended* and *fullness* (Table 4). Additionally, it tended to be rated higher in *woody* and *must/dusty*, but lower in *roasted, burnt, ashy* and *bitter taste* than other samples.

Differences in sensory characteristics across the eight brewed black coffee samples were mainly due to differences in the roast level of the coffee beans. Generally, intensities of *coffee ID, roasted, burnt, acrid, bitter aromatic, smoky, ashy, dark chocolate, creosote/tar, bitter taste* and *fullness* tended to increase with an increased roast level from light to dark (Table 4). On the other hand, *sour aromatic* and *sour taste* intensities tended to decrease with the degree of roasting. The increases in *coffee ID, roasted, burnt, acrid, smoky* and *ashy* notes with an increased roast level have been reported by other researchers [3,33,46,47], while the decreases in *sour* characters with an increased roast level were in agreement with the studies of Bhumiratana et al. [33] and Akiyama et al. [47]. Roasting is one of the important factors that affects the sensory properties of coffee [48]. During the roasting process, coffee beans are exposed to a high temperature, and sugars in the beans undergo a caramelization reaction. Once the beans are heated to 205 °C, thermal decompositions and chemical changes occur, resulting in development and degradation of various volatile compounds, such as carbon dioxide, aldehydes, ketones, ethers, acetic acid, methanol, oils, and glycerol [3]. Thus, the temperature and condition during the roasting process are among the main factors that control the complexity of coffee aroma and flavor. An exception was observed for sample 562. Although it was labelled as a light roasted coffee, its *coffee ID, roasted* and *burnt* intensities were as high as those of other medium and dark roasted coffees. Apart from a roast level, other factors, such as the growing region/condition and processing methods from coffee cherries to green coffee beans [33] could have an impact on the aroma and flavor of coffee.

3.2. Consumer Liking, Purchase Intent and Emotions

Sample 712 received the highest overall liking score; however, it was not significantly different ($p > 0.05$) from samples 139 and 311 (Table 5). These three samples were scored higher than six (like slightly) on a 9-point hedonic scale. On the contrary, sample 187 received the lowest ($p \leq 0.05$) liking score that was in the range of 'dislike slightly' to 'neither like nor dislike', while scores of samples 849, 745, 968 and 562 were in the middle range of the scale (5.6–5.7) and were not significantly different ($p > 0.05$) from one another. The purchase intent also was significantly ($p \leq 0.05$) different among samples and followed a similar trend of liking scores.

Table 5. Overall liking score and purchase intent of eight black coffee samples, as evaluated by 100 consumers.

Coffee Samples (Roast Level *)	Overall Liking Scores	% Consumers Who Would Purchase the Sample
712 (d)	6.5 ± 1.6^{a}	75^{a}
139 (m)	6.2 ± 1.6^{a}	$63^{a,b,c}$
311 (m)	6.1 ± 1.6^{a}	$69^{a,b}$
849 (m)	5.7 ± 1.7^{b}	$53^{b,c}$
745 (d)	5.7 ± 1.7^{b}	$53^{b,c}$
968 (m)	5.6 ± 1.7^{b}	$51^{b,c}$
562 (l)	5.6 ± 1.8^{b}	48^{c}
187 (l)	4.4 ± 1.9^{c}	26^{d}

Liking scores were the average values of 100 consumers. * roast level: l = light, m = medium and d = dark. a,b,c,d Values within columns with different superscripts are significantly different ($p \leq 0.05$).

Results indicated that liking scores and purchase intent tended to increase as roast levels increased from light to dark. However, an exception was observed for sample 745. Although the sample was a dark roasted coffee, it was rated lower in liking and purchase intent than other medium roasted coffee samples. Previously, Bhumiratana et al. [21,33] found that liking scores of one consumer cluster ($n = 10$) toward brewed coffee tended to decrease with increased roast levels ($n = 10$); however, no consistent trend was observed for other consumer clusters ($n = 84$) regarding the effect of roast levels.

Results based on Cochran's Q test revealed that differences existed across the eight coffee samples for 17 out of 20 emotion terms. The non-discriminating emotion terms were *bored, joyful* and *relaxed*. A symmetrical correspondence analysis (CA) map (Figure 2) shows the positioning of each coffee sample in the emotion space that explains 75.2% of total variability. The top three most liked samples (712, 139 and 311) were on the left quadrants and were explained by the emotion terms *active, alert, awake, energetic, enthusiastic, feel good, good mood, happy, impressed, jump start, pleased, refreshed, vigorous* and *wistful*. Next, on the right of these samples were samples 745, 968 and 562, which received liking scores in a neutral range. Although sample 849 was also rated in a neutral range for overall liking, its position in the CA map was closer to the top three most liked sample. By contrast, sample 187 that was disliked by consumers was on the right quadrant and was anchored by the terms *disappointed, grouchy* and *unfulfilled*. Results clearly showed that positive emotions were elicited by coffee samples that received liking scores in positive and neutral ranges of the scale, with generally higher frequency counts of positive emotions for liked samples than for neutrally liked samples, while negative emotions were evoked by disliked coffee samples. These results were supported by previous works on coffee [21,26,33] and other food categories, such as sweetener [49], salty snack, yogurt, cheese [29], beef [50] and fruit and vegetable juices [15,51] that samples with higher liking scores were more strongly associated with positive emotions, while less liked samples were more strongly associated with negative emotions.

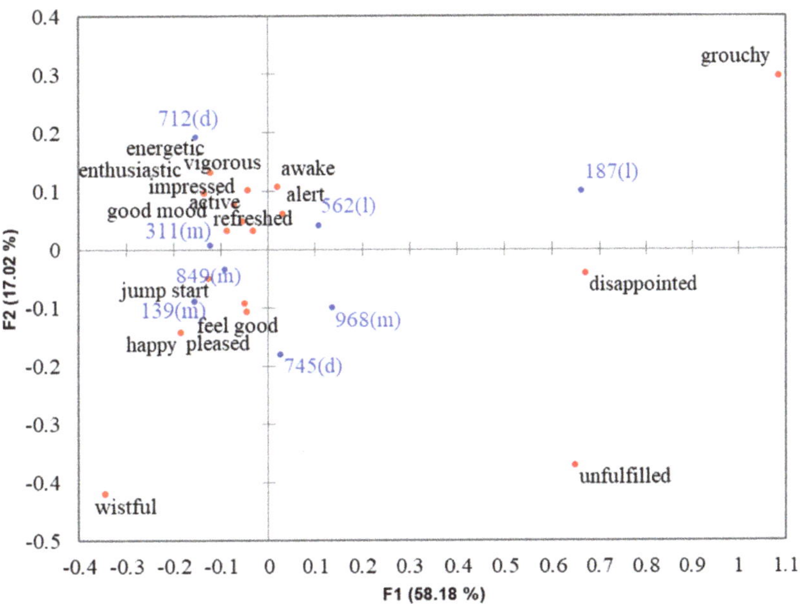

Figure 2. Correspondence analysis (CA) map showing the positioning of eight brewed black coffee samples in the emotion space. • = emotion terms, • = samples with roast level in parenthesis (l = light, m = medium and d = dark).

Product liking sometimes does not correlate well with emotions [52]. Even though liking scores of samples 712, 139 and 311 were not significantly different (p >0.05), sample 712 was described to stimulate positive, high-energy emotions such as *alert, awake* and *energetic*, more frequently ($p \leq 0.05$) than the other two samples. Additionally, the feeling of *good mood* was mentioned more frequently ($p \leq 0.05$) when drinking sample 849 compared to sample 745, although both samples received similar ($p > 0.05$) liking scores. Thus, the measurement of emotions elicited using a consumer-defined lexicon provided more discrimination across coffee samples than the hedonic measure, despite the fact that CATA emotion measures were only monitored in terms of presence/absence. The results concurred with those observed in the literature [15,21,32,53–56].

3.3. Sensory Drivers of Liking, Purchase Intent and Emotions

Results from PLSR analysis (Figure 3) revealed sensory drivers of liking, purchase intent and emotions toward brewed black coffee based on 100 Thai coffee users. The first two components generated by PLSR explained 72.8% of total variability found in sensory attributes (x, explanatory variables) and 77.2% of total variability found in overall liking, purchase intent (%) and emotions (proportion of frequency counts) (y, dependent variables). Only the attributes with standardized β-coefficients in the PLSR models (data not shown) greater than 0.05 in absolute value were interpreted as the sensory drivers of these parameters [57]. It was found that liking, purchase intent and positive emotions including *active, alert, awake, energetic, enthusiastic, feel good, good mood, happy, jump start, impressed, pleased, refreshed* and *vigorous* were mainly driven by *coffee ID, roasted, balanced/blended* and *fullness. Ashy* and *pipe tobacco* also stimulated liking, purchase intent and almost all positive emotions except *good mood.* The *bitter taste* of brewed black coffee not only promoted liking, but also evoked an *impressed* feeling and positive high-energy feelings, including *alert, awake, energetic, jump start, refreshed* and *vigorous.* Unexpectedly, *rubber* was positively associated with liking, purchase intent and the feelings of *alert, awake, energetic, impressed, jump start, pleased, refreshed* and *vigorous.* This attribute was present in only one sample (sample 712) at very low intensity (Table 4). The consumers also related an *overall sweet* note to a *happy* feeling. *Longevity* of coffee flavor during tasting and after swallowing made the consumers feel *alert, awake* and *impressed.*

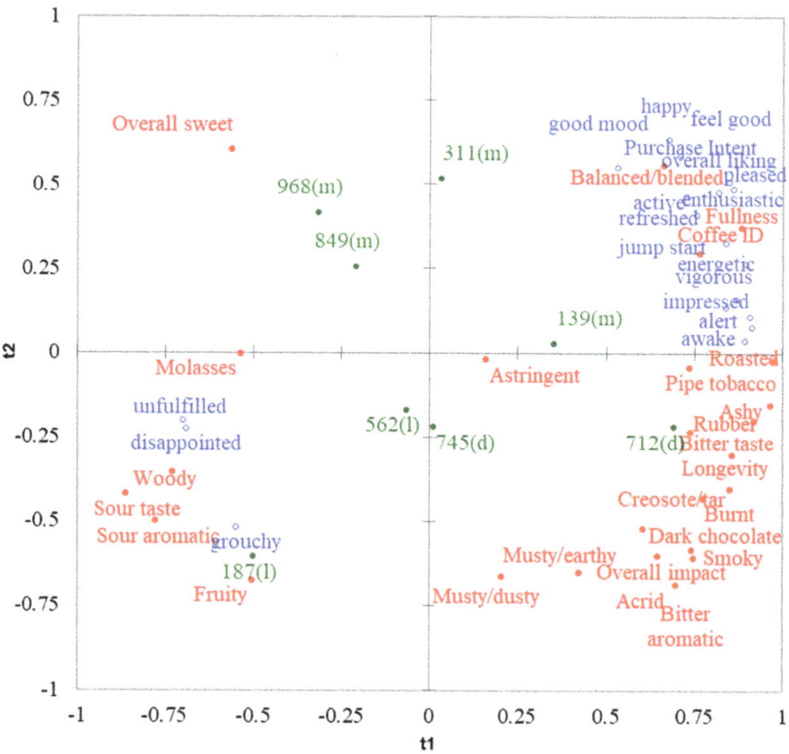

Figure 3. Partial least square regression map showing positioning of eight brewed black coffee samples in the sensory and emotion space. • = sensory attributes, o = overall liking/purchase intent/emotion terms, • = samples with roast level in parenthesis (l = light, m = medium and d = dark).

On the contrary, *sour aromatic, sour taste, fruity, woody, musty/earthy, musty/dusty* and *molasses* negatively affected liking and purchase intent. In addition, the presence of these attributes generally stimulated negative emotions and/or suppressed positive emotions. Specifically, *sour aromatic, sour taste, fruity* and *woody* evoked *disappointed, grouchy* and *unfulfilled* feelings and suppressed all positive feelings. The presence of the *musty/earthy* flavor induced *grouchy*, and decreased *active, enthusiastic, feel good, good mood, pleased* and *happy* emotions. Similarly, a *musty/dusty* note elicited *grouchy* and decreased *feel good, good mood* and *happy* feelings. Although *molasses* did not evoke any negative emotions, it decreased the feelings of *pleased, feel good, refresh* and *jump start. Acrid* was also associated with the decrease of *happy* feeling. Therefore, these attributes were not desirable in the view of brewed black coffee users.

The positive impact *of Coffee ID* found in this study agreed with common knowledge that the coffee aroma/flavor generally induces positive feelings and is a key driver of coffee consumption [48,58]. Moreover, our findings on the positive effects of *roasted, pipe tobacco*, and *bitter taste* and the negative effects of *sour aromatic* and *sour taste* on emotions were consistent with those reported by Bhumiratana et al. [32] who determined sensory drivers for emotions toward brewed coffee using sensory descriptive data evaluated by trained panelists and emotion data evaluated by 94 US consumers. They found that *roasted* aroma and flavor aroused the feelings of *jump start, satisfied, boosted* and *special*, while *tobacco* flavor made the consumers feel *jolted* and *content*. In addition, *bitter taste* was reported to evoke *energetic* and *productive* feelings, while *acidity*, which corresponded to *sour aromatic* and *sour taste* in the current study, was found to arouse an *off-balance* feeling. However, a contradictory result was observed for the *coffee* aroma. Bhumiratana et al. [32] found the *coffee* aroma to associate with negative emotions (*bored, disgusted, annoyed* and *disappointed*) and attributed that to different understandings toward the term '*coffee*' aroma in the views of consumers and trained panelists. Other attributes found to play an important role on consumer acceptance, purchase intent and emotions toward brewed black coffee in the current study (i.e., *balanced/blended, fullness, rubber, overall sweet, fruity, woody, musty/earthy, musty/dusty, molasses* and *acrid*) were not evaluated in the study of Bhumiratana et al. [32]. Another study by Hu and Lee [26] found that Korean and Chinese consumers did not like coffee with a strong *bitter taste*, which was contradictory to our findings. Disagreement between the two studies could be due to differences in terms of coffee sample context and consumer groups. In the current study, all samples being tested were black coffee, and all of the consumers participating in the test were black coffee users. Thereby, these consumers preferred coffee samples with a strong bitter taste, while in the study of Hu and Lee [26], various types of coffee, including the sweetened all-in-one type were evaluated, and participants were general coffee users.

It seems that most of the sensory attributes that drive acceptance, purchase intent and positive emotions while suppressing negative emotions for Thai consumers who drink brewed black coffee are the characteristics of dark roast coffees, which are the results of the roasting process. Product developers and coffee manufacturers could use the information obtained from this research in creating or modifying the sensory profile of coffee to increase consumer acceptance.

4. Conclusions

This research determined the sensory characteristics of brewed black coffee that affected consumer acceptance, purchase intent and emotions using data obtained from descriptive analysis and a consumer test. Results indicated that *coffee ID, roasted, ashy, pipe tobacco, bitter taste, rubber, overall sweet, balanced/blended, fullness* and *longevity* were the key sensory attributes driving liking, purchase intent and positive emotions, such as *active, alert, awake, energetic, enthusiastic, feel good, happy, jump start, impressed, pleased, refreshed* and *vigorous*. While *sour aromatic, sour taste, fruity, woody, musty/earthy, musty/dusty* and *molasses* decreased liking, purchase intent and positive emotions and stimulated negative emotions, such as *disappointed, grouchy* and *unfulfilled*, thus they were undesirable attributes. An increased roast level from light to dark tended to increase the intensities of desirable attributes (e.g., *coffee ID, roasted, ashy, bitter taste* and *fullness),* while decreasing the intensities of undesirable attributes (e.g., *sour aromatic* and *sour taste*). This information could be valuable for product developers and the coffee industry for creating or modifying the sensory profile of brewed black coffee to increase consumer acceptance. This research also expanded the pre-existing lexicons for brewed coffee by providing new sensory terms and references that could be used in sensory research.

Author Contributions: Conceptualization, A.P., S.S., P.C. and W.P.; methodology, A.P. and S.S.; software, S.S. and W.P.; validation, A.P. and S.S.; formal analysis, S.S., P.C. and W.P.; investigation, A.P. and S.S.; resources, A.P. and S.S.; data curation, A.P. and S.S.; writing—original draft preparation, A.P. and S.S.; writing—review and editing, S.S., P.C. and W.P.; visualization, A.P. and S.S.; supervision, S.S., P.C. and W.P.; project administration, S.S.; funding acquisition, S.S. All authors have read and agreed to the published version of the manuscript.

Funding: This research received no external funding.

Institutional Review Board Statement: This research was a collaborative effort between Kasetsart University, Thailand and Louisiana State University (LSU), Agricultural Center. The study was conducted according to the guidelines of the Declaration of Helsinki, and the use of human subjects in research was approved by the Institutional Review Board, approval number IRBAG-21-0063 (Title: Consumer Acceptance, Emotion, and Purchase Intent of New and Healthier Food Products).

Informed Consent Statement: All participants agreed to participate in this study.

Data Availability Statement: The data presented in this study are available on request from the corresponding author.

Acknowledgments: We acknowledge Edgar IV Chambers and a trained panelist of Center for Sensory Analysis and Consumer Behavior, Kansas State University for help with panel training on brewed coffee flavor. Sarisuk Sittiketgorn and the graduate students (Siriporn Siralertmukul, Teepanee Pinitjun and Sayamon Netsuwan) are thanked for assistance with the lab work.

Conflicts of Interest: The authors declare no conflict of interest.

References

1. Coffee. Available online: https://www.statista.com/outlook/cmo/hot-drinks/coffee/worldwide (accessed on 1 November 2021).
2. Labbe, D.; Ferrage, A.; Rytz, A.; Pace, J.; Martin, N. Pleasantness, emotions and perception induced by coffee beverage experience depend on the consumption motivation (hedonic or utilitarian). *Food Qual. Prefer.* **2015**, *44*, 56–61. [CrossRef]
3. Bhumiratana, N.; Adhikari, K.; Chambers, E., IV. Green coffee beans to brewed coffee: Evolution of coffee aroma. *LWT-Food Sci. Technol.* **2011**, *44*, 2185–2192. [CrossRef]
4. Di Donfrancesco, B.; Guzman, N.G.; Chambers, E., IV. Comparison of results from cupping and descriptive sensory analysis of Columbian brewed coffee. *J. Sens. Stud.* **2014**, *29*, 301–311. [CrossRef]
5. Sanchez, K.; Chambers, E., IV. How does product preparation affect sensory properties? An example with coffee. *J. Sens. Stud.* **2015**, *30*, 499–511. [CrossRef]
6. Chapko, M.J.; Seo, H.S. Characterizing product temperature-dependent sensory perception of brewed coffee beverages: Descriptive sensory analysis. *Food Res. Int.* **2019**, *121*, 612–621. [CrossRef]
7. Muñoz, A.E.; Hernández, S.S.; Tolosa, A.R.; Burnollo, S.P.; Herrera, M.O. Evaluation of differences in the antioxidant capacity and phenolic compounds of green and roasted coffee and their relationship with sensory properties. *LWT-Food Sci. Technol.* **2020**, *128*, 109547. [CrossRef]
8. Chambers, E., IV; Sanchez, K.; Phan, U.X.T.; Miller, R.; Civille, G.V.; Di Donfrancesco, B. Development of a "living" lexicon for descriptive sensory analysis of brewed coffee. *J. Sens. Stud.* **2016**, *31*, 465–480. [CrossRef]
9. Samant, S.S.; Chapko, M.J.; Seo, H.S. Predicting consumer liking and preference based on emotional responses and sensory perception: A study with basic taste solutions. *Food Res. Int.* **2017**, *100*, 325–334. [CrossRef]
10. King, S.C.; Meiselman, H.L. Development of a method to measure consumer emotions associated with foods. *Food Qual. Prefer.* **2010**, *21*, 168–177. [CrossRef]
11. Jiang, Y.; King, J.M.; Prinyawiwatkul, W. A review of measurement and relationships between food, eating behavior and emotion. *Trends Food Sci. Technol.* **2014**, *36*, 15–28. [CrossRef]
12. Nestrud, M.A.; Meiselman, H.L.; King, S.C.; Lesher, L.L.; Cardello, A.V. Development of EsSense25, a short version of the EsSense Profile®. *Food Qual. Prefer.* **2016**, *48*, 107–117. [CrossRef]

13. Thomson, D.M.H.; Crocker, C.; Marketo, C.G. Linking sensory characteristics to emotions: An example using dark chocolate. *Food Qual. Prefer.* **2010**, *21*, 1117–1125. [CrossRef]
14. Spinelli, S.; Masi, C.; Dinnella, C.; Zoboli, G.P.; Monteleone, E. How does it make you feel? A new approach to measuring emotions in food product experience. *Food Qual. Prefer.* **2014**, *37*, 109–122. [CrossRef]
15. Ng, M.; Chaya, C.; Hort, J. Beyond liking: Comparing the measurement of emotional response using EsSense Profile and consumer defined check-all-that-apply methodologies. *Food Qual. Prefer.* **2013**, *28*, 193–205. [CrossRef]
16. Ferrarini, R.; Carbognin, C.; Casarotti, E.M.; Nicolis, E.; Neneini, A.; Meneghini, A.M. The emotional response to wine consumption. *Food Qual. Prefer.* **2010**, *21*, 720–725. [CrossRef]
17. Chaya, C.; Eaton, C.; Hewson, L.; Vázquez, R.F.; Fernández-Ruiz, V.; Smart, K.A.; Hort, J. Developing a reduced consumer-led lexicon to measure emotional response to beer. *Food Qual. Prefer.* **2015**, *45*, 100–112. [CrossRef]
18. Garzaro, M.; Raimondo, L.; Pecorari, G.; Sensini, M.; Riva, G.; Palmo, A.; Giordano, C. Digestibility, palatability and emotional status after ingestion of an iced dessert: Analysis of subjective responses in 100 healthy volunteers. *J. Biol. Regul. Homeost.* **2011**, *25*, 101–107.
19. Wardy, W.; Jack, A.R.; Chonpracha, P.; Alonso, J.R.; King, J.M.; Prinyawiwatkul, W. Gluten-free muffins: Effects of sugar reduction and health benefit information on consumer liking, emotion, purchase intent. *Int. J. Food Sci. Technol.* **2018**, *53*, 262–269. [CrossRef]
20. Ardoin, R.; Prinyawiwatkul, W. Consumer perceptions of insect consumption: A review of western research since 2015. *Int. J. Food Sci. Technol.* **2021**, *56*, 4942–4958. [CrossRef]
21. Bhumiratana, N.; Adhikari, K.; Chambers, E., IV. The development of an emotion lexicon for coffee-drinking experience. *Food Res. Int.* **2014**, *61*, 83–92. [CrossRef]
22. Kanjanakorn, A.; Lee, J. Examining emotions and comparing the EsSense Profile® and the Coffee-drinking Experience in coffee drinkers in the natural environment. *Food Qual. Prefer.* **2017**, *56*, 69–79. [CrossRef]
23. van Zyl, H.; Meiselman, H.I. The roles of culture and language in designing emotion lists: Comparing the same language in different English and Spanish speaking countries. *Food Qual. Prefer.* **2015**, *41*, 201–213. [CrossRef]
24. Jaeger, S.R.; Cardello, A.V.; Schutz, H.G. Emotion questionnaires: A consumer-centric perspective. *Food Qual. Prefer.* **2013**, *30*, 229–241. [CrossRef]
25. Gmuer, A.; Guth, J.N.; Runte, M.; Siegrist, M. From emotion to language: Application of a systematic, linguistic-based approach to design a food-associated emotion lexicon. *Food Qual. Prefer.* **2015**, *40*, 77–86. [CrossRef]
26. Hu, X.; Lee, J. Emotions elicited while drinking coffee: A cross-cultural comparison between Korean and Chinese consumers. *Food Qual. Prefer.* **2019**, *76*, 160–168. [CrossRef]
27. Pinsuwan, A.; Suwonsichon, S.; Chompreeda, P.; Prinyawiwatkul, W. Development of Thai emotion lexicon for coffee using word association and check-all-that-apply methods. *J. Food Sci. Agric. Technol.* **2018**, *4*, 46–52.
28. Mojet, J.; Dürrschmid, K.; Danner, L.; Jöchl, M.; Heiniö, R.L.; Holthuysen, N.; Köster, E. Are implicit emotion measurements evoked by food unrelated to liking? *Food Res. Int.* **2015**, *76*, 224–232. [CrossRef]
29. Jaeger, S.R.; Lee, P.Y.; Xia, Y.; Chheang, S.L.; Roigard, C.M.; Ares, G. Using the emotion circumplex to uncover sensory drivers of emotional associations to products: Six case studies. *Food Qual. Prefer.* **2019**, *77*, 89–101. [CrossRef]
30. Jaeger, G.; Schlich, P.; Tijjsen, I.; Yao, J.; Visalli, M.; de Graaf, C.; Steiger, M. Temporal dominance of emotions: Measuring dynamics of food-related emotions during consumption. *Food Qual. Prefer.* **2014**, *37*, 87–99. [CrossRef]
31. Jaeger, S.R.; Spinelli, S.; Ares, G.; Monteleone, E. Linking product-elicited emotional associations and sensory perceptions through a circumplex model based on valence and arousal: Five consumer studies. *Food Res. Int.* **2018**, *109*, 626–640. [CrossRef]
32. Spinelli, S.; Monteleone, E.; Ares, G.; Jaeger, S.R. Sensory drivers of product-elicited emotions are moderated by liking: Insights from consumer segmentation. *Food Qual. Prefer.* **2019**, *78*, 103725. [CrossRef]
33. Bhumiratana, N.; Wolf, M.; Chambers, E., IV; Adhikari, K. Coffee-drinking and emotions: Are there key sensory drivers for emotions? *Beverages* **2019**, *5*, 27. [CrossRef]
34. Coffee-Thailand. Available online: https://www.statista.com/outlook/cmo/hot-drinks/coffee/thailand (accessed on 1 November 2021).
35. Wake up and sell the coffee. Available online: https://www.bangkokpost.com/business/1631922/wake-up-and-sell-the-coffee (accessed on 1 November 2021).
36. Back to Black ... and 10 Other Coffee Trends for 2020. Available online: https://www.bighospitality.co.uk/Article/2020/02/25/Back-to-black-and-10-other-coffee-trends-for-2020 (accessed on 7 December 2021).
37. ISO 6668. *Green Coffee–Preparation of Samples for Use in Sensory Analysis*; ISO: Genève, Switzerland, 2008.
38. Heymann, H.; Machado, B.; Torri, L.; Robinson, A.L. How many judges should one use for sensory descriptive analysis? *J. Sens. Stud.* **2012**, *27*, 111–122. [CrossRef]
39. Maximo-Gacula, J.R.; Rutenbeck, D. Sample size in consumer test and descriptive analysis. *J. Sens. Stud.* **2006**, *21*, 129–145. [CrossRef]
40. Drake, M.A. Sensory analysis of dairy foods. *J. Dairy Sci.* **2007**, *90*, 4925–4937. [CrossRef]
41. Pujchakarn, T.; Suwonsichon, S.; Suwonsichon, T. Development of a sensory lexicon for a specific subcategory of soy sauce: Seasoning soy sauce. *J. Sens. Stud.* **2016**, *31*, 443–452. [CrossRef]

42. King, S.C.; Snow, J.; Meiselman, H.L.; Sainsbury, J.; Carr, B.T.; McCafferty, D.; Serrano, D.; Gillette, M.; Millard, L.; Li, Q. Development of a questionnaire to measure consumer wellsense associated with foods: The WellSense Profile™. *Food Qual. Prefer.* **2015**, *39*, 82–94. [CrossRef]
43. Ares, G.; Etchemendy, E.; Antúnez, L.; Vidal, L.; Giménez, A.; Jaeger, S.R. Visual attention by consumers to check-all-that-apply questions: Insights to support methodological development. *Food Qual. Prefer.* **2014**, *32*, 210–220. [CrossRef]
44. Sheskin, D.J. *Handbook of Parametric and Nonparametric Statistical Procedures*, 5th ed.; CRC Press: Boca Raton, FL, USA, 2020.
45. Suwonsichon, S. The importance of sensory lexicons for research and development of food products: A Review. *Foods* **2019**, *8*, 27. [CrossRef]
46. Czerny, M.; Mayer, F.; Grosch, W. Sensory study on the character impact odorants of roasted Arabica coffee. *J. Agric. Food Chem.* **1999**, *47*, 695–699. [CrossRef]
47. Akiyama, M.; Murakami, K.; Hirano, Y.; Ikeda, M.; Iwatsuki, K.; Wada, A.; Tokuno, K.; Onishi, M.; Iwabuchi, H. Characterization of headspace aroma compounds of freshly brewed Arabica coffee and studies on a characteristic aroma compound of Ethiopia coffee. *J. Food Sci.* **2008**, *73*, C335–C346. [CrossRef]
48. Illy, A.; Viani, R. *Espresso Coffee: The Science of Quality*, 2nd ed.; Elsevier Academic Press: London, UK, 2005.
49. Leitch, K.A.; Dincan, S.E.; O'Keefe, S.; Rudd, R.; Gallagher, D.L. Characterizing consumer emotional response to sweeteners using an emotion terminology questionnaire and facial expression analysis. *Food Res. Int.* **2015**, *76*, 283–292. [CrossRef]
50. Borgogno, M.; Cardello, A.V.; Favotto, S.; Piasentier, E. An emotional approach to beef evaluation. *Meat Sci.* **2017**, *127*, 1–5. [CrossRef] [PubMed]
51. Samant, S.S.; Seo, H.S. Using both emotional responses and sensory attribute intensities to predict consumer liking and preference toward vegetable juice products. *Food Qual. Prefer.* **2019**, *73*, 75–85. [CrossRef]
52. Gurdian, C.E.; Torrico, D.D.; Li, B.; Tuuri, G.; Prinyawiwatkul, W. Effect of disclosed information on product liking, emotional profile and purchase intent: A case of chocolate brownies containing edible-cricket protein. *Foods* **2021**, *10*, 1769. [CrossRef]
53. King, S.C.; Meiselman, H.L.; Carr, T. Measuring emotions associated with foods in consumer testing. *Food Qual. Prefer.* **2010**, *21*, 1114–1116. [CrossRef]
54. King, S.C.; Meiselman, H.L.; Carr, T. Measuring emotions associated with foods: Important elements of questionnaire and test design. *Food Qual. Prefer.* **2013**, *28*, 8–16. [CrossRef]
55. Cardello, A.V.; Meiselman, H.L.; Schutz, H.G.; Craig, C.; Given, Z.; Lesher, L.L.; Eicher, S. Measuring emotional responses to foods and food names using questionnaire. *Food Qual. Prefer.* **2012**, *24*, 243–250. [CrossRef]
56. Jaeger, S.R.; Hedderley, D.I. Impact of individual differences in emotional intensity and private body consciousness on EsSense Profile® responses. *Food Qual. Prefer.* **2013**, *27*, 54–62. [CrossRef]
57. Lykomitros, D.; Fogliano, V.; Capuano, E. Drivers of preference and perception of freshness in roasted peanuts (*Arachis* spp.) for European consumers. *J. Food Sci.* **2018**, *83*, 1103–1115. [CrossRef]
58. Seo, H.S.; Hirano, M.; Shibato, M.; Rakwal, R.; Hwang, I.K.; Masuo, Y. Effect of coffee bean aroma on the rat brain assessed by sleep deprivation: A selected transcript and 2D get-based proteome analysis. *J. Agric. Food Chem.* **2008**, *56*, 4665–4673. [CrossRef]

foods

MDPI

Article

Effects of Tasting and Ingredient Information Statement on Acceptability, Elicited Emotions, and Willingness to Purchase: A Case of Pita Chips Containing Edible Cricket Protein

Cristhiam E. Gurdian [1], Damir D. Torrico [2], Bin Li [3] and Witoon Prinyawiwatkul [1,*]

1 School of Nutrition and Food Sciences, Louisiana State University, Agricultural Center, Baton Rouge, LA 70803, USA; cgurdi3@lsu.edu
2 Department of Wine, Food and Molecular Biosciences, Faculty of Agriculture and Life Sciences, Lincoln University, Lincoln 7647, New Zealand; damir.torrico@lincoln.ac.nz
3 Department of Experimental Statistics, Louisiana State University, Agricultural Center, Baton Rouge, LA 70803, USA; bli@lsu.edu
* Correspondence: wprinya@lsu.edu

Abstract: Sustainable and nutritious alternatives are needed to feed the ever-increasing world population. The successful incorporation of edible-cricket protein (ECP) into foods needs deeper consumer insights. Treatments (plain, Italian, and Cajun pita chips containing 6.9% *w/w* ECP) were evaluated by subjects for overall liking (OL), emotions, and purchase intent (PI) in three different moments: (1) before tasting, (2) after tasting/before ECP statement, and (3) after tasting/after ECP statement. Attributes' liking scores were evaluated only after tasting/before ECP statement. Liking scores (mixed-effects ANOVA), emotions, and PI across moments within treatments/across treatments within moments were evaluated. Emotion-based penalty-lift analyses for OL within moments were assessed using two-sample *t*-tests ($p < 0.05$). Random forest model analyzed after-tasting informed PI and variables' importance. Although formulations' OL and PI were similar across moments, plain and Italian chips had higher after-tasting (before and after ECP statement) OL than the Cajun chips. Moments indirectly affected OL via emotions elicitation. Valence and activation/arousal emotions discriminated across moments for the plain treatment whereas valence and mostly activation/arousal terms discriminated across moments for the Italian and Cajun treatments, respectively. For either formulation or moment, "interested" and "adventurous" positively affected OL. Before and after-tasting attribute liking, "satisfied," and "enthusiastic" emotions were critical in predicting after-tasting informed PI.

Keywords: emotion-based liking; alternative protein; consumer behavior; purchase intent; perceptions; holistic sensory analysis

Citation: Gurdian, C.E.; Torrico, D.D.; Li, B.; Prinyawiwatkul, W. Effects of Tasting and Ingredient Information Statement on Acceptability, Elicited Emotions, and Willingness to Purchase: A Case of Pita Chips Containing Edible Cricket Protein. *Foods* **2022**, *11*, 337. https://doi.org/10.3390/foods11030337

Academic Editor: Derek V. Byrne

Received: 15 December 2021
Accepted: 19 January 2022
Published: 25 January 2022

1. Introduction

The expected increase in the global population accompanied by increasing rates of the depletion of natural resources, such as water and land, urgently calls for innovative changes in the actual food supply [1]. In addition, consumers' demand for more sustainable food products that still meet their expectations regarding nutrient and hedonic profiles continues to rise [2]. Edible cricket protein (ECP) is considered among the most acceptable insect-derived ingredients in the Western world [3,4]. Besides imparting the desired functionality as a protein ingredient, ECP imparts a higher quality nutritional profile [5] to foods without significantly changing the sensory acceptability [6,7]. However, achieving this goal is highly dependable on how ECP products are perceived in terms of their sensory characteristics. Moreover, ECP products' acceptability can be affected by the disclosed information, demographic factors (e.g., gender, age, and race [1,8,9]), and other consumer-niche variables [10]. The level at which ECP can be incorporated into foods varies greatly depending on the food application [1,6,7,11–13]. Changes in the flavor profile is the most cited limiting constraint/factor for ECP [5], followed by modifications in color and/or texture [1,6,7,11,13]. In a recent study, Ardoin et al. [14] found that 15% of ECP in crackers produced a 20% rejection rate in consumers and thus recommended 7.9% as the upper level for its addition into foods.

Several studies have suggested strategies (e.g., familiarity tasting approach, influencers adoption broadcasting, invisible ingredients, educational sessions, etc.) to improve the adoption of insect-derived ingredients [8,9,15]. In addition, consumers' acceptability can be influenced by extrinsic factors [16,17], including packaging [18], benefits statements [19], serving inputs [20], and vestibular sensations [21]. Moreover, ECP incorporations faces psychological constraints [1,8], which are accentuated in Western cultures [22]. Food neophobia [4] and disgust sensitivity have been identified as the major limitations of foods containing insect-derived ingredients in Western societies [23–25]. Despite this, food neophobia can be overcome through familiarization of the consumers with entomophagy mainly through repeated exposure to successful tasting experiences. However, suppressing the disgust sensation towards edible insect ingredients represents a more challenging task for new product formulators.

In Western cultures, the ultimate success of foods containing edible insect ingredients will indeed be driven by consumer behaviors. Hence, it has been proposed that, instead of attempting to convince unwilling consumers to experience edible insects, the efforts and strategies in products formulated with edible insects should be aimed at improving the eating experience for potential early adopters [9] and identifying the market niche for these types of products [1,8]. Potential adopters of products formulated with ECP or other insect-derived ingredients seem to have a consistent pattern of sensations and emotions when experiencing products containing edible insects. Hence, these so-called sensation-seeker perceptions, expectations, liking, and behaviors should then be further studied to obtain meaningful insights for the incorporation of products containing edible insects in Western cultures.

Previous research involving consumers' attitudes towards edible insects suggests that food-evoked emotions in addition to hedonic ratings, product liking, and other descriptive and experimental components shall be considered to achieve holistic models that accurately predict and interpret consumer behaviors [8,26,27]. Therefore, the aim of this study was to explore the effect of tasting and informing an ECP statement on the hedonic perceptions, evoked sentiments, and PI of three pita chips formulations containing 6.9% *w/w* ECP (plain, Italian, and Cajun) in three moments: (1) before tasting, (2) after tasting/before ECP statement, and (3) after tasting/after ECP statement). Likewise, the formulations' effect on hedonic perceptions and emotional patterns as they relate to consumer behaviors (PI) was investigated. The associations among variables (consumer, product, and experimental) were explored to provide meaningful insights for the development of novel foods containing ECP.

2. Materials and Methods

2.1. Pita Chips Preparation

Whole wheat Gold Medal flour (General Mills Sales, Inc., Minneapolis, MN, USA), Morton lite salt (Morton Salt, Inc., Chicago, IL, USA), Great Value double-acting baking powder, purified drinking water, and Sam's Choice Italian style herb (basil, marjoram, oregano, rosemary, sage, and thyme) grinder seasoning (Great Value, Wal-Mart Stores, Inc., Bentonville, AR, USA), Slap Ya Mama low sodium Cajun seasoning (Walker & Sons, Inc., Ville Platte, LA, USA), and McCormick sundried tomato basil pasta sauce and seasoning mix (McCormick & Co., Inc., Hunt Valley, MD, USA) were purchased at Walmart Supercenter (Baton Rouge, LA, USA). Organic triple filtered coconut oil (Trader Joe's, Monrovia, CA, USA) was purchased at Trader Joe's grocery store and Thailand unique microwave-dried edible cricket protein (ECP, JR Unique Foods Ltd., Udon Thani, Thailand) made of 100% farmed and powdered house crickets (*Acheta domesticus*) was purchased online from www.amazon.com (accessed on 28 August 2020). Treatments' seasonings (plain, Italian, and Cajun) were selected in consensus from a focus group session with n = 15 pita chips consumers and a skilled moderator considering available information from recipes, magazines, and personal cooking experience followed by a preliminary trial evaluating overall acceptability using a 9-point hedonic scale and open-ended questions to relate back to product development in bench (e.g., what would you change in this product?) with the same subjects. Batches of treatments (plain, Italian, and Cajun pita chip formulations) were prepared by hand mixing the ingredients (Table 1), followed by kneading to produce dough, refrigerated (4 °C) resting for 30 min, rolling, and shaping (triangular shape). Then, raw chips were placed in 45.7 cm × 66 cm aluminum trays and baked in a pre-heated OV310G mini rotating rack oven (Baxter Mfg, a Division of ITW FEG, LLC, Orting, WA, USA) at 325 °F for 20 min. Baked treatments (plain, Italian, and Cajun pita chips containing 6.9% *w/w* ECP) were stored in separate containers at room temperature overnight (12 h approximately) until the consumer test was performed. The ECP concentration (6.9% *w/w*) added to the treatments was selected based on the recommendations from previous research [1,5,7,11–13] and a preliminary trial with n = 20 subjects, who were consumers of pita chips and indicated overall acceptability scores of at least 5.5 on a 9-point hedonic scale.

Table 1. Formulation of pita chip treatments †.

Ingredients	Plain		Italian		Cajun	
	Amount (g)	Amount (%)	Amount (g)	Amount (%)	Amount (g)	Amount (%)
Whole wheat flour	36.78	45.98%	35.60	44.50%	34.68	43.35%
Purified drinking water	33.04	41.30%	33.04	41.30%	33.04	41.30%
ECP	5.52	6.90%	5.52	6.90%	5.52	6.90%
Coconut oil	3.67	4.59%	3.67	4.59%	3.67	4.59%
Lite salt	0.70	0.88%	0.70	0.88%	0.70	0.88%
Baking powder	0.29	0.36%	0.29	0.36%	0.29	0.36%
Cajun seasoning	-		-		2.10	2.63%
Sundried tomato basil seasoning	-		0.92	1.15%	-	
Italian-style herb seasoning	-		0.26	0.33%	-	

† Treatments are described in Figure 1. ECP = Edible cricket protein.

Figure 1. Graphical description of treatments (plain, Italian, and Cajun pita chip formulations containing 6.9 % *w/w* ECP). ECP = Edible cricket protein.

2.2. *Sensory Evaluation*

2.2.1. Panelists

The research protocol for this study was approved by the Louisiana State University (LSU) Agricultural Center Institutional Review Board (IRB # HE 18-9). Untrained participants (n = 84) 18 years of age and older (Table 2) were recruited from the LSU campus, Baton Rouge, LA, USA, on 10 November 2020 and subsequently screened. The number of participants for this study was low because the study took place during the early stage of the COVID-19 pandemic and social distance in addition to other restrictions and precautions prevented the study from recruiting more participants. The screening criteria consisted of: (1) no allergies or adverse reactions toward any ingredient of the samples or unsalted crackers, (2) willingness to taste products containing edible cricket protein (ECP), (3) no indication of conditions that may impair their performance as panelists, and (4) being regular consumers (at least once per month) of pita chips.

Table 2. Demographic profile of participants from the consumer study.

Demographic Variables	Levels	n	%
Gender	Female	41	48.81%
	Male	43	51.19%
Age group	18–22	45	53.57%
	23–29	24	28.57%
	30–39	10	11.90%
	40–49	2	2.38%
	50–59	1	1.19%
	≥60	2	2.38%
Race	Asian	5	5.95%
	African American	22	26.19%
	Latino	14	16.67%
	Caucasian	41	48.81%
	Other	2	2.38%
Previously consumed products containing edible insects	Yes	33	39.29%
	No	51	60.71%

2.2.2. Consumer Study

On the day of the study, pita chip treatments (plain, Italian, and Cajun formulations) were placed (6 chips per treatment) inside Great Value (5.1 × 4.7 in) clear-plastic square snack zipper bags (Wal-Mart Stores, Inc., Bentonville, AR, USA) labeled with three-digit random codes. All participants received the three treatments in different moments (before tasting, after tasting before ECP statement, and after tasting after ECP statement) in one session. The consumer study was conducted in sanitized (before each subject's session) partitioned booths. Each booth was separated from another by a 6 ft distance to comply with the COVID-19 precautions established in place at LSU. The sensory booths were equipped with white lights and the room was kept at 25 °C.

2.2.3. Questionnaire

Participants' responses were collected with the online Qualtrics software version 11.2020 (Qualtrics, Provo, UT, USA) accessed on 1–10 November 2020. Before evaluating the treatments (Figure 1), subjects agreed with and signed a consent form included in the approved research protocol. Then, demographic information (gender, age, and race) and whether a previous edible insect consumption had occurred was recorded from the participants. The three pita chip treatments were presented together, but panelists were instructed to evaluate them in a monadic sequential order as indicated on the screen. Participants were asked to clean their palate with unsalted crackers and water before tasting the first sample and in between samples. The treatments' evaluation followed a balanced randomized complete block design

For a given treatment, first, panelists were instructed to evaluate the sample before tasting (moment 1) for: (1) aroma (smell) and overall liking (OL) with a 9-point hedonic scale (1 = dislike extremely and 9 = like extremely), (2) purchase intent (PI) with a binomial (yes or no) scale, and (3) evoked emotions with a check-all-that-apply (CATA) task using twenty-five emotion terms obtained from the Essense25 list [28]. Then, panelists were instructed to taste the sample (moment 2) and evaluate crunchiness, overall flavor, OL (9-point hedonic scale), PI (binomial scale), and evoked emotions using CATA (Essense25). Finally, the questionnaire displayed the following statement regarding the ECP used in the formulations: "Edible insects are safe to eat and are considered a sustainable source of high-quality protein and other nutrients. Edible insect production has less negative environmental impact than traditional livestock production. An estimated 2 billion people worldwide consume edible insects", and panelists were instructed to evaluate again the sample (moment 3) for OL (9-point hedonic scale), PI (binomial scale), and evoked emotions using CATA (Essense25).

2.3. Data Analysis

The evaluation of the pita chip treatments (Figure 1) followed a balanced randomized complete block design (panelists as blocks). The R software version 4.0.3 (RStudio, Inc., Boston, MA, USA) [29] and the XLSTAT version 2019.3.1 (Addinsoft, New York, NY, USA) statistical software version 2020 [30] with $\alpha = 0.05$ significance level were used for the data analysis. The effect of demographics, formulation (plain, Italian, and Cajun), moment (before tasting, after tasting before ECP statement, and after tasting after ECP statement), two-way interactions (gender (females vs. males) * previous edible insect consumption (yes vs. no) and previous edible insect consumption * formulation), and interactions up to three-way (gender * formulation * moment) on sensory likings (excluding moment) and on overall liking (OL) were investigated with multi-way analysis of variance (ANOVA) in a mixed-effects model having panelists as a random effect and Tukey's HSD post-hoc test. Two-sided Cochran's Q test followed by asymptotic McNemar test for post hoc multiple pairwise comparisons [31] with p-value adjusted by false discovery rate [32] was used to compare the frequencies of PI = "Yes" across moments (before tasting, after tasting before ECP statement, and after tasting after ECP statement) within treatments (plain, Italian, and Cajun) and across treatments within moments. Check-all-that-apply (CATA) binary data from evoked emotions were analyzed according to Meyners et al. [33] and Ares et al. [34] across moments within treatments and across treatments within moments. First, Cochran's Q tests determined the overall and individual effect of moments within treatments and treatments within moments in emotions list distribution and each emotion term frequency distribution, respectively. Subsequently, all pairwise comparisons were performed for moments as well as treatment groups following the Marascuilo and McSweeney method [35]. For each treatment, emotions, moments, and PI were then visualized with a correspondence analysis based on chi-square distances. For each moment (pooling treatments together), the relationship (drivers/inhibitors) between evoked emotions and product liking was studied through penalty-lift analysis of OL. Overall-liking mean impact was calculated as the mean OL difference from present vs. absent categories for each emotion with a 20% population threshold [36]. This difference was then standardized, and its significance ($p < 0.05$) was tested with a two-sample t-test. Finally, a random forest (RF) algorithm ensembled and combined $n = 1000$ decision trees to predict after-tasting (after ECP statement) PI [37] using mtry = 6 features out of 36 in the random selection at each splitting node of the $n = 1000$ decision trees. Formulation, population variables, OL from each moment, and after-tasting sensory likings and emotions were input to the RF algorithm using full data as the main interest on model interpretation. The misclassification rate for RF was calculated using the out-of-bag observations, which provide a cross-validation-like estimation of the prediction accuracy, and the classifier's performance was plotted on the receiver operating characteristic (ROC) curve. The variables' contribution for the prediction of the response variable in an RF model can be determined by two approaches: (1) the mean decrease in prediction accuracy after permuting each predictor on the out-of-bag samples, and (2) the mean decrease in the Gini index, which measures node impurity for classification trees, from splitting on the variable and averaging over the trees in the RF ensemble.

3. Results and Discussion

3.1. Overall Significance of Main Effects on Product Liking

The significance of the main effects and their two and three-way interactions of interest on the sensory acceptability of treatments (Figure 1) are presented in the analysis of variance (ANOVA; Table 3). Formulation and its two-way interaction with gender were significant ($p < 0.05$) effects for all the sensory attributes. Disregarding all other effects, the liking scores for aroma (plain = 5.83, Italian = 6.63, Cajun = 6.22), crunchiness (plain = 6.43, Italian = 6.59, Cajun = 6.08), overall flavor (plain = 5.72, Italian = 5.97, Cajun = 5.28), and overall liking (OL, plain = 5.97, Italian = 6.13, Cajun = 5.42) were different depending on the formulation (Figures 2 and 3). Gender levels (female vs. male) influenced the way subjects rated their liking depending on the formulation levels (plain, Italian, Cajun) for aroma, crunchiness, overall flavor, and OL (Table 3). Females' liking scores for aroma were higher than those of males but the liking scores for crunchiness, overall flavor, and OL were higher for males than for females (data not shown). These results agree with the findings from other studies involving edible cricket protein (ECP) for which females exhibited lower taste thresholds and blind acceptability upon tasting for products formulated with ECP than males [1,38]. However, under the conditions of this study, females may have exhibited a higher rejection threshold for the aroma or smells (before tasting) imparted by ECP than males. Previous edible insect consumption significantly ($p < 0.05$) interacted with the formulation effect causing the treatments' aroma (before tasting) acceptability to be mediated by previous edible insect experience (Table 3). Overall, subjects who had experienced edible insects had a lower aroma liking score for all the treatments except for the Cajun formulation (plain with edible insects experience = 5.69 vs. plain without edible insects experience = 5.96; Italian with edible insects experience = 6.54 vs. Italian without edible insects experience = 6.73; Cajun with edible insects experience = 6.58 vs. Cajun without edible insects experience = 5.87; data not shown in Tables or Figures). Findings from the aroma liking for the Cajun formulation agree with other studies reporting higher sensory likings for products formulated with edible insects when subjects had experienced them before [8,25,39].

Table 3. ANOVA † table for sensory acceptability ‡ of treatments §.

Effects	Aroma		Crunchiness		Overall Flavor		Overall Liking *	
	F Value	Pr > F	F Value	Pr > F	F Value	Pr > F	F Value	Pr > F
Gender	1.81	0.18	1.61	0.21	3.30	0.07	1.70	0.20
Age	0.87	0.51	0.26	0.93	0.79	0.56	0.70	0.63
Race	0.39	0.82	0.66	0.62	0.81	0.52	0.60	0.66
Previous edible insect consumption	0.08	0.78	0.07	0.79	0.01	0.91	0.15	0.70
Formulation	31.63	<0.01	11.88	<0.01	12.24	<0.01	16.56	<0.01
Moment	-	-	-	-	-	-	2.00	0.14
Gender * Previous edible insect consumption	0.45	0.51	0.17	0.68	0.02	0.88	0.00	0.99
Gender * Formulation	4.93	0.01	5.10	0.01	3.36	0.04	3.07	0.05
Gender * Moment	-	-	-	-	-	-	0.32	0.73
Previous edible insect consumption * Formulation	14.23	< 0.01	1.98	0.14	0.04	0.96	0.54	0.58
Formulation * Moment	-	-	-	-	-	-	1.37	0.24
Gender * Formulation * Moment	-	-	-	-	-	-	0.29	0.88

† ANOVA = Analysis of variance 2 genders (female and male), 6 age groups (18–22, 23–29, 30–39, 40–49, 50–59, ≥60 years old), 5 races (Asian, African American, Latino, Caucasian, Other), 2 levels of previous edible insect consumption (yes and no), 3 formulations (plain, Italian, and Cajun), and 3 levels of moment (before tasting, after tasting, and after edible cricket (ECP) protein statement). ‡ Liking data from $n = 84$ consumers were collected using a 9-point hedonic scale (1 = dislike extremely, 9 = like extremely) and analyzed by a mixed-effects model with panelists as a random effect. § Treatments are described in Figure 1. * Overall liking determined at three moments (before tasting, after tasting, and after ECP statement).

3.2. Effects of Formulation on Sensory Acceptability of Treatments

Figure 2 shows the sensory acceptability of the treatments (Figure 1) for aroma (before tasting) and after-tasting (before ECP statement) crunchiness and overall flavor. The Italian formulation had the highest ($p < 0.05$) aroma liking followed by the plain formulation, which obtain a significantly ($p < 0.05$) higher aroma liking than the Cajun formulation. However, for crunchiness and overall flavor likings, the Italian and plain treatments presented similar ratings, which were significantly ($p < 0.05$) higher than that for the Cajun treatment. These results suggest that the seasonings used for snacks with incorporations of ECP have an effect on consumers' acceptability. More extravagant or accentuated flavors like the spices contained in the Cajun treatment may become more acceptable in a later stage of the life cycle of products formulated with ECP but are not yet attractive in an introductory phase [40]. In addition, the observed results show the positive effect of introducing new ingredients, such as ECP, through familiar products [1,11,41] and flavors (flavor notes imparted by the Cajun treatment may not be familiar to pita chips consumers).

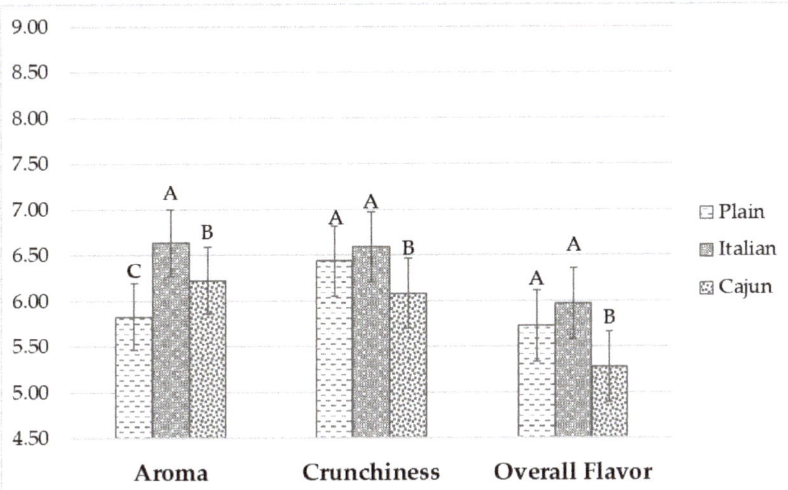

Figure 2. Treatments' sensory acceptability bar chart for the before-tasting (aroma) and after-tasting (before edible cricket protein (ECP) statement) moments (crunchiness and overall flavor). Data are liking ratings least square means and standard errors from $n = 84$ consumers. Treatments are described in Figure 1. Different uppercase letters indicate significantly ($p < 0.05$) different liking scores (Tukey's means separation) across treatments.

3.3. Effects of Moment on Overall Product Liking and Purchase Intent

Figure 3 depicts the overall liking (OL) and purchase intent (PI) of the treatments in the before-tasting and after-tasting (before and after the ECP statement) moments. Although moment did not significantly ($p > 0.05$) affect the overall liking (OL) nor the purchase intent (PI) for any of the treatments, the OL within the after tasting moments (before and after the ECP statement) differed across the treatments. Plain and Italian formulations presented higher ($p > 0.05$) OL ratings than the Cajun formulation in both after-tasting moments. These results agree with the observed higher liking ratings for these two formulations for after-tasting (before ECP statement) crunchiness and overall flavor, which suggests that the liking of these two attributes [11] influenced ultimate product liking [20] more than other sensory attributes or product benefit statements. No PI differences across formulations were observed within each moment either. This suggests that other orthogonal variables to the sensory dimension, such as emotions and sensations, may also affect PI and consumer behavior [26,42–44]. The non-significant effect of the

ECP statement in the OL and PI of treatments observed in this study agrees with the findings from other studies investigating the psychological traits behind the reluctance to consume edible insects [45–47]. Possibly, for products formulated with edible insects, environmental [24,48] or health-related statements about entomophagy are insufficient to significantly improve their sensory liking profile [46,49]. However, the OL, which was not disconfirmed upon tasting, and PI achieved by the plain and Italian treatments indicate an overall positive marketing potential [50] as they represent a new concept for the Western consumers, and their formulations could be further optimized to achieve higher sensory liking [1].

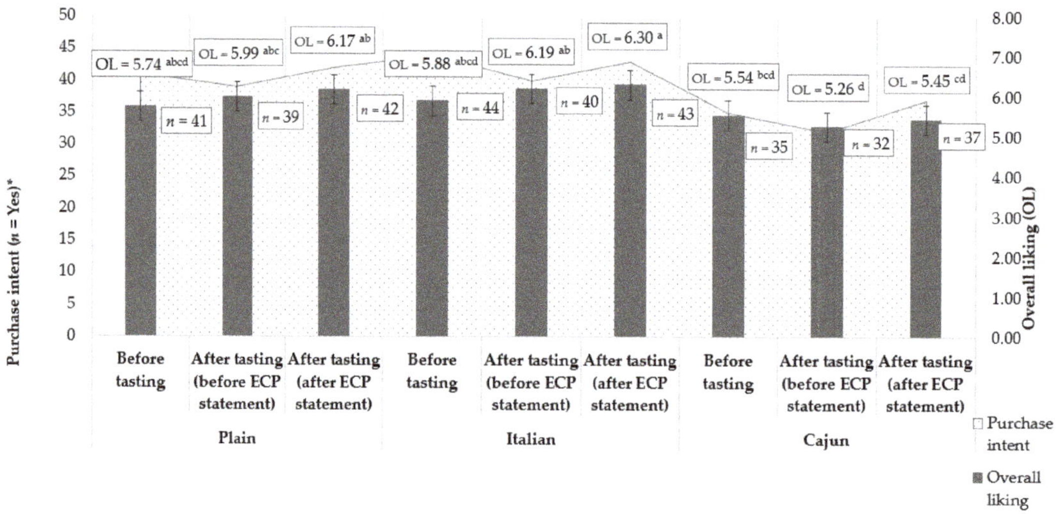

Figure 3. Treatments overall liking (OL; grey bars) and purchase intent (PI; trend line) chart comparing formulations and moments (before tasting, after tasting, and after edible cricket protein (ECP) statement). Data are OL least square means and standard errors/frequencies of PI = Yes from n = 84 consumers. Treatments are described in Figure 1. Different lowercase letters indicate significantly ($p < 0.05$) different OL scores (Tukey's means separation) across treatments and moments. * No significant ($p > 0.05$) difference in PI frequencies across treatments within a given moment or across moments within a given treatment (Cochran's Q test followed by asymptotic McNemar test for post hoc multiple pairwise comparisons and p-value adjustment by false discovery rate).

3.4. Discriminative Effects of Moments and Formulation on Treatments' Emotional Profile

The emotional profile of treatments was segmented and evaluated across moments within formulations and across formulations within moments (Tables 4–6).

Table 4. Emotional profile † elicited by the plain treatment ‡ across moments.

Emotions	Plain		
	Before Tasting	After Tasting (Before ECP Statement)	After Tasting (After ECP Statement)
Active	3 [a]	4 [a]	6 [a]
Adventurous	38 [a,(A)]	24 [b]	31 [a,b]
Aggressive	0 [b]	5 [a]	3 [a,b,(B)]
Bored	9 [a]	10 [a]	12 [a]
Calm	17 [a]	13 [a]	13 [a]
Disgusted	8 [a]	14 [a]	11 [a]
Enthusiastic	9 [a]	5 [a,(B)]	8 [a,(B)]
Free	2 [a]	0 [a]	2 [a]
Good	9 [b]	18 [a]	14 [a,b]
Good-natured	7 [a]	6 [a]	11 [a]
Guilty	3 [a]	0 [a]	0 [a]
Happy	5 [a]	5 [a]	8 [a]
Interested	46 [a]	28 [b]	27 [b]
Joyful	3 [a]	1 [a]	2 [a]
Loving	0 [a]	0 [a]	0 [a]
Mild	11 [a(A,B)]	12 [a]	12 [a]
Nostalgic	3 [a]	0 [a]	0 [a]
Pleasant	7 [a]	14 [a]	10 [a]
Satisfied	4 [b]	16 [a,(A,B)]	17 [a]
Safe	5 [a]	5 [a]	6 [a]
Tame	5 [a]	6 [a]	4 [a]
Understanding	4 [b]	7 [a,b]	12 [a]
Warm	0 [a]	1 [a,(B)]	2 [a,(B)]
Wild	8 [a]	5 [a]	6 [a]
Worried	14 [a]	9 [a]	7 [a]

† Frequency of emotions for the plain treatment across moments (before tasting, after tasting, and after edible cricket protein (ECP) statement) from $n = 84$ consumers analyzed by two-sided Cochran's Q test with Marascuilo and McSweeney procedure. Different lowercase/uppercase letters within a row represent significant ($p < 0.05$) differences across moments for the plain treatment/treatments (Tables 5 and 6) within a given moment. ‡ Treatments are described in Figure 1.

Table 5. Emotional profile † elicited by the Italian treatment ‡ across moments.

Emotions	Italian		
	Before Tasting	After Tasting (Before ECP Statement)	After Tasting (After ECP Statement)
Active	6 [a]	9 [a]	6 [a]
Adventurous	29 [a,(B)]	24 [a]	26 [a]
Aggressive	2 [a]	3 [a]	2 [a,(B)]
Bored	9 [a]	6 [a]	6 [a]
Calm	14 [a]	10 [a]	8 [a]
Disgusted	10 [a]	13 [a]	12 [a]
Enthusiastic	11 [a]	11 [a,(A,B)]	8 [a,(B)]
Free	5 [a]	3 [a]	3 [a]
Good	12 [a]	19 [a]	17 [a]
Good-natured	6 [b]	9 [a,b]	14 [a]
Guilty	3 [a]	0 [a]	0 [a]
Happy	8 [a]	11 [a]	12 [a]
Interested	37 [a]	27 [a]	38 [a]
Joyful	2 [a]	3 [a]	1 [a]
Loving	0 [a]	1 [a]	0 [a]
Mild	19 [a,(A)]	13 [a]	11 [a]
Nostalgic	2 [a]	1 [a]	1 [a]
Pleasant	5 [b]	12 [a]	9 [a,b]
Satisfied	5 [b]	17 [a,(A)]	14 [a]
Safe	6 [a]	7 [a]	5 [a]
Tame	3 [a]	3 [a]	3 [a]
Understanding	4 [b]	4 [b]	10 [a]
Warm	0 [a]	2 [a,(B)]	3 [a,(A,B)]
Wild	6 [a]	9 [a]	6 [a]
Worried	10 [a]	6 [a]	5 [a]

† Frequency of emotions for the Italian treatment across moments (before tasting, after tasting, and after edible cricket protein (ECP) statement) from $n = 84$ consumers analyzed by two-sided Cochran's Q test with Marascuilo and McSweeney procedure. Different lowercase/uppercase letters within a row represent significant ($p < 0.05$) differences across moments for the Italian treatment/treatments (Tables 4 and 6) within a given moment.
‡ Treatments are described in Figure 1.

Table 6. Emotional profile † elicited by the Cajun treatment ‡ across moments.

Emotions	Cajun		
	Before Tasting	After Tasting (Before ECP Statement)	After Tasting (After ECP Statement)
Active	4 [a]	5 [a]	6 [a]
Adventurous	40 [a,(A)]	25 [b]	27 [b]
Aggressive	1 [b]	9 [a]	9 [a,(A)]
Bored	5 [a]	8 [a]	6 [a]
Calm	11 [a]	7 [a]	8 [a]
Disgusted	14 [a]	19 [a]	16 [a]
Enthusiastic	11 [a]	13 [a,(A)]	17 [a,(A)]
Free	4 [a]	0 [a]	2 [a]
Good	11 [a]	14 [a]	12 [a]
Good-natured	7 [a]	6 [a]	8 [a]
Guilty	2 [a]	1 [a]	0 [a]
Happy	3 [a]	4 [a]	5 [a]
Interested	41 [a]	27 [b]	32 [a,b]
Joyful	4 [a]	2 [a]	3 [a]
Loving	0 [a]	0 [a]	0 [a]
Mild	10 [a,(B)]	13 [a]	12 [a]
Nostalgic	2 [a]	1 [a]	1 [a]
Pleasant	4 [a]	6 [a]	5 [a]
Satisfied	8 [a]	6 [a,(B)]	9 [a]
Safe	7 [a]	3 [a]	5 [a]
Tame	6 [a]	3 [a]	2 [a]
Understanding	5 [a]	7 [a]	9 [a]
Warm	1 [b]	9 [a,(A)]	8 [a,b,(A)]
Wild	7 [a]	11 [a]	10 [a]
Worried	12 [a]	9 [a,b]	4 [b]

† Frequency of emotions for the Cajun treatment across moments (before tasting, after tasting, and after edible cricket protein (ECP) statement) from n = 84 consumers analyzed by two-sided Cochran's Q test with Marascuilo and McSweeney procedure. Different lowercase/uppercase letters within a row represent significant ($p < 0.05$) differences across moments for the Cajun treatment/treatments (Tables 4 and 5) within a given moment. ‡ Treatments are described in Figure 1.

3.4.1. Emotional Profiles across Moments

Figure 4 depicts the symmetric plot of elicited emotions and moments (before tasting, after tasting before ECP statement, and after tasting after ECP statement) for the plain treatment. Tasting had a more pronounced effect than the ECP statement in discriminating the evoked emotions for the plain formulation. For the before-tasting moment, the "worried" (extreme activation/arousal) and "interested" (moderate activation/arousal) emotions were associated with the plain treatment. Possibly subjects' worry arose from a safety (health-related context) concern because of the limited information regarding edible insects' regulations and processes that guarantee their safe use as ingredients for human foods [8,51–54]. Yet, their curiosity and interest may have been triggered by the fact that the treatments contained ECP and the tasting experience they were about to have [24,55,56].

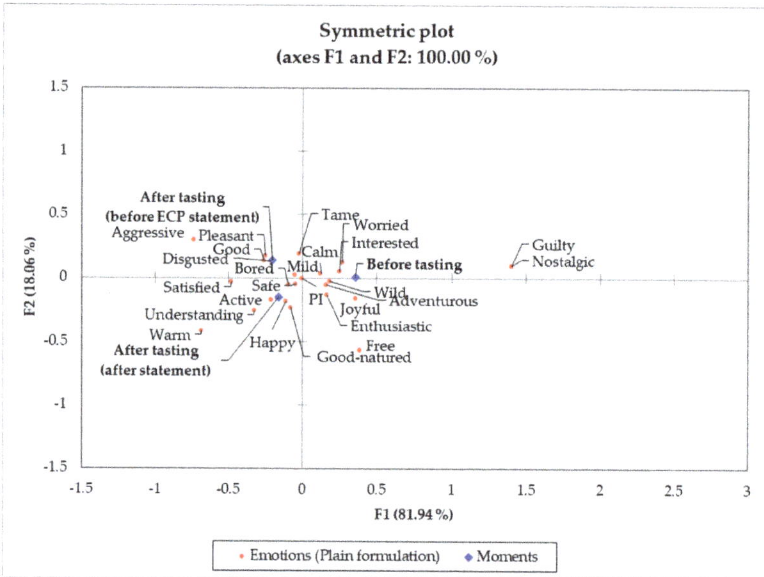

Figure 4. Correspondence analysis (chi-squared distance) symmetric plot visualizing the plain treatment moments (before tasting, after tasting, and after edible cricket protein (ECP) statement) and emotions from n = 84 consumers. Treatments are described in Figure 1.

Both after-tasting moments (before and after the ECP statement) were associated with positive-valence emotions such as "good" and "pleasant" (before the ECP statement) and "happy" and "good-natured" (after the ECP statement). However, the "disgusted" term was also highly associated with the after-tasting before ECP statement moment while the "active" high activation/arousal term was more associated with the after-tasting after ECP statement moment for the plain treatment (Figure 4).

Ultimate product acceptability is positively associated with positive-valence emotions [26,27,49] and negatively associated with "disgust" feeling, which can vary across genders [57]. For novel products formulated with edible insects, sensation-seeking emotions, such as "active" should be elicited as they are important predictors of product acceptability [24,58], and previous research has emphasized that the context for launching products formulated with ECP should encompass novelty, adventure, and wild features [8,43].

On the other hand, the "safe" (positive valence and low activation/arousal) and "bored" (low activation/arousal) were closely positioned to both after tasting moments (before and after the ECP statement). The pattern of emotions occurrence belonging to both positive and negative valences in the pleasantness dimension and both high and low activation/arousal dimension suggests that consumers belonged to different segments [59]. However, the pattern of emotions should be interpreted with caution as not all terms occurrences may be significant.

Table 4 shows the frequency distribution for the plain treatment emotions. Both valence and activation/arousal emotions equally contributed to the differentiation across moments, having only six ("adventurous," "aggressive," "good," "interested," "satisfied," and "understanding") significant emotions. The "adventurous" term frequency decreased upon tasting, but, after showing the ECP statement, it increased to a similar level than before tasting [60]. Although the "aggressive" emotion was more elicited ($p < 0.05$) after tasting (before the ECP statement) than before tasting, the compared frequencies are sparse and may not indicate any practical differences. On the other hand, the ECP statement was not effective in achieving the initial interest level for the plain treatment, which decreased upon tasting. This could reflect a partial (if not total) disagreement with the ECP statement [61].

Regarding emotions lying in the pleasantness dimension, "satisfied" and "understanding" were more elicited ($p < 0.05$) after tasting before the ECP statement and after tasting after the ECP statement, respectively, than before tasting [62]. However, for the "good" emotion term, tasting significantly increased its frequency of occurrence, but disclosing the ECP statement after tasting caused a slight reduction in its frequency.

Figure 5 depicts the symmetric plot of elicited emotions and moments (before tasting, after tasting before ECP statement, and after tasting after ECP statement) for the Italian treatment. Again, the effect of tasting was larger than that of the ECP statement. For the before tasting moment, the "worried" (extreme activation/arousal), "mild," "calm," and "bored" (low to moderate activation/arousal), "free," and "nostalgic" [63] (positive) emotions were associated with the Italian treatment. While the worry and concern about the safety of ECP are very likely to negatively affect product liking and consumption [64], the Italian formulation may benefit from the low to moderate activation/arousal elicited emotions as they have been found to increase appetite and food intake [65]. The other positive emotions have an overall positive effect on product liking and consumption that has been well documented, although some studies report a weak contribution of the "nostalgic" emotion [27,42].

For both after-tasting moments (before and after the ECP statement), an association with positive-valence emotions such as "good" and "happy" but also with the negative "disgusted" term was observed for the Italian treatment (Figure 5), suggesting that elicited emotions for this treatment varied across subjects, possibly because there were consumers from different segments [59]. On the other hand, the Italian formulation was strongly associated with the sensation-seeking emotions "wild," "aggressive," and "active" with regard to both dimensions (F1 and F2). This reflects a positive effect of tasting for the Italian treatment as these emotions lie on the high activation/arousal dimension, which, together with liking and emotions that belong to the pleasantness dimension, have strong predictive power for consumer behavior and positively affect product consumption willingness [8,9,24,58].

Table 5 contains the frequency of the emotional responses for the Italian treatment. Only positive-valence emotions ("good-natured," "pleasant," "satisfied," and "understanding") were significant in discriminating across moments. Overall, tasting and the ECP statement had a positive effect on the Italian treatment as positive-valence terms were more evoked ($p < 0.05$) upon tasting than before tasting ("pleasant" and "satisfied") [62] or upon disclosing the ECP statement ("good-natured" and "understanding") [8].

Figure 6 contains the symmetric plot of evoked emotions and moments (before tasting, after tasting before ECP statement, and after tasting after ECP statement) for the Cajun treatment. Like the other treatments, the tasting effect was larger than the ECP statement effect in the emotional profile across moments. For the before-tasting moment, the Cajun treatment was mainly associated with the "nostalgic" term, which has been categorized as a driver of sensory pleasure [63] but also as a neutral term [42], and to a lesser extent with the "adventurous" term, indicative of an active and energetic state [8,64].

Regarding Cajun treatment's after-tasting moments, the before ECP statement was highly associated with both high and low to moderate activation/arousal emotions ("wild" and "bored", respectively) [66] and with the positive-valence "pleasant" term and, to a lower extent, with the negative "disgusted" term (Figure 6). As previously mentioned, this could reflect the need for further consumer segmentation [59], but it could also indicate mixed feelings (consumer ambivalence) [67] toward the entomophagy concept, which is still an unfamiliar practice for Western cultures [68]. Except for "disgusted", the close association with the other terms may be beneficial for the Cajun treatment as evoking sensations and feelings that belong to the activation/arousal and pleasantness dimensions is a positive effect for novel foods and ingredients such as ECP [8], and "bored" does not necessarily negatively impact food consumption [65]. On the other hand, when the ECP statement was disclosed, the "active" and "enthusiastic" emotions belonging to the high activation/arousal dimension characteristic of the sensation seekers [69] were highly associated with the Cajun formulation as was the positive-valence "happy" emotion. This suggests that the ECP statement had a positive effect maintaining the sensation-seeking and pleasant emotions while distancing from the "disgusted" negative emotion. Disgust feeling is among the top constraints to entomophagy in Western cultures [24,70]; therefore, it is a key sensation yet to be further investigated to find ways to minimize or prevent it from being elicited in foods formulated with edible insect ingredients [71,72]. However, the significance of each term frequency will determine their ultimate effect in discriminating among the treatments, moments, and their impact on product liking.

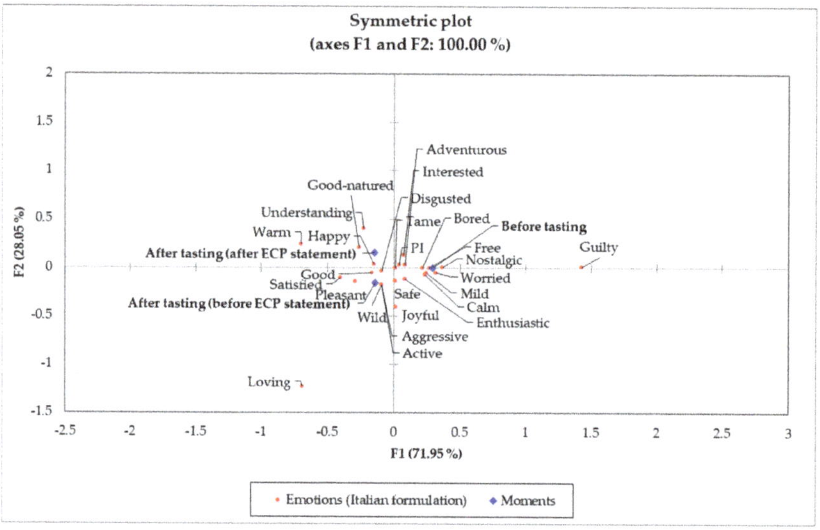

Figure 5. Correspondence analysis (chi-squared distance) symmetric plot visualizing the Italian treatment moments (before tasting, after tasting, and after edible cricket protein (ECP) statement) and emotions from n = 84 consumers. Treatments are described in Figure 1.

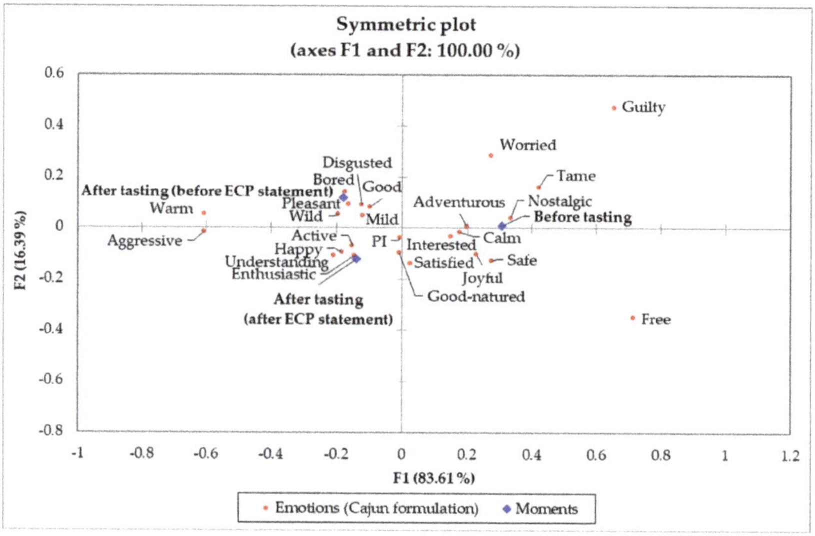

Figure 6. Correspondence analysis (chi-squared distance) symmetric plot visualizing the Cajun treatment moments (before tasting, after tasting, and after edible cricket protein (ECP) statement) and emotions from n = 84 consumers. Treatments are described in Figure 1.

Table 6 presents the frequency of emotions for the Cajun treatment exploring the effects of tasting and ECP statement. As for the other treatments, emotions distribution significantly ($p < 0.05$) varied across moments, but, for this formulation, mostly activation/arousal terms discriminated across moments. The frequency of the "adventurous" emotion decreased ($p < 0.05$) upon tasting compared to before tasting regardless of the ECP statement, but, for the "aggressive" term, it was increased upon tasting regardless of the ECP statement. The "adventurous" emotion is characteristic of the sensation seekers [69] market niche and belongs to the activation/arousal dimension [73,74], which has been suggested as the appropriate context for the introduction of ECP [8]. Yet, the "aggressive" term has been classified as an arousal [64], neutral [42], and as a negative [73] term. Hence, this effect needs to be further investigated for its relationship with product liking and the willingness to consume or to purchase the product. A positive effect of the ECP statement was observed for the "interested" and "worried" emotions. The first became less evoked ($p < 0.05$) upon tasting than before tasting but increased to a similar level than before tasting after disclosing the ECP statement. The second became less frequent ($p < 0.05$) when the ECP statement was disclosed when compared to the before-tasting moment. Finally, tasting increased the "warm" emotion frequency, which remained constant after disclosing the ECP statement.

However, this may represent the "warm" taste sensation imparted from the spiciness of the Cajun formulation rather than a "warm" feeling elicited by this treatment.

3.4.2. Emotional Profiles across Formulations

For the before-tasting moment, the overall hypothesis testing whether the distribution of emotions differed across formulations could not be rejected at the $p = 0.05$ confidence level. However, the Italian formulation presented a significantly lower frequency of the "adventurous" emotion than the other two formulations. "Adventurous" term is an important emotional attribute ideally evoked before and after tasting because it belongs to the sensation-seeking emotions, which have strong predictive power and are drivers of liking for novelty products formulated with ECP [24,43,64].

For both after-tasting moments (before and after the ECP statement), the distribution of emotions varied ($p < 0.05$) across formulations. For the after-tasting (before ECP statement) moment, the "enthusiastic" term belonging to the pleasantness and activation/arousal dimensions was significantly ($p < 0.05$) lower for the plain treatment when compared to the Cajun treatment while the "satisfied" positive-valence emotion was significantly lower for the Cajun treatment when compared to the Italian treatment. Possibly the variety of flavors and hot/spicy sensations from the Cajun formulation contrasted with the plain formulation flavor, making the subjects feel more "enthusiastic" about the Cajun flavor notes [74]. The observed decreased frequency in the "satisfied" term for the Cajun treatment suggests that this emotion is an important driver of product liking, possibly leading to the observed lower OL for the Cajun formulation in the after-tasting (before ECP) moment [8]. In addition, the "warm" emotion was significantly ($p < 0.05$) more elicited for the Cajun treatment when compared to the plain and Italian formulations in both after-tasting moments (before and after the ECP statement) but, as previously mentioned, this could reflect the "warm" sensations imparted from the spicy flavor notes in the Cajun formulation. On the other hand, for the after-tasting (after ECP statement) moment, the plain and Italian formulations had lower ($p < 0.05$) frequencies of the "aggressive", "enthusiastic" (active and pleasant) [75], and "warm" (pleasant) [42] terms than the Cajun formulation. However, as previously mentioned, the "aggressive" term does not have a definite valence, its impact on product liking varies greatly across studies, and the "warm" term may just reflect the spiciness perception for the Cajun formulation. Hence, the observed differences in the "enthusiastic" emotion frequency across formulations may be attributed to a more varied flavor profile imparted by the Cajun treatment when compared to the plain and Italian treatments.

3.5. Elicited Emotions and Product Liking

The effect of emotions elicited at each moment (before tasting, after tasting before ECP statement, and after tasting after ECP statement) on overall product liking (disregarding the formulation effect as it was shown to be minimal within each moment emotional profile) was investigated by calculating overall liking (OL) mean impacts.

Evoked emotions in the before-tasting moment and their respective OL mean impact (increase or decrease) from all the three treatments (Figure 1) are presented in Figure 7.

For the before-tasting moment, only the "interested" emotion caused a significant ($p < 0.001$) increase in the treatments' OL, whereas "good," "adventurous, and "interested" were all significant ($p < 0.001$) drivers for the after-tasting (before ECP statement) OL of the treatments (Figure 8A). These results agree with other studies highlighting the importance of evoking sensation-seeking emotions and positive-valence emotions for expected and actual product liking in foods containing edible insects [8]. The increased experience of positive-valence emotions upon tasting, such as "good", has been widely documented in research with foods [76], but the continued frequency across moments of significant sensation-seeking emotions seems to be specific for foods, such as spicy foods [77], and those lying in the context of novelty and adventure, like edible insects. Unexpectedly, when the ECP statement was shown, "good" was no longer a significant driver of product liking because its frequency (pooled across treatments) was reduced, falling below the user-defined 20% selection threshold (Figure 8B). Possibly, although the ECP statement contained environmental and health benefits information associated with entomophagy, subjects may have experience disagreement with some (if not all) the information communicated [61]. On the other hand, it is also possible that the statement triggered a reminder of mental associations that decreased the "good" emotion of the participants from this study [78–80].

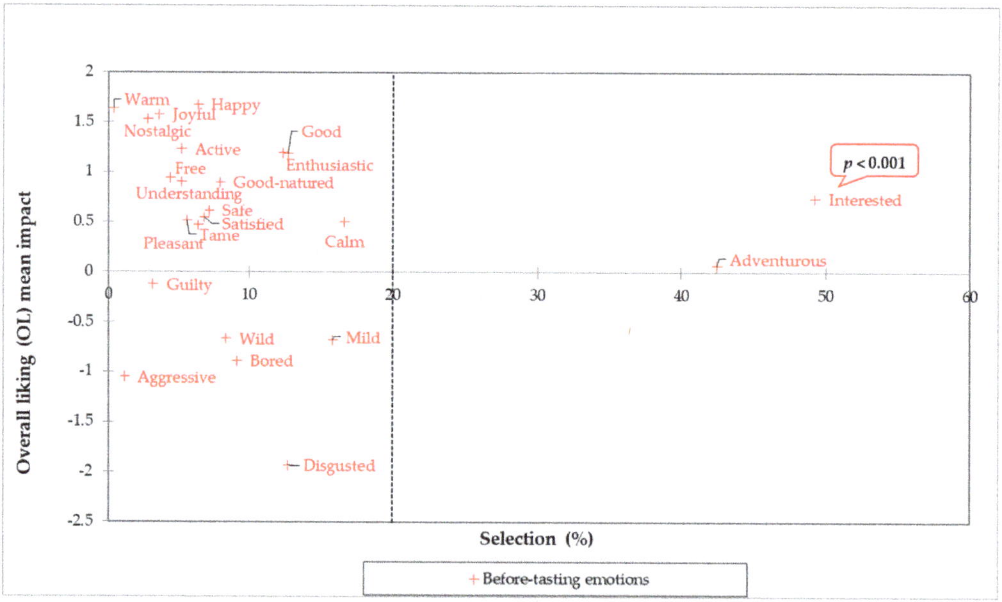

Figure 7. Treatments' before-tasting overall liking (OL) mean impact (mean OL difference from present vs. absent categories for each emotion with a 20% population threshold size) vs. significant ($p < 0.05$, 2-sample t-test) emotions in the before-tasting moment (%) from $n = 84$ consumers. Before-tasting emotions and OL from treatments (Figure 1) were pooled together for the analysis.

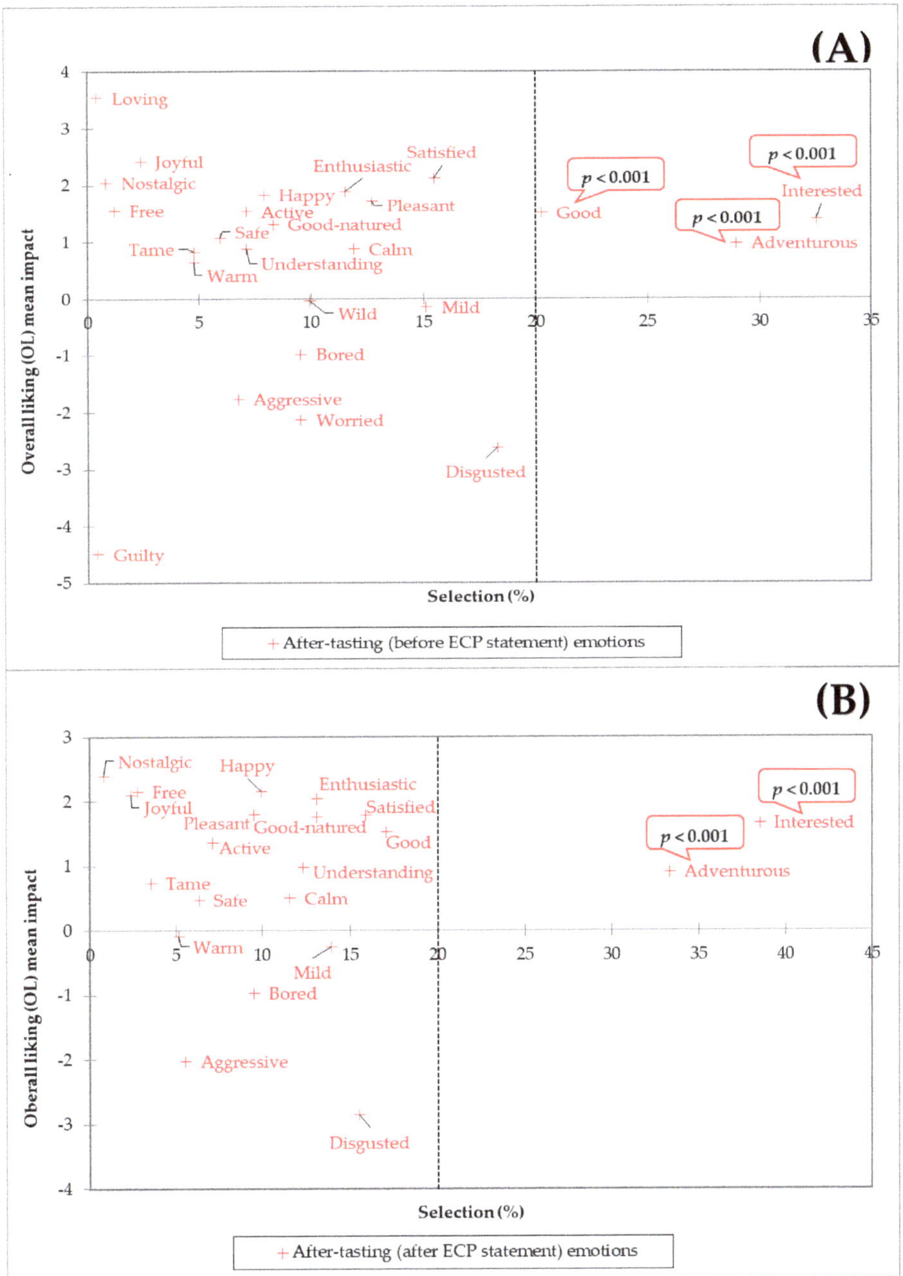

Figure 8. Treatments' after-tasting overall liking (OL) mean impact (mean OL difference from present vs. absent categories for each emotion with a 20% population threshold size) vs. significant ($p < 0.05$, 2-sample t-test) emotions in the after-tasting moment (%) from n = 84 consumers (**A**) before edible cricket protein (ECP) statement and (**B**) after ECP statement. After-tasting emotions and OL from treatments (Figure 1) were pooled together for the analysis.

3.6. Purchase Intent (PI) Prediction and Variables' Importance

The holistic approach of incorporating emotions and other variables that may be orthogonal to the product liking dimension has proven to be an efficient tool to provide information beyond liking and to better understand consumers' behavior [27,42,64]. A random forest (RF) classifier was used in this study to predict the after-tasting (after ECP claim) PI based on demographic variables, formulations, sensory likings, OL from the before and after-tasting (before and after ECP statement) moments, and elicited emotions in the after-tasting (after ECP statement) moment. Figure 9 displays the classifier's performance in terms of its sensibility and specificity at different classification thresholds. The obtained area under the curve (AUC = 0.91) shows that the model accurately discriminates among consumers willing to purchase and those not willing to purchase the product. Moreover, this model that obtained an out-of-bag accuracy of 84.52% provided input variables' relative importance for the PI prediction as illustrated in Figure 10. After-tasting (before and after ECP statement) OL, actual overall flavor liking [39,81], positive-valence "satisfied" emotion [26,82], and negative-valence "disgusted" [23,58,83] emotion were common top-10 predictors for the after-tasting informed PI as determined by the mean decrease in classification accuracy and the mean decrease in node impurity when the variable is permuted and split, respectively. In addition, high and low activation/arousal emotions ("enthusiastic," "calm," "bored," and "worried") [84] and the "good" positive-valence emotion [85] were important inputs to obtain an accurate PI prediction, whereas before-tasting aroma and OL, after-tasting (before ECP statement) crunchiness liking [8], race [44], and age [45,86] were important variables to achieve a higher node purity. According to the RF partial dependence plots (data not shown), the odds of PI = Yes increase for either pita chip formulation when the overall acceptability (regardless of tasting and communication of the ECP statement), flavor, and crunchiness likings increase. In addition, the more elicited the after-tasting (after ECP statement) "disgusted" is and the less elicited the after-tasting (after ECP statement) "interested," "wild," "understanding," "safe," and "good-natured" are, the lower the probability that either pita chip formulation will be purchased. These results emphasize the importance of yielding an adequate liking profile upon tasting (especially an overall flavor and texture liking) and eliciting pleasant and sensation-seeking emotions but also encourage deeper research to find creative solutions that minimize the disgust sensation that has an overall negative impact from all points of view for novelty products formulated with ECP.

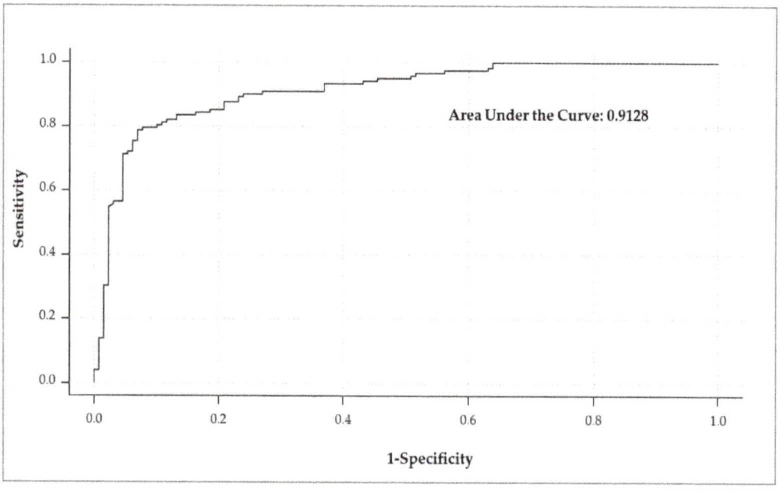

Figure 9. Receiver operating characteristic (ROC) curve illustrating the area under the curve (AUC) for the random forest classifier.

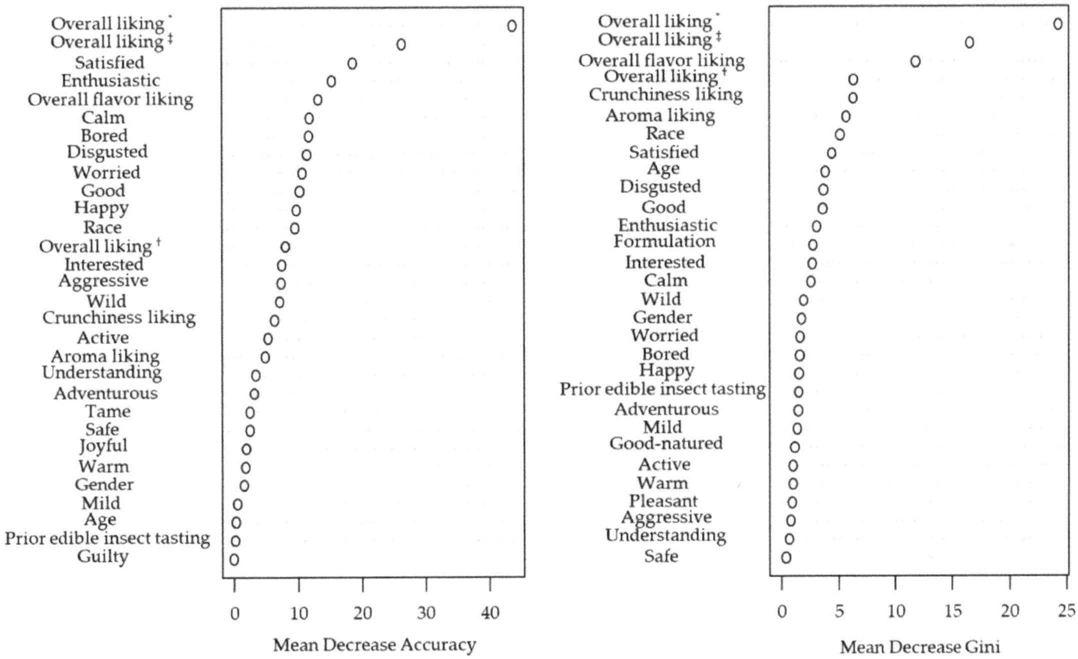

Figure 10. Random forest classifier variables importance plots for after-tasting (after edible cricket protein (ECP) statement) purchase intent (PI) prediction. † Before-tasting moment; ‡ after-tasting (before ECP statement) moment; * after-tasting (after ECP statement) moment. Emotions included in the model were from the after-tasting (after ECP statement) moment.

4. Study Limitations

This research faced many COVID-19 pandemic restrictions in place at the time of the study that led to limited recruitment of participants (n = 84). Hence, no consumer segmentation was performed for the analysis, and the demographic distribution of the subjects is neither balanced nor represents the actual distribution of the US population. Therefore, the findings from this study shall be interpreted with caution, and no inferences should be made on the entire population. For future studies, we recommend testing more seasonings and increasing the sample size, exploring gender and previous edible insect consumption segmentations as these two effects presented a significant interaction with the treatments of this study. In addition, we recommend performing a consumer-based descriptive panel to unearth insights and perceptions that may also affect liking, emotions, and consumption of products formulated with edible cricket protein (ECP). Finally, we would like to emphasize that statistical significance does not always imply practical significance. The former measures the likelihood of a difference (e.g., treatment effect) due to random chance while the latter can tell whether the observed effect is large enough to be "useful" in the reality.

5. Conclusions

Although communicating the associated benefits of the consumption of edible cricket protein (ECP) has proven to be beneficial in previous studies, the ECP statement used in this study did not significantly affect the overall liking of the treatments. Similar to the pattern observed for the after-tasting (before ECP statement) OL, after communicating the ECP statement, the plain and Italian treatments presented similar OL ratings, which were significantly higher than for the Cajun treatment. However, tasting and communicating the ECP statement affected the treatments' emotional profiles more than formulation, which in turn affected the overall product liking in the different moments. This suggests that, although expectations, disconfirmations, and product claims may not have a direct effect on product liking, they may still indirectly affect the overall product acceptability via emotional elicitation. In this study, the plain treatment had both valence and activation/arousal terms discriminate across moments, but only valence and mostly activation/arousal emotions discriminated across moments for the Italian and Cajun treatments, respectively. Minimal differences were observed across the formulations' emotional profiles; rather larger discrimination was observed across moments within each formulation. This research found that evoking "interested" and "adventurous" emotions plays a significant role as drivers of product liking regardless of the formulation and moment. On the other hand, our results showed that the PI can be improved if the sensory profile of products containing ECP is optimized and through the elicitation of pleasant emotions upon tasting while decreasing the "disgusted" feeling. This research may serve as a guide to optimize novel product development incorporating ECP to foods that are appealing for the market niche they are intended to.

Author Contributions: Conceptualization: C.E.G., D.D.T., B.L. and W.P.; performing the research work: C.E.G.; data analysis and interpretation: C.E.G., D.D.T., B.L. and W.P.; writing—original draft preparation: C.E.G.; writing—review and editing: C.E.G., D.D.T., B.L. and W.P.; supervision: W.P.; project administration and resource/funding acquisition: W.P. All authors have read and agreed to the published version of the manuscript.

Funding: This work was partially supported by the USDA National Institute of Food and Agriculture Hatch project (Accession No. 1022270; Hatch Project No. LAB94473) and the LSU Agricultural Center.

Institutional Review Board Statement: The study was conducted according to the guidelines of the Declaration of Helsinki and approved by the Institutional Review Board of Louisiana State University Agricultural Center (IRB # HE 18-9).

Informed Consent Statement: Informed consent was obtained from all subjects involved in the study.

Data Availability Statement: The data that support the findings of this study are available from the corresponding author upon reasonable request.

Acknowledgments: We acknowledge the graduate and undergraduate students under the direct supervision of the corresponding author in the Sensory Services Lab, LSU Agricultural Center for assistance with the lab work.

Conflicts of Interest: All authors declare no conflict of interest. The funders had no role in the design of the study; in the collection, analyses, or interpretation of data; in the writing of the manuscript; or in the decision to publish the results.

References

1. Gurdian, C.E.; Torrico, D.D.; Li, B.; Tuuri, G.; Prinyawiwatkul, W. Effect of Informed Conditions on Sensory Expectations and Actual Perceptions: A Case of Chocolate Brownies Containing Edible-Cricket Protein. *Foods* **2021**, *10*, 1480. [CrossRef] [PubMed]
2. Chen, Z.; Gurdian, C.; Sharma, C.; Prinyawiwatkul, W.; Torrico, D.D. Exploring Text Mining for Recent Consumer and Sensory Studies about Alternative Proteins. *Foods* **2021**, *10*, 2537. [CrossRef] [PubMed]
3. Fischer, A.R.; Steenbekkers, L.B. All insects are equal, but some insects are more equal than others. *Br. Food J.* **2018**, *120*, 852–863. [CrossRef] [PubMed]
4. Barton, A.; Richardson, C.D.; McSweeney, M.B. Consumer attitudes toward entomophagy before and after evaluating cricket (*Acheta domesticus*)-based protein powders. *J. Food Sci.* **2020**, *85*, 781–788. [CrossRef]
5. Osimani, A.; Milanović, V.; Cardinali, F.; Roncolini, A.; Garofalo, C.; Clementi, F.; Pasquini, M.; Mozzon, M.; Foligni, R.; Raffaelli, N.; et al. Bread enriched with cricket powder (*Acheta domesticus*): A technological, microbiological and nutritional evaluation. *Innov. Food Sci. Emerg. Technol.* **2018**, *48*, 150–163. [CrossRef]
6. Smarzyński, K.; Sarbak, P.; Musiał, S.; Jeżowski, P.; Piątek, M.; Kowalczewski, P.Ł. Nutritional analysis and evaluation of the consumer acceptance of pork pâté enriched with cricket powder—Preliminary study. *Open Agric.* **2019**, *4*, 159–163. [CrossRef]
7. Biró, B.; Sipos, M.A.; Kovács, A.; Badak-Kerti, K.; Pásztor-Huszár, K.; Gere, A. Cricket-Enriched Oat Biscuit: Technological Analysis and Sensory Evaluation. *Foods* **2020**, *9*, 1561. [CrossRef]

8. Gurdian, C.E.; Torrico, D.D.; Li, B.; Tuuri, G.; Prinyawiwatkul, W. Effect of Disclosed Information on Product Liking, Emotional Profile and Purchase Intent: A Case of Chocolate Brownies Containing Edible-Cricket Protein. *Foods* **2021**, *10*, 1769. [CrossRef]
9. Ardoin, R.; Prinyawiwatkul, W. Consumer perceptions of insect consumption: A review of western research since 2015. *Int. J. Food Sci. Technol.* **2021**, *56*, 4942–4958. [CrossRef]
10. Palczak, J.; Blumenthal, D.; Rogeaux, M.; Delarue, J. Sensory complexity and its influence on hedonic responses: A systematic review of applications in food and beverages. *Food Qual. Prefer.* **2019**, *71*, 66–75. [CrossRef]
11. Ardoin, R.; Marx, B.D.; Boeneke, C.; Prinyawiwatkul, W. Effects of cricket powder on selected physical properties and US consumer perceptions of whole-wheat snack crackers. *Int. J. Food Sci. Technol.* **2021**, *56*, 4070–4080. [CrossRef]
12. Pambo, K.O.; Okello, J.J.; Mbeche, R.M.; Kinyuru, J.N.; Alemu, M.H. The role of product information on consumer sensory evaluation, expectations, experiences and emotions of cricket-flour-containing buns. *Food Res. Int.* **2018**, *106*, 532–541. [CrossRef]
13. Duda, A.; Adamczak, J.; Chełmińska, P.; Juszkiewicz, J.; Kowalczewski, P. Quality and Nutritional/Textural Properties of Durum Wheat Pasta Enriched with Cricket Powder. *Foods* **2019**, *8*, 46. [CrossRef]
14. Ardoin, R.; Romero, R.; Marx, B.; Prinyawiwatkul, W. Exploring New and Modified Rejection-Type Thresholds Using Cricket Snack Crackers. *Foods* **2020**, *9*, 1352. [CrossRef]
15. Delicato, C.; Schouteten, J.J.; Dewettinck, K.; Gellynck, X.; Tzompa-Sosa, D.A. Consumers' perception of bakery products with insect fat as partial butter replacement. *Food Qual. Prefer.* **2020**, *79*, 103755. [CrossRef]
16. Verastegui-Tena, L.; Schulte-Holierhoek, A.; van Trijp, H.; Piqueras-Fiszman, B. Beyond expectations: The responses of the autonomic nervous system to visual food cues. *Physiol. Behav.* **2017**, *179*, 478–486. [CrossRef]
17. Chonpracha, P.; Ardoin, R.; Gao, Y.; Waimaleongora-ek, P.; Tuuri, G.; Prinyawiwatkul, W. Effects of Intrinsic and Extrinsic Visual Cues on Consumer Emotion and Purchase Intent: A Case of Ready-to-Eat Salad. *Foods* **2020**, *9*, 396. [CrossRef]
18. Wardy, W.; Chonpracha, P.; Chokumnoyporn, N.; Sriwattana, S.; Prinyawiwatkul, W.; Jirangrat, W. Influence of package visual cues of sweeteners on the sensory-emotional profiles of their products. *J. Food Sci.* **2017**, *82*, 500–508. [CrossRef]
19. Wardy, W.; Jack, A.R.; Chonpracha, P.; Alonso, J.R.; King, J.M.; Prinyawiwatkul, W. Gluten-free muffins: Effects of sugar reduction and health benefit information on consumer liking, emotion, and purchase intent. *Int. J. Food Sci. Technol.* **2018**, *53*, 262–269. [CrossRef]
20. Gurdian, C.E.; Torrico, D.D.; Li, B.; Prinyawiwatkul, W. Effect of serving plate types and color cues on liking and purchase intent of cheese-flavored tortilla chips. *Foods* **2021**, *10*, 886. [CrossRef]
21. Biswas, D.; Szocs, C.; Abell, A. Extending the boundaries of sensory marketing and examining the sixth sensory system: Effects of vestibular sensations for sitting versus standing postures on food taste perception. *J. Consum. Res.* **2019**, *46*, 708–724. [CrossRef]
22. Fuentes, S.; Wong, Y.Y.; Gonzalez Viejo, C. Non-invasive Biometrics and Machine Learning Modeling to Obtain Sensory and Emotional Responses from Panelists during Entomophagy. *Foods* **2020**, *9*, 903. [CrossRef] [PubMed]
23. La Barbera, F.; Verneau, F.; Amato, M.; Grunert, K. Understanding Westerners' disgust for the eating of insects: The role of food neophobia and implicit associations. *Food Qual. Prefer.* **2018**, *64*, 120–125. [CrossRef]
24. Lammers, P.; Ullmann, L.M.; Fiebelkorn, F. Acceptance of insects as food in Germany: Is it about sensation seeking, sustainability consciousness, or food disgust? *Food Qual. Prefer.* **2019**, *77*, 78–88. [CrossRef]
25. Sogari, G.; Menozzi, D.; Mora, C. The food neophobia scale and young adults' intention to eat insect products. *Int. J. Consum. Stud.* **2019**, *43*, 68–76. [CrossRef]
26. Gutjar, S.; Dalenberg, J.R.; de Graaf, C.; de Wijk, R.A.; Palascha, A.; Renken, R.J.; Jager, G. What reported food-evoked emotions may add: A model to predict consumer food choice. *Food Qual. Prefer.* **2015**, *45*, 140–148. [CrossRef]
27. Gutjar, S.; de Graaf, C.; Kooijman, V.; de Wijk, R.A.; Nys, A.; Ter Horst, G.J.; Jager, G. The role of emotions in food choice and liking. *Food Res. Int.* **2015**, *76*, 216–223. [CrossRef]
28. Nestrud, M.A.; Meiselman, H.L.; King, S.C.; Lesher, L.L.; Cardello, A.V. Development of EsSense25, a shorter version of the EsSense Profile®. *Food Qual. Prefer.* **2016**, *48*, 107–117. [CrossRef]
29. R Core Team. *R: A Language and Environment for Statistical Computing*; R Foundation for Statistical Computing: Vienna, Austria, 2020.
30. Addinsoft. XLSTAT Statistical and Data Analysis Solution. Available online: https://www.xlstat.com (accessed on 15 June 2021).
31. Smith, M.Q.P.; Ruxton, G.D. Effective use of the McNemar test. *Behav. Ecol. Sociobiol.* **2020**, *74*, 1–9.
32. Benjamini, Y.; Hochberg, Y. Controlling the false discovery rate: A practical and powerful approach to multiple testing. *J. R. Stat. Soc. Ser. B Methodol.* **1995**, *57*, 289–300. [CrossRef]
33. Meyners, M.; Castura, J.C.; Carr, B.T. Existing and new approaches for the analysis of CATA data. *Food Qual. Prefer.* **2013**, *30*, 309–319. [CrossRef]
34. Ares, G.; Dauber, C.; Fernández, E.; Giménez, A.; Varela, P. Penalty analysis based on CATA questions to identify drivers of liking and directions for product reformulation. *Food Qual. Prefer.* **2014**, *32*, 65–76. [CrossRef]
35. Sheskin, D.J. *Handbook of Parametric and Nonparametric Statistical Procedures*, 5th ed.; CRC Press: Boca Raton, FL, USA, 2020.
36. Carabante, K.M.; Ardoin, R.; Scaglia, G.; Malekian, F.; Khachaturyan, M.; Janes, M.E.; Prinyawiwatkul, W. Consumer acceptance, emotional response, and purchase intent of rib-eye steaks from grass-fed steers, and effects of health benefit information on consumer perception. *J. Food Sci.* **2018**, *83*, 2560–2570. [CrossRef]
37. Breiman, L. Random forests. *Mach. Learn.* **2001**, *45*, 5–32. [CrossRef]

38. Castro Delgado, M.; Chambers IV, E.; Carbonell-Barrachina, A.; Noguera Artiaga, L.; Vidal Quintanar, R.; Burgos Hernandez, A. Consumer acceptability in the USA, Mexico, and Spain of chocolate chip cookies made with partial insect powder replacement. *J. Food Sci.* **2020**, *85*, 1621–1628. [CrossRef]
39. Wendin, K.; Nyberg, M. Factors influencing consumer perception and acceptability of insect-based foods. *Curr. Opin. Food Sci.* **2021**, *40*, 67–71. [CrossRef]
40. Schwarz, F.H.; Marcus, M.; Schwarz, F.J. *Soy Protein and National Food Policy*, 1st ed.; CRC Press: New York, NY, USA, 2019.
41. Megido, R.C.; Gierts, C.; Blecker, C.; Brostaux, Y.; Haubruge, É.; Alabi, T.; Francis, F. Consumer acceptance of insect-based alternative meat products in Western countries. *Food Qual. Prefer.* **2016**, *52*, 237–243. [CrossRef]
42. Jiang, Y.; King, J.M.; Prinyawiwatkul, W. A review of measurement and relationships between food, eating behavior and emotion. *Trends Food Sci. Technol.* **2014**, *36*, 15–28. [CrossRef]
43. Tuccillo, F.; Marino, M.G.; Torri, L. Italian consumers' attitudes towards entomophagy: Influence of human factors and properties of insects and insect-based food. *Food Res. Int.* **2020**, *137*, 109619. [CrossRef]
44. Kauppi, S.-M.; Pettersen, I.N.; Boks, C. Consumer acceptance of edible insects and design interventions as adoption strategy. *Int. J. Food Des.* **2019**, *4*, 39–62. [CrossRef]
45. Hartmann, C.; Shi, J.; Giusto, A.; Siegrist, M. The psychology of eating insects: A cross-cultural comparison between Germany and China. *Food Qual. Prefer.* **2015**, *44*, 148–156. [CrossRef]
46. Schäufele, I.; Albores, E.B.; Hamm, U. The role of species for the acceptance of edible insects: Evidence from a consumer survey. *Br. Food J.* **2019**, *121*, 2190–2204. [CrossRef]
47. Sogari, G.; Bogueva, D.; Marinova, D. Australian consumers' response to insects as food. *Agriculture* **2019**, *9*, 108. [CrossRef]
48. Berger, S.; Bärtsch, C.; Schmidt, C.; Christandl, F.; Wyss, A.M. When utilitarian claims backfire: Advertising content and the uptake of insects as food. *Front. Nutr.* **2018**, *5*, 88. [CrossRef]
49. Schouteten, J.J.; De Steur, H.; De Pelsmaeker, S.; Lagast, S.; Juvinal, J.G.; De Bourdeaudhuij, I.; Verbeke, W.; Gellynck, X. Emotional and sensory profiling of insect-, plant-and meat-based burgers under blind, expected and informed conditions. *Food Qual. Prefer.* **2016**, *52*, 27–31. [CrossRef]
50. Ardoin, R.; Prinyawiwatkul, W. Product appropriateness, willingness to try and perceived risks of foods containing insect protein powder: A survey of U.S. consumers. *Int. J. Food Sci. Technol.* **2020**, *55*, 3215–3226. [CrossRef]
51. Imathiu, S. Benefits and food safety concerns associated with consumption of edible insects. *NFS J.* **2020**, *18*, 1–11. [CrossRef]
52. Mishyna, M.; Chen, J.; Benjamin, O. Sensory attributes of edible insects and insect-based foods–Future outlooks for enhancing consumer appeal. *Trends Food Sci. Technol.* **2020**, *95*, 141–148. [CrossRef]
53. Dobermann, D.; Swift, J.; Field, L. Opportunities and hurdles of edible insects for food and feed. *Nutr. Bull.* **2017**, *42*, 293–308. [CrossRef]
54. Belluco, S.; Halloran, A.; Ricci, A. New protein sources and food legislation: The case of edible insects and EU law. *Food Secur.* **2017**, *9*, 803–814. [CrossRef]
55. Ali, L.; Ali, F. Perceived risks related to unconventional restaurants: A perspective from edible insects and live seafood restaurants. *Food Control* **2022**, *131*, 108471. [CrossRef]
56. Legendre, T.S.; Baker, M.A. Legitimizing edible insects for human consumption: The impacts of trust, risk–benefit, and purchase activism. *J. Hosp. Tour. Res.* **2020**, 1096348020914375. [CrossRef]
57. King, S.C.; Meiselman, H.L.; Carr, B.T. Measuring emotions associated with foods in consumer testing. *Food Qual. Prefer.* **2010**, *21*, 1114–1116. [CrossRef]
58. Ruby, M.B.; Rozin, P.; Chan, C. Determinants of willingness to eat insects in the USA and India. *J. Insects Food Feed* **2015**, *1*, 215–225. [CrossRef]
59. Spinelli, S.; Monteleone, E.; Ares, G.; Jaeger, S.R. Sensory drivers of product-elicited emotions are moderated by liking: Insights from consumer segmentation. *Food Qual. Prefer.* **2019**, *78*, 103725. [CrossRef]
60. Van Huis, A.; Van Gurp, H.; Dicke, M. *The Insect Cookbook: Food for a Sustainable Planet*; Columbia University Press: New York, NY, USA, 2014.
61. Shelomi, M. Why we still don't eat insects: Assessing entomophagy promotion through a diffusion of innovations framework. *Trends Food Sci. Technol.* **2015**, *45*, 311–318. [CrossRef]
62. Tan, H.S.G.; Fischer, A.R.; van Trijp, H.C.; Stieger, M. Tasty but nasty? Exploring the role of sensory-liking and food appropriateness in the willingness to eat unusual novel foods like insects. *Food Qual. Prefer.* **2016**, *48*, 293–302. [CrossRef]
63. Chrea, C.; Grandjean, D.; Delplanque, S.; Cayeux, I.; Le Calvé, B.; Aymard, L.; Velazco, M.I.; Sander, D.; Scherer, K.R. Mapping the semantic space for the subjective experience of emotional responses to odors. *Chem. Senses* **2009**, *34*, 49–62. [CrossRef]
64. Cardello, A.V.; Meiselman, H.L.; Schutz, H.G.; Craig, C.; Given, Z.; Lesher, L.L.; Eicher, S. Measuring emotional responses to foods and food names using questionnaires. *Food Qual. Prefer.* **2012**, *24*, 243–250. [CrossRef]
65. Macht, M. How emotions affect eating: A five-way model. *Appetite* **2008**, *50*, 1–11. [CrossRef]
66. Mollahosseini, A.; Hasani, B.; Mahoor, M.H. Affectnet: A database for facial expression, valence, and arousal computing in the wild. *IEEE Trans. Affect. Comput.* **2017**, *10*, 18–31. [CrossRef]
67. Ruth, J.A.; Brunel, F.F.; Otnes, C.C. Linking thoughts to feelings: Investigating cognitive appraisals and consumption emotions in a mixed-emotions context. *J. Acad. Mark. Sci.* **2002**, *30*, 44–58. [CrossRef]

68. Batat, W.; Peter, P. The healthy and sustainable bugs appetite: Factors affecting entomophagy acceptance and adoption in Western food cultures. *J. Consum. Mark.* **2020**, *37*, 291–303. [CrossRef]
69. Norbury, A.; Husain, M. Sensation-seeking: Dopaminergic modulation and risk for psychopathology. *Behav. Brain Res.* **2015**, *288*, 79–93. [CrossRef]
70. Liu, A.-J.; Li, J.; Gómez, M.I. Factors influencing consumption of edible insects for Chinese consumers. *Insects* **2020**, *11*, 10. [CrossRef]
71. Berger, S.; Christandl, F.; Bitterlin, D.; Wyss, A.M. The social insectivore: Peer and expert influence affect consumer evaluations of insects as food. *Appetite* **2019**, *141*, 104338. [CrossRef]
72. Baker, M.A.; Shin, J.T.; Kim, Y.W. An exploration and investigation of edible insect consumption: The impacts of image and description on risk perceptions and purchase intent. *Psychol. Mark.* **2016**, *33*, 94–112. [CrossRef]
73. Mora, M.; Urdaneta, E.; Chaya, C. Effect of personality on the emotional response elicited by wines. *Food Qual. Prefer.* **2019**, *76*, 39–46. [CrossRef]
74. Scott, N.O.; Burgess, B.; Tepper, B.J. Perception and liking of soups flavored with chipotle chili and ginger extracts: Effects of PROP taster status, personality traits and emotions. *Food Qual. Prefer.* **2019**, *73*, 192–201. [CrossRef]
75. Scherer, K.R.; Shuman, V.; Fontaine, J.; Soriano Salinas, C. The GRID meets the Wheel: Assessing emotional feeling via self-report. In *Components of Emotional Meaning: A Sourcebook*; Fontaine, J.J.R., Scherer, K.R., Soriano, C., Eds.; Oxford University Press: Oxford, UK, 2013.
76. Gmuer, A.; Guth, J.N.; Runte, M.; Siegrist, M. From emotion to language: Application of a systematic, linguistic-based approach to design a food-associated emotion lexicon. *Food Qual. Prefer.* **2015**, *40*, 77–86. [CrossRef]
77. Byrnes, N.K.; Hayes, J.E. Behavioral measures of risk tasking, sensation seeking and sensitivity to reward may reflect different motivations for spicy food liking and consumption. *Appetite* **2016**, *103*, 411–422. [CrossRef] [PubMed]
78. Balzan, S.; Fasolato, L.; Maniero, S.; Novelli, E. Edible insects and young adults in a north-east Italian city an exploratory study. *Br. Food J.* **2016**, *118*, 318–326. [CrossRef]
79. Martins, Y.; Pliner, P. "Ugh! That's disgusting!": Identification of the characteristics of foods underlying rejections based on disgust. *Appetite* **2006**, *46*, 75–85. [CrossRef] [PubMed]
80. Verneau, F.; La Barbera, F.; Kolle, S.; Amato, M.; Del Giudice, T.; Grunert, K. The effect of communication and implicit associations on consuming insects: An experiment in Denmark and Italy. *Appetite* **2016**, *106*, 30–36. [CrossRef]
81. Cunha, L.M.; Ribeiro, J.C. Sensory and consumer perspectives on edible insects. In *Edible Insects in the Food Sector*; Springer: Cham, Switzerland, 2019; pp. 57–71.
82. Kim, J.-Y.; Prescott, J.; Kim, K.-O. Emotional responses to sweet foods according to sweet liker status. *Food Qual. Prefer.* **2017**, *59*, 1–7. [CrossRef]
83. Ruby, M.B.; Rozin, P. Disgust, sushi consumption, and other predictors of acceptance of insects as food by Americans and Indians. *Food Qual. Prefer.* **2019**, *74*, 155–162. [CrossRef]
84. Kusumasondjaja, S.; Tjiptono, F. Endorsement and visual complexity in food advertising on Instagram. *Internet Res.* **2019**. [CrossRef]
85. Tepsongkroh, B.; Jangchud, K.; Jangchud, A.; Chonpracha, P.; Ardoin, R.; Prinyawiwatkul, W. Consumer perception of extruded snacks containing brown rice and dried mushroom. *Int. J. Food Sci. Technol.* **2020**, *55*, 46–54. [CrossRef]
86. Verbeke, W. Profiling consumers who are ready to adopt insects as a meat substitute in a Western society. *Food Qual. Prefer.* **2015**, *39*, 147–155. [CrossRef]

foods
MDPI

Article

Emotion and Wellness Profiles of Herbal Drinks Measured Using Different Questionnaire Designs

Pannapa Hanmontree [1], Witoon Prinyawiwatkul [2] and Amporn Sae-Eaw [1,*]

1 Department of Food Technology, Faculty of Technology, Khon Kaen University, Khon Kaen 40002, Thailand; pannapa@kkumail.com
2 School of Nutrition and Food Sciences, Louisiana State University, Agricultural Center, Baton Rouge, LA 70803, USA; wprinya@lsu.edu
* Correspondence: sampor@kku.ac.th; Tel.: +66-43-362-131

Abstract: The emotion and wellness profiles of herbal drinks were assessed using six different questionnaire designs. The questionnaire designs were constructed from two formats of questionnaire items, including words and sentences, and three types of measuring scales, including a rating scale (5-point intensity; 1 = 'not at all', 5 = 'extremely'), a checklist scale (check-all-that-apply, CATA), and a combination of CATA and rating scales (rate-all-that-apply, RATA; 5-point intensity; 1 = 'slightly', 5 = 'extremely'). The 39 emotional terms of the EsSense Profile® and the 45 wellness terms of the WellSense Profile™ were translated into Thai, then screened for relevance to herbal drinks. The seven positive emotional terms (active, energetic, good, happy, polite, satisfied, and warm), three negative emotional terms (bored, disgusted, and worried), and five wellness terms (comforted, healthy, invigorated, relaxed, and refreshed) were selected and included in the questionnaire. A central location test was performed to determine the emotion and wellness profiles of five herbal drinks: roselle (*Hibiscus sabdariffa*) drink, chrysanthemum (*Chrysanthemum morifolium Ramat*) drink, ginger (*Zingiber officinale)* drink, Jubliang (a mixture of eight herbs) drink, and Krachai Dam (*Kaempferia parviflora)* drink. For herbal drinks, measuring emotion and wellness with a questionnaire using full sentences did not show increased benefit over questionnaires using words alone. All three measuring methods—a rating scale, CATA, and RATA—produced similar emotion and wellness profiles. However, each method has different advantages and limitations, which researchers should carefully consider.

Keywords: emotion; wellness; herbal drink; product discrimination; questionnaire designs

Citation: Hanmontree, P.; Prinyawiwatkul, W.; Sae-Eaw, A. Emotion and Wellness Profiles of Herbal Drinks Measured Using Different Questionnaire Designs. *Foods* **2022**, *11*, 348. https://doi.org/10.3390/foods11030348

Academic Editor: Han-Seok Seo

Received: 10 December 2021
Accepted: 22 January 2022
Published: 26 January 2022

Publisher's Note: MDPI stays neutral with regard to jurisdictional claims in published maps and institutional affiliations.

1. Introduction

In the last decade, emotional responses have been more frequently used in consumer research and food-product development because the use of liking data alone may not effectively predict product success in the market. Collecting emotional responses that offer insight into consumer experience can provide additional useful information for product development [1]. Several findings have noted that emotions influence consumers' eating behavior and decision-making processes [2–4]. Understanding the emotions elicited by their products could be beneficial for food industries when considering packaging design, branding, and advertising [2].

Several methods for measuring emotions associated with foods have been developed and reported, such as EsSense Profile® [5], consumer-defined check-all-that-apply (CD-CATA) [6], EmoSemio [7], and EmoSensory® Wheel [8]. Furthermore, the health and wellness perception of food products has become an important aspect of consumers' purchasing decisions. The WellSense Profile™, a questionnaire measuring wellness associated with foods from the consumer perspective, was recently developed [9] and was adapted in recent research to measure organic food-related wellbeing [10] in relation to the Krachai

Dam drink [11]. Accordingly, the emotion and wellness responses of consumers should be considered in food-product development. Although the EsSense profile® and the WellSense Profile™ methods have been widely used, the application of these methods in in countries with different cultures and languages should be appropriately performed. For this reason, cross-cultural studies of emotion elicited by food have gained attention [12–14]. However, research investigating the impact of questionnaire design and measurement scales on emotion and wellness profiles is scarce. In Thailand, methods for measuring emotional response elicited by food products have not been widely applied, so more research is required.

Herbs have been recognized as having health benefits [15]. Herbs and their extracts contain different bioactive compounds that can provide therapeutic effects, such as reducing cardiovascular problems, enhancing immune functions, and preventing cancer [16]. In Thailand, herbs have been used as medicine and food for a long time. Herbal drinks are becoming more popular, especially among health-conscious consumers, since these beverages are prepared with natural ingredients. The results obtained in a study by Jabeur et al. [17] highlighted the potential of *Hibiscus sabdariffa* L. (Hs), also known as roselle, as a source of bioactive and natural coloring ingredients destined for food and pharmaceutical industries. Roselle extracts showed antibacterial, antioxidant, anticholesterol, antidiabetic and antihypertensive effects, among others [18]. In recent years, the flowers of medicinal chrysanthemum (*Chrysanthemum morifolium Ramat*) were shown to act as common materials in functional and healthy tea or beverages due to their unique flavor, color, and health benefits, such as detoxification, improving liver function, decreasing inflammation, and improving eyesight [19]. Ginger (*Zingiber officinale*) rhizomes are commonly used in foods and beverages for their characteristic pungency and piquant flavor and have exhibited various pharmacological effects, such as antioxidant, anti-inflammatory, gastroprotective, antibacterial, and antidiabetic properties [20]. Jubliang is a mixture of eight herbs, namely *Bombax ceiba* L., *Chrysanthemum morifolium*, *Imperata cylindrical* (L.) P. Beauv., *Lophatherum gracile* Brogn., *Nelumbo nucifera* Gaertn, *Oroxylum indicum* L., *Pragmites communis* Trin, and *Prunella vulgaris*. Jubliang extract is a good source of water-soluble antioxidants, phenolic compounds, and antimutagens [21]. *Kaempferia parviflora*, or Krachai Dam (KD), is a Thai herb that belongs to the *Zingiberaceae* family. Its rhizomes have been reputed to have beneficial medicinal effects, owing to its major methoxyflavones, including anticancer, cardioprotective, neuroprotective, antioxidative, antimicrobial, and transdermal-permeable activity [22].

There are many kinds of commercial herbal drinks in Thailand, some of which are very popular; others are not popular, even though they have health-promotion benefits. There are many factors that have influence consumer perception and buying decisions. An important factor is the emotion and wellness responses of consumers. Additionally, different questionnaire designs may influence the results of emotion and wellness assessments. We hypothesized that using a questionnaire designed with full sentences and a rating-scale measurement method would have a greater influence on emotion and wellness profiles than that with words and a check-all-that-apply (CATA) measurement method. Therefore, in this research, we aimed to examine and compare the emotion and wellness responses associated with herbal drinks using questionnaires constructed according to two formats—words vs. full sentences—and three types of measurement methods, including a rating scale, a checklist scale (CATA), and a combination of the two scales (i.e., RATA).

2. Materials and Methods

2.1. Herbal Drink Samples

Five herbal drink samples commercially available in Thailand were used, including roselle drink, chrysanthemum drink, ginger drink, Jubliang drink, and Krachai Dam drink (Table 1). Roselle drink, chrysanthemum drink, Jubliang drink, and Krachai Dam drink were ready-to-drink products. After purchased, they were kept at 4 °C until used for consumer testing. Samples were presented to the participants in random 3-digit labelled cups at approximately 10 °C, except ginger drink, which was served at approximately

70 °C. The ginger drink was prepared immediately prior to sampling by adding 18 g of instant ginger powder to 150 mL of hot water and kept in an insulated bottle.

Table 1. The description of five herbal drink samples.

Herbal Drinks	Main Ingredients	Serving Condition/Temperature
Roselle drink Ready to drink, 200 mL/box	Roselle extract 94%, sucrose 6%	Cold/10 °C
Chrysanthemum drink Ready to drink, 350 mL/bottle	Chrysanthemum extract 90.00%, fructose syrup 4.00%, sucrose 4.00%	Cold/10 °C
Ginger drink Instant powder [1], 18 g/sachet	Ginger extract 8.03%, sugar 10.47%, vitamin B1 0.0003%, vitamin B2 0.0003%, vitamin B6 0.0009%	Hot/70 °C
Jubliang drink Ready to drink, 300 mL/box	Jubliang extract 94.68%, fructose 4.0%, sugar 1.0%	Cold/10 °C
Krachai Dam drink Ready to drink, 90 mL/bottle	Krachai Dam extract 20%, fructose 18%, honey 1.5%, citric acid 0.5%, vitamin C 0.12%, vitamin A 0.012%, vitamin B6 0.004%	Cold/10 °C

[1] ginger drink was prepared by diluting 18 g of instant ginger powder in 150 mL of hot water.

2.2. Screening of Emotion and Wellness Terms for Relevance to Herbal Drinks

The research protocol concerned with the use of human subjects was approved by the Khon Kaen University Ethics Committee for Human Research (KKUEC), Thailand (approval code, HE611230). Emotion and wellness terms were selected from the EsSence Profile® [5] and WellSense Profile™ tests [9]. All terms were translated into Thai by a group of experienced linguists (n = 3) and back-translated into English by another experienced group (n = 3) for meaning validation (Table 2). Thai emotion and wellness terms relevant to five herbal drinks (Roselle drink, Chrysanthemum drink, Ginger drink, Jubliang drink, and Krachai Dam drink) were screened using online CATA questionnaires. Thai consumers (n = 1678; 43.1% males and 56.9% females) who regularly consume herbal drinks participated voluntarily. Participants were asked to select the emotion and wellness terms that they associate with when thinking about herbal drinks that they had consumed. Terms with at least 20% frequency of use were recommended as a criterion [2,5,23,24]. However, terms with least 15% frequency of use were adopted for development of the EmoSensory® Wheel, while negative emotion terms with a lower frequency of use (≤10%) were used to obtain a balance of positive and negative terms [8]. For our current research, terms with at least 18% frequency of use were used. First, the selected terms from all of five herbal drinks were collected. Then, the duplicated terms were removed before inclusion in the questionnaire. Based on the results, seven positive emotion terms (active, energetic, good, happy, polite, satisfied, and warm) and five wellness terms (comforted, healthy, invigorated, relaxed, and refreshed) were selected. To control the balance on both of positive and negative terms, three negative emotion terms (bored, disgusted, and worried) were added to the questionnaire (Table 3). Therefore, a total of 15 emotion and wellness terms were included in the questionnaire to measure the consumer responses elicited by the five herbal drinks.

2.3. Testing the Test Designs for Measurement of Emotion and Wellness Responses Elicited by Herbal Drinks

The proposed test designs for measurement of emotion and wellness responses were composed of 2 factors. Factor 1 was composed of two formats of questionnaire items, including words and full sentences (see the attached questionnaire file). The sentence format was used to reduce ambiguity within the questionnaire because including a context increases understanding of the specified emotion and wellness state. The full sentences were created using selected emotion and wellness terms, in consultation with a focus group of consumers (n = 29, average age = 45.76 ± 7.83 years old) (adapted from [7]). Factor 2 was composed of three types of measuring scales, including a rating scale (5-point intensity; 1 = 'not at all', 5 = 'extremely'), a checklist scale (CATA), and a combination (rate-all-that-apply, RATA; 5-point intensity; 1 = 'slightly', 5 = 'extremely'). Thus, the questionnaire designs comprised six test forms (Table 4).

Table 2. Thai translation of emotion and wellness terms obtained from the EsSense Profile® and the WellSense Profile™.

EsSense Profile® (39 terms)					
Active	รู้สึกกระฉับกระเฉง	Glad	ยินดี/ดีใจ	Pleased	พึงพอใจ
Adventurous	ชอบผจญภัย	Good	ดี	Polite	สุภาพ
Affectionate	รักใคร่	Good-natured	มีใจเมตตา	Quiet	เงียบ
Aggressive	ก้าวร้าว	Guilty	รู้สึกผิด	Satisfied	พึงพอใจ
Bored	เบื่อ	Happy	มีความสุข	Secure	ปลอดภัย
Calm	สงบ	Interested	รู้สึกสนใจ	Steady	คงเส้นคงวา
Daring	กล้าหาญ	Joyful	ร่าเริง	Tame	จืดชืด/ไม่น่าสนใจ
Disgusted	น่ารังเกียจ/ขยะแขยง	Loving	รู้สึกรัก	Tender	นุ่มนวล
Eager	กระตือรือร้น	Merry	รู้สึกครึกครื้น	Understanding	มีความเข้าอกเข้าใจ
Energetic	รู้สึกมีพลัง	Mild	อ่อนโยน	Warm	อบอุ่น
Enthusiastic	กระตือรือร้น	Nostalgic	ระลึกถึงความหลัง	Whole	รู้สึกสมบูรณ์ครบถ้วน
Free	เป็นอิสระ	Peaceful	สงบ	Wild	คึกคะนอง
Friendly	เป็นมิตร	Pleasant	รู้สึกเพลิดเพลิน	Worried	รู้สึกกังวล
WellSense Profile™ (45 terms)					
Accepted	ได้รับการยอมรับ	Energetic	เต็มไปด้วยพลัง	Resilient	ยืดหยุ่น
Accomplished	บรรลุเป้าหมาย	Fatigued	เหนื่อยล้า	Rested	ได้พักผ่อน
Active	รู้สึกกระฉับกระเฉง	Focused	ตั้งใจจดจ่อ	Sad	เศร้า
Affectionate	รักใคร่	Friendly	เป็นมิตร	Satisfied	พึงพอใจ
Alert	ตื่นตัว	Fulfilled	รู้สึกได้รับการเติมเต็ม	Secure	ปลอดภัย
Approachable	เข้าถึงได้	Grateful	กตัญญูรู้คุณ	Sociable	ชอบเข้าสังคม
Attentive	เอาใจใส่	Happy	มีความสุข	Stimulated	ถูกกระตุ้น
Calm	สงบ	Healthy	มีสุขภาพดี	Stressed	เครียด
Comforted	รู้สึกสบาย	Invigorated	มีชีวิตชีวา	Supported	ได้รับการสนับสนุน
Compassionate	เห็นอกเห็นใจ	Joyful	ร่าเริง	Unfulfilled	ยังไม่ได้รับการเติมเต็ม
Concerned for others	มีความห่วงใยต่อผู้อื่น	Lonely	เหงา	Uninspired	ไม่มีแรงบันดาลใจ
Connected	ผูกพัน	Loved	รู้สึกได้รับความรัก	Uninterested	ไม่สนใจ
Creative	สร้างสรรค์	Peaceful	สงบ	Tense	เครียด
Curious	อยากรู้อยากเห็น	Refreshed	สดชื่น	Tired	เหนื่อยล้า
Disconnected	ไม่ผูกพัน	Relaxed	รู้สึกผ่อนคลาย	Whole	รู้สึกสมบูรณ์ครบถ้วน

The impact of the test designs (Test 1 to Test 6) was evaluated between 15 and 30 June 2018. Consumers were recruited by an accidental sampling method. According to Meilgaard et al. [25], 50–300 responses should be collected per location for the central location test (CLT). In this study, about 200 Thai consumers were recruited for each test design (a total of 1252; 49.3% males and 50.7% females) from 4 regions of Thailand (Table 4), with almost equal proportions of four age groups participating in the CLT. To be selected, consumers were required to be familiar with and frequently consumers of herbal drinks. Each consumer was randomly assigned to one of the six test designs. Five herbal drinks were coded with random 3-digit numbers. Each herbal drink sample (20 mL) was poured into a clear plastic cup and served under the corresponding condition/temperature described in Table 1. All five samples were served in a balanced order. All consumers received five herbal drink samples for tasting, along with water to cleanse their palates between samples. First, they were asked to taste and rate their overall liking of each sample using a 9-point hedonic scale (1 = extremely dislike, 9 = extremely like). Secondly, they indicated their emotion and wellness responses to each sample using the scales, as described above. Consumers were seated separately and were instructed not to interact with other participants.

Table 3. Selected emotional and wellness terms relevant to herbal drinks.

Term Categories	English	Thai
Positive emotional terms	Active	รู้สึกกระฉับกระเฉง
(7 terms)	Energetic	รู้สึกมีพลัง
	Good	รู้สึกดี
	Happy	รู้สึกมีความสุข
	Polite	รู้สึกสุภาพ
	Satisfied	รู้สึกพึงพอใจ
	Warm	รู้สึกอบอุ่น
Negative emotional terms	Bored	รู้สึกเบื่อ
(3 terms)	Disgusted	รู้สึกน่ารังเกียจ
	Worried	รู้สึกกังวล
Wellness terms	Comforted	รู้สึกสบาย
(5 terms)	Healthy	มีสุขภาพดี
	Invigorated	มีชีวิตชีวา
	Relaxed	รู้สึกผ่อนคลาย
	Refreshed	รู้สึกสดชื่น

Table 4. Demographic characteristics of participants for the six test designs.

Test Design	Test 1	Test 2	Test 3	Test 4	Test 5	Test 6	Total
Emotion and wellness items	Words	Sentences	Words	Sentences	Words	Sentences	
Measurement method	Rating	Rating	CATA	CATA	RATA	RATA	
Number of participants	209	208	209	207	208	211	1252
Gender (%) [1]							
Male	49.8	49.5	49.3	49.3	49.0	48.8	49.3
Female	50.2	50.5	50.7	50.7	51.0	51.2	50.7
Age (%) [1]							
18–21 years	24.9	24.5	24.4	26.6	25.2	27.5	25.6
22–40 years	24.9	26.0	25.8	24.2	26.5	24.6	25.2
41–59 years	25.4	25.5	25.8	25.1	24.4	24.2	25.1
≥60 years	24.8	24.0	24.0	24.1	23.9	23.7	24.1
Region of Thailand (%) [1]							
Northern	20.1	20.2	21.1	18.8	20.7	19.4	20.1
Central	22.5	20.7	19.1	24.2	25.0	26.6	23.0
Northeastern	34.0	36.0	33.5	36.7	31.7	34.6	34.4
Southern	23.4	23.1	26.3	20.3	22.6	19.4	22.5

[1] Percentage within each test design.

2.4. Statistical Analysis

Analysis of variance (ANOVA) and Duncan's multiple range test (DMRT) were performed on the overall liking scores to determine significant overall and pairwise differences among the five herbal drinks and six test designs.

For rating scores, the mean score of each emotion and wellness term was calculated from consumer responses. For CATA data, a frequency count was calculated by counting the number of consumers who chose the terms for each sample. RATA data were analyzed using one of two approaches [26]: frequency of selection (RATA frequency) or weighted frequency of selection (RATA score). RATA frequency was counted from the number of consumers who chose the terms for each sample. RATA scores were calculated by summing up the scores that the consumers rated for each selected term (0–5 scoring; 0 = 'not selected', 1 = 'slightly', and 5 = 'extremely') [27] and dividing by the number of participating consumers.

Both rating and RATA scores (when treated as a continuous outcome) were analyzed using ANOVA to determine the emotion and wellness terms with significant differences among the five herbal drinks. Next, the significant terms were analyzed using DMRT for multiple comparisons [28]. Principal component analysis (PCA) was performed to identify the emotion and wellness terms that contributed to overall differences among the herbal drinks and to classify the products based on emotion and wellness responses. Overall product differences were determined using multivariate analysis of variance (MANOVA) based on emotion and wellness responses, followed by discriminant analysis (DA) to identify the important terms of group discrimination.

CATA and RATA frequency counts were analyzed following the recommendation of Meyners et al. (2016) [27]. Cochran's Q test was carried out to determine the differences among herbal drink samples for each emotion or wellness response, and the sign test was used for pairwise comparison. Correspondence analysis (CA) based on Chi-square distances was used to identify the emotion and wellness terms that correlated to overall differences among the sampled herbal drinks. DA was performed to identify the important terms of group discrimination.

All statistical analyses were performed at 5% significance level using the SPSS 19 software (SPSS Inc., Chicago, IL, USA), except DA, which was carried out using XLSTAT® Basic version 2019 software (Addinsoft; New York, NY, USA).

3. Results and Discussion

3.1. Overall Liking of Herbal Drink Samples

Regarding the overall liking scores of the five herbal drinks from six test designs, the data were analyzed by ANOVA and DMRT (Table 5). In each test, significant differences in overall liking were observed ($p < 0.05$) among the five herbal drinks, indicating that consumers liked the five herbal drinks to varying degrees. In general, roselle and chrysanthemum drinks were similarly well-liked (7.11–7.49) and were found to be more acceptable than the three other herbal drinks (6.48–6.87 for ginger and Jubliang; 6.62–7.06 for Krachai Dam). For overall liking evaluation, different groups of consumers participated in each test design. Homogeneity of hedonic perception should be determined before comparing the different questionnaire designs. When comparing the overall liking score of each herbal drink among different consumers from the six test-design groups, no significant differences were found. This implies that the consumer population in this study was homogenous in terms of hedonic perception of the test samples. Thus, the subsequent emotion and wellness results of the six test designs could be compared since overall liking and emotion terms may be correlated, depending on product, product category, demographics, and psychographics [5].

3.2. Emotion and Wellness Profiles of Herbal Drinks from Different Test Designs

As seen from Tables 6 and 7, the type of herbal drink significantly affected most emotion (except "good", "happy", "bored", "worried", and "disgusted" in some cases) and all wellness responses of consumers. Generally, negative emotion responses were rated lower than positive emotion and wellness responses. These findings confirm and extend the positive asymmetry of emotions. According to the theory of positive asymmetry, previous research findings reported that consumers experience primarily positive emotions in response to food and food names [29–31]. In the case of beverages, Calve-Porral et al. [32] found positive asymmetry of emotions in beverage consumption, regardless of the type of beverage. Thus, our findings are in agreement with previous studies.

Table 5. The mean overall liking scores of the five herbal drinks in each test design.

Test Design [1]	Roselle	Chrysanthemum	Ginger	Jubliang	Krachai Dam	F-Value [3]	p-Value [3]
1	7.11 [a] ± 1.33	7.32 [a] ± 1.35	6.48 [c] ± 2.03	6.81 [b] ± 1.28	6.65 [bc] ± 1.82	11.867	0.000
2	7.14 [a] ± 1.25	7.29 [a] ± 1.23	6.55 [b] ± 1.98	6.76 [b] ± 1.39	6.62 [b] ± 1.79	12.870	0.000
3	7.13 [a] ± 1.38	7.41 [a] ± 1.45	6.59 [b] ± 2.19	6.81 [b] ± 1.41	6.85 [b] ± 1.83	9.983	0.000
4	7.13 [b] ± 1.40	7.46 [a] ± 1.21	6.48 [c] ± 2.08	6.48 [c] ± 1.73	6.74 [c] ± 1.81	19.866	0.000
5	7.18 [a] ± 1.33	7.35 [a] ± 1.42	6.81 [b] ± 2.04	6.87 [b] ± 1.51	7.06 [ab] ± 1.74	4.841	0.001
6	7.23 [a] ± 1.28	7.49 [a] ± 1.22	6.86 [b] ± 1.96	6.79 [b] ± 1.59	6.64 [c] ± 2.01	13.335	0.000
F-value [2]	0.179	0.753	1.439	1.478	1.184	-	-
***p*-value** [2]	0.970	0.584	0.208	0.195	0.315	-	-

Mean ± SD from 1252 consumer responses based on a 9-point hedonic scale. [a–c] Different letters in the same row indicate significant differences across samples according to Duncan's multiple range test ($p < 0.05$). [1] Test 1 rating, words; Test 2 rating, sentences; Test 3 CATA, words; Test 4 CATA, sentences; Test 5 RATA, words; Test 6 RATA, sentences. [2] F-value and p-value in ANOVA were used to assess whether there is a difference among the six test designs. [3] F-value and p-value in ANOVA were used to assess whether there is a difference among the five herbal drinks.

Table 6. Mean ± standard deviations of overall liking [1] and emotion/wellness responses elicited by five herbal drinks in rating [2] and RATA [3] questionnaires.

Attributes	Test Samples					*p*-Value
	R	C	G	J	K	
			Test design 1: Rating-Words (n = 209)			
Overall liking	7.11 ± 1.33 a	7.32 ± 1.35 a	6.48 ± 2.03 c	6.81 ± 1.28 b	6.65 ± 1.82 bc	<0.001
Emotion						
Active	2.70 ± 1.29 a	2.37 ± 1.33 bc	2.24 ± 1.36 c	2.46 ± 1.33 b	2.89 ± 1.38 a	<0.001
Energetic	2.28 ± 1.32 b	2.36 ± 1.35 b	2.67 ± 1.34 a	2.32 ± 1.32 b	2.76 ± 1.36 a	<0.001
Good	2.90 ± 1.31	2.91 ± 1.37	2.80 ± 1.41	2.72 ± 1.33	2.89 ± 1.34	0.311
Happy	2.62 ± 1.31 a	2.71 ± 1.32 a	2.26 ± 1.34 c	2.56 ± 1.26 ab	2.42 ± 1.37 bc	<0.001
Polite	2.04 ± 1.26 b	2.36 ± 1.29 a	2.09 ± 1.28 b	2.31 ± 1.14 a	2.12 ± 1.33 b	<0.001
Satisfied	3.03 ± 1.25 a	2.84 ± 1.34 ab	2.66 ± 1.33 b	2.73 ± 1.23 b	2.67 ± 1.44 b	0.003
Warm	1.90 ± 1.15 b	2.04 ± 1.31 b	2.88 ± 1.34 a	2.00 ± 1.15 b	2.02 ± 1.26 b	<0.001
Bored	1.36 ± 0.81 b	1.33 ± 0.74 bc	1.49 ± 0.88 a	1.33 ± 0.77 bc	1.25 ± 0.66 c	<0.001
Disgusted	1.30 ± 0.81 ab	1.31 ± 0.80 a	1.37 ± 0.83 a	1.28 ± 0.71 ab	1.19 ± 0.58 b	0.020
Worried	1.42 ± 1.02	1.36 ± 0.86	1.47 ± 0.90	1.36 ± 0.78	1.40 ± 0.79	0.403
Wellness						
Comforted	2.30 ± 1.31 b	2.74 ± 1.33 a	2.63 ± 1.40 a	2.68 ± 1.22 a	2.32 ± 1.33 b	<0.001
Healthy	2.40 ± 1.26 c	2.51 ± 1.33 bc	3.00 ± 1.32 a	2.43 ± 1.27 bc	2.62 ± 1.35 b	<0.001
Invigorated	3.01 ± 1.30 a	2.60 ± 1.43 b	2.26 ± 1.35 c	2.51 ± 1.33 b	3.00 ± 1.25 a	<0.001
Relaxed	2.30 ± 1.28 c	2.94 ± 1.21 a	2.84 ± 1.30 ab	2.72 ± 1.27 b	2.45 ± 1.38 c	<0.001
Refreshed	3.47 ± 1.18 a	3.19 ± 1.31 bc	2.28 ± 1.32 d	3.00 ± 1.19 c	3.24 ± 1.15 b	<0.001
			Test design 2: Rating-Sentences (n = 208)			
Overall liking	7.14 ± 1.25 a	7.29 ± 1.23 a	6.55 ± 1.98 b	6.76 ± 1.39 b	6.62 ± 1.79 b	<0.001
Emotion						
Active	2.58 ± 1.24 ab	2.41 ± 1.26 b	2.18 ± 1.24 c	2.43 ± 1.38 ab	2.61 ± 1.28 a	<0.001
Energetic	2.32 ± 1.16 b	2.28 ± 1.20 b	2.38 ± 1.31 b	2.22 ± 1.15 b	2.73 ± 1.27 a	<0.001
Good	2.97 ± 1.24 a	2.98 ± 1.35 a	2.57 ± 1.33 c	2.75 ± 1.22 bc	2.81 ± 1.34 ab	<0.001
Happy	2.63 ± 1.24 b	2.85 ± 1.26 a	2.35 ± 1.27 c	2.55 ± 1.30 b	2.51 ± 1.34 bc	<0.001

Table 6. *Cont.*

Attributes	Test Samples					p-Value
	R	C	G	J	K	
Polite	2.08 ± 1.17 c	2.49 ± 1.31 a	2.03 ± 1.15 c	2.27 ± 1.18 b	2.00 ± 1.17 c	<0.001
Satisfied	2.94 ± 1.26 a	2.89 ± 1.26 a	2.49 ± 1.22 b	2.78 ± 1.26 a	2.74 ± 1.38 a	<0.001
Warm	2.11 ± 1.19 b	2.19 ± 1.23 b	2.81 ± 1.30 a	2.08 ± 1.16 b	2.08 ± 1.20 b	<0.001
Bored	1.47 ± 0.94 a	1.29 ± 0.68 b	1.34 ± 0.71 ab	1.43 ± 0.83 a	1.43 ± 0.87 a	0.034
Disgusted	1.31 ± 0.74 bc	1.24 ± 0.62 c	1.42 ± 0.85 a	1.27 ± 0.65 c	1.39 ± 0.85 ab	0.002
Worried	1.40 ± 0.83	1.37 ± 0.81	1.38 ± 0.78	1.43 ± 0.83	1.50 ± 0.86	0.119
Wellness						
Comforted	2.54 ± 1.27 c	3.07 ± 1.28 a	2.54 ± 1.31 c	2.75 ± 1.17 b	2.43 ± 1.37 c	<0.001
Healthy	2.53 ± 1.26 b	2.56 ± 1.29 b	2.99 ± 1.26 a	2.47 ± 1.30 b	2.63 ± 1.26 b	<0.001
Invigorated	3.08 ± 1.15 a	2.88 ± 1.32 b	2.27 ± 1.23 c	2.70 ± 1.25 b	2.87 ± 1.28 b	<0.001
Relaxed	2.46 ± 1.26 b	2.79 ± 1.27 a	2.56 ± 1.28 b	2.51 ± 1.29 b	2.25 ± 1.30 c	<0.001
Refreshed	3.32 ± 1.12 a	3.52 ± 1.16 a	2.45 ± 1.27 c	3.03 ± 1.19 b	3.06 ± 1.19 b	<0.001
Test design 5: RATA-Words (n = 208)						
Overall liking	7.18 ± 1.33 a	7.35 ± 1.42 a	6.81 ± 2.04 b	6.87 ± 1.51 b	7.06 ± 1.74 ab	<0.001
Emotion						
Active	1.83 ± 1.73 b	1.40 ± 1.81 c	1.33 ± 1.73 c	1.67 ± 1.85 b	2.20 ± 1.97 a	<0.001
Energetic	1.68 ± 1.79 ab	1.53 ± 1.85 bc	1.60 ± 1.82 bc	1.39 ± 1.75 c	1.91 ± 1.89 a	0.001
Good	2.35 ± 1.80 a	2.15 ± 1.96 ab	1.96 ± 1.86 c	1.90 ± 1.83 c	2.05 ± 1.94 ab	0.023
Happy	1.88 ± 1.77 a	1.95 ± 1.92 a	1.37 ± 1.71 b	1.61 ± 1.78 b	1.50 ± 1.83 b	<0.001
Polite	1.42 ± 1.79 b	1.72 ± 1.79 a	1.13 ± 1.63 c	1.28 ± 1.64 bc	1.08 ± 1.62 c	<0.001
Satisfied	2.26 ± 1.80 a	2.06 ± 1.92 ab	1.79 ± 1.76 b	2.25 ± 1.79 a	2.00 ± 1.96 ab	0.008
Warm	1.15 ± 1.56 b	1.20 ± 1.66 b	2.22 ± 1.98 a	0.97 ± 1.52 b	1.08 ± 1.66 b	<0.001
Bored	0.47 ± 1.01 a	0.41 ± 1.04 a	0.44 ± 0.99 a	0.34 ± 0.89 ab	0.24 ± 0.80 b	0.004
Disgusted	0.26 ± 0.81 b	0.26 ± 0.86 b	0.41 ± 1.01 a	0.20 ± 0.75 b	0.21 ± 0.76 b	0.009
Worried	0.56 ± 1.22 a	0.41 ± 1.05 b	0.40 ± 1.09 b	0.30 ± 0.87 b	0.38 ± 0.95 b	0.014
Wellness						
Comforted	1.71 ± 1.80 b	2.05 ± 1.87 a	1.59 ± 1.82 bc	1.72 ± 1.74 b	1.35 ± 1.79 c	<0.001
Healthy	1.73 ± 1.78 bc	1.56 ± 1.86 c	2.36 ± 1.94 a	1.59 ± 1.78 c	1.90 ± 1.84 b	<0.001
Invigorated	2.39 ± 1.71 a	1.69 ± 1.89 b	1.38 ± 1.74 c	1.84 ± 1.86 b	2.20 ± 1.90 a	<0.001
Relaxed	2.06 ± 1.79 ab	2.22 ± 1.85 a	2.03 ± 1.85 ab	1.90 ± 1.82 b	1.54 ± 1.81 c	<0.001
Refreshed	3.20 ± 1.46 a	2.96 ± 1.74 ab	1.86 ± 1.90 d	2.54 ± 1.71 c	2.67 ± 1.86 bc	<0.001
Test design 6: RATA-Sentences (n = 211)						
Overall liking	7.23 ± 1.28 a	7.49 ± 1.22 a	6.86 ± 1.96 b	6.79 ± 1.59 b	6.64 ± 2.01 b	<0.001
Emotion						
Active	1.97 ± 1.76 a	1.71 ± 1.79 b	1.49 ± 1.68 b	1.54 ± 1.67 b	1.93 ± 1.76 a	<0.001
Energetic	1.78 ± 1.69 b	1.47 ± 1.68 c	1.68 ± 1.63 bc	1.51 ± 1.68 c	2.07 ± 1.70 a	<0.001
Good	2.56 ± 1.70 a	2.43 ± 1.74 ab	2.40 ± 1.71 ab	2.26 ± 1.71 c	2.17 ± 1.76 c	0.039
Happy	2.06 ± 1.72 a	2.14 ± 1.81 a	1.65 ± 1.74 b	2.08 ± 1.69 a	1.41 ± 1.67 c	<0.001
Polite	1.36 ± 1.59 b	1.62 ± 1.63 a	1.29 ± 1.62 b	1.48 ± 1.61 ab	1.31 ± 1.70 b	0.008
Satisfied	2.57 ± 1.69 a	2.42 ± 1.77 a	2.28 ± 1.72 ab	2.37 ± 1.65 a	2.05 ± 1.84 b	0.007
Warm	1.40 ± 1.66 b	1.31 ± 1.64 b	2.39 ± 1.75 a	1.27 ± 1.65 b	1.27 ± 1.62 b	<0.001
Bored	0.47 ± 1.04 ab	0.34 ± 0.91 b	0.48 ± 1.14 ab	0.38 ± 0.95 b	0.54 ± 1.18 a	0.039
Disgusted	0.35 ± 0.96	0.23 ± 0.79	0.39 ± 1.10	0.25 ± 0.79	0.32 ± 0.93	0.089
Worried	0.34 ± 0.90 c	0.36 ± 1.03 c	0.44 ± 1.16 bc	0.54 ± 1.15 ab	0.64 ± 1.18 a	0.001
Wellness						
Comforted	1.94 ± 1.76 b	2.13 ± 1.79 ab	1.98 ± 1.74 b	2.27 ± 1.66 a	1.61 ± 1.72 c	<0.001
Healthy	1.86 ± 1.76 c	1.74 ± 1.78 c	2.64 ± 1.71 a	1.72 ± 1.78 c	2.19 ± 1.80 b	<0.001
Invigorated	2.52 ± 1.67 a	1.88 ± 1.78 b	1.58 ± 1.71 c	2.02 ± 1.73 b	2.36 ± 1.74 a	<0.001
Relaxed	1.80 ± 1.73 bc	2.14 ± 1.73 a	2.00 ± 1.61 ab	1.75 ± 1.65 c	1.50 ± 1.71 d	<0.001
Refreshed	3.09 ± 1.44 a	3.07 ± 1.52 a	1.93 ± 1.81 c	2.79 ± 1.36 b	2.90 ± 1.55 ab	<0.001

[1] Mean ± SD of overall liking scores based on a 9-points hedonic scale (1 = extremely dislike to 5 = extremely like). [2] Mean ± SD of rating scores based on a 5-points scale (1 = not at all to 5 = extremely). [3] Mean ± SD of RATA rating scores based on a 5-points scale (1 = slightly to 5 = extremely). a–d Mean values in the same row followed by different letters are significantly different ($p < 0.05$). Note: R = Roselle drink; C = Chrysanthemum drink; G = Ginger drink; J = Jubliang drink; K = Krachai Dam drink.

Table 7. Mean ± standard deviations of overall liking [1] and frequency counts of emotion/wellness responses elicited by five herbal drinks in CATA [2] and RATA [2] questionnaires.

Attributes	Test Samples					Cochran's Q	p-Value
	R	C	G	J	K		
Test design 3: CATA-Words (n = 209)							
Overall liking	7.13 ± 1.38 a	7.41 ± 1.45 a	6.59 ± 2.19 b	6.81 ± 1.41 b	6.85 ± 1.83 b		<0.001
Emotion							
Active	58	41	42	48	95	68.962	<0.001
Energetic	39	41	74	45	68	45.305	<0.001
Good	79	91	74	81	84	4.603	0.331
Happy	46	56	40	45	52	8.082	0.089
Polite	28	48	36	31	27	16.994	0.002
Satisfied	69	79	52	66	63	13.338	0.010
Warm	18	24	100	27	26	222.749	<0.001
Bored	14	21	23	18	13	14.667	0.005
Disgusted	14	13	24	13	16	21.500	<0.001
Worried	24	19	23	21	26	3.476	0.482
Wellness							
Comforted	45	81	62	79	43	42.904	<0.001
Healthy	43	54	106	65	67	71.609	<0.001
Invigorated	84	62	39	70	93	52.529	<0.001
Relaxed	60	104	77	86	60	40.209	<0.001
Refreshed	156	143	66	137	114	119.685	<0.001
Test design 4: CATA-Sentences (n = 207)							
Overall liking	7.13 ± 1.40 b	7.46 ± 1.21 a	6.48 ± 2.08 c	6.48 ± 1.73 c	6.74 ± 1.81 c		<0.001
Emotion							
Active	53	47	40	42	90	68.784	<0.001
Energetic	44	33	64	30	72	65.502	<0.001
Good	85	108	79	89	81	17.227	0.002
Happy	64	82	46	53	53	36.553	<0.001
Polite	20	47	32	41	16	45.342	<0.001
Satisfied	72	73	49	70	67	14.916	0.005
Warm	22	35	82	41	33	106.995	<0.001
Bored	18	19	22	24	21	4.750	0.314
Disgusted	17	16	25	23	22	14.571	0.006
Worried	15	18	20	21	22	5.500	0.240
Wellness							
Comforted	75	102	76	95	66	27.069	<0.001
Healthy	61	115	69	63	68	65.954	<0.001
Invigorated	81	77	48	79	105	46.724	<0.001
Relaxed	71	77	77	58	47	24.281	<0.001
Refreshed	160	141	77	123	124	89.387	<0.001
Test design 5: RATA-Words (n = 208)							
Overall liking	7.18 ± 1.33 a	7.35 ± 1.42 a	6.81 ± 2.04 b	6.87 ± 1.51 b	7.06 ± 1.74 ab		<0.001
Emotion							
Active	116	83	83	102	124	52.089	<0.001
Energetic	104	90	97	86	115	21.243	<0.001
Good	137	120	118	114	118	10.919	0.027
Happy	116	111	87	100	89	25.585	<0.001
Polite	86	106	71	87	69	39.768	<0.001
Satisfied	133	116	113	135	113	15.948	0.003
Warm	78	78	125	64	66	94.646	<0.001
Bored	39	31	36	28	18	21.387	<0.001
Disgusted	22	20	32	15	16	17.864	0.001
Worried	40	30	27	24	32	10.590	0.032
Wellness							
Comforted	104	120	96	109	82	30.556	<0.001
Healthy	106	91	131	100	115	34.691	<0.001

Table 7. *Cont.*

Attributes	Test Samples					Cochran's Q	*p*-Value
	R	C	G	J	K		
Invigorated	145	98	86	110	127	69.516	<0.001
Relaxed	125	128	122	116	95	21.713	<0.001
Refreshed	179	161	109	152	147	81.199	<0.001
	Test design 6: RATA-Sentences (n = 211)						
Overall liking	7.23 ± 1.28 a	7.49 ± 1.22 a	6.86 ± 1.96 b	6.79 ± 1.59 b	6.64 ± 2.01 b		<0.001
Emotion							
Active	127	109	98	103	126	34.400	<0.001
Energetic	118	99	116	101	137	41.902	<0.001
Good	157	149	151	144	139	6.643	0.156
Happy	133	130	108	135	95	56.124	<0.001
Polite	96	112	88	105	88	22.218	<0.001
Satisfied	158	147	144	151	125	20.266	<0.001
Warm	95	90	147	84	89	121.892	<0.001
Bored	38	27	35	31	39	7.353	0.118
Disgusted	26	18	26	20	24	4.935	0.294
Worried	28	26	31	42	53	28.398	<0.001
Wellness							
Comforted	123	132	127	148	108	36.416	<0.001
Healthy	119	113	158	107	137	64.692	<0.001
Invigorated	158	120	104	129	147	65.381	<0.001
Relaxed	116	137	136	118	99	39.635	<0.001
Refreshed	184	180	122	181	175	96.239	<0.001

[1] Mean ± SD of overall liking scores based on a 9-point hedonic scale (1 = extremely dislike, 5 = extremely like). [2] Frequency counts of selected emotion/wellness terms were analyzed using Cochran's Q test. a–c Mean values in the same row followed by different letters are significantly different ($p < 0.05$). Note: R = roselle drink; C = chrysanthemum drink; G = ginger drink; J = Jubliang drink; K = Krachai Dam drink.

When comparing the two formats of questionnaire items—words vs. full sentences—some minor differences, in terms of the number of significant emotion and wellness terms, were observed, depending on the measurement scale used. The number of significant terms for test 1 (rating-words) vs. test 2 (rating-sentences) was 13 vs. 14 for test 5 (RATA/scores-words) vs. test 6 (RATA/scores-sentences) was 15 vs. 14, for test 3 (CATA-words) vs. test 4 (CATA-sentences) was 12 vs. 13, and for test 5 (RATA/frequency-words) vs. test 6 (RATA/frequency-sentences) was 15 vs. 12. A greater difference was observed between words and sentences in the RATA frequency test format. However, overall, sentence format may not reduce ambiguity over the word-only format in the questionnaire. For terms unfamiliar to consumers (such as "polite"), including a context may help to increase understanding of the specified emotion or wellness state. However, the emotion and wellness terms used in the questionnaire in this study were pre-screened by consumers who regularly consume the herbal drink products. Consumers selected the terms that they were familiar with; therefore, a full sentence in the questionnaire may not provide much more clarity under the conditions of the current study.

The results from three types of measurement methods, including a rating scale, CATA, and RATA, were compared. The RATA data consisted of two types: RATA frequency and RATA score. Hence, the RATA scores were compared with rating scores, and RATA frequencies were compared with CATA frequencies.

To compare rating scores (test 1) and RATA scores (test 5) from the word-format questionnaire, mean scores of emotion and wellness responses were plotted on a spider web, as shown in Figure 1. The overall profiles were visually similar, but the ratings from the rating scores were higher than those from the RATA scores. The rating-words profile (test 1) in Figure 1A reveals that significant product differences were observed for 13 terms ($p < 0.05$), except for "worried" (1.36–1.47 scores) and "good" (2.72–2.91 scores) terms (Table 6). Interestingly, the "refreshed" term showed higher scores of four herbal drinks (roselle, Jubliang, chrysanthemum, and Krachai Dam drink) (3.47, 3.00, 3.19, and 3.24, respectively) than other emotion and wellness terms. Meanwhile, the score of the "warm" term (2.88) was higher for ginger drink than other terms, with a maximum mean-difference value of 0.98. King et al. [29] suggested that a mean emotional difference of ≥ 0.2 units (on a 5-point scale) may be of practical value. For the rata-words profile (test 5) in Figure 1B, all 15 terms had significant differences among the five products. Therefore, using the RATA scale with words (Test 5) showed better discrimination ability than the rating scale with words (15 vs. 13) in terms of improving the number of significantly different terms.

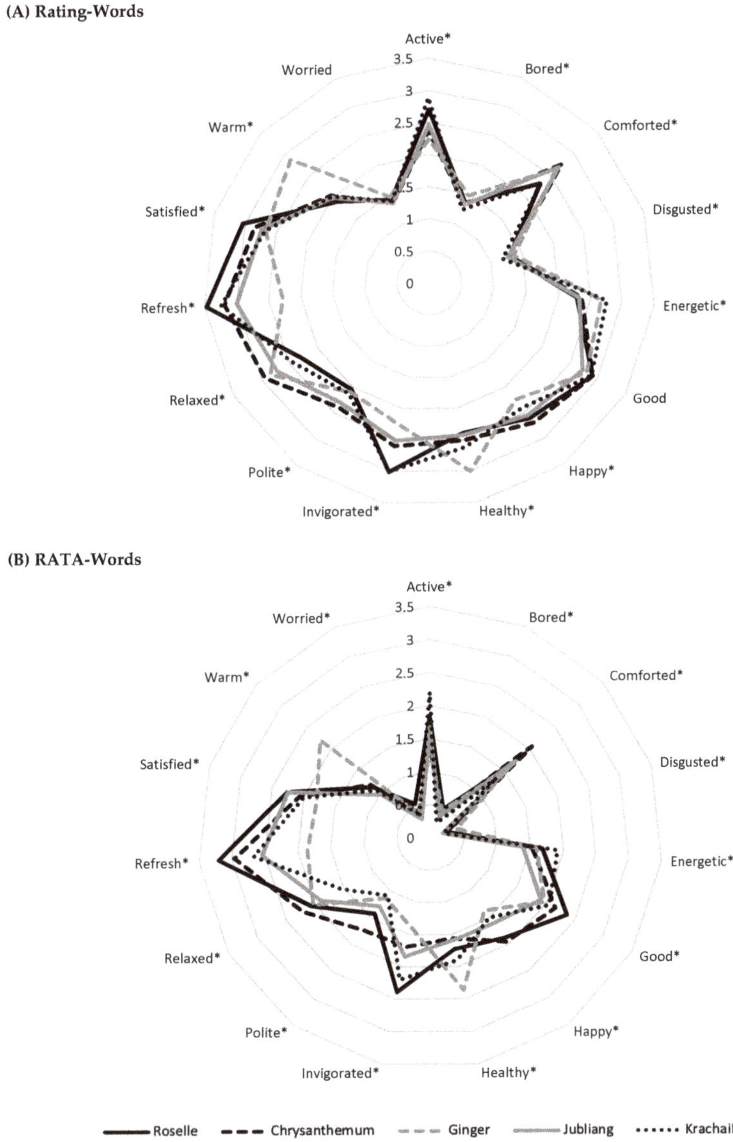

Figure 1. Emotion and wellness profiles of five herbal drinks from the (**A**) Rating-Words (score 1–5; $n = 209$) and (**B**) RATA-Words (score 0–5; $n = 208$) questionnaires. * indicates significant differences among the five herbal drinks ($p < 0.05$).

To compare the CATA frequency (test 3) and RATA frequency (test 5) from the word-format questionnaire, emotion and wellness responses, in terms of percentage of the frequency count, were plotted on a line chart, as shown in Figure 2. The two profiles were visually similar, but the frequency counts of the RATA-Words format was higher than those of the CATA-Words format. For CATA-Words in Figure 2A, the "refreshed" term was popular among consumers for roselle (74.64%), Jubliang (65.55%), chrysanthemum (68.42%), and Krachai Dam drink (54.55%), while "good", "happy", and "worried" terms were not significantly different among the five herbal drinks ($p > 0.05$; Table 7). For RATA-Words in Figure 2B, all 15 terms were significantly different among the five products. Thus, the RATA-Words show a slight improvement in the number of significantly different terms. Therefore, using the RATA scale with words (Test 5) showed slightly better discrimination ability than the CATA scale with words (15 vs. 12) in terms of increasing the number of significant terms.

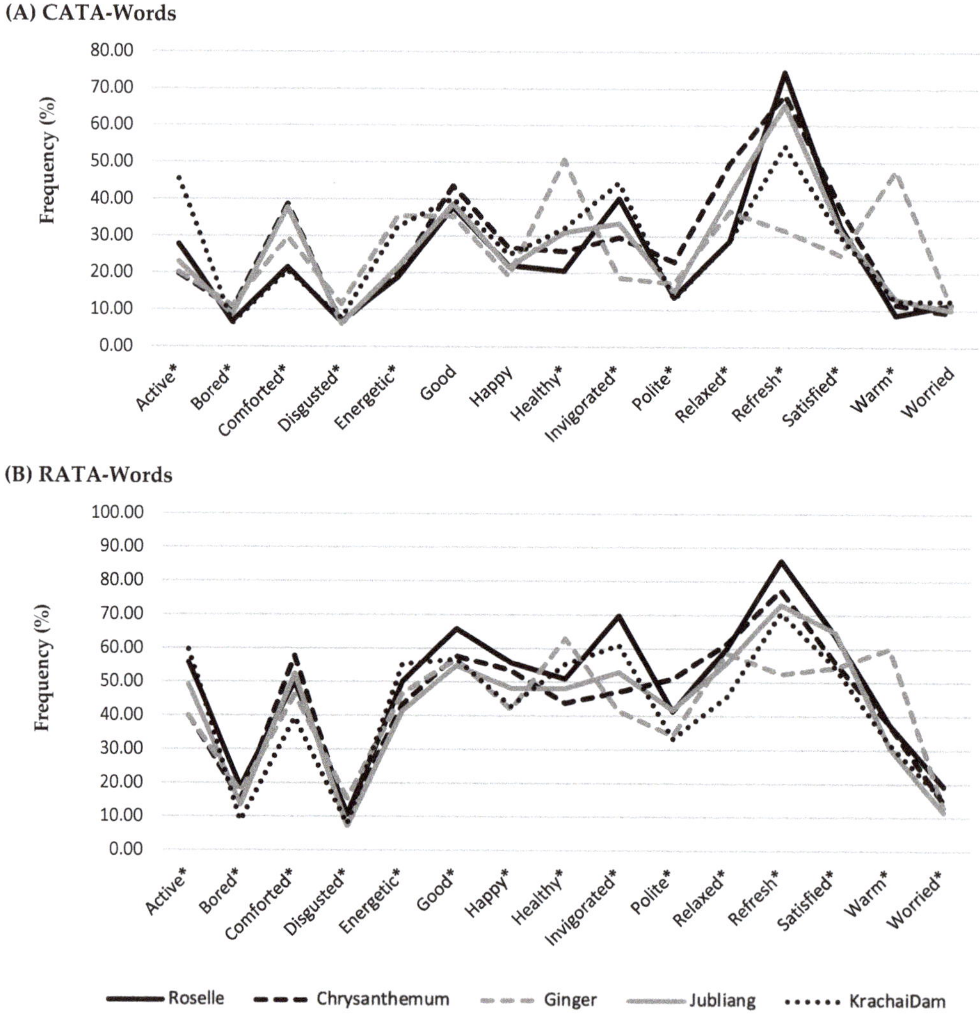

Figure 2. Emotion and wellness profiles of the five herbal drinks from the (**A**) CATA-Words (test 3; $n = 209$) and (**B**) RATA-Words (test 5; $n = 208$). * indicates significant differences among five herbal drinks ($p < 0.05$).

Soares et al. [33] mentioned that the word-association (WA) method is used in consumer and marketing research to provide insights on perceptions and attitudes to achieve a deeper understanding of what consumers really think and feel about a product. Many researchers have studied the benefit of herbs, such as roselle, ginger and chrysanthemum, and found that they are a good source of vitamins and minerals and rich in phenolic compounds known to have free-radical-scavenging abilities [34–38]. In this study, consumers selected the "active" and "energetic" emotion terms, as well as the "invigorated", "refreshed", and "relaxed" wellness terms for these five herbal drink products. It is possible that some participants realized the important benefits of these herbal drink; thus, they chose those emotion and wellness terms based on their prior knowledge and consumption experiences. It is known that characteristics of an individual participant may affect the emotion profile of foods [2]; these characteristics include personal preference, past experience, frequency of consumption [30], and culture [39].

3.3. Discriminating Herbal Drinks Based on Emotion and Wellness Responses

To determine the overall product differences and which emotion and wellness attributes were mainly responsible for product discrimination, MANOVA and DA were performed (Table 8). When considering the MANOVA results of rating scores and RATA scores, significant differences were observed among the five herbal drinks in terms of emotion and wellness responses ($p < 0.0001$). Based on the first canonical dimension (Can 1), DA identified the "warm" emotion and the "refreshed" wellness terms as the two most discriminating terms among the herbal drinks based on questionnaire responses. However, one additional wellness term ("healthy") was identified in Can 1 for the CATA frequency questionnaire. When considering Can 2, "active" and "comforted" terms significantly contributed to overall product differences. However, RATA scores identified one more wellness term ("relaxed"), while rating scores and CATA frequencies identified two more wellness terms ("relaxed" and "invigorated"). In summary, DA of CATA data revealed the greatest number of terms that discriminated the products, including two positive emotion terms ("warm" and "active") and all five wellness terms ("refreshed", "healthy", "comforted", "relaxed", and "invigorated") with greatest explained variance (93.6%). This finding (Table 8) is not in agreement with that previously reported based on the results from Tables 6 and 7. One possible explanation is that the former data analysis took into account intercorrelations among emotion and wellness attributes, while the latter did not.

PCA was performed to investigate the relationship between the five herbal drinks and emotion and wellness responses. PCA biplots of rating scores and RATA scores were compared (Figure 3A,B) with the first two PCs (PC1 and PC2), explaining 78.98% and 77.50% of total variance, respectively. For rating-words data, the PCA biplots identified four groups of herbal drinks: I-ginger drink, II-Krachai Dam drink, III-roselle drink, and IV-chrysanthemum-Jubliang drink (Figure 3A). Ginger drink and roselle drink corresponded with PC1 but in the opposite quadrant, while Krachai Dam drink and chrysanthemum-Jubliang drink corresponded with PC2 but in the opposite quadrant as well. Ginger drink correlated with "warm", "healthy", "disgusted", and "bored". Roselle drink correlated with "refreshed", "invigorated", and "active". Krachai Dam drink correlated with "energetic" and "worried". Chrysanthemum and Jubliang drinks correlated with "polite" and "comforted".

For RATA-words data, the PCA biplots revealed three groups of herbal drinks: I-ginger drink, II-Krachai Dam drink, and III-roselle, chrysanthemum, and Jubliang drinks (Figure 3B). Ginger drink and roselle drink corresponded with PC1, and Krachai Dam drink and Chrysanthemum drink corresponded with PC2, in opposite directions. Jubliang drink corresponded with PC3 (not shown in Figure 3), as it could not be explained using PC1 and PC2. Ginger drink correlated with "warm", "healthy", and "disgusted". Roselle drink correlated with "refreshed", "invigorated", "happy", "satisfied", and "polite". Krachai Dam drink correlated with "energetic" and "active". Chrysanthemum drink correlated with "polite", "comforted", and "bored". Obviously, the "warm" emotion term and the "refreshed" wellness term were the most important terms differentiating the hot ginger drink from the cold herbal drinks (Table 8 and Figure 3).

Table 8. Canonical structure (r's) [1] describing group differences among the five herbal drinks based on emotion and wellness responses obtained from the rating, CATA, and RATA questionnaires.

Responses	Rating Scores [2]		RATA Scores [2]		CATA Frequencies [2]		RATA Frequencies [2]	
	Can 1	Can 2	Can 1	Can 2	Can 1	Can 2	Can 1	Can 2
Emotion								
Active	0.181	**−0.319**	0.148	**−0.402**	0.109	−0.531	0.177	**−0.349**
Energetic	−0.142	−0.232	−0.001	−0.191	−0.210	−0.234	0.003	−0.224
Good	0.037	−0.039	0.079	0.025	0.059	0.044	0.055	−0.011
Happy	0.162	0.159	0.174	0.173	0.069	0.014	0.159	0.147
Polite	0.032	0.228	0.134	0.287	−0.017	0.210	0.139	0.284
Satisfied	0.117	0.028	0.148	0.008	0.120	0.111	0.114	0.035
Warm	**−0.457**	−0.020	**−0.465**	0.137	**−0.648**	0.036	**−0.461**	0.114
Bored	−0.122	0.041	−0.027	0.161	−0.071	0.127	−0.043	0.140
Disgusted	−0.072	0.092	−0.139	0.071	−0.115	−0.044	−0.165	0.053
Worried	−0.049	−0.060	0.030	0.010	−0.005	−0.097	0.061	−0.053
Wellness								
Comforted	−0.091	**0.316**	0.075	**0.316**	−0.013	**0.409**	0.080	**0.302**
Healthy	−0.259	−0.088	−0.267	−0.112	**−0.348**	−0.045	−0.222	−0.177
Invigorated	0.282	**−0.315**	0.264	−0.281	0.248	**−0.303**	0.293	−0.284
Relaxed	−0.166	**0.355**	−0.003	**0.313**	−0.004	**0.399**	−0.016	0.259
Refreshed	**0.525**	−0.110	**0.441**	0.051	**0.493**	0.192	**0.485**	0.059
% Cumulative variance explained	62.5	89.0	61.7	90.7	68.5	93.6	54.4	87.5
MANOVA; Wilks' Lambda p-value	<0.0001		<0.0001					

[1] Based on the pooled within-group variance with $p < 0.0001$ for rating score, RATA score, CATA frequency, and RATA frequency. Can1 and Can2 refer to the first and second canonical discrimination function, respectively. [2] Rating scores from Rating-Words question, RATA scores; RATA frequencies from RATA-Words question; CATA frequencies from CATA-Words question.

The CA symmetric plots of CATA frequency and RATA frequency were compared (Figure 4A,B). The first two dimensions (Dim1 and Dim2) accounted for 94.79% and 89.60% of variance, respectively. CA configuration revealed four groups of herbal drinks: I-ginger drink, II-Krachai Dam drink, III-roselle drink, and IV-chrysanthemum-Jubliang drink. The ginger drink and the other four herbal drinks were clearly separated, on the opposite sides of Dim1 in both CA configurations. However, the CA configurations of CATA data vs. RATA data revealed some slightly different relationships between herbal drinks and emotion and wellness profiles. For CATA data (Figure 4A), ginger drink correlated with "warm", "healthy", and "disgusted". Roselle drink correlated with "refreshed" and "invigorated". Krachai Dam drink correlated with "active". Chrysanthemum and Jubliang drinks correlated with "polite", "relaxed", and "comforted". For RATA data (Figure 4B), ginger drink correlated with "warm" and "disgusted". Roselle drink correlated with "refreshed" and "invigorated". Krachai Dam drink correlated with "energetic" and "active". Chrysanthemum and Jubliang drinks correlated with "polite", "comforted", "relaxed", and "happy".

King et al. [31] evaluated the impact of rating scale and CATA on measurement of emotional responses. They found that the rating scale provided differentiation for more attributes at lower levels of emotional response, while CATA provided greater differentiation at higher levels of emotional frequency for a few select emotions. The rating scale was more sensitive than CATA, but both were found to be acceptable approaches, depending on the objective of the test [31]. In addition, Ng et al. [6] evaluated the effectiveness of the CATA approach compared to intensity scaling for the EsSense® Profile. They reported that the qualitative nature of the data obtained from CATA limited the extent of statistical analysis, making it difficult to draw clear inferential conclusions from the data obtained with the EsSense Profile®. Therefore, using a combination of both approaches, RATA was proposed.

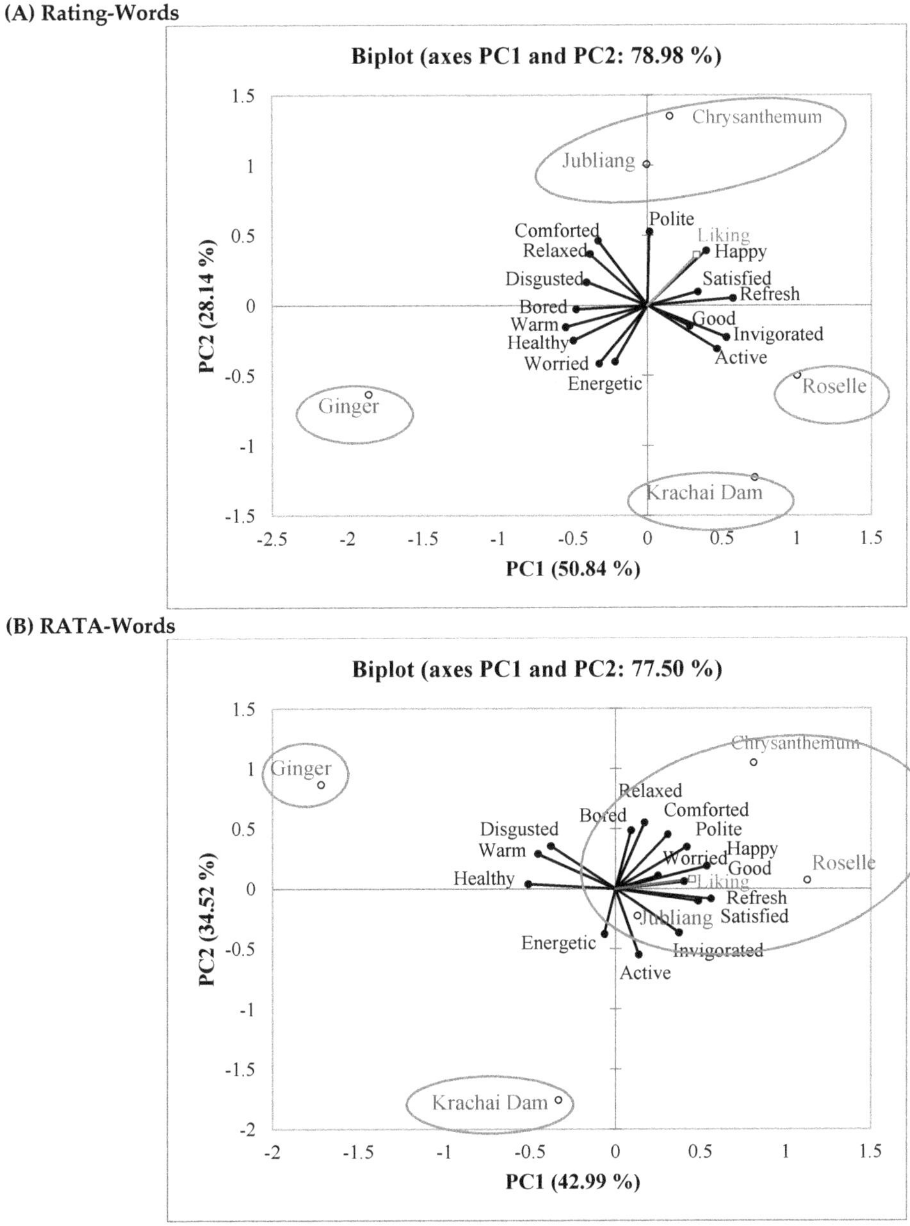

Figure 3. PCA biplots of emotion and wellness responses of the five herbal drinks in the (**A**) Rating-Words and (**B**) RATA-Words questionnaires. Symbols (○) and (•) indicate the position of herbal drinks and emotion-wellness terms, respectively.

(A) CATA-Words

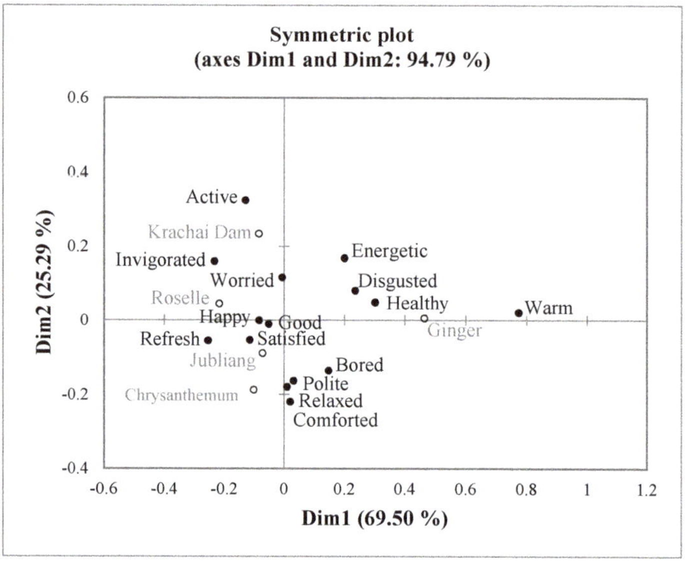

(B) RATA-Words

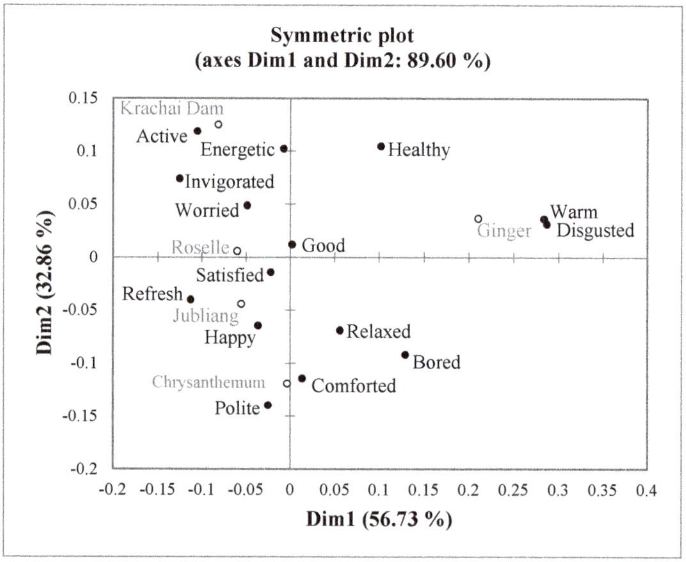

Figure 4. CA symmetric plots of emotion and wellness responses of the five herbal drinks in the (**A**) CATA-Words and (**B**) RATA-Words questionnaires. Symbols (○) and (•) indicate the position of herbal drinks and emotion and wellness terms, respectively.

Meyners et al. [27] mentioned that compared to CATA questions, the use of a RATA variant was found to increase the number of attribute terms selected to describe samples and led to a slight increase in the percentage of terms for which significant differences among samples were identified. Jaeger et al. [40] compared the CATA and RATA question formats using an emoji questionnaire. They found that neither CATA nor RATA emoji-questions were regarded by consumers as difficult or tedious. Their recommendation for method selection was to use CATA emoji questions when samples have distinct emotional associations, whereas RATA seems to better discriminate between samples with more similar emotional profiles. However, in this study, the CA configurations of CATA data vs. RATA data revealed only slightly different relationships between herbal drinks and emotion and wellness profiles (Figure 4).

3.4. Overall Discussion

Overall, all types of data (rating, RATA, and CATA) showed similar results in terms of emotion and wellness profiles of the five herbal drinks studied. The "warm" and "healthy" terms were important and relevant to ginger drink, and the negative emotion term "disgusted" was elicited. "Refreshed" and "invigorated" were the important wellness terms for roselle drink. "Active", "energetic", and "worried" emotion terms were often elicited by Krachai Dam drink. Additionally, "comforted", "relaxed", and "bored" were significant terms for describing the chrysanthemum and Jubliang drinks. Although chrysanthemum drink had the highest overall liking scores (Table 5), some consumers expressed a "bored" emotion. It is a challenge of product developers to reduce the negative emotions associated with products. Elicited negative emotion terms should be carefully considered when developing or improving products. In the case of herbal drinks, ginger drink was associated with "disgusted" by some consumers, possibly because of its natural spicy flavor. Krachai Dam drink was associated with "worried" because it is not well-known by many Thai consumers and it normally imparts a pungent and sour flavor. "Bored" was elicited by chrysanthemum, possibly because it is a well-known traditional product that has not changed for a long time (i.e., not much product innovation applied to this drink).

When the three types of measuring methods—a rating scale, CATA question, and RATA question—were compared, they all produced similar emotion and wellness profiles, though there were advantages and limitations for each method. For a rating scale, the participants needed more time to complete the questionnaire than for other methods. However, a rating scale is appropriate for measuring intensity of attributes. The PCA and CA of rating scores were able to differentiate five herbal drinks into four groups with a higher percentage explained variance (78.98%) when compared with the RATA score (three groups; 77.50%). The CATA questionnaire was the most consumer-friendly method and was found to be appropriate for use for measurement of the presence or absence of the selected attributes. DA of CATA data identified the important terms discriminating the overall differences among the herbal tested drinks with the highest percentage of explained variance (93.6% total for Can1 and Can2). Additionally, CA of CATA data separated the five herbal drinks into four groups, similar to those of the rating data, with a higher percentage of explained variance (94.79%) when compared with the RATA frequency (four groups, 89.60%). However, the RATA score and RATA frequency from the RATA-Words question seemed to show better discrimination ability in term of improving the number of significant terms among the five herbal drinks.

4. Conclusions

In this study, we demonstrated the effects of two factors in designing a questionnaire for measurement of emotion and wellness responses elicited by herbal drinks consumed by Thai consumers. The first factor was questionnaire items in the form of words vs. full sentences. Results indicate that measuring emotion and wellness using full sentences did not provide a clear benefit over using words alone. When using familiar terms clearly understood by consumers, a full sentence in the questionnaire is not needed. For the second factor, three types of measuring methods, including a rating scale, CATA question, and RATA question were compared for their ability to discriminate among five herbal drinks. All three measuring methods produced similar emotion and wellness profiles. However, each method has advantages and limitations that researchers should carefully consider. Overall, this study provides some useful options in terms of questionnaire-item formats and measurement methods in the design of questionnaire to measuring consumers' emotion and wellness responses to herbal drinks. The information elucidated in this study is applicable to other food products and beverages. In the future, the impact of health-benefit statements on consumer perception and purchase intent with regard to herbal drinks formulated in our lab will be assessed by product tasting and with an appropriate sample size representing the target population. We also plan to collaborate with a medical school for human clinical trials to determine the health benefits of these herbal drinks.

Author Contributions: Conceptualization, W.P. and A.S.-E.; methodology, W.P., A.S.-E. and P.H.; formal analysis, A.S.-E. and P.H.; investigation, A.S.-E. and P.H.; writing—original draft preparation, P.H.; writing—review and editing, A.S.-E. and W.P.; visualization, P.H.; supervision, A.S.-E. and W.P.; project administration, A.S.-E.; funding acquisition, A.S.-E. and P.H. All authors have read and agreed to the published version of the manuscript.

Funding: This research receives funding from Sakon Nakhon Rajabhat University (grant no. 0213).

Institutional Review Board Statement: The study was reviewed and approved by the Ethics Committee of Khon Kaen University (HE611230, approved on 7 June 2018).

Informed Consent Statement: Informed consent was obtained from all subjects involved in the study.

Data Availability Statement: The data presented in this study are available in the manuscript.

Acknowledgments: The authors express their grateful appreciation to Sakon Nakhon Rajabhat University for financial support. The department of Food Technology, Faculty of Technology, Khon Kaen University, Thailand and the School of Nutrition and Food Sciences, Louisiana State University, Agricultural Center, Baton Rouge, LA, USA, are also gratefully acknowledged for their support. We would also like to thank the consumers who participated in this research.

Conflicts of Interest: The authors declare no conflict of interest. The funders had no role in the design of the study; in the collection, analyses, or interpretation of data; in the writing of the manuscript, or in the decision to publish the results.

References

1. Spinelli, S.; Niedziela, M. Chapter 5 Emotion Measurements and Application to Product and Packaging Development. In *Integrating the Packaging and Product Experience in Food and Beverages: A Road-Map to Consumer Satisfaction*; Burgess, P., Ed.; Woodhead: Publishing, UK, 2016; pp. 77–119.
2. Jiang, Y.; King, J.M.; Prinyawiwatkul, W. A review of measurement and relationships between food, eating behavior and emotion. *Trends Food Sci. Technol.* **2014**, *36*, 15–28. [CrossRef]
3. Köster, E.P.; Mojet, J. From mood to food and from food to mood: A psychological perspective on the measurement of food-related emotions in consumer research. *Food Res. Int.* **2015**, *76*, 180–191. [CrossRef]
4. Volz, K.G.; Hertwig, R. Emotions and decisions: Beyond conceptual vagueness and the rationality muddle. *Perspect. Psychol. Sci.* **2016**, *11*, 101–116. [CrossRef] [PubMed]
5. King, S.C.; Meiselman, H.L.; Carr, B.T. Measuring emotions associated with foods in consumer testing. *Food Qual. Prefer.* **2010**, *21*, 1114–1116. [CrossRef]
6. Ng, M.; Chaya, C.; Hort, J. Beyond liking: Comparing the measurement of emotional response using EsSense Profile and consumer defined check-all-that-apply methodologies. *Food Qual. Prefer.* **2013**, *28*, 193–205. [CrossRef]
7. Spinelli, S.; Masi, C.; Dinnella, C.; Zoboli, G.P.; Monteleone, E. How does it make you feel? A new approach to measuring emotions in food product experience. *Food Qual. Prefer.* **2014**, *37*, 109–122. [CrossRef]
8. Schouteten, J.J.; Steur, H.D.; Pelsmaeker, S.D.; Lagast, S.; Bourdeaudhuij, I.D.; Gellynck, X. An integrated method for the emotional conceptualization and sensory characterization of food products: EmoSensory® Wheel. *Food Res. Int.* **2015**, *78*, 96–107. [CrossRef]
9. King, S.C.; Snow, J.; Meiselman, H.L.; Sainsbury, J.; Carr, B.T.; McCafferty, D.; Serrano, D.; Gillette, M.; Millard, L.; Li, Q. Development of a questionnaire to measure consumer wellness associated with foods: The WellSense Profile™. *Food Qual. Prefer.* **2015**, *39*, 82–94. [CrossRef]
10. Apaolaza, V.; Hartmann, P.; D'Souza, C.; López, M. Eat organic—Feel good? The relationship between organic food consumption, health concern and subjective wellbeing. *Food Qual. Prefer.* **2018**, *63*, 51–62. [CrossRef]
11. Hanmontree, P.; Sae-Eaw, A. Krachai Dam (*Kaempferia parviflora*) drinks: Physicochemical properties, consumer acceptance, purchase intent, and emotion and wellness responses. *Ital. J. Food Sci.* **2020**, *32*, 631–644.
12. Jang, J.A.; Oh, J.E.; Na, Y.; Yeo, G.E.; Cho, M.S. Emotions evoked by colors and health functionality information of colored rice: A cross-cultural study. *Foods* **2021**, *10*, 231. [CrossRef] [PubMed]
13. Hu, X.; Lee, J. Emotions elicited while drinking coffee: A cross-cultural comparison between Korean and Chinese consumers. *Food Qual. Prefer.* **2019**, *76*, 160–168. [CrossRef]

14. Gunaratne, T.M.; Viejo, C.G.; Fuentes, S.; Torrico, D.D.; Gunaratne, N.M.; Ashman, H.; Dunshea, F.R. Development of emotion lexicons to describe chocolate using the Check-All- That-Apply (CATA) methodology across Asian and Western groups. *Food Res. Int.* **2019**, *115*, 526–534. [CrossRef] [PubMed]
15. Farzaneh, V.; Carvalho, I.S. A review of the health benefit potentials of herbal plant infusions and their mechanism of actions. *Ind. Crops Prod.* **2015**, *65*, 247–258. [CrossRef]
16. Craig, W.J. Health-promoting properties of common herbs. *Am. J. Clin. Nutr.* **1999**, *70*, 491s–499s. [CrossRef]
17. Jabeur, I.; Pereira, E.; Barros, L.; Calhelha, R.C.; Soković, M.; Oliveira, M.B.P.P.; Ferreira, I.C.F.R. *Hibiscus sabdariffa* L. as a source of nutrients, bioactive compounds and colouring agents. *Food Res. Int.* **2017**, *100*, 717–723. [CrossRef]
18. Da-Costa-Rocha, I.; Bonnlaender, B.; Sievers, H.; Pischel, I.; Heinrich, M. *Hibiscus sabdariffa* L.—A phytochemical and pharmacological review. *Food Chem.* **2014**, *165*, 424–443. [CrossRef]
19. Shi, X.-F.; Chu, J.-Z.; Zhang, Y.-F.; Liu, C.-Q.; Yao, X.-Q. Nutritional and active ingredients of medicinal chrysanthemum flower heads affected by different drying methods. *Ind. Crops Prod.* **2017**, *104*, 45–51. [CrossRef]
20. Pratap, S.R.; Gangadharappa, H.V.; Mruthunjaya, K. Ginger: A potential neutraceutical, an updated review. *Int. J. Pharmacogn. Phytochem. Res.* **2017**, *9*, 1227–1238. [CrossRef]
21. Kruawan, K.; Kangsadalampai, K. Antioxidant activity, phenolic compound contents and antimutagenic activity of some water extract of herbs. *Thai J. Pharm. Sci.* **2006**, *30*, 28–35.
22. Chen, D.; Li, H.; Li, W.; Feng, S.; Deng, D. *Kaempferia parviflora* and Its Methoxyflavones: Chemistry and Biological Activities. *Evid.-Based Complement. Alternat. Med.* **2018**, *2018*, 4057456. [CrossRef] [PubMed]
23. Poonnakasem, N.; Pujols, K.D.; Chaiwanichsiri, S.; Laohasongkram, K.; Prinyawiwatkul, W. Different oils and health benefit statements affect physicochemical properties, consumer liking, emotion, and purchase intent: A case of sponge cake. *J. Food Sci.* **2016**, *81*, S165–S173. [CrossRef] [PubMed]
24. Wardy, W.; Jack, A.R.; Chonpracha, P.; Alonso, J.R.; King, J.M.; Prinyawiwatkul, W. Gluten-free muffins: Effects of sugar reduction and health benefit information on consumer liking, emotion, and purchase intent. *Int. J. Food Sci. Technol.* **2018**, *53*, 262–269. [CrossRef]
25. Meilgaard, M.C.; Civille, G.V.; Carr, B.T. *Sensory Evaluation Techniques*, 5th ed.; CRC Press: Boca Raton, FL, USA, 2016; p. 589.
26. Ares, G.; Bruzzone, F.; Vidal, L.; Cadena, R.S.; Giménez, A.; Pineau, B.; Hunter, D.C.; Paisley, A.G.; Jaeger, S.R. Evaluation of a rating-based variant of check-all-that-apply questions: Rate-all-that-apply (RATA). *Food Qual. Prefer.* **2014**, *36*, 87–95. [CrossRef]
27. Meyners, M.; Jaeger, S.R.; Ares, G. On the analysis of Rate-All-That-Apply (RATA) data. *Food Qual. Prefer.* **2016**, *49*, 1–10. [CrossRef]
28. Ngoenchai, P.; Alonso, J.R.; Suwonsichon, T.; Suwonsichon, S.; Prinyawiwatkul, W. Effects of visual cues on consumer expectation, emotion and wellness responses, and purchase intent of red chili powders. *J. Food Sci.* **2019**, *84*, 3018–3026. [CrossRef]
29. King, S.C.; Meiselman, H.L. Development of a method to measure consumer emotions associated with foods. *Food Qual. Prefer.* **2010**, *21*, 168–177. [CrossRef]
30. Cardello, A.V.; Meiselman, H.L.; Schutz, H.G.; Craig, C.; Given, Z.; Lesher, L.L.; Eicher, S. Measuring emotional responses to foods and food names using questionnaires. *Food Qual. Prefer.* **2012**, *24*, 243–250. [CrossRef]
31. King, S.C.; Meiselman, H.L.; Carr, B.T. Measuring emotions associated with foods: Important elements of questionnaire and test design. *Food Qual. Prefer.* **2013**, *28*, 8–16. [CrossRef]
32. Calvo-Porral, C.; Rivaroli, S.; Orosa-González, J. The asymmetric emotional associations to beverages: An approach through the theory of positive asymmetry. *Foods* **2021**, *10*, 794. [CrossRef]
33. Soares, E.K.B.; Esmerino, E.A.; Ferreira, M.V.S.; da Silva, M.A.A.P.; Freitas, M.Q.; Cruz, A.G. What are the cultural effects on consumers' perceptions? A case study covering coalho cheese in the Brazilian northeast and southeast area using word association. *Food Res. Int.* **2017**, *102*, 553–558. [CrossRef] [PubMed]
34. Cao, X.; Xiong, X.; Xu, Z.; Zeng, Q.; He, S.; Yuan, Y.; Wang, Y.; Yang, X.; Su, D. Comparison of phenolic substances and antioxidant activities in different varieties of chrysanthemum flower under simulated tea making conditions. *J. Food Meas. Charact.* **2020**, *14*, 1443–1450. [CrossRef]
35. Chen, S.; Liu, J.; Dong, G.; Zhang, X.; Liu, Y.; Sun, W.; Liu, A. Flavonoids and caffeoylquinic acids in *Chrysanthemum morifolium* Ramat flowers: A potentially rich source of bioactive compounds. *Food Chem.* **2021**, *344*, 128733. [CrossRef] [PubMed]
36. Ghafoor, K.; Juhaimi, F.A.; Özcan, M.M.; Uslu, N.; Babiker, E.E.; Ahmed, I.A.M. Total phenolics, total carotenoids, individual phenolics and antioxidant activity of ginger (*Zingiber officinale*) rhizome as affected by drying methods. *Food Sci. Technol.* **2020**, *126*, 109354. [CrossRef]
37. Nguyen, Q.V.; Chuyen, H.V. Processing of herbal tea from roselle (*Hibiscus sabdariffa* L.): Effects of drying temperature and brewing conditions on total soluble solid, phenolic content, antioxidant capacity and sensory quality. *Beverages* **2020**, *6*, 2. [CrossRef]
38. Riaz, G.; Chopra, R. A review on phytochemistry and therapeutic uses of *Hibiscus sabdariffa* L. *Biomed. Pharmacother.* **2018**, *102*, 575–586. [CrossRef]
39. Olegario, L.S.; Estevéz, M.; González-Mohino, A.; Madruga, M.S.; Ventanas, S. Cross-cultural emotional response to food stimuli: Influence of consumption context. *Food Res. Int.* **2021**, *142*, 110194. [CrossRef]
40. Jaeger, S.R.; Lee, S.M.; Kim, K.-O.; Chheang, S.L.; Roigard, C.M.; Ares, G. CATA and RATA questions for product-focused emotion research: Five case studies using emoji questionnaires. *Food Qual. Prefer.* **2018**, *68*, 342–348. [CrossRef]

foods

MDPI

Article

Communicating Function and Co-Creating Healthy Food: Designing a Functional Food Product Together with Consumers

Petar Vrgović [1,*], Milica Pojić [2], Nemanja Teslić [2], Anamarija Mandić [2], Aleksandra Cvetanović Kljakić [3], Branimir Pavlić [3], Alena Stupar [2], Mladenka Pestorić [2], Dubravka Škrobot [2] and Aleksandra Mišan [2]

1 Faculty of Technical Sciences, University of Novi Sad, Trg Dositeja Obradovića 7, 21000 Novi Sad, Serbia
2 Institute of Food Technology, University of Novi Sad, Bulevar Cara Lazara 1, 21000 Novi Sad, Serbia; milica.pojic@fins.uns.ac.rs (M.P.); nemanja.teslic@fins.uns.ac.rs (N.T.); anamarija.mandic@fins.uns.ac.rs (A.M.); alena.tomsik@fins.uns.ac.rs (A.S.); mladenka.pestoric@fins.uns.ac.rs (M.P.); dubravka.skrobot@fins.uns.ac.rs (D.Š.); aleksandra.misan@fins.uns.ac.rs (A.M.)
3 Faculty of Technology, University of Novi Sad, Bulevar Cara Lazara 1, 21000 Novi Sad, Serbia; amod@uns.ac.rs (A.C.K.); bpavlic@uns.ac.rs (B.P.)
* Correspondence: vrgovic@uns.ac.rs; Tel.: +381-641-715-897

Abstract: Functional food is lately an interesting topic from the new product development perspective; complex motivation and expectations of consumers regarding it present a challenge when new products are designed. Co-creation is an interesting alternative to the standard practices by the R & D departments since it directly involves consumers in the various stages of the creation process. This work aims to describe experiences of engaging consumers in different development stages of a functional food product within a project realized at a food research institute. Four consecutive studies were conducted: the first study explored current trends in Serbia regarding the way consumers use functional food and are informed about it; the second study described development of a raspberry seeds extract with antioxidant and anti-proliferative activity confirmed in vitro; the third study tested the same extract in a sample of consumers, validating its usability in food products; and the fourth study described a co-creation session with 18 participants, during which a number of activities were realized to stimulate idea generation. Rather than the final product idea itself, this work is valuable because of detailed insights into the various phases of the co-creation process. It is shown that consumers and food researchers can together engage in the new food product development process as long as the communication between them is rich and with mutual understanding.

Keywords: co-creation; communication; functional food; NADES; raspberry

Citation: Vrgović, P.; Pojić, M.; Teslić, N.; Mandić, A.; Kljakić, A.C.; Pavlić, B.; Stupar, A.; Pestorić, M.; Škrobot, D.; Mišan, A. Communicating Function and Co-Creating Healthy Food: Designing a Functional Food Product Together with Consumers. *Foods* **2022**, *11*, 961. https://doi.org/10.3390/foods11070961

Academic Editor: Witoon Prinyawiwatkul

Received: 15 February 2022
Accepted: 18 March 2022
Published: 26 March 2022

1. Introduction

The link between diet and general health has always been visible, but it is becoming increasingly important in the new age of rising health care costs, rising living standards, and drastically changing perceptions of food; such demands may be achieved by formulating functional foods [1]. Although there is no strict definition of functional food, it can generally be said that it is food that, in addition to the basic nutritional value, has a positive effect on one or more target functions in the body [2]. Such foods include foods that are enriched with nutrients naturally present in them (by increasing the concentrations of these components) [3] or new components that are not naturally found in those foods [3,4]; further, they can be modified [3] or improved products [5].

Consumers around the world are now more than ever health conscious and more educated about what goes into their food; health is one of the most important motivators that influence consumers' food choices [6,7], and health motivation significantly influences the way consumers perceive and process information displayed on the food package [8]. However, the relationship between consumers' interest in their own health and the purchase of a food product is not straightforward, as different emotions arise when different food groups are mentioned to the consumers and thus mediate said relationship. In the case of functional food, although it prompts a generally positive affect with the customers, it still induces more fear compared to regular and organic food [9], and it is even documented to evoke several different emotions simultaneously from both the "positive" and "negative" categories [10].

Such a complex relationship between consumers' intentions, underlying emotions, and attitudes towards functional food [11] implies that acceptance rates of novel functional food products may not be easily projected and could be quite worrisome [12]. It is not uncommon for a significant percentage of novel food products to fail on the market because there was a lack of response to the complex needs of consumers [13,14]. New product development (NPD) in the food industry is usually a closed, intra-organizational process with little or no contact with the end-users during the early stages—companies frequently rely on their own intuition, wisdom, or competence to come up with ideas for new products, and they become interested in consumers' impressions only once they introduce a new product to the market [15]. Although a range of methods is available that can aid companies in generating new product ideas based on input from marketing research [16], these methods are still just indirectly voicing end-users' perspective, as consumers usually are contacted when the company's interest in obtaining a glimpse of the market is present or during the product-testing phase. Aiming to bridge this gap, some companies embark on a journey to create new food products together with the consumers from the start by shifting towards consumer-led product development, which is saturated with rich market communication between R&D/production departments (voice of the company) and end-consumers (voice of the consumer) [14]: the outcome of the creative process could be significantly improved if consumer requirements are integrated into the NPD objectives from the start [17]. Co-creating a new food product (NFP) together with potential end-users brings a number of advantages [18] and is recently perceived as a vital factor in NPD [19], where rich, two-way communication with the consumers in both early stages and the latter promotion stages is crucial for a product's market success [20].

Customer co-creation—Defined as a "process in which customers consciously and actively engage in a firm's innovation process, taking over innovation activities traditionally executed by the firm" [21] (p. 664)—Offers an interesting and effective approach to radicalize the NPD process, to humanize it [22], and to make it more spontaneous and playful [23]. In the last decade, a number of publications have appeared that describe successful cases of co-creating NFP with consumers. Kemp [24] described the case of Pepsico's Mountain Dew® "DEWmocracy™" campaign where consumers were co-creators of a new variant of a nonalcoholic beverage, with a series of crowdsourcing activities and even a story-based, interactive on-line game. Barone et al. [20] co-created healthier meat products on the basis of plant-based ingredients with lower fat and salt content. Prior to the co-creation session, consumers' opinions, perceptions, preferences, and needs for healthier meat products were explored. Banović et al. [19] demonstrated that application of projective and creative techniques is beneficial for consumers' creation of new ideas for aquaculture products during the early stages of NPD process, while Bettiga and Ciccullo [25] described a number of Italian food companies that co-created their products (such as gluten-free pasta, nuts biscuits, frozen pizza, and homogenized meat) by involving customers in different parts of their NPD processes.

A couple of recently published papers provide deeper insights into the factors of the co-creation process and their possible effect. Hoppe et al. [18] demonstrated that the main driver for consumer participation in innovative activities in the food industry is the innovativeness of an individual—Substantial curiosity about new food products will probably lead to significant willingness to participate in the NPD activities. Jacobsen et al. [26] examined the trust that consumers have in co-created food products and concluded that if the targeted consumer groups do not possess significant knowledge and understanding of the co-creation process, then the information about a new food product should be communicated by the consumer co-creators (their peers that were engaged in NFP development) rather than by the food company that places the product on the market. In this context, where functional foods are becoming a major focus of NPD, requiring a more open and flexible approach [27] and understanding of consumer needs (both explicit and latent) are vital for the food science and the food industry [28]; the objective of the present work was to explore and describe one research institute's complex process of co-creating ideas for functional food products together with consumers. In the framework of a national project realized in the Republic of Serbia, which aimed to create a functional food based on an extract rich in bioactive compounds from sustainable sources by utilizing deep eutectic solvents, a multidisciplinary team of researchers engaged in a series of studies that relied heavily on rich communication with consumers and their active involvement. This team, composed of a number of scientists from the fields of food science, pharmaceutical engineering, chemistry, and psychology, designed and conducted a few interconnected studies to explore the width with which it was possible to approach the NPD process without the standard constraints of a commercial endeavor, allowing deeper insights into consumers' needs and their creative potential.

Serbia is one of the largest world producers of raspberries (*Rubus idaeus* L.) [29], and the processing of fresh raspberries into jams, fruit pulp, juices, and other similar products generates a large quantity of seeds, which are currently underutilized in Serbia. These raspberry seeds are often discarded as waste although they contain a considerable number of bioactive compounds, such as polyphenols, tocopherols, ω-3, and ω-6 fatty acids [30]. Due to relatively high content of bioactive compounds, high availability, and low market price in Serbia, raspberry seeds were identified as a sustainable alternative source of ellagic acid and other polyphenols with antioxidant and anti-proliferative properties [31]. Ellagic acid is connected with a broad spectrum of biological effects, such as antioxidant, anti-carcinogenic, anti-obesity, anti-inflammatory and anti-angiogenic, anti-neurodegenerative, and hepatoprotective activity. In order to recover valuable compounds from industrial side-streams, sustainable extraction processes, which do not have a negative impact on the environment, have to be developed. This means, among other things, that the usage of toxic organic solvents has to be avoided at first place, but instead, the application of "green solvents", such as Natural Deep Eutectic Solvents (NADES) [32], has to be considered.

This paper aims to describe experiences and results obtained through the complex process of creating a NFP idea based on raspberry seeds extract in order to provide evidence for embarking on an interesting journey of co-creating a functional food product together with a wide array of potential consumers, thus developing a novel and healthier food product. It contributes to the literature by identifying current trends and potential for communicating about health and functional ingredients [26,33] and also by answering the call to provide insights into the new methods and approaches to generate new ideas about food together with customers [34].

2. Materials and Methods

The research protocols for all studies that involved human subjects were approved by the Ethics Committee of the Institute of Food Technology in Novi Sad (Ethical approval, Ref. No. 175/I/2-3 of 3 June 2021).

2.1. Study 1: Consumer Preferences

The first study had an exploratory approach, utilizing a custom-made survey to collect information about consumers' current needs, preferences, and practices related to their diet, sources of information about functional food, and ultimately, their choices and consumption of the said food. The aim of this study was to provide information that would later be used in co-creating the NFP. A concise, multi-layered survey (presented in Appendix A) was designed to assess consumers' current needs, preferences, and practices related to their type of diet, lifestyle, sources of information about functional food, and their choices and consumption of the said food. The survey items were identified during a number of research team's analysis sessions, where relevant publications were consulted to provide potential questions that could obtain needed information for the research project. The first version of the survey was tested on a pilot sample (N = 20) from general population in Serbia, to ensure all items were understandable and relevant.

By utilizing an online survey platform Google Forms to collect responses during June, July, and October of 2021, a convenient sample of 613 valid cases was obtained from the population of adult citizens from the Republic of Serbia (sample properties are presented in Table 1). Since various methods were used to promote the survey and attract potential participants (from the campaign in the university campus to different social media networks and channels), it was not possible to calculate the response rate precisely; therefore, the convenience sample used in this study should be observed with some caution regarding generalizability of results.

Table 1. Sample properties from Study 1.

Sample Property					
Gender	Female 71.10%			Male 28.90%	
Age	Minimum 18		Median 32		Maximum 76
Working status	Permanently employed 38%	University students 34.6%	Occasionally employed 9.6%	Unemployed 7.3%	Retired 3.6%

Statistical analysis: Excel 2016 (Microsoft Corporation, Redmond, WA, USA) and SPSS Version 26 (Armonk, NY, USA) software were used for data statistics in Study 1. Descriptive statistics, Frequencies, Crosstabs, and Multiple response commands were used to describe the sample and survey participants' responses in order to reach the explorative goals of the survey.

2.2. Study 2: Consumer Ranking Test of Beverage Developed on the Basis of Raspberry Seed Extract

The second study examined results of a consumer ranking test for a soft drink made with a novel extract compared to a handful of other samples, both commercial and experimental. The intention of this study was not to speculate about the final product; rather, this study's aim was to validate the extract regarding its potential acceptance on the market, making sure that it is likeable for use. A raspberry seed extract rich in compounds with in vitro anti-cancer activity and in vitro antioxidant activity by extraction with edible, "green" solvent (eutectic mixture consisting of citric acid, water and betaine—a novel and healthier component for food products) was developed. Details about this process have been recently published elsewhere [31]; however, for better understanding of the entire co-creation concept, some of previously published results are briefly included in the present study. In order to test the idea that produced "green" raspberry seed extract could be used as an ingredient in food products without fear that it has any off-putting properties, in the next phase of research, it was decided to create a soft drink, such as iced tea, in which the extract would serve as a carrier of bioactive components, acidifier, and colorant and to investigate how consumers like the product.

As a result of the development process, a NADES (citric acid: betaine: water 3:1:3) extract containing 7.93 ± 0.15 mg/100 g ellagic acid was obtained. Apart from that, its in vitro antioxidant and in vitro anti-proliferative activity was confirmed [31]. In summary, raspberry seeds, side-products of raspberry processing, were used to obtain an extract rich in compounds with in vitro anti-cancer activity by extraction with edible, "green" solvent, an eutectic mixture consisting of citric acid, water and betaine. In addition to the high content of ellagic acid, the extract contains ~12.5 g betaine, which is also a bioactive compound that can serve as an osmolyte that assists in cellular water homeostasis [35]. Furthermore, due to its edible nature, the obtained NADES extract is ready to use and can have various technological roles in products, such as a source of antioxidants, acidifier (citric acid), and colorant.

Five types of similar beverage samples were included in the ranking test. Three samples were commercially available on the local market: peach-flavored iced tea (IT1), black elder-flavored iced tea (IT2), and forest fruit-flavored iced tea (IT3). The remaining two tested samples were NR and NT.

The NR sample was prepared by mixing 0.5 g of NADES (citric acid: betaine: water 3:1:3), 6 g of honey, 93.5 mL of raspberry tea, and 300 μL of blueberry fruit base. Raspberry tea was prepared with 2 tea bags, which were added to 1 L of boiling water and left for 10 min prior to further use.

The NT sample was prepared by mixing 2 g of NADES (proline–glycerol–water; molar ratio 1:2:1), 5 g of honey, 93 mL of wild thyme tea, 200 μL citron aroma, and 200 μL of pomegranate fruit base. Wild thyme tea was prepared with 5 g of plant material, which was added to 1 L of boiling water and left for 10 min prior to further use. This sample was included since the research project under which these activities were performed had an open starting point regarding the development of an extract; from the list of six relevant herbal materials, the research team narrowed down the extraction processes to raspberry and thyme. However, it will later be noticed that proline had a distinctive odor that was not pleasant for most in the consumer test.

This study was conducted in the Serbian city of Novi Sad within the 2021 "Researchers' Night Festival", a large and renowned cultural event in Serbia. The experiment took place in two consecutive days in one of the event facilities. A convenience sample of subjects was used: they were essentially passers-by taking part in the larger city event. In total, 93 subjects, willing to take part in testing, without aversion to the ingredients used for drink, and without any known food allergy took part in the study. Demographic information collected were gender (58 women and 35 men); age, which spanned unevenly from 18 to over 50 years old (age interval 18–30, $n = 32$; 31–40, $n = 27$; 36–50: $n = 26$; over 50, $n = 13$); and frequency of iced tea drinking (at least once a week, $n = 21$ (22.6%); at least once a month, $n = 41$ (44.1%); and never, $n = 31$ (33.3%)). The consumers were previously informed about the products and testing procedures, as described in the project approved by the Ethics Committee of the Institute of Food Technology in Novi Sad, and which referred to the survey-collaborative session. Before analysis, consumers were also instructed on the procedures describing the basic steps of the tests. Each ranking test was split for the attributes of color, odor, sweetness, sourness, flavor, and overall likeability. The lowest rank (1) corresponded to the least likeability, whereas the highest rank (5) corresponded to the most likeability. No ties were allowed; however, consumers were allowed to retaste any sample.

Samples from the refrigerator at 4 °C were distributed to consumers in 40 mL, odorless plastic bottles marked with random three-digit codes. Each consumer received five samples in a monadic sequence, with a balanced serving order (given by the XLSTAT software (Addinsoft, New York, NY, USA) to minimize systematic carry-over and position effects. For mouth rinsing, water was provided after each sample.

Statistical analysis: Excel (Microsoft Corporation, Redmond, WA, USA) and XLSTAT 2012.2.02 (Addinsoft, NY, USA) software were used for data statistics and significant difference analysis in Study 2. Rank data were analyzed using Friedman's nonparametric test and multiple comparisons methods based on sum ranks according to ISO 8587 (ISO, 2006). Following statistically significant effects ($p < 0.05$), a post hoc LSD test was used to compare products.

2.3. Study 3: Co-Creating Ideas for Functional Food Product with Raspberry Seeds Extract

The third study brings the focus back to the consumers, as it compiles the results from the previous studies and, based on them, employs a co-creative approach to designing functional food products based on the developed raspberry seeds extract. In order to democratize this process and involve different stakeholders, a workshop was organized on the premises of the Institute of Food Technology, in Novi Sad, Serbia, in September 2021.

This workshop aimed to utilize the creative potential of subjects who find this topic to be relevant from their perspective by including them as workshop participants and active co-creators of the NFP ideas. Evaluation of the innovative potential and emotional profile of generated product ideas was then conducted.

2.3.1. The Co-Creation Session

The design of the workshop, presented in detail in Appendix B, Table A1, was inspired by Ind and Coates's [23] understanding of the co-creation process as a spontaneous and playful activity rather than a strongly rational and controlled approach dictated by a company's management—These authors suggest that every person can be creative enough as long as there is proper motivation and relevant knowledge about certain subject. Application of Amabile's [36] general model of creativity helped to achieve this in the present study by influencing the three proposed components of creativity:

Motivation—During the workshop preparation, the organizers scouted for the individuals who were personally interested in food design and experimentation—The ones who saw themselves as highly curious and who observed creativity as a playful and enjoyable activity [37]; thus, this study used a convenience sample. Participants (N = 18) were invited individually, based on their interests and profession, to be included as one of the three groups of stakeholders: food and agricultural researchers (10 participants), food technologists and entrepreneurs in the food industry (3 participants), and end-consumers (5 participants). These participants were, during the workshop, split into four groups.

Expertise—Technical and domain-specific knowledge is necessary for successful creative activity. Ideas do not come out of nothing [38]—They are products of our past experience and new insights, where priming one's mind with information relevant to the topic prior to the ideation session yields higher productivity and idea originality [39]. People who "brainstorm" about some topic have more success in creating quality ideas the more they know about the subject in question, so it is important to make sure they have some relevant expertise or to engage in interdisciplinary collaboration with the ones who already possess said expertise [40]. Therefore, at the start of the workshop, all participants were briefed by an expert about the purpose of the whole project but also about basic principles and benefits of using "green solvents" and then about relevant information obtained from Study 1—How the general population consumes and informs themselves about functional food and other topics, presented in Appendix B, Table A1.

Creative-thinking skills—The third component of creativity describes the importance of the way people approach problems in front of them. The more these skills are guided, improved, and trained, the better the creative output [41]; this means that, for co-creative sessions, it is not enough to gather a number of participants together, expecting them to spontaneously come about new ideas after the presentation of a challenge—Co-creation events need to be properly facilitated and guided in order to fulfil the desired goals [22]. In the present study, the workshop facilitator employed a number of techniques to stimulate participants' creative thinking during different stages of the workshop. This first included a combination of brain purge activity [42] and the SCAMPER technique [43,44]. Second, since playing is the most important activity to unleash creative potential in children and adults alike [23], and since it is established that gamification can contribute to a pleasurable participation experience and that it can enhance task absorption [45], sets of Lego Serious Play® were distributed to the participants, allowing them to fully engage in creative thinking and acting in order to develop their NFP ideas and the stories around them.

2.3.2. Idea Evaluation—Originality, Feasibility, Effectiveness

The list of ideas produced during the workshop session was screened by a research associate in order to filter out entries that were incomprehensible, ambiguous, irrelevant, or redundant. The final list of ideas was then presented to a panel of three judges, two of them being experienced food technologists and researchers and the third one being a food technologist working in the food industry. All three judges were blind to the study procedure. They independently evaluated ideas on their dimensions of originality (the extent to which the idea was novel and out of the ordinary), feasibility (the extent to which the idea was precise and the ease with which it could be implemented), and effectiveness (the extent to which the idea helps to solve the problem) on separate 5-point scales ranging from 1 to 5. This approach was based on widely used Diehl and Stroebe's idea quality-scoring method [46] for the first two dimensions, while the effectiveness dimension was added since some authors insist that it should also be observed when assessing idea quality [47]. These evaluations were then compared to determine inter-rater agreement percentages. Finally, two of the most promising ideas (the ones with the highest cumulative scores from all three judges) were selected for the next step—Evaluation of consumer emotions.

2.3.3. Idea Development and Evaluation on Emotional Dimensions

The two ideas that received highest total scores from the three judges were further developed by three members of the research team using transcripts from the Lego Serious Play® session and insights from Studies 1 and 2. This resulted in a concise description of these two NFP ideas, together with information about the health function of the added raspberry seeds extract (in the following part, these two ideas will have their titles marked with an "f" letter: P1f and P2f). Next, these descriptions were censored to remove information related to the raspberry seeds extract functions in order to explore if the information about function altered the consumers' emotional response (in the following part, these two ideas will have their titles marked without an "f" letter: P1 and P2). In this way, two pairs of NFP ideas were created: the first pair with a brief information about how the product carries additional health function due to the extract being present and the second pair without the said information, mentioning the extract without its potential health benefits.

A brief list of emotions was constructed to measure emotional response from the consumers when they are presented with the two NFP ideas. This list was based on Laros and Steenkamp's [9] hierarchical model of emotions in consumer behavior related to functional food and other types of food, which was built upon the previously established Richins's Consumption Emotion Set [48]; only the list of "basic emotions" that Laros and Steenkamp proposed was used. However, since the same authors suggested that their list should not be taken for granted, advising other researchers to modify the list if they believe there is a good reason to do so, it was decided to enrich this list with two additional emotions that Barrena et al. [10] observed when measuring consumers' emotional response regarding functional food surprise and worry.

An online survey was then designed to measure consumers' emotional reactions to the NFP idea descriptions, using Google Forms. After a short description of the research project and the participants' consent form, this survey asked participants for a few basic pieces of information about themselves. Study participants were next randomized into one of two groups: group 1 received information about additional health function, while in group 2, information about additional health function was omitted. On the next survey page, participants were presented with a short description of the first NFP idea, followed by the following instruction:

> "Read carefully this description of a food product. Think for a moment about how that description makes you feel. After that, mark on a scale from 1 to 5 how much you currently feel each of these feelings, in relation to your reading of the product description. (Mark 1 means 'I don't feel this emotion at all', Mark 5 means 'I feel this emotion very strongly'). Then move on to the next question."

The order of emotions presented for evaluation for each food product idea was randomized for each participant in order to reduce potential bias. Next, participants were presented with the next survey page, which provided a description of the second NFP idea, followed by the same instruction and the task with a list of ten emotions, again randomized. After this, the survey ended, and the participants were thanked for their contribution. In this way, for each of the two NFP ideas, two groups of participants were established: the ones who expressed felt emotions after reading product description that contained information about health function of the raspberry seeds extract and the ones who expressed felt emotions after reading product description that did not contain information about the extract's health function. This survey was conducted in three days during February 2022, and since various methods were used in parallel to promote the survey and attract potential participants, it was not possible to calculate the response rate precisely.

Statistical analysis: Excel 2016 (Microsoft Corporation, Redmond, WA, USA) and SPSS Version 26 (Armonk, NY, USA) software were used for data statistics in Study 3. Descriptive statistics were used as well as Mann–Whitney U test due to the data not being normally distributed.

3. Results and Discussion

3.1. Study 1: Consumer Preferences

Participants from the sample first described their general health and their diet. A total of 70.1% of the participants believe that their diet is average regarding its health status, 12.6% believe that their diet is not very healthy, while 15.2% describe their diet as very healthy; 2.1% did not answer this question. Regarding the diet regime, 90.4% of the participants describe their diet as conventional in generally consuming all food groups, 2.6% stated that their diet is vegetarian, 1.0% as vegan, and 5% reported that their diet is special due to a medical condition or their personal decision. Dietary supplements are regularly consumed by 37.2% of the people from the sample, and these supplements are mostly consumed in the form of pills and capsules (69.7%), then powders (12.3%), ready-made drinks (8.3%), drops (6.1%), and syrups (3.5%).

Observing the self-reported frequency of functional food consumption (appropriate description was written in the survey, ensuring that study participants understood this term properly), the following percentages were calculated for each of the suggested answers: never 17.8%, seasonal 11.8%, several times a month 22.2%, several times a week 29.6%, and daily 18.6%. Out of those who consume functional food at least several times during one month, 35.3% do not care whether the food product is of domestic origin, 26.9% prefer domestic producers, while 37.3% are more interested in the reputation of the manufacturer and the product itself.

Participants who consume functional food several times a week or on daily basis had option to rate different aspects of functional food packages regarding the importance for their decision for purchase on a Likert 1–5 scale: this set of questions was answered by 162 participants (26.4% of the whole sample), and the highest score was recorded for the "description of the beneficial effect" (mean $\mu = 3.42$, standard deviation $\sigma = 1.261$), next for the "description of ingredients" ($\mu = 3.27$, $\sigma = 1.230$), followed by "package design" ($\mu = 3.08$, $\sigma = 1.191$), while the least important aspect of the food package was the "name of the product" ($\mu = 2.63$, $\sigma = 1.228$). When asked how they inform themselves about the functions of food/diet products (multiple answers could be marked, hence the percentages surpass 100 in sum), the respondents stated that they mostly use packaging/declaration of food products (53.7% of cases) and internet portals (53.4%); all other sources of information are less frequently utilized (Figure 1).

When asked how they inform themselves about new products (food supplements and functional foods) on the market (multiple answers could be marked, hence the percentages surpass 100 in sum), social networks were the dominant channel of communication (45.5%), and relatives and friends were also sources of news regarding this topic (24.6%), while store staff (23.4%) all other sources of information were less frequently utilized (Figure 2). If the respondents consume food supplements frequently (which was reported by 29.2% of the study participants), the most dominant reasons were (multiple answers could be marked, hence the percentages surpass 100 in sum) preserving health and improving immunity (59.5%) and improving nutrition (46.5%), followed by other reasons presented in Figure 3. Respondents prefer to consume these products mostly in a form that they can prepare themselves (effervescent tablets, tea, instant drinks, etc.) (30.1%) and pills and capsules (26.5%), followed by already-prepared beverages (juice, water, bottled tea, shot, etc.) (14.3%), powdered or grainy or mushy meals (11.0%), powders (7.4%), bars (7.4%), and drops (3.3%).

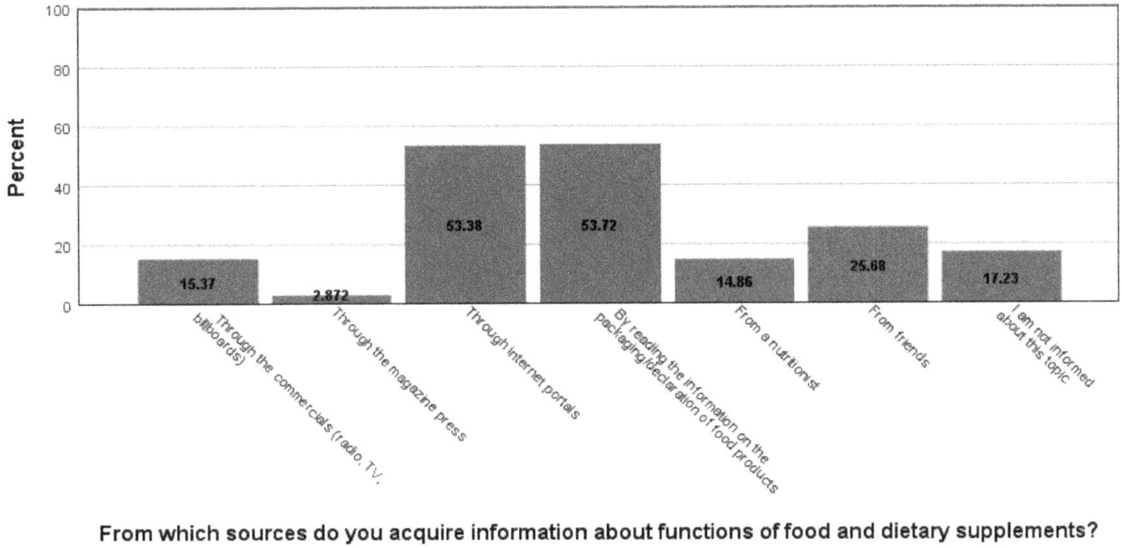

Figure 1. Sources of information about functions of food and dietary supplements.

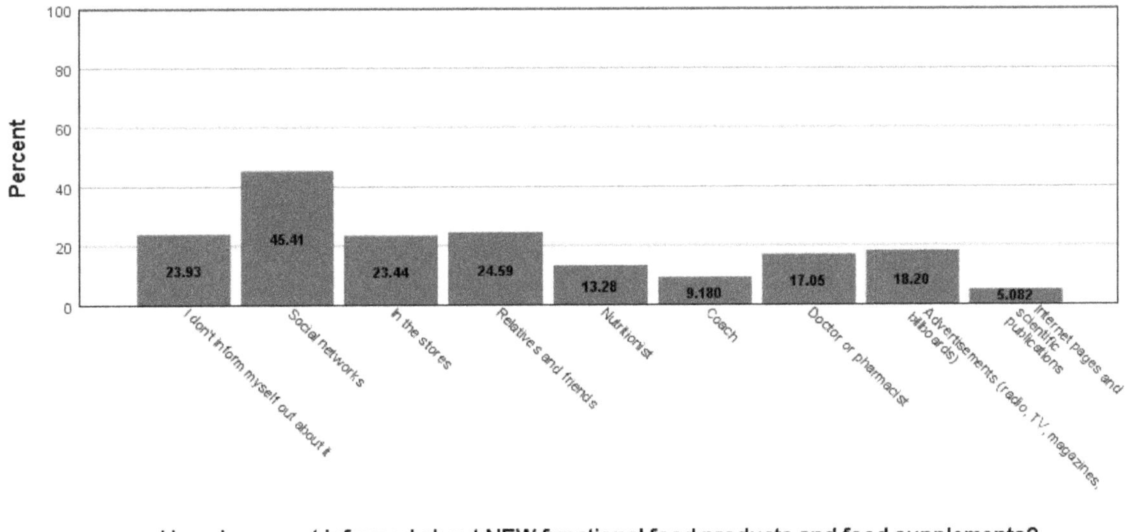

Figure 2. Sources of information about new functional food products and food supplements.

Study participants observed scientific evidence of efficiency as the most important factor when choosing a functional food product (28.9%). Composition is also important (21.7%), while taste (11.8%) and other people's recommendations (7.7%) are not so important for those who do consume functional food products. The most important findings from this study were compiled into a concise report to be presented orally to the Study 3 participants.

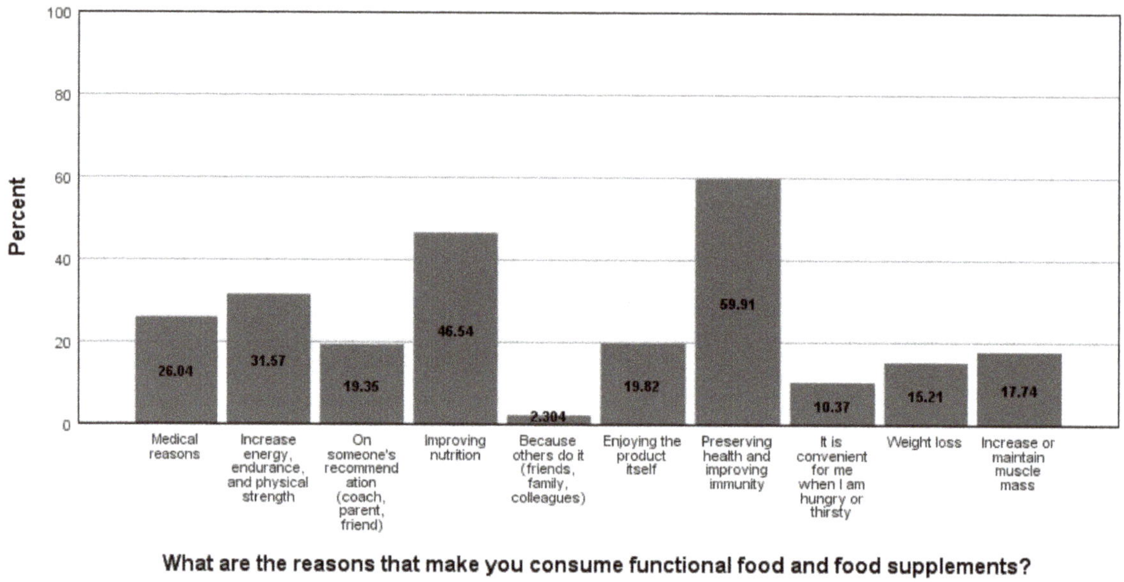

Figure 3. Reasons for functional food and food supplements consumption.

3.2. *Study 2: Consumer Ranking Test of Beverage Developed on the Basis of Raspberry Seed Extract*

Due to the simplicity of the procedure, the ranking test as a traditionally discriminative test was used in a number of consumer studies [49–52]. Therefore, the test applied in this research aimed at gaining the first insight into consumer preferences related to analyzed beverage samples. Moreover, the obtained results provided us with the information on how attractive consumers found two newly created NADES products (NT and NR) in relation to other similar iced tea products on the market. Figure 4 shows the results of the ranking test performed on five beverage samples. For all ranked attributes, significant ($p < 0.05$) differences were evidenced among five samples, indicating that consumers' preferences contributed to obtaining differences between the ranked samples.

For the color, it can be clearly seen that NR was the best ranked compared to the remaining test samples. At the same time, the color of NT products was the least acceptable in relation to the samples of iced tea samples from the market. Furthermore, consumers showed a smaller difference in likeability of NT sample for odor property compared to IT1 as the best ranked sample as well as for sourness of NR sample compared to IT3 and IT1, the best ranked samples. According to the results (Figure 4), consumers demonstrated better preferences for IT1 and IT3 samples, indicating that the preference between them differed significantly in relation to sweetness, flavor, and overall likeability. However, results obtained for NR are promising since this sample was superior to other commercial samples in terms of color, while it was comparable to IT2 in terms of overall likeability, sweetness, and flavor property.

3.3. *Study 3: Co-Creating Ideas for Functional Food Product with Raspberry Seeds Extract*

During the workshop, participants individually and in groups produced a total of 46 starting ideas related to raspberry seeds extract as a component in a food product—Many of these ideas were modifications or improvisations based on ideas previously suggested by other participants, which was expected with the usage of brain purge and SCAMPER ideation techniques. Upon initial inspection, it was possible to categorize generated ideas into one of the following groups based on the form of the final product with the extract:

1. Soft drinks such as iced tea, energy drink, or hot chocolate beverage;

2. A component for application in producing sweet pastries and cakes;
3. Sauce or a marinade for meat or fish;
4. A final food product that is ready and sufficient on its own for consumption.

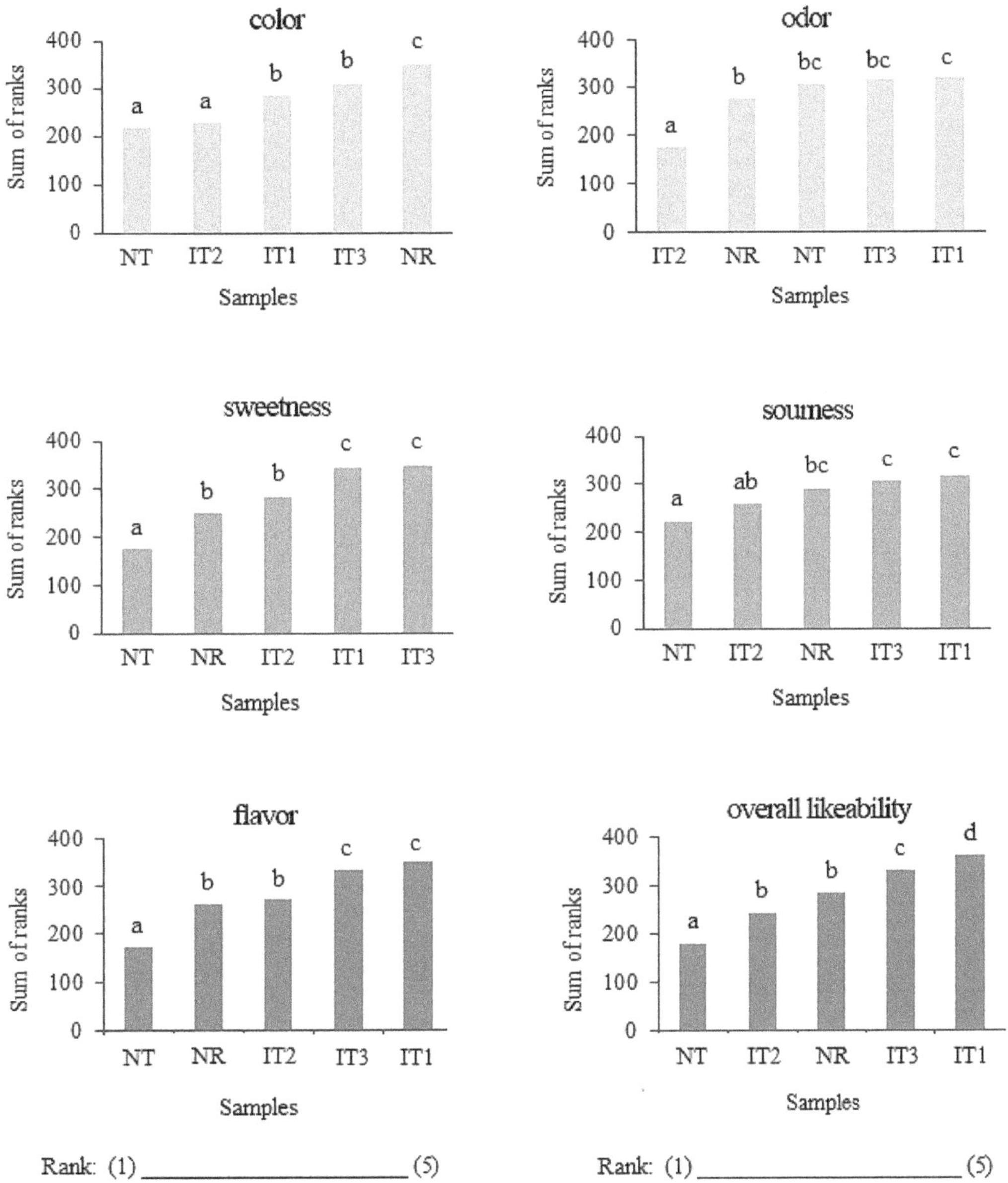

Figure 4. Results of preference ranking test of beverage samples. IT1, peach-flavored iced tea; IT2, black elder-flavored iced tea; IT3, forest fruit-flavored iced tea; NT, NADES thyme extract-based soft drink; NR, NADES raspberry seeds extract-based soft drink. Values marked with the same letter are not statistically different ($p < 0.05$).

After eliminating ideas that obviously lacked novelty (market is already saturated with similar products) and after merging similar and redundant ideas, the final list of 15 ideas was compiled. This list was then presented to the three independent judges for originality, feasibility, and effectiveness evaluation. Defining the three judges' ratings as being in agreement whenever their evaluations fell within one point of each other (for example, agreement was recorded if an idea received marks 4, 4, and 3, and disagreement was recorded if an idea received marks such as 5, 4, and 3) [46,53], it was found that all three judges were in agreement in 6 out of 15 ideas regarding the originality evaluation, in 9 out of 15 ideas regarding the feasibility evaluation, and again in 9 out of 15 ideas regarding the effectiveness evaluations. More generally, if it was counted that all three judges had agreement on at least two out of the three idea-quality dimensions regarding every idea, their evaluations were marked as an agreement for 9 out of 15 ideas. Results of judges' evaluations on three idea-quality dimensions are presented in Table 2, with their detailed evaluations in Appendix C, Table A2.

Table 2. Final list of NFP ideas and sums of the judges' evaluation points. Judges' agreement was recorded when at least two of the three idea-quality dimensions were identical.

Idea	Judges Had Agreement	Sum for the Originality Points	Sum for the Feasibility Points	Sum for the Effectiveness Points	Total Sum of Points
1. Fruit yogurt with raspberry seeds extract. Ingredient for fruit cake or ice cream, not for direct consumption	No	11	12	12	35
2. Ice cream with raspberry seeds extract	Agreement	9	15	10	34
3. Butter with raspberry seeds extract for cake and pastry	No	13	11	11	35
4. Instant ice cream with raspberry seeds extract	Agreement	8	11	11	30
5. Rice pudding with raspberry seeds extract	Agreement	11	15	12	38
6. Healthy jelly cube-jelly honey with raspberry seeds extract	No	13	9	9	31
7. Sweet noodles with raspberry seeds extract	No	12	10	9	31
8. Meat sausage with the addition of raspberry seeds extract	No	11	8	8	27
9. Meat dressing with raspberry seeds extract for roasted meat	Agreement	11	13	10	34
10. Marinade for meat and fish with raspberry seeds extract, before cooking	No	8	9	9	26
11. Vegan soy sausage with raspberry seeds extract	Agreement	14	13	12	39
12. Hot chocolate drink with raspberry seeds extract	Agreement	10	14	11	35
13. Topping for cake, pancake, or ice cream with raspberry seeds extract	Agreement	9	15	12	36
14. Raspberry seeds extract as a liquid additive for sweet dough and cake	Agreement	7	11	8	26
15. Stuffing for muffin and donut with raspberry seeds extract	Agreement	8	14	12	34

Ideas number 11 (vegan soy sausage with raspberry seeds extract) and 5 (rice pudding with raspberry seeds extract) received the highest number of points on all three dimensions of idea quality combined, and since both of them had satisfying agreement between the judges, these two ideas were chosen for further development and analysis of emotional response. As described earlier, they were further developed by the research team to produce concise descriptions of a NFP idea. These descriptions were then formulated with (containing letter "f' at the title code) and without (not containing letter "f" at the title code) information about functional descriptions for the raspberry seeds extract, resulting in the following four descriptions:

(1) P1f Vegan soy sausage with raspberry seeds extract

This product does not contain raw materials of animal origin and is suitable for vegetarians, vegans, and fasting days. The product is made of soy and does not contain artificial preservatives, colors, and flavors. It is enriched with natural raspberry seed extract that contains ellagic acid, a compound with potential anti-cancer activity and antioxidant activity. It also contains betaine, which lowers high levels of homocysteine in the body, which contributes to reducing the risk of cardiovascular, cerebral, peripheral vascular, and neurodegenerative diseases. It has a pleasant, slightly salty taste and is designed to imitate several sensory properties of animal meat, such as appearance, taste, texture, and mild pink color, which makes it a delicious alternative to meat products. The product is a rich source of protein. It does not contain GMOs.

(2) P1 Vegan soy sausage with raspberry seeds extract

This product does not contain raw materials of animal origin and is suitable for vegetarians, vegans, and fasting days. The product is made of soy and does not contain artificial preservatives, colors, and flavors. It is enriched with natural raspberry seed extract. It has a pleasant, slightly salty taste and is designed to imitate several sensory properties of animal meat, such as appearance, taste, texture, and mild pink color, which makes it a delicious alternative to meat products. The product is a rich source of protein. It does not contain GMOs.

(3) P2f Rice pudding with raspberry seeds extract and vegetable milk

This product does not contain raw materials of animal origin and is suitable for vegetarians, vegans, and fasting days. The product is made from rice and vegetable milk and does not contain gluten, lactose, milk protein, or artificial preservatives, colors, and flavors. It is enriched with natural raspberry seed extract that contains ellagic acid, a compound with potential anti-cancer activity and antioxidant activity. It also contains betaine, which lowers high levels of homocysteine in the body, which contributes to reducing the risk of cardiovascular, cerebral, peripheral vascular, and neurodegenerative diseases. It has a pleasant, slightly sweet taste and a slightly pink color.

(4) P2 Rice pudding with raspberry seeds extract and vegetable milk

This product does not contain raw materials of animal origin and is suitable for vegetarians, vegans, and fasting days. The product is made from rice and vegetable milk and does not contain gluten, lactose, milk protein, or artificial preservatives, colors, and flavors. It is enriched with natural raspberry seed extract. It has a pleasant, slightly sweet taste and a slightly pink color.

The online survey about consumers' emotional reactions to NFP ideas was successfully completed by 126 adult participants; age span was from 18 to 70, with a mean of 40.74; 88 participants were female (69.8 percent). Regarding consumption of functional food, 33 participants (26.2%) stated that they never consume it, 76 participants (60.3%) stated that they consume it to a smaller extent, and 17 participants (13.5%) to a large extent. Regarding consumption of dietary supplements, 37 participants (29.4%) stated that they never consume them, 46 participants (36.5%) stated that they consume them to a smaller extent, and 43 participants (34.1%) to a large extent. Most of the participants, i.e., 107 (84.9%), described their diet as conventional; 13 participants (10.3%) described their diet as vegetarian or vegan; and 6 participants (4.8%) stated that they have a special diet due to health concerns or personal choice.

This survey showed that study participants feel emotions with a positive affect, as defined by Laros and Steenkamp (contentment, happiness, and love) [9] to a larger extent in the case of rice pudding with raspberry seeds extract than for the vegan soy sausage with raspberry seeds extract (Figure 5). This finding is expected since rice pudding is a common dessert in Serbian households (as elsewhere) although usually prepared with cow's milk and only sometimes with addition of fruit extracts. Rice pudding, thanks to its mild taste and high nutritional value, is an option frequently offered to small children and, at least in Serbia, a common dessert allowed for consumption when a child is having stomach problems. These facts surely cause this product to be associated with a set of pleasant emotions, such as relief and comfort, which may lead to the emotion of contentment and even love, making it a product with sentimental value [9] for many. Happiness is also an expected reaction regarding the rice pudding, as sweet taste has a metaphoric association with this emotion [54]. Since vegetable milk is a relatively a close alternative to cow's milk, and since fruit extracts similar to the raspberry seeds extract are often used in various desserts in the Serbian cuisine, this NFP idea was expectedly followed with pleasant emotions, as described.

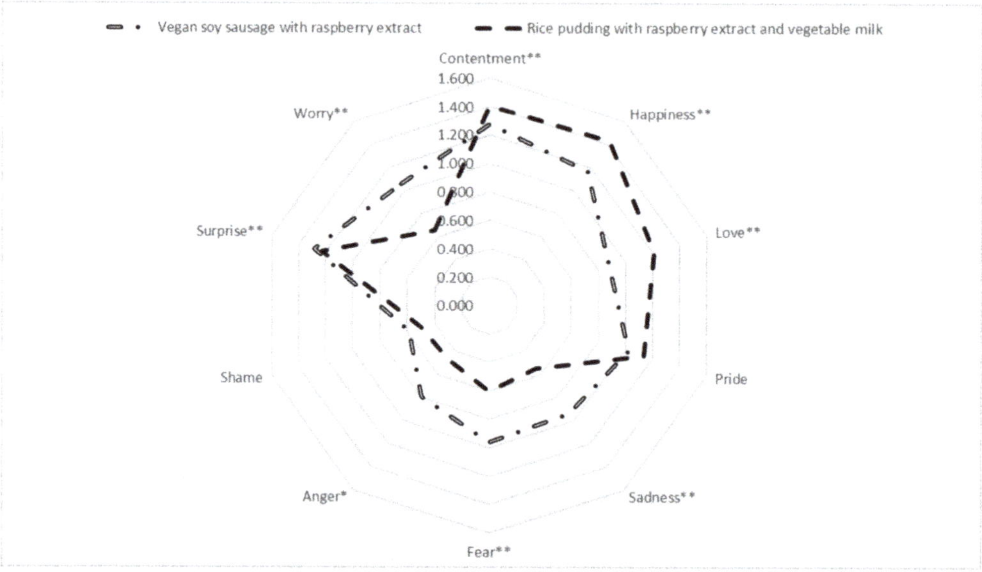

Figure 5. Radar chart of emotions felt while reading about the two product ideas. Emotions marked with * showed differences significant at 0.05 level; those marked with ** showed differences significant at 0.01 level.

The same cannot be said for the vegan sausage, which is surely not perceived as a known, sweet, or comforting product by most consumers in Serbia. The description of vegan soy sausage with raspberry seeds extract evoked emotions with a negative affect to a larger extent (sadness, fear, anger) probably because the sausage as a form of food is mostly associated with pork meat in Serbia. The pork sausage is one of the most frequently used meat products in Serbia, and different parts of the country have their own regional varieties [55,56], making this product a part of national heritage and pride for many. The notion of a vegan product being associated with a name of a food product that is usually based on meat may thus bring unpleasant emotions and negative views from an average consumer [57] at least in Serbia, where vegan food is not consumed by many. This may also explain the significantly higher level of worry felt while reading about vegan sausage, as it introduces soy filling to a product shape that is usually filled with spicy meat. An additional reason for the elevated level of this emotion may be found in the fact that raspberry seeds extract is added in the already controversial product, where raspberries are almost exclusively perceived as something utilized in sweet food products on the Serbian market. These unpleasant emotions may additionally be exacerbated by the situation in which the raspberry is considered by many as one of the most recognizable Serbian food products but never associated with savory food or in combination with the word sausage. Emotions of pride and shame did not differ between the two product ideas (Figure 5); explanation for this may lie in the fact that the emotion of pride generally occurs when a consumer feels superior compared to another person regarding some achievement or a property [9], while the emotion of shame occurs when an individual perceives a failure of some aspect of the global self [58]—None of which apply here since the NFP ideas were not generated by the emotional evaluation study participants.

One additional observation may be interesting. After the survey completion, the research team received a few informal comments by some study participants regarding their interest in the study and the way it was conducted. One interesting comment by a handful of participants stood out: they asked why the emotion of "disgust" was not present as an option, as they stated that they were "disgusted by the idea of a vegan soy sausage with raspberry seeds extract". While we understand the skepticism of some regarding a food product that may sound controversial in a country where pork products are a part of national identity, while vegan products are not widely consumed and frowned upon by many, we provide the following explanation on why the emotion of disgust was not included in this study, in addition to the methodological argument that the studies that we relied on ignored the said emotion [9,10,48]. Shaver provided detailed evidence on basic human emotions and commented that, in the modern age, the emotion of disgust has shifted away from a primary physical and emotional reaction (with distinct facial expression and supposed links to innate reactions to bad tastes and smells) to a type of anger akin to contempt [59] (pp. 1069). In other words, the word "disgust" in an adult's emotional lexicon has dominantly transformed from an emotional reaction into an evaluative statement that aims to disprove the object of disgust.

Interestingly, for both products, there were no significant differences found regarding the intensity of emotions felt in relation to whether the information about functional effect of the raspberry seeds extract was disclosed or not. Only when the levels of emotions were observed, regardless of the food product (for both products together), a significant difference was found on one dimension, contentment, showing that participants who did not receive information about functional effect felt more content than the ones who did receive the said information (mean ranks 137.15 and 117.98, respectively, with Mann–Whitney U value of 6647.000, $Z = -2;144$, $p = 0.032$). A possible explanation could be that mentions of various diseases (cancer, cardiovascular, cerebral, peripheral vascular, and neurodegenerative diseases) as part of the information about functional properties of the extract upset the readers, thus making them feel less contempt. Finding that information about functional effect does not generally affect emotional response is somewhat contradictory to the finding from Study 1, where it was found that people who regularly consume functional food rated "description of the beneficial effect" as the most important factor for their decision to buy functional food. A possible explanation would be that Study 4 did not include motivational aspects (cognitive and conative) of functional food consumption, asking participants to only rate their emotions regardless of their current needs for a food with additional function.

4. Conclusions

This work, complex in its approach, aimed to explore and describe a food research institute's process of co-creating a functional food product together with a wide array of potential consumers. It contributes to the contemporary literature by providing an overview of methods and actions for co-creation of NFP from the perspective of a research institute as the main actor rather than a commercial company. Although the current literature provides examples of NFP co-creation in the food industry, there are not any records available regarding these activities in a non-commercial context. The present work documents various activities that could be also utilized by other organizations elsewhere when the creative potential of consumers, researchers, and producers is combined. Acknowledging the importance of communication during the NFP co-creation [26], it was this multidisciplinary team's intention to engage in action research together with relevant and diverse stakeholders in order to revalorize raspberry seeds as a source of nutrients by utilizing natural compounds and to produce more than just an extract—to produce an idea for a healthy, functional food product.

The first study described current needs of consumers in Serbia regarding functional food. It was found that consumers do use functional food products to a good extent, mostly with intention to preserve their health but also to improve their diet or for reasons related to sports or health issues. For a number of study participants, information about functional food and about scientific evidence of efficiency is quite important—product information is especially important for dietary considerations, and these are probably conscious food shoppers [60]. Social networks and personal communication with friends and relatives are the main channels for information about new functional food products, which is in line with previous research that suggested a more personal approach when advertising NFP [26]. In the second study, the NADES raspberry seeds extract showed satisfactory ratings in the consumers ranking test, qualifying it for usage in the next step. Finally, the third study depicted activities related to next stages of the NFP design process—instead of R&D departments or food technology specialists being responsible for this step, it was decided to gather a number of interested stakeholders and to explore potential usages for the raspberry seeds extract together with them. Through a playful set of activities, a number of interesting NFP ideas emerged; an evaluation panel selected two of these ideas as the most promising, and these two ideas were further developed into NFP descriptions subject to an evaluation of emotional reactions by a sample of the general public in Serbia. The idea of a rice pudding with raspberry seeds extract and vegetable milk showed most potential and evoked mostly pleasant emotions within a sample of consumers, which was explained with similarities of the proposed dish with a dessert that is cherished in Serbia as well as around the world, being more acceptable to an average consumer than the other food product idea, the vegan sausage.

A number of limitations, such as convenience samples in three studies from just one country, relatively small samples in Studies 2 and 3, a limited set of resources, and just one co-creation session as well as relatively modest input from the industry, are definitely present. However, aiming to show that open communication and creativity stimulation have a significant role in NFP development, this work will hopefully inspire other researchers and companies to create a stake in co-creating functional food products together with their consumers. This research approach was, surely, far more explorative than confirmative in its nature. The study presents experiences of a research institute's project to design novel food within a research project. Therefore, the results of our studies can hardly be compared to the traditional R&D practices, and it would be troublesome to benchmark the practicalities and benefits of this approach, as it is not commercial in nature. The final limitation of this work is that the final NFP idea was not validated in terms of actual sensory quality or consumer acceptance, the reason for which lies in the fact that the aim of the project was limited to the development of a functional extract and its potential usage, without activities related to final product validation.

Author Contributions: Conceptualization, P.V. and A.M. (Aleksandra Mišan); methodology, P.V., M.P. (Mladenka Pestorić), N.T., B.P., A.M. (Aleksandra Mišan) and D.Š.; software, P.V., N.T. and D.Š.; validation, M.P. (Mladenka Pestorić), M.P. (Milica Pojić) and A.S.; formal analysis, M.P. (Milica Pojić), A.S. and A.C.K.; resources, A.M. (Anamarija Mandić), B.P. and N.T.; data curation, A.S., M.P. (Mladenka Pestorić) and N.T.; writing—original draft preparation, P.V., M.P. (Milica Pojić), N.T. and A.C.K.; writing—review and editing, P.V., A.C.K., N.T., B.P., D.Š., A.M. (Anamarija Mandić) and A.S.; visualization, P.V. and M.P. (Mladenka Pestorić); supervision, M.P. (Milica Pojić); project administration, A.M. (Anamarija Mandić) and A.M. (Aleksandra Mišan); funding acquisition, A.M. (Aleksandra Mišan). All authors have read and agreed to the published version of the manuscript.

Funding: This research was funded by the Science Fund of the Republic of Serbia, grant number 6060592.

Institutional Review Board Statement: The study was conducted in accordance with the Declaration of Helsinki, and approved by the Ethics Committee of the Institute of Food Technology in Novi Sad (protocol code Ref. No. 175/I/2-3, 3 June 2021) for studies involving humans.

Informed Consent Statement: Informed consent was obtained from all subjects involved in the study.

Data Availability Statement: Not applicable.

Acknowledgments: Authors would like to thank the Science Fund of the Republic of Serbia, PROMIS, Grant No. 6060592, DEStiny; IDEJE, Grant No. 7750168, BioUtilize and the Ministry of Education, Science, and Technological Development of the Republic of Serbia, Grant No. 451-03-68/2022-14/200222.

Conflicts of Interest: The authors declare no conflict of interest. The funders had no role in the design of the study; in the collection, analyses, or interpretation of data; in the writing of the manuscript, or in the decision to publish the results.

Appendix A

Survey used in Study 1, as downloaded from Google Forms

Sex: (Mark only one):

- ☐ Female
- ☐ Male
- ☐ Undisclosed

Birth date: ____________________

Education level (Mark only one):

- ☐ Elementary school
- ☐ High school
- ☐ Bachelor's
- ☐ Master's
- ☐ Ph.D.

Employment status (Mark only one):

- ☐ Unemployed
- ☐ Occasional employment or fixed-term work
- ☐ Permanently employed
- ☐ Self-employed (entrepreneur)
- ☐ Retired
- ☐ In disability pension
- ☐ Student who is not employed
- ☐ Other: ________________________________

How would you rate your daily diet? (Mark only one)

- ☐ Not very healthy
- ☐ Average healthy
- ☐ Very healthy
- ☐ I cannot tell for sure

What diet regime do you have? Mark only one.

- ☐ Conventional—I mostly consume all food groups
- ☐ Vegetarian
- ☐ Vegan
- ☐ Special diet due to allergy/intolerance to certain food ingredients or other reason or decision

☐ Other: ______________________________

If you suffer from an acute and/or chronic disease or if you would be in such a situation in the future, would you use functional food products or dietary supplements to solve health problems? Mark only one.

☐ YES, unless that way means giving up my usual way of life and diet
☐ YES, even if that way implies a change in the current way of life
☐ YES, I already use some functional food products or dietary supplements to solve health problems
☐ NO, I do not believe that functional food products or dietary supplements may help to solve health problems.

Do you consume some of the following products and how often? (check all that apply, mark only one per row)

☐ (Daily Several times a week Several times a month Seasonal Never)
☐ Enriched food and beverages, vitamins, minerals, probiotics,
☐ Fiber (enriched mineral water or juices, fortified dairy products, muesli
☐ with vitamins, etc.)
☐ Food products and beverages with reduced salt, sugar, and/or fat content
☐ Sports drinks (protein drinks or electrolytes)
☐ Sports bars (energy and protein)
☐ Protein supplements (concentrates and isolates)
☐ Multivitamin complexes
☐ Multimineral complexes
☐ Energy drinks
☐ Dietary supplements based on plants and herbs extracts (tablets, powders, drops, etc.)
☐ Products intended for weight loss for people on a regulated diet
☐ Products for people with intolerance to gluten, lactose, etc.

How do you inform yourself about the functions of food/diet products? (Multiple answers can be marked)

☐ Through the commercials (radio, TV, billboards)
☐ Through the magazine press
☐ Through the internet portals
☐ By reading the information on the packaging/declaration of food products
☐ From a nutritionist
☐ From friend
☐ I am not informed about this topic

What is the most important factor for you when it comes to choosing a functional food product? (Mark only one)

☐ I do not use functional food products
☐ Taste/aroma
☐ Scientific evidence of efficiency
☐ Composition
☐ Recommendation from somebody

Do you read food labels/declaration? (Mark only one)

☐ YES
☐ NO
☐ Sometimes
1. I want to read, but they are often illegible, small
☐ I would if they were written in a more understandable way

Do you trust the information provided by the manufacturer in the declaration? (Mark only one)

☐ YES

- ☐ NO
- ☐ Depends on the manufacturer
- ☐ I don't think about it

If you consume products and supplements that can be classified into categories, such as functional/dietary/"healthy food", state all the reasons why? If not, skip this question. (Multiple answers can be marked)

- ☐ Medical reasons
- ☐ Increase energy, endurance, and physical strength
- ☐ On someone's recommendation (coach, parent, friend)
- ☐ Improving nutrition
- ☐ Because others do it (friends, family, colleagues)
- ☐ Enjoying the product itself
- ☐ Preserving health and improving immunity
- ☐ It is convenient for me when I am hungry or thirsty
- ☐ Weight loss
- ☐ Increase or maintain muscle mass
- ☐ Due to allergies/sensitivity/intolerance to food or other special dietary needs
- ☐ Other: ______________________________

In what form do you prefer to consume products and supplements that can be classified into categories, such as functional/dietary/"healthy food", if you consume them? If not, skip this question. (Mark only one)

- ☐ Already-prepared beverage (juice, water, bottled tea, shot, etc.)
- ☐ Beverage that I make myself (effervescent tablets, tea, instant drinks, etc.)
- ☐ Powders
- ☐ Tablets and capsules
- ☐ Drops
- ☐ Bars
- ☐ Powdered, grainy, or mushy meal

When buying functional/dietary products and "healthy food", do you give preference to domestic producers? If you do not buy them, skip this question. (Mark only one)

- ☐ YES
- ☐ NO
- ☐ It depends on the product and the reputation of the manufacturer
- ☐ I do not pay attention to the origin of the product

How do you inform yourself out about NEW products (food supplements and functional foods) on the market? (Multiple answers can be marked)

- ☐ I don't inform myself out about it
- ☐ Social networks
- ☐ In the stores
- ☐ Relatives and friends
- ☐ Nutritionist
- ☐ Coach
- ☐ Doctor or pharmacist
- ☐ Advertisements (radio, TV, magazines, billboards)
- ☐ Internet pages and scientific publications

Do you compensate the missing nutrients by using some of the following products? (Multiple answers can be marked)

- ☐ Dietary supplements
- ☐ Teas
- ☐ Enriched food products
- ☐ I can't find an adequate way

If you use dietary supplements, in what form do you prefer to use them? If you do not use them, skip this question. (Multiple answers can be marked)

- ☐ Tablet and capsule
- ☐ Powders
- ☐ Syrup
- ☐ Drops
- ☐ Ready-made drinks

If you describe yourself as a fairly regular consumer of any kind of functional food, to which extent do the following aspects of food package attract your attention and influence your intention to buy that product?

On a 1–5 scale, where 1 means "not at all", and 5 means "to a large extent"

- ☐ Package design
- ☐ Name of the food product
- ☐ Description of the health effect brought by the product
- ☐ Description of the contents

* Questions for health issues

Do you have any of the following health issues for a substantial period of time? (Multiple answers can be marked)

- ☐ Higher blood pressure
- ☐ Atherosclerosis
- ☐ High blood sugar
- ☐ High triglyceride levels in blood
- ☐ Overweight
- ☐ Hormone imbalance
- ☐ I haven't tested my health in a long time

What do you use for the illness treatment if you have any)? (Multiple answers can be marked)

- ☐ Classic drugs prescribed by a specialist doctor
- ☐ Alternative medicines (e.g., homeopathic medicines)
- ☐ Teas, drops, aerosols, etc.
- ☐ Functional products and dietary supplements for the prevention and treatment of these diseases
- ☐ Balanced diet and healthy lifestyle

Do you intentionally consume functional foods and/or dietary supplements to ingest compounds that have a positive effect on your health (antioxidants, vitamins, minerals, probiotics, fiber, etc.)? (Mark only one)

- ☐ YES
- ☐ NO
- ☐ Only sometimes.

Appendix B

Table A1. Workshop structure and activities.

Duration (Minutes)	Workshop Unit	Description and Activities
20	Workshop introduction	Presenting the purpose of the whole project but also the basic principles and benefits of using "green solvents" Presenting basic results from Study 1 about how the information about functional food is being communicated between companies and consumers and about what are relevant consumer needs Individual introduction by each participant, team members, and the facilitator Setting the basic rules of conduct and managing participants' expectations from the workshop
20	Warming-up and Ice-breaking	Investigate in teams of four (using available resources on the internet) and report on potential benefits of ellagic acid and betaine, which are found in the raspberry seeds extract—participants are allowed to use their mobile phones and search online for information
10	Getting acquainted with the raspberry seeds extract	Two team members describe process of obtaining the raspberry seeds extract from Study 2. Next, they demonstrate how to mix the extract with other components, how to test, and how to record results and observations
40	Experimenting with raspberry seeds extract usage	In pairs, participants go through the process of making a soft drink with the raspberry seeds extract and a number of other available materials: water, yoghurt, milk, rice milk, honey, and fruit aromas Presentation of each pair's best solutions
60	Ideation about potential products with raspberry seeds extract as a functional component	Informing participants about basic rules of brainwriting/brainstorming sessions necessary for fruitful creativity Allow each participant to write a few ideas that spontaneously come into their minds about how to use the extract in NFP—this allows to "purge" participants' minds from the most obvious ideas, which are frequently underdeveloped and redundant Use a combination of brain purge and SCAMPER methods for idea stimulation, with participants working in groups of 4 or 5—every person writes an idea on top of a sheet of paper (about a possible functional food product with the raspberry seeds extract for one of the interest groups of their choice: people with chronic health issues, people who are interested in a healthy diet, university students, athletes), hands it down to the person sitting right, and the next person needs to transform the received idea with one of the SCAMPER methods that the facilitator calls for (substitute, combine, adopt, modify, put to other use, eliminate, rearrange). The paper is then passed to the next person sitting right, and the new recipient must transform the last idea written on the paper; five iterations will happen, resulting in a number of paper sheets with a few ideas on them—approximately 20 ideas in total for each group.
30	Creating a story about the product	Every group decides on which idea from the ones they have produced as a group is the best for all members, and after reaching a consensus, they create a story of production, promotion, and usage of the NFP with the use of Lego Serious Play® sets, presenting their story to other workshop participants

Appendix C

Table A2. Individual scores for each idea by the three judges, records of their agreements, and summary scores.

	Originality			**Feasibility**			**Effectiveness**			**Agreement**			**Sums of Scores**			
idea	o1	o2	o3	f1	f2	f3	e1	e2	e3	originality	feasibility	effectiveness	sum o	sum f	sum e	total sum
	4	4	3	5	4	3	5	3	4	Yes t	No	No	11	12	12	35
2	2	3	4	5	5	5	3	3	4	No	Yes	Yes	9	15	10	34
3	5	5	3	5	2	4	5	3	3	No	No	No	13	11	11	35
4	2	3	3	3	4	4	4	3	4	Yes	Yes	Yes	8	11	11	30
5	3	4	4	5	5	5	4	4	4	Yes	Yes	Yes	11	15	12	38
6	5	5	3	2	3	4	3	3	3	No	No	Yes	13	9	9	31
7	4	5	3	2	4	4	3	3	3	No	No	Yes	12	10	9	31
8	5	5	1	5	2	1	4	3	1	No	No	No	11	8	8	27
9	3	4	4	5	4	4	3	3	4	Yes	Yes	Yes	11	13	10	34
10	3	4	1	5	3	1	5	3	1	No	No	No	8	9	9	26
11	5	5	4	5	4	4	5	3	4	Yes	Yes	No	14	13	12	39
12	2	4	4	4	5	5	5	3	3	No	Yes	No	10	14	11	35
13	2	3	4	5	5	5	4	4	4	No	Yes	Yes	9	15	12	36
14	1	4	2	3	4	4	2	3	3	No	Yes	Yes	7	11	8	26
15	2	3	3	5	5	4	4	4	4	Yes	Yes	Yes	8	14	12	34

References

1. Roberfroid, M.B. Global view on functional foods: European perspectives. *Br. J. Nutr.* **2002**, *88*, S133–S138. [CrossRef] [PubMed]
2. Roberfroid, M.B. Defining functional foods. In *Functional Foods Concept to Product*; Gibson, G.R., Williams, C.M., Eds.; Woodhead Publishing Limited: Cambridge, UK; CRC Press LLC: Boca Raton, FL, USA, 2000; pp. 9–25.
3. Spence, J.T. Challenges related to the composition of functional foods. *J. Food Compos. Anal.* **2006**, *19*, S4–S6. [CrossRef]
4. Siró, I.; Kápolna, E.; Kápolna, B.; Lugasi, A. Functional food. Product development, marketing and consumer acceptance—A review. *Appetite* **2008**, *51*, 456–467. [CrossRef] [PubMed]
5. Lewis, N.M.; Seburg, S.; Flanagan, N.L. Enriched Eggs as a Source of N-3 Polyunsaturated Fatty Acids for Humans1. *Poult. Sci.* **2000**, *79*, 971–974. [CrossRef] [PubMed]
6. Chen, P.-J.; Antonelli, M. Conceptual Models of Food Choice: Influential Factors Related to Foods, Individual Differences, and Society. *Foods* **2020**, *9*, 1898. [CrossRef] [PubMed]
7. Rankin, A.; Bunting, B.P.; Poínhos, R.; van der Lans, I.A.; Fischer, A.R.H.; Kuznesof, S.; Almeida, M.D.V.; Markovina, J.; Frewer, L.J.; Stewart-Knox, B.J. Food choice motives, attitude towards and intention to adopt personalised nutrition. *Public Health Nutr.* **2018**, *21*, 2606–2616. [CrossRef]
8. Visschers, V.H.M.; Hess, R.; Siegrist, M. Health motivation and product design determine consumers' visual attention to nutrition information on food products. *Public Health Nutr.* **2010**, *13*, 1099–1106. [CrossRef]
9. Laros, F.J.M.; Steenkamp, J.-B.E.M. Emotions in consumer behavior: A hierarchical approach. *J. Bus. Res.* **2005**, *58*, 1437–1445. [CrossRef]
10. Barrena, R.; García, T.; Sánchez, M. The effect of emotions on purchase behaviour towards novel foods. An application of Means–End chain methodology. *Agrekon* **2017**, *56*, 173–190. [CrossRef]
11. Goetzke, B.; Nitzko, S.; Spiller, A. Consumption of organic and functional food. A matter of well-being and health? *Appetite* **2014**, *77*, 96–105. [CrossRef]
12. Urala, N.; Lähteenmäki, L. Consumers' changing attitudes towards functional foods. *Food Qual. Prefer.* **2007**, *18*, 1–12. [CrossRef]
13. Dijksterhuis, G. New product failure: Five potential sources discussed. *Trends Food Sci. Technol.* **2016**, *50*, 243–248. [CrossRef]
14. Costa, A.I.A.; Jongen, W.M.F. New insights into consumer-led food product development. *Trends Food Sci. Technol.* **2006**, *17*, 457–465. [CrossRef]
15. Harmsen, H.; Grunert, K.G.; Declerck, F. Why did we make that cheese? An empirically based framework for understanding what drives innovation activity. *RD Manag.* **2000**, *30*, 151–166. [CrossRef]

16. Grunert, K.G.; Verbeke, W.; Kügler, J.O.; Saeed, F.; Scholderer, J. Use of consumer insight in the new product development process in the meat sector. *Meat Sci.* **2011**, *89*, 251–258. [CrossRef]
17. Rudolph, M.J. The food product development process. *Br. Food J.* **1995**, *97*, 3–11. [CrossRef]
18. Hoppe, A.; De Barcellos, M.D.; Perin, M.G.; Jacobsen, L.F.; Lähteenmäki, L. Factors influencing consumers' willingness to participate in new food product development activities. *Br. Food J.* **2018**, *120*, 1195–1206. [CrossRef]
19. Banović, M.; Krystallis, A.; Guerrero, L.; Reinders, M.J. Consumers as co-creators of new product ideas: An application of projective and creative research techniques. *Food Res. Int.* **2016**, *87*, 211–223. [CrossRef]
20. Barone, A.M.; Banovic, M.; Asioli, D.; Wallace, E.; Ruiz-Capillas, C.; Grasso, S. The usual suspect: How to co-create healthier meat products. *Food Res. Int.* **2021**, *143*, 110304. [CrossRef]
21. Gemser, G.; Perks, H. Co-creation with customers: An evolving innovation research field. *J. Prod. Innov. Manag.* **2015**, *32*, 660–665. [CrossRef]
22. Roberts, D.L.; Darler, W. Consumer Co-Creation: An Opportunity to Humanise the New Product Development Process. *Int. J. Mark. Res.* **2017**, *59*, 13–33. [CrossRef]
23. Ind, N.; Coates, N. The meanings of co-creation. *Eur. Bus. Rev.* **2013**, *25*, 86–95. [CrossRef]
24. Kemp, S.E. Consumers as part of food and beverage industry innovation. In *Open Innovation in the Food and Beverage Industry*; Garcia-Martinez, M., Ed.; Woodhead Publishing: Cambridge, UK, 2013; pp. 109–138. ISBN 978-0-85709-595-4.
25. Bettiga, D.; Ciccullo, F. Co-creation with customers and suppliers: An exploratory study. *Bus. Process Manag. J.* **2019**, *25*, 250–270. [CrossRef]
26. Jacobsen, L.F.; Tudoran, A.A.; Martinez, M.G. Examining trust in consumers as new food co-creators: Does the communicator matter? *Food Qual. Prefer.* **2020**, *86*, 104004. [CrossRef]
27. Khan, R.S.; Grigor, J.; Winger, R.; Win, A. Functional food product development—Opportunities and challenges for food manufacturers. *Trends Food Sci. Technol.* **2013**, *30*, 27–37. [CrossRef]
28. Azanedo, L.; Garcia-Garcia, G.; Stone, J.; Rahimifard, S. An Overview of Current Challenges in New Food Product Development. *Sustainability* **2020**, *12*, 3364. [CrossRef]
29. FAOSTAT. Available online: https://www.google.com/url?q=https://www.fao.org/faostat/en/%23data/QCL&sa=D&source=docs&ust=1644825938084357&usg=AOvVaw21W8alQ3KuyudCk0jZ8GER (accessed on 13 December 2021).
30. Marić, B.; Abramović, B.; Ilić, N.; Krulj, J.; Kojić, J.; Perović, J.; Bodroža-Solarov, M.; Teslić, N. Valorization of red raspberry (Rubus idaeus L.) seeds as a source of health beneficial compounds: Extraction by different methods. *J. Food Process. Preserv.* **2020**, *44*, e14744. [CrossRef]
31. Teslić, N.; Santos, F.; Oliveira, F.; Stupar, A.; Pojić, M.; Mandić, A.; Pavlić, B.; Kljakić, A.C.; Duarte, A.R.C.; Paiva, A.; et al. Simultaneous Hydrolysis of Ellagitannins and Extraction of Ellagic Acid from Defatted Raspberry Seeds Using Natural Deep Eutectic Solvents (NADES). *Antioxidants* **2022**, *11*, 254. [CrossRef]
32. Mišan, A.; Nađpal, J.; Stupar, A.; Pojić, M.; Mandić, A.; Verpoorte, R.; Choi, Y.H. The perspectives of natural deep eutectic solvents in agri-food sector. *Crit. Rev. Food Sci. Nutr.* **2020**, *60*, 2564–2592. [CrossRef]
33. Polizer Rocha, Y.J.; de Noronha, R.L.F.; Trindade, M.A. Relations between consumer's concern with own health and their perception about frankfurters with functional ingredients. *Meat Sci.* **2019**, *155*, 91–101. [CrossRef]
34. Galler, M.; Gonera, A.; Varela, P. Children as food designers: The potential of co-creation to make the healthy choice the preferred one. *Int. J. Food Des.* **2020**, *5*, 125–131. [CrossRef]
35. Yancey, P.H.; Clark, M.E.; Hand, S.C.; Bowlus, R.D.; Somero, G.N. Living with Water Stress: Evolution of Osmolyte Systems. *Science* **1982**, *217*, 1214–1222. [CrossRef] [PubMed]
36. Amabile, T.M. *How to Kill Creativity*; Harvard Business School Publishing: Brighton, MA, USA, 1998.
37. Füller, J. Refining Virtual Co-Creation from a Consumer Perspective. *Calif. Manag. Rev.* **2010**, *52*, 98–122. [CrossRef]
38. Baer, J. Creativity Doesn't Develop in a Vacuum. *New Dir. Child Adolesc. Dev.* **2016**, *2016*, 9–20. [CrossRef] [PubMed]
39. Rietzschel, E.F.; Nijstad, B.A.; Stroebe, W. Relative accessibility of domain knowledge and creativity: The effects of knowledge activation on the quantity and originality of generated ideas. *J. Exp. Soc. Psychol.* **2007**, *43*, 933–946. [CrossRef]
40. Baer, J. The Importance of Domain-Specific Expertise in Creativity. *Roeper Rev.* **2015**, *37*, 165–178. [CrossRef]
41. Im, H.; Hokanson, B.; Johnson, K.K.P. Teaching Creative Thinking Skills: A Longitudinal Study. *Cloth. Text. Res. J.* **2015**, *33*, 129–142. [CrossRef]
42. VanGundy, A.B. *101 Activities for Teaching Creativity and Problem Solving*; Pfeiffer: San Francisco, CA, USA, 2005; ISBN 0-7879-7402-1.
43. Eberle, R.F. Developing Imagination Through Scamper. *J. Creat. Behav.* **1972**, *6*, 199–203. [CrossRef]
44. Serrat, O. The SCAMPER Technique. In *Knowledge Solutions: Tools, Methods, and Approaches to Drive Organizational Performance*; Serrat, O., Ed.; Springer Singapore: Singapore, 2017; pp. 311–314. ISBN 978-981-10-0983-9.
45. Chou, Y.-K. *Actionable Gamification: Beyond Points, Badges, and Leaderboards*; Octalysis Media: Milpitas, CA, USA, 2015; ISBN 1511744049.
46. Diehl, M.; Stroebe, W. Productivity loss in brainstorming groups: Toward the solution of a riddle. *J. Pers. Soc. Psychol.* **1987**, *53*, 497–509. [CrossRef]
47. Barki, H.; Pinsonneault, A. Small Group Brainstorming and Idea Quality: Is Electronic Brainstorming the Most Effective Approach? *Small Gr. Res.* **2001**, *32*, 158–205. [CrossRef]
48. Richins, M.L. Measuring emotions in the consumption experience. *J. Consum. Res.* **1997**, *24*, 127–146. [CrossRef]

49. Chauhan, J.; O'Mahony, M. Use of a signal detection ranking analysis to measure preference for commercial and 'health modified' cakes. *J. Sens. Stud.* **1993**, *8*, 69–75. [CrossRef]
50. Lee, H.-S.; O'Mahony, M. Sensory evaluation and marketing: Measurement of a consumer concept. *Food Qual. Prefer.* **2005**, *16*, 227–235. [CrossRef]
51. Rodrigue, N.; Guillet, M.; Fortin, J.; Martin, J.-F. Comparing information obtained from ranking and descriptive tests of four sweet corn products. *Food Qual. Prefer.* **2000**, *11*, 47–54. [CrossRef]
52. Torbica, A.; Škrobot, D.; Janić Hajnal, E.; Belović, M.; Zhang, N. Sensory and physico-chemical properties of wholegrain wheat bread prepared with selected food by-products. *LWT* **2019**, *114*, 108414. [CrossRef]
53. Kohn, N.W.; Paulus, P.B.; Choi, Y. Building on the ideas of others: An examination of the idea combination process. *J. Exp. Soc. Psychol.* **2011**, *47*, 554–561. [CrossRef]
54. Zhou, Y.; Tse, C.-S. Sweet taste brings happiness, but happiness does not taste sweet: The unidirectionality of taste-emotion metaphoric association. *J. Cogn. Psychol.* **2021**, 1–23. [CrossRef]
55. Živković, D.; Radulović, Z.; Aleksić, S.; Perunović, M.; Stajić, S.; Stanišić, N.; Radović, C. Chemical, sensory and microbiological characteristics of Sremska sausage (traditional dry-fermented Serbian sausage) as affected by pig breed. *Afr. J. Biotechnol.* **2012**, *11*, 3858–3867. [CrossRef]
56. Ikonić, P.; Peulić, T.; Jokanović, M.; Šojić, B.; Škaljac, S.; Popović, S.; Šarić, L.; Novaković, A.; Tomović, V.; Vasilev, D. Evaluation of the physicochemical, biochemical and microbiological characteristics of three Serbian traditional dry-fermented sausages. *J. Food Sci. Technol.* **2021**, *58*, 3215–3222. [CrossRef]
57. Cooper, K.; Dedehayir, O.; Riverola, C.; Harrington, S.; Alpert, E. Exploring Consumer Perceptions of the Value Proposition Embedded in Vegan Food Products Using Text Analytics. *Sustainability* **2022**, *14*, 2075. [CrossRef]
58. Tagney, J.P.; Dearing, R.L. *Shame and Guilt (Emotions and Social Behavior)*; The Guilford Press: New York, NY, USA, 2003.
59. Shaver, P.; Schwartz, J.; Kirson, D.; O'connor, C. Emotion knowledge: Further exploration of a prototype approach. *J. Pers. Soc. Psychol.* **1987**, *52*, 1061–1086. [CrossRef] [PubMed]
60. Szakály, Z.; Szente, V.; Kövér, G.; Polereczki, Z.; Szigeti, O. The influence of lifestyle on health behavior and preference for functional foods. *Appetite* **2012**, *58*, 406–413. [CrossRef] [PubMed]

foods MDPI

Article

Nature Ambience in a Lunch Restaurant Has the Potential to Evoke Positive Emotions, Reduce Stress, and Support Healthy Food Choices and Sustainable Behavior: A Field Experiment among Finnish Customers

Saara Vanhatalo [1], Hilkka Liedes [2,*] and Kyösti Pennanen [1]

[1] VTT Technical Research Centre of Finland Ltd., 02150 Espoo, Finland; saara.vanhatalo@vtt.fi (S.V.); kyosti.pennanen@gmail.com (K.P.)
[2] VTT Technical Research Centre of Finland Ltd., 90590 Oulu, Finland
* Correspondence: hilkka.liedes@vtt.fi

Abstract: Laboratory experiments have indicated that exposure to restorative ambiences in food environments can lead to beneficial outcomes for consumers, but there is little evidence if this positive effect holds true in real-life consumption conditions. Therefore, the aim of this study was to analyze the effects of lunch restaurant ambience on customers' emotional responses, stress recovery, food choices, and generation of plate waste. The expectation was that ambience inducing positive emotional responses would lead to alleviated stress, healthier food choices, and reduced plate waste. A field experiment with a baseline and two experimental ambiences ('nature ambience' to induce positive emotions and 'fast food ambience' to induce less positive emotions) including visual and auditory stimuli was conducted in a lunch restaurant for one week per ambience. Emotional responses, and objective and subjective stress were measured from a subgroup of participants (n = 32). Food choices and plate waste were measured for all customers (n = 1610–1805 depending on the study week). During 'nature ambience' week, customers more often chose vegetarian dishes and generated less plate waste. The results on emotional responses and stress recovery were partially in line with the expectations. The study provides real-life evidence that restaurant ambience modification could lead to beneficial consequences for customers.

Keywords: food environment; stress; emotions; food choice; environmental sustainability; field experiment

Citation: Vanhatalo, S.; Liedes, H.; Pennanen, K. Nature Ambience in a Lunch Restaurant Has the Potential to Evoke Positive Emotions, Reduce Stress, and Support Healthy Food Choices and Sustainable Behavior: A Field Experiment among Finnish Customers. *Foods* **2022**, *11*, 964. https://doi.org/10.3390/foods11070964

Academic Editor: Witoon Prinyawiwatkul

Received: 8 February 2022
Accepted: 24 March 2022
Published: 26 March 2022

1. Introduction

Modern lifestyles in terms of mental workload, rush, and imbalances between work and personal life threaten people's psychological and physical welfare, predisposing them to stress, unhealthy food choices and unsustainable behaviors, posing significant adverse effects on individuals and societies in terms of public health and the environmental sustainability of the food system. Stress alone, defined as a condition in which the perceived demands are not met by the resources of an individual [1], predisposing individuals to anxiety, depression and unhealthy lifestyle [2], was responsible for financial costs of 26 billion euros in the EU-15 in 2014 [3], with 62% of workers experiencing stress at least once a week [4]. In a similar vein, unhealthy eating habits have led to the trebling of global obesity rates since the mid-1970s, making obesity and being overweight larger killers than being underweight for most of the world's population [5]. On top of these, unsustainable food consumption behaviors are partially responsible for the increased global CO_2 emissions and therefore contribute to climate change [6,7].

The development of food environments where consumers make food choices and consume food offers one potential tool to deal with stress and improve eating habits both from the health and sustainability perspectives. There is plenty of evidence that food environment and context have effects on various aspects of consumers' food consumption, such as (un)healthy food choices [8,9], food preference [10], and food product acceptance [11]. One potential explanation for the food environment's effect on consumers' behavior lies in its capability to generate different emotional responses in consumers [12]. This follows the idea of appraisal theories, which postulate that a person's evaluation of an object (e.g., food environment) will lead to an emotional reaction with subsequent psychological (e.g., increased or decreased stress) and behavioral consequences (e.g., increased or decreased healthy food choices, more or less environmentally sustainable behavior) [13]. In general, it has been suggested that a positive emotional reaction will lead to more beneficial behaviors from the individual's perspective, such as higher satisfaction [14] and increased quality evaluations of healthy snacks [12], while negative appraisals might lead to unbeneficial outcomes such as higher preference for products containing sugar [15].

Emotions have been shown to influence stress recovery, healthy food choices and sustainable behaviors. Restorative physical environments such as nature have been found to contribute to stress recovery by reducing negative mood states and enhancing positive emotions [1]. Already short nature visits [16] and even exposure to nature views on TV have been observed to generate positive emotional reactions and reduce stress [17]. In line with this, positive emotional states have also been observed to trigger healthy food choices [18]. Emotional reactions caused by the environment have an effect on sustainable behaviors as well. For instance, it has been found that exposure to restorative nature videos inducing positive emotional responses in comparison to videos about built environments inducing negative affective reactions promotes sustainable behaviors [19], and positive emotions in general are connected to sustainable behaviors [20]. In the sustainable food consumption context, negative emotional states are found to be among the factors explaining, for example, food wasting behavior [21]. These observations imply that stress recovery, healthy food choices and sustainable food behaviors could be promoted through a food environment that induces positive emotional reactions in people.

A notable part of the working age population consumes daily meals in lunch restaurants, defined as restaurants that mainly serve lunch and provide other food services, such as breakfast or cafeteria services, during office hours and are located on-site or in close proximity of business centers or major business areas. In 2015, it was estimated that approximately 17% of EU workers ate their lunch in on-site catering environments [22], making them attractive venues for the development of food environments to generate positive emotional responses and promote stress recovery, healthier food choices and sustainable food behaviors. Indeed, the restaurant servicescape (the physical environment of a service organization where the service transaction occurs) has been shown to influence consumers' emotions and satisfaction [23]. Although the scientific evidence on the influence of the food environment on stress and food behaviors in a real food consumption environment is sparse [8], it is proposed that behaviors could be influenced by illumination, colors, odors and temperature modifications [24]. From this perspective, recent studies done in restaurant dining contexts have shown that multisensory lunch conditions including nature themed images, sounds and odors indeed enhance positive emotions among the study participants, but no effect on food choices was found [25], and that auditory stimuli that was perceived as "healthy music" increased healthy food choices [26]. However, the above-mentioned studies were carried out in laboratory settings instead of field conditions.

The aim of this study is to analyze the effects of lunch restaurant ambience on customers' emotional states, stress recovery, food choices and sustainable food behavior in a real-life consumption context in Finland. In comparison to earlier studies, the current study intends to make a contribution by adopting the logic of appraisal theories in the development of two experimental ambiences ('nature' to extract positive emotions and 'fast food' to induce less positive emotions) and by applying them in real-life conditions instead of a laboratory setting. The main expectation of the study is that the 'nature' ambience aimed at inducing more positive emotional responses leads to alleviated stress, healthier food choices and less plate waste in comparison to those ambiences which induce less positive emotional reactions.

2. Materials and Methods

2.1. Study Design

The study was conducted in TestEat, a research restaurant located in Helsinki, Finland. Although the restaurant is designed for research purposes, it is commercially operated by a large international food service provider and serves lunch for 300–400 customers every day from Monday to Friday throughout the year. The study ran during a seven-week period in spring 2019 with three actual study weeks (see overview of the study design in Figure 1).

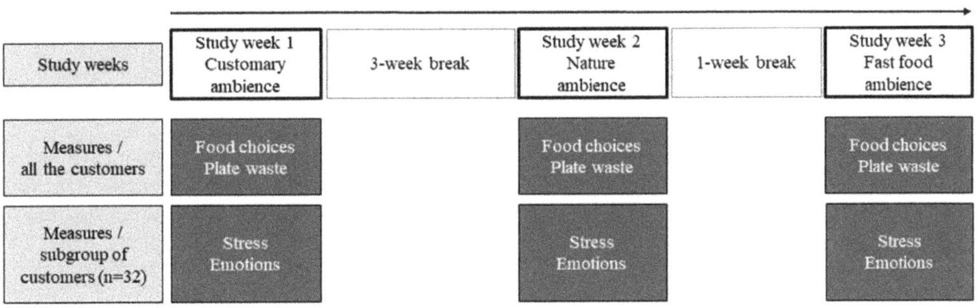

Figure 1. Overall study design with the study weeks, duration of breaks between the study weeks, measures for all customers, and measures for the subgroups of customers.

There was a three-week break between study week 1 and study week 2 and a one-week break between study weeks 2 and 3. The reason for the unequal duration of the breaks was an attempt to avoid various holiday seasons (Finnish skiing holiday, Easter, 1st of May) during the study period, which might have had an effect on the restaurant menu, clientele amount and composition. The visual and auditory ambience of the restaurant was the customary one during the first study week, serving as the baseline measure, and it was modified during the following two study weeks (study week 2: 'nature ambience', with the aim to induce more positive emotional responses; study week 3: 'fast food ambience', with the aim to induce less positive emotional responses). The lunch menu, restaurant, and arrangement of the buffet line were kept constant during the study weeks. Photos of the arrangements in the buffet line and restaurant environment were taken every day in the first study week before the lunch service and were used every day in the coming study weeks to guarantee coherent arrangements between the weeks. No campaigns (price, marketing), which could influence customers' food choices, took place during the study weeks or during the weeks between the study weeks. Food choices and the amount of plate waste of all the customers, as well as emotional responses to the restaurant ambience and the objective and subjective stress of a subgroup of customers (n = 32) were measured in all study weeks.

2.2. Participants

The sample for the study consisted of all the customers who purchased lunch in the restaurant during the study weeks (food choices and plate waste measurements) and a subgroup of customers (n = 32) who were recruited for the study and whose emotional responses to the eating situation, as well as objective and subjective stress, were measured throughout the study.

2.2.1. Subgroup of Customers Subjected to Emotion, Objective, and Subjective Stress Measures

Recruitment of the subgroup of customers was carried out in collaboration with the human resources department of a major employer located in the same building as the restaurant. Invitations to join the study were distributed through the company's intranet, weekly newsletter and information screens located in the research restaurant and other locations in the office building. As the objective stress measures were based on heart rate variability, only healthy adults (exclusion criteria: chronic diseases influencing the functioning of the heart or disturbances of the mind potentially causing unknown variation in stress) were included in the study in order to control the stress measures. In addition, only those who ate lunch at least 3–4 days a week in the research restaurant were invited to participate in the study to ensure frequent exposure to the ambiences and to obtain as much relevant data as possible for the emotion and stress measures. Volunteers were guided to fill in an online screening form and those who met the inclusion criteria were invited to participate in the study. At the info visit, the volunteers were given a detailed written description of the study and the information was summarized verbally by the study personnel. After that, the participants signed informed consent forms, and received study IDs and activity bracelets (Garmin, Vivosmart 4) for objective stress measurement. Detailed written instructions about the use of the activity bracelets were provided. The participants were instructed to wear the bracelet in the non-dominant hand superior to the carpal bones and to attach the bracelet firmly, but not tightly, during the working hours in the study weeks. Before the study, the participants were instructed to start using the activity bracelet 3 days in advance to get comfortable with the use. One participant withdrew from the study due to workplace change. Thirty-two participants (21 females, 11 males, 45.6 $\pm$ 9.2 years) completed the study. After the study, the participants were given three movie tickets and a voucher (total worth 50 euros) to compensate for the time and effort used for the study.

2.2.2. All Customers Subjected to Food Choice and Plate Waste Measures

The other study population consisted of all the restaurant clientele during the study weeks. They were mainly office workers of the office building in which the restaurant is located, their visitors, as well as members of the general population living in the neighborhood. The sample sizes for each study week were calculated according to the transactions recorded in the restaurant sales data. In study week 1 (baseline week), 1610 lunch purchases were recorded, followed by study week 2 ('nature ambience' week) with 1714 purchases and study week 3 ('fast food ambience' week) with 1805 purchases. The customers were not informed about the ongoing study. However, the restaurant personnel were instructed to provide an information leaflet about the study with the researchers' contact information for further queries to those customers who might ask about the changes in the restaurant during the study. At the end, no customers contacted the research group.

The ethical committee of VTT Technical Research Centre of Finland Ltd evaluated and approved the study design and experimental procedure (121739/Opportunity_FE_4.0_2019, approved 16.1.2019). The study was conducted according to the ethical principles of good research and clinical practice described in the declaration of Helsinki.

2.3. *Ambience Design*

The visual appearance, soundscape, and brightness of the general lighting in the restaurant were modified to create two unobtrusive and immersive experimental ambiences: 'nature ambience' to induce a positive emotional response and 'fast food ambience' to induce a less positive emotional response. The choice of the ambiences was based on previous research indicating that nature ambiences (including visual appearance and soundscape) are perceived restorative [16,17], induce positive emotional reactions in humans [12,27,28] and thus potentially contribute to stress recovery, healthy food choices and sustainable behaviors (i.e., plate waste in this study). On the contrary, exposure to colors with longer wavelengths (red, yellow, orange) in abstract forms is considered as arousing [29] and fast-paced instrumental music has been shown to lead to sympathetic arousal [30] and to delay recovery from stress [31]. In terms of lighting, it has been proposed that warm ambient light has a relaxing effect on humans while bright general lighting contributes to arousal and activation [24].

A pre-test was conducted to confirm that the images and soundtracks (images purchased from Adobe Stock library under standard license and sounds from the FMA Free Music Archive and from the YouTube Audio Library under the Creative Commons license) representing 'nature ambience' and 'fast food ambience' induced opposite emotional responses. Eighteen volunteers were recruited from the staff of VTT Technical Research Centre of Finland Ltd., Espoo, Finland. They were presented with four distinct nature images as image pairs constituting eight pairs of nature images and four red-yellow-orange abstract images as image pairs constituting eight pairs of abstract images in randomized order on paper. In addition, they were played two soundtracks with birds singing and three soundtracks with fast-paced instrumental music. The volunteers were asked to evaluate the calmness (positive emotion) and restlessness (negative emotion) evoked by the image pairs and soundtracks one after each other on a visual analogue scale (VAS) from 0 to 10, with higher values indicating calmer (or more restless) feeling.

The evaluations of calmness evoked by the nature or abstract image pairs ranged from 2.6 ± 2.2 to 8.9 ± 0.9 on a scale from 0 to 10 in which 0 = does not evoke calmness at all and 10 = evokes calmness a lot. Evaluations of restlessness ranged from 1.0 ± 1.1 to 6.9 ± 1.9 on a scale from 0 to 10 in which 0 = does not evokes restlessness at all and 10 = evokes restlessness a lot. The image pair evoking the least intensive restlessness (1.0 ± 1.1) and the most intensive calmness (8.9 ± 0.9) was chosen to represent 'nature ambience'. The pair evoking the most intense restlessness (6.9 ± 1.8) and the least intensive calmness (2.6 ± 2.1) was chosen to represent 'fast food ambience' (Figure 2). There were statistically significant differences between the chosen 'nature ambience' and 'fast food ambience' image pairs regarding evoked calmness and restlessness ($p < 0.001$ for both).

Figure 2. Image pairs chosen for the experimental ambiences.

Birdsong soundtrack 1 scored 8.4 ± 1.9 in calmness on a scale from 0 to 10 in which 0 = does not evoke calmness at all and 10 = evokes calmness a lot and 1.5 ± 1.8 in restlessness. Birdsong soundtrack 2 received calmness evaluations of 3.4 ± 1.9 and restlessness evaluations of 5.9 ± 2.1 on a scale from 0 to 10 in which 0 = does not evokes restlessness at all and 10 = evokes restlessness a lot. Based on this result, birdsong soundtrack 1 was chosen to be used for 'nature ambience'. Fast-paced music soundtrack 1 scored 4.3 ± 2.6 in calmness and 5.5 ± 3.0 in restlessness. Fast-paced music soundtrack 2 received a calmness evaluation of 4.5 ± 2.3 and a restlessness evaluation of 5.2 ± 2.5. Fast-paced music soundtrack 3 scored 5.2 ± 1.7 in calmness and 3.9 ± 2.5 in restlessness. The soundtracks did not differ regarding the evoked calmness. The soundtrack evoking the most intensive restlessness was chosen to be used in 'fast food ambience' (soundtrack 1). There were statistically significant differences between birdsong soundtrack 1 and fast-paced music soundtrack 1 regarding evoked calmness and restlessness ($p < 0.001$ for both).

The practical application of ambiences in the lunch restaurant included three elements: (1) visual images projected on the walls ('nature ambience': nature images; 'fast food ambience': red-yellow-orange abstract images), (2) soundscape ('nature ambience': birdsong; 'fast food ambience': fast-paced instrumental music), and (3) general lighting ('nature ambience': slightly dimmed; 'fast-food ambience': full lightning). The visual ambiences in the restaurant were created by presenting four large still image pairs (approx. 6 m^2 each) on the restaurant walls by using multiple ultra-high-definition projectors (Figure 3). The volume of the birdsong soundtrack in the 'nature ambience' week and the fast-paced instrumental music soundtrack during the 'fast food ambience' week was adjusted to 46–47 dB. During the 'nature ambience' week, general lighting was slightly dimmed (to retain constant lighting throughout the week, the correct setting was marked in the dimmer) while during 'fast food ambience' week, full lighting was on all the time [24]. The curtains were kept closed during all study weeks to exclude the possible influence of changing seasonal and weather conditions on customer behavior.

Figure 3. Illustration of 'nature ambience' projections in different parts of the lunch restaurant.

2.4. *Measures*

2.4.1. Emotions

The subgroup of participants ($n = 32$) was asked to evaluate their emotional states just after lunch through an online questionnaire administered in QuestBack software. The link to the emotion questionnaire was sent to the participants by email at 11 a.m. each study day. The intensities of the positive (contentment, joy, calmness) and negative (worry, discontentment) emotions were measured on a scale from 1–5 (1 = not at all, 2 = a little, 3 = somewhat, 4 = a lot, 5 = very much) adapted from the Consumption Emotions Set (CES) scale [32]. Based on the measures, composite variables for the analyses were formed and Cronbach's alphas were calculated to check the reliability of the formed variables for positive emotions ('customary ambience' week $\alpha = 0.71$, 'nature ambience' week $\alpha = 0.83$, 'fast food ambience' week $\alpha = 0.91$) and negative emotions ('customary ambience' week $\alpha = 0.77$, 'nature ambience' week $\alpha = 0.66$, 'fast food ambience' week $\alpha = 0.92$).

2.4.2. Objective and Subjective Stress

Objective and subjective stress were measured from the subgroup of participants ($n = 32$) recruited for the study. Heart rate variability (HRV) based data were collected through Garmin Vivosmart 4 activity bracelets as an objective indicator of stress [33]. The data were recorded for the entire working day, but for the analysis, data from 180 min before and 180 min after lunch were used. The activity bracelet measured stress level on a scale from 0 to 100 (0–25 = resting state, 26–50 = low stress, 51–75 = medium stress, 76–100 = high stress). The stress data provided by the bracelet were first synchronized with a Garmin Connect application in the participants' mobile phones and then synchronized with Garmin servers. From there, the data was transferred through the Garmin Health API to study servers from where the research group extracted the data.

Subjective stress was measured with a questionnaire, which participants were instructed to fill in right before and after lunch. A link to the pre-lunch stress questionnaire was sent to the participants by email each study day at 10 a.m. The link to the post-lunch stress questionnaire was sent to the participants by email at 11 a.m. each study day. In the questionnaire, the participants were asked to evaluate their level of positive stress (excitement, motivation, vitality) and negative stress (anxiety, restlessness, nervousness) on a scale from 1 to 5 (1 = not at all, 2 = a little, 3 = somewhat, 4 = a lot, 5 = very much). Composite variables for analyses were developed and Cronbach's alphas were calculated to check the reliability of the formed variables for positive stress before lunch ('customary ambience' week $\alpha = 0.79$, 'nature ambience' week $\alpha = 0.77$, and 'fast food ambience' week $\alpha = 0.86$), positive stress after lunch ('customary ambience' week $\alpha = 0.84$, 'nature ambience' week $\alpha = 0.90$, 'fast food ambience' week $\alpha = 0.80$), negative stress before lunch ('customary ambience' week $\alpha = 0.72$, 'nature ambience' week $\alpha = 0.87$, 'fast food ambience' week $\alpha = 0.72$), and negative stress after lunch ('customary ambience' week $\alpha = 0.80$, 'nature ambience' week $\alpha = 0.80$, 'fast food ambience' week $\alpha = 0.79$).

2.4.3. Food Choices and Plate Waste

During the study, the lunch restaurant offered five different lunch options each day, including two vegetarian warm dish options and three warm meat-based dish options (fish, red meat or chicken being the main protein component in these dishes). In addition to the indicated main components, a typical dish also included salad from the buffet line, a carbohydrate component (e.g., rice, pasta, potato), bread, drinks (e.g., water, milk, juice), dessert, coffee, or tea. According to sales data, the number of purchased vegetarian dish options for each week were summed up to represent healthier and more sustainable meal choices [6]. The rest of the choices were combined to represent meat-based and less healthy and less sustainable meal choices and summed up for each week.

For the plate waste measure, the restaurant personnel weighed the total amount of plate waste at the end of the lunch service in each study day. To increase the reliability of the weighing, the personnel were trained by the researchers before the data collection. In addition, they were provided written instructions and a template onto which the weighing results were marked every day. The average amount of plate waste per customer per day was calculated based on the total weighed plate waste divided by the number of customers in each day.

2.5. Data Analysis

2.5.1. Pre-Processing of the Stress Data

As the lunch time for the participants in the recruited subgroup varied, the time points for "before lunch" and "after lunch" were defined for each participant for each study day using the time stamp of their answers to the pre-lunch and post-lunch questionnaires. For the study, it was crucial that the questionnaire answers were given just before or right after the lunch to get an understanding of the momentary stress and emotional states and the potential effects of lunch ambience on those. Therefore, the time stamp data were pre-screened to check that the times were valid. The following principles were used to pre-screen the data: (1) If the pre-lunch questionnaire was filled in 15 min before or during the opening time of the lunch restaurant and the post-lunch questionnaire was filled in by a latest of 90 min after the opening time and the duration of lunch (15–90 min) seemed reasonable, the time stamps were used as such. (2) If there was only one time stamp available and the timing was reasonable (as earlier), the time stamp was used as such. (3) If the first time stamp was more than 15 min before the opening time of the restaurant, the time stamp was replaced with a time stamp where 1 h was subtracted from the post-lunch questionnaire time stamp. (4) If the interval between time stamps was less than 15 min, but the timing of the post-lunch questionnaire was reasonable (not more than 90 min after the restaurant opening time), the post-lunch time stamp was used and the pre-lunch stamp was discarded.

Through the validated time stamps, objective stress values for the 180 min preceding the pre-lunch questionnaire and 180 min after the post-lunch questionnaire were identified and extracted from the activity bracelet data. Garmin Vivosmart 4 provides stress values every 3 min at its best, but movement by the user and low signal quality may hinder data collection and lead to missing values. To remove noise and impute missing values, the stress values were filtered with a Gaussian-weighted moving average filter with a window length of five samples, which corresponds to 12 min. The moving average filter slides a window over the samples and calculates an average over the samples in the window. The Gaussian-weighted filter gives less weight to samples which lie further from the center of the window.

To study the evolution of stress values over time, the average and standard deviation of the filtered stress values for each time point were calculated for each study week based on all available data. For each participant, median values for stress were defined for periods of 180 min before lunch and for 1–90 min and 91–180 min after lunch. At least 30% of the potential data points per participant had to be available to calculate the median stress values.

2.5.2. Statistical Analyses

Repeated measures ANOVA with Bonferroni correction was used to study the differences in emotions and subjective stress between the study weeks for the subgroup of participants. The objective stress of the subgroup between the study weeks was analyzed with a paired samples *t*-test. For these analyses, the 180 min before lunch stress measures were averaged over all study weeks for all study participants and compared with both 1–90 min after lunch and 91–180 min after lunch stress measures. Correlations between emotions, subjective stress and objective stress were analyzed using Spearman's rank correlation coefficient. A two proportions Z-test was used to analyze the differences in vegetarian and meat-based food choices of all customers between the study weeks and one-way ANOVA was used to analyze the amount of plate waste between the study weeks. Due to the real-life nature of the experiment, the criterion for statistical significance was set to a more liberal p-value of ≤ 0.1. The data were analyzed using IBM SPSS Statistics software (Version 24, IBM Corp, Chicago, IL, USA) and MATLAB R2020a (The MathWorks, Inc., Natick, MA, USA).

3. Results

3.1. Emotions

Lunch experience in the 'nature ambience' evoked stronger positive emotions than that in the 'fast food ambience' (Table 1). However, no difference between 'customary ambience' and 'nature ambience' or 'fast food ambience' was detected. Eating lunch in 'fast food ambience' evoked stronger negative emotions than eating lunch in the 'customary ambience'. No difference between 'customary ambience' and 'nature ambience' was detected in negative emotions.

Table 1. Positive and negative emotions evoked by the lunch experience on a scale in which 1 = low intensity and 5 = high intensity. Data from the subgroup of participants (n = 32). The different letters next to the means indicate statistically significant difference ($p \leq 0.1$) between the study weeks.

	Ambience				
	Customary	**Nature**	**Fast Food**	**F (df)**	***p* Value**
Positive emotions, mean (SD)	3.2 ± 0.5^{ab}	3.3 ± 0.7^{b}	3.0 ± 0.7^{a}	3.267 (2)	0.052
Negative emotions, mean (SD)	1.3 ± 0.4^{a}	1.6 ± 0.7^{ab}	1.6 ± 0.6^{b}	4.376 (2)	0.024

3.2. *Objective and Subjective Stress*

3.2.1. Objective Stress

Figure 4 shows how objective stress varied during the six hours of the working day around lunchtime (180 min before lunch and 180 min after lunch). Similar patterns were detected for all study weeks. In the morning hours, stress decreases until lunch. After lunch, there is a clear increase in measured stress. After the post-lunch peak, the stress slowly decreases towards the end of the working day. In general, stress levels in all weeks seem low.

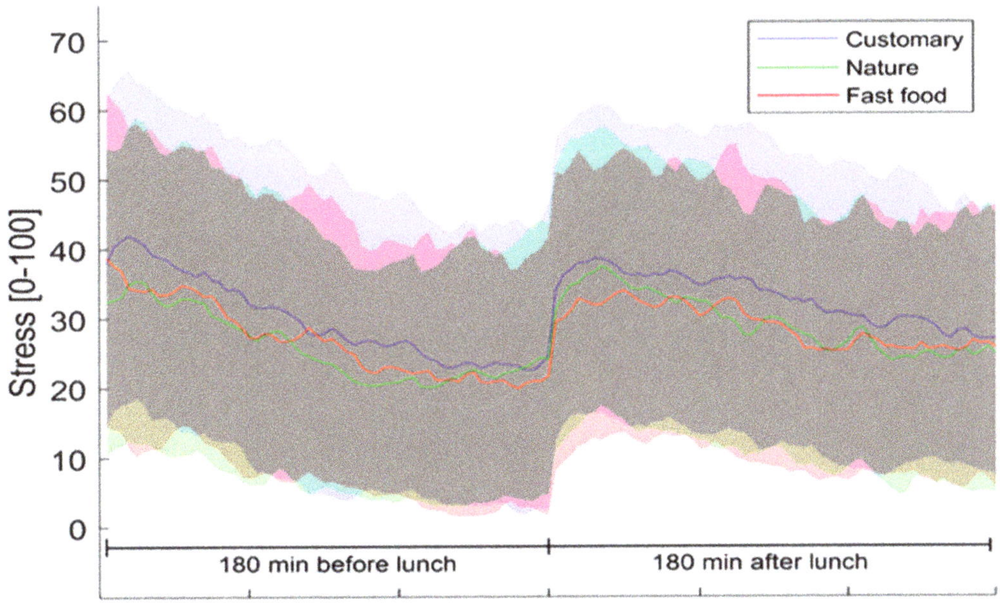

Figure 4. Evolution of objective stress over time in different ambiences on a scale in which 0–25 = resting state, 26–50 = low stress, 51–75 = medium stress, 76–100 = high stress. Solid lines show averages and shaded areas show standard deviations. Data from the subgroup of participants (n = 32).

For the objective stress analyses, the before lunch stress levels were averaged over all study weeks for all participants to achieve a more valid baseline measure as the descriptive statistics before the lunch showed some unexpected variation between the study weeks (Figure 4). Figure 5a shows that stress was increased in each study week 1–90 min after lunch compared to 180 before lunch (all weeks combined), ('customary ambience', t(31) = −5.21, p = 0.000; 'nature ambience', t(31) = −4.62, p = 0.000; 'fast food ambience', t(31) = −3.82, p = 0.001). During the 'nature ambience' week, participants' stress decreased 91–180 min after lunch compared to 180 min before lunch, t(31) = 2.15, p = 0.04. During the 'customary ambience' week, stress was still higher in 91–180 min period after the lunch in comparison to 180 min before lunch t(31) = −1.81, $p = 0.08$. (Figure 5b).

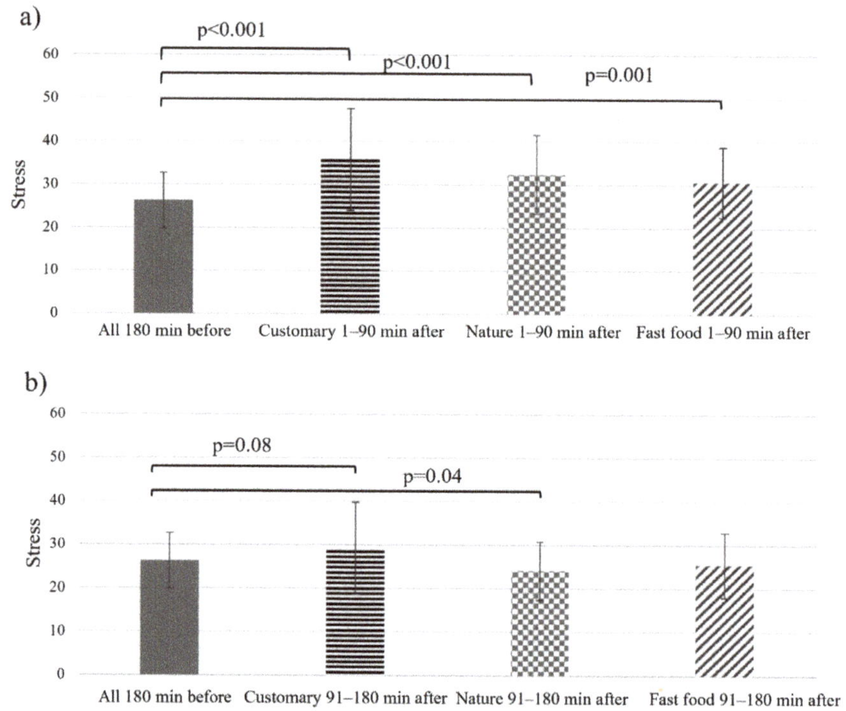

Figure 5. Averages of median objective stress 180 min before lunch (all weeks combined) and (**a**) 1–90 min after lunch and (**b**) 91–180 min after lunch. Scale: 0–25 = resting state, 26–50 = low stress, 51–75 = medium stress, 76–100 = high stress. Error bars are standard deviations of the medians. Data from the subgroup of participants ($n = 32$).

3.2.2. Subjective Stress

Subjective positive stress right before or after the lunch did not differ between the study weeks (Table 2). Subjective negative stress after the lunch was higher during the 'fast food ambience' week in comparison to the 'customary ambience' week.

Table 2. Subjective positive and negative stress just before and just after lunch during the study weeks measured on a scale in which 1 = low intensity and 5 = high intensity. Data from the subgroup of participants ($n = 32$). Different letters next to the means indicate statistically significant difference ($p \leq 0.1$) between the study weeks.

	Ambience				
	Customary	**Nature**	**Fast Food**	**F (df)**	***p* Value**
Subjective positive stress					
Just before lunch	2.8 ± 0.7	2.9 ± 0.7	2.8 ± 0.7	0.268 (2)	0.766
Just after lunch	2.8 ± 0.7	2.8 ± 0.7	2.7 ± 0.6	0.328 (2)	0.722
Subjective negative stress					
Just before lunch	2.1 ± 0.6	2.2 ± 0.6	2.2 ± 0.6	0.756 (2)	0.475
Just after lunch	1.9 ± 0.6 ^{a}	2.1 ± 0.7 ^{ab}	2.2 ± 0.6 ^{b}	3.337 (2)	0.046

3.2.3. Correlations between Objective and Subjective Stress and Emotions

Table 3 presents correlations between objective stress, subjective stress, and emotions. In most cases, no statistically significant correlations were found between objective stress and subjective stress or emotions except for between objective stress and subjective negative stress before lunch during the 'customary ambience' week and after lunch during the 'nature ambience' week. Subjective stress correlated with emotions. Positive stress and positive emotions had positive correlations, negative stress and negative emotions had positive correlations, negative stress and positive emotions had negative correlations, and positive stress and negative emotions had negative correlations.

Table 3. Spearman's rank correlations between objective stress, subjective stress and emotions evoked by the ambiences. Data from the subgroup of participants (n = 32). ** $p \leq 0.01$, * $p \leq 0.1$, ^{ns} non-significant correlation.

	Customary	Nature	Fast Food
	r	r	r
Before lunch			
Objective stress			
Subjective positive stress	−0.05 ^{ns}	0.07 ^{ns}	−0.08 ^{ns}
Subjective negative stress	0.23 *	0.06 ^{ns}	−0.03 ^{ns}
After lunch			
Objective stress			
Subjective positive stress	−0.04 ^{ns}	−0.15 ^{ns}	−0.13 ^{ns}
Subjective negative stress	0.13 ^{ns}	0.24 *	−0.05 ^{ns}
Positive emotions	−0.10 ^{ns}	−0.21 ^{ns}	−0.11 ^{ns}
Negative emotions	0.10 ^{ns}	0.21 ^{ns}	0.15 ^{ns}
Subjective positive stress			
Positive emotions	0.30 **	0.47 **	0.36 **
Negative emotions	−0.23 *	−0.19 *	−0.14 ^{ns}
Subjective negative stress			
Positive emotions	−0.47 **	−0.42 **	−0.44 **
Negative emotions	0.37 **	0.44 **	0.38 **

3.3. Food Choices

Meat-based warm dish options were clearly the most popular dish options during all the study weeks (Figure 6). However, vegetarian lunch options were slightly more popular (18.6 %) during the 'nature ambience' week compared to the 'customary ambience' week (16.3 %), Z = −1.74, p = 0.08. No differences were detected between the 'nature ambience' week and the 'fast food ambience' week (17.2 %), Z = −1.08, p = 0.28 or between the 'customary ambience' and the 'fast food ambience' weeks, Z = 0.70, p = 0.48.

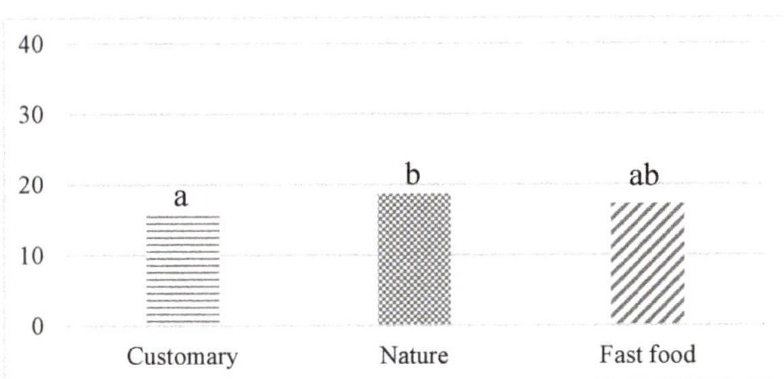

Figure 6. Relative share of vegetarian dishes of all lunch dish choices during the weeks with 'customary ambience', 'nature ambience' and 'fast food ambience'. Data from all customers who purchased lunch during the study weeks ('customary ambience' week = 1610, 'nature ambience week' = 1714, 'fast food ambience' week = 1805). Different letters above the bars indicate statistically significant difference ($p \leq 0.1$) between the study weeks.

3.4. Plate Waste

On average, 23.8 ± 3.2 g, 14.6 ± 3.4 g, and 19.7 ± 7.2 g plate waste per customer was produced during the 'customary ambience', 'nature ambience' and 'fast food ambience' weeks, respectively, with some differences between the weeks [$F(2,12) = 4.279$, $p = 0.04$] (Figure 7). The Tukey post hoc test showed that the amount of plate waste was lower during the 'nature ambience' week in comparison to the 'customary ambience' week ($p = 0.03$). No statistically significant differences were detected between the 'customary ambience' and 'fast food ambience' weeks ($p = 0.42$) or the 'nature ambience' and 'fast food ambience' weeks ($p = 0.28$).

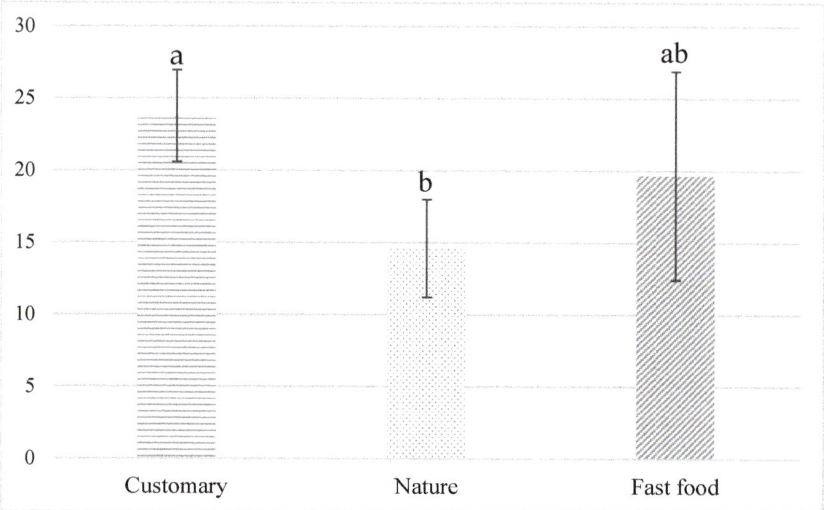

Figure 7. Average plate waste in grams per customer during the 'customary ambience', 'nature ambience' and 'fast food ambience' weeks. Data from each day of the study weeks ($n = 15$). Error bars are standard deviations of the five days of the study week. Different letters above the bars indicate statistically significant difference ($p \leq 0.1$) between the study weeks.

4. Discussion

The main expectation of this study was that different ambiences in a lunch restaurant can cause variation in customers' positive or negative emotional responses and therefore influence stress recovery, food choices and sustainable behaviors. The results provided support to some of these assumptions. Partially in line with the expectations, the 'nature ambience' evoked stronger positive emotions than 'fast food ambience' and 'fast food ambience' evoked stronger negative emotions than the 'customary ambience'. In terms of objective stress, a slight decrease in heart rate variability-based stress 91–180 min after lunch in 'nature ambience' was detected. No differences emerged in subjective stress between the study weeks except for a slight increase in negative after-lunch stress during the 'fast food ambience' week in comparison to the 'customary ambience' week. The correlations between positive and negative emotions and subjective stress were as expected, but surprisingly no correlations were found between emotions and objective stress or between objective and subjective stress. The data on food choices and plate waste showed that in 'nature ambience', the share of vegetarian dish choices was elevated and customers produced less plate waste in comparison to 'customary ambience'.

In general, the results are to some extent in line with the earlier literature on restorative environments and their effect on emotional responses and stress recovery [1,16,17,27–30]. The 'nature ambience' developed for this study generated more positive emotional reactions than the 'fast food ambience'. However, no differences were observed between 'nature ambience' and 'customary ambience'. Although the effects of eating environment or context on consumers' emotional responses have been demonstrated in many previous studies carried out in laboratory conditions [25,34,35], the differences between the ambiences in our field study were small despite the drastic differences observed in the pre-test. This might be because in a lunch restaurant setting, there are a multitude of other stimuli such as the lunch companion, restaurant layout, and rush that influence the situation simultaneously and might hide the experimental ambience effect. Possibly the effect could have been enhanced by bringing more elements such as plants or decorations in the restaurant. Both experimental ambiences could have brought a novelty aspect to the restaurant clientele as well, which could have hindered the desired effects on emotions. It is also possible that the customers were familiar with the customary ambience and changes to it in the first study week ('nature ambience') might not have been considered only as positive, which might explain why no differences in emotional responses were observed between the 'customary ambience' and 'nature ambience'. Finally, emotional responses were measured only from a relatively small sample of study participants. With a larger sample, the results may have been more conclusive.

Stress during working time was evaluated objectively with a wrist device that measures heart rate variability (HRV) and subjectively through questionnaires. 'Nature ambience' was expected to induce more positive emotional responses and therefore decrease stress. In each of the study weeks, objective stress slightly decreased during morning hours, increased clearly after lunch and gradually decreased after the post-lunch peak towards the end of the day. Measured stress increased in all study weeks when comparing the median stress of all weeks before lunch (180 min) to 1–90 min after lunch. This result can be explained by a meal-induced sympathetic activation, which is a natural consequence of eating. The activation was found to start soon (approx. in 15 min) after eating and last at least 90 min [36]. The properties of a meal (size, macronutrients) and individual characteristics (weight, age) influence the extent of meal-induced sympathetic activation as well. Coffee drinking [37], which is typical for Finnish people after lunch and moving back to office from the lunch restaurant most probably increased sympathetic activation as well. For these reasons, the stress values were observed not only straight after lunch but also 91–180 min after lunch when the meal-induced activation was expected to finish. In that time period (91–180 min after the lunch), objective stress in 'nature ambience' decreased compared to the combined pre-lunch stress from all weeks. This indicates that eating in 'nature ambience' might result in slightly decreased stress during the afternoon hours. However, the objective stress measures should be interpreted with caution. This is due to the fact that no variation was detected in subjective stress measurements between the study weeks. In addition, neither emotion measures nor subjective stress measures correlated with the objective stress measurements, which is illogical especially as the correlations between emotion measures and subjective stress measures were logical and according to expectations. These observations raise questions of how reliably HRV measurements with a wristband actually measure stress, how the data could be interpreted, and what type of conclusions can be drawn based on such data.

The customers' lunch choices and plate waste were recorded throughout the study weeks. The results show that the relative share of vegetarian dish choices was marginally higher and the amount of plate waste lower during the restorative 'nature ambience' week than the 'customary ambience' week. These are desirable changes since vegetable rich options are generally healthy and plate waste is unwanted from both ecological and financial perspectives. Previous research has shown that environmental cues are able to prime food choices. For example, a lunch buffet with green plants and odor of herbs nudged customers to take less food from the buffet line [38]. Even just an exposure to a nature poster increased healthy choices [8]. There is also evidence from laboratory experiments that exposure to nature videos leads to more sustainable behavior [19]. It can be assumed that the 'nature ambience' in this study might have activated either a mental concept of a healthy diet or pro-environmental behavior leading to healthier and more sustainable choices. Whether these are separate or connected mechanisms in our study is debatable and requires further enquiries. For instance, the activation of a mental concept of a healthy diet might have led customers to choose smaller portions, a significant predictor of the amount of plate waste [39,40], which could have led to less plate waste, suggesting that the discovered results on food waste are mediated by the activated mental concept related to health. On the other hand, there is evidence that visual exposure to nature images or videos has a direct effect on sustainable behaviors as well [19]. However, the latter studies have not been carried out in a food consumption context and thus do not provide direct explanations on the matter. Finally, gender and food type [40,41] have also shown to influence the amount of plate waste. In the current study, the food types were controlled over the study weeks, but there is no data on the potential changes in gender composition of the restaurant clientele between the study weeks, which could have had an effect on the study results. To sum up, although the exact mechanisms behind the ambience effect on food choices and plate waste remain unclear, the study provides evidence that ambience modification in real life can have at least a moderate effect on healthier and more sustainable food behaviors.

4.1. Suggestions for Further Research

In the current study design, the expectation was that positive emotional responses to the eating environment would have beneficial effects on stress, food choices and plate waste. It must be noted that there is recent evidence that negative emotional appraisals might have a beneficial effect especially on sustainable food behaviors. For instance, a recent study found that a feeling of guilt leads to the reduction of household food waste [42]. Similar findings were introduced in a study related to meat consumption from a sustainability perspective [43]. These are interesting findings for further research aiming at understanding how eating environment modification affects food consumption behaviors.

This study focused on positive and negative emotions on a general level. However, it has been proposed that certain specific positive emotions such as joy and pride could lead to, for instance, environmentally sustainable behaviors [20]. Further research could consider priming specific emotional responses instead of the simpler valence of the emotions. Understanding which specific emotions are connected to healthy and environmentally sustainable behaviors would support the development of practical applications to facilitate consumers' healthy and sustainable food consumption.

The ambience modification was implemented through visual and auditory changes of the environment. Further research could test other measures of ambience modification such as olfactory ambience modification. Technological advancements in virtual reality tools could also offer interesting opportunities to develop innovative eating ambiences.

4.2. Limitations of the Study

Five main limitations of the study deserve discussion. Firstly, the effect of the experimental ambiences on customers' emotions was weaker than expected based on the literature and pre-test results. Stronger manipulation might have produced stronger effects on emotions, stress, food choices and amount of plate waste. In hindsight, it would have been useful to pre-test the chosen experimental images and sounds in the research restaurant environment instead of a more controlled paper and pen approach. Secondly, the curtains of the restaurant were kept closed during the study weeks to minimize the effect of changing weather and season on the study. Some of the customers might have been discontent with curtains being closed and it might have influenced how the participants perceived the lunch experience. Thirdly, there was a considerably large number of missing values (48% of the filtered data) in HRV based stress data, which needed to be balanced in the data pre-screening phase (see details in Section 2.5.1). Missing data was most probably due to participants' movements and low signal quality. Similar type of wristband (Empatica) based HRV measurements have been shown to function well when the user is seated and in rest but not in dynamic conditions [44]. An HRV detection system with electrodes attached to the skin would have provided higher quality data, but in the context of a three-week study it would have been too burdensome for the participants. Consumer wearables have gained popularity during the past years also in the research field, but in practice their use is not optimal [45]. Fourthly, the sample subjected to emotion, subjective and objective stress measurements (n = 32) was relatively small, limiting the power of the study. Therefore, the results should be treated with caution and indicative at best. Despite the recruitment efforts, no more participants were attracted. The interest might have been low as participation in the study required long-term commitment by the participants and almost daily visits to the restaurant during the study weeks. Finally, there was imprecision in the timing when the participants filled in the pre and post lunch questionnaires. The participants were sent the questionnaires twice a day, but they had to remember to fill them in at the right time (right before and right after lunch). Some data were lost because of incredible time stamps. A more accurate way to collect the data would have been to send a notification to the participant at the moment they were entering or leaving the restaurant.

Despite the limitations, the study has strengths. The main strength of this study is that it was conducted in a real-life environment, unlike many previous studies that have observed the impact of different stimuli on emotions, stress, food choices and sustainable behaviors in laboratory settings. In addition, this study approached the study topics with various methods including subjective measures, objective physiological measures (heart rate variability) as well as objective observations on food choices and amount of plate waste. In addition, the number of participants especially regarding food choices and plate waste (all the customers) was high, allowing strong conclusions.

5. Conclusions

A field experiment in a lunch restaurant was carried out to understand the effects of two experimental visual and auditory ambiences, 'nature ambience' intended to evoke positive emotions and 'fast food ambience' intended to evoke less positive emotions, on customers' emotional states, recovery from stress, food choices and plate waste. The ambiences were designed to generate varying emotional responses with potential subsequent effects on the other measures and presented to customers through large projections on the walls and via loudspeakers. The results of the study indicate that 'nature ambience' in a lunch restaurant has at least a moderate positive effect on customers' emotional states, vegetarian lunch choices and plate waste. The results also showed that under 'nature ambience', stress recovery 91–180 min after lunch was marginally elevated. However, the stress results should be interpreted with caution. To conclude, the study provides real-life evidence that restaurant visual and auditory ambience modification might lead to beneficial consequences among the clientele.

Author Contributions: Conceptualization, S.V. and K.P.; methodology, S.V. and K.P.; formal analysis, S.V., H.L. and K.P.; investigation, S.V.; writing—original draft preparation, S.V., H.L. and K.P.; writing—review and editing, H.L. and K.P.; supervision, K.P.; project administration, K.P.; funding acquisition, K.P. All authors have read and agreed to the published version of the manuscript.

Funding: This work was funded by VTT Technical Research Centre of Finland Ltd under the Food Economy 4.0 programme and Academy of Finland [grant number: 316805].

Institutional Review Board Statement: The study was conducted according to the guidelines of the Declaration of Helsinki, and approved by the Ethical Committee of VTT Technical Research Centre of Finland Ltd (121739/Opportunity_FE_4.0_2019, approved 16.1.2019).

Informed Consent Statement: Informed consent was obtained from all participants from whom the questionnaire and stress data were collected.

Data Availability Statement: Data is available by request from the authors. The public availability of the data is restricted due to privacy and ethical reasons.

Acknowledgments: The authors acknowledge the human resources department experts as well as IT experts of IBM Finland for helping in recruiting the study participants and delivering objective stress data to the research group, Compass Group Finland Oy and the staff of the TestEat restaurant for keeping the study setting constant, recording food choices and measuring plate waste, Antti Väätänen and Timo Urhemaa for supporting the technical arrangements of the ambiences and Lauri Vanhatalo for processing the objective stress data.

Conflicts of Interest: The authors declare no conflict of interest. The funders had no role in the design of the study; in the collection, analyses, or interpretation of data; in the writing of the manuscript, or in the decision to publish the results.

References

1. Berto, R. The Role of Nature in Coping with Psycho-Physiological Stress: A Literature Review on Restorativeness. *Behav. Sci.* **2014**, *4*, 394–409. [CrossRef] [PubMed]
2. Schneiderman, N.; Ironson, G.; Siegel, S. Stress and Health: Psychological, Behavioral, and Biological Determinants. *Annu. Rev. Clin. Psychol.* **2005**, *1*, 607–628. [CrossRef] [PubMed]
3. Hassard, J.; Teoh, K.R.H.; Visockaite, G.; Dewe, P.; Cox, T. The Cost of Work-Related Stress to Society: A Systematic Review. *J. Occup. Health Psychol.* **2018**, *23*, 1–17. [CrossRef] [PubMed]
4. Yildirmaz, A.; Klein, S. The Workforce View 2020 | Volume Two post-COVID-19. 2020. Available online: https://www.adpri.org/assets/the-workforce-view-2020-post-covid-19/ (accessed on 9 November 2021).
5. WHO. Obesity and Overweight. Available online: https://www.who.int/news-room/fact-sheets/detail/obesity-and-overweight (accessed on 9 June 2021).
6. EAT-Lancet Commission. Summary Report of the EAT-Lancet Commission. 2019. Available online: https://eatforum.org/eat-lancet-commission/eat-lancet-commission-summary-report/ (accessed on 12 October 2021).
7. Henchion, M.; Hayes, M.; Mullen, A.M.; Fenelon, M.; Tiwari, B. Future Protein Supply and Demand: Strategies and Factors Influencing a Sustainable Equilibrium. *Foods* **2017**, *6*, 53. [CrossRef]
8. Stöckli, S.; Stämpfli, A.E.; Messner, C.; Brunner, T.A. An (Un)healthy Poster: When Environmental Cues Affect Consumers' Food Choices at Vending Machines. *Appetite* **2016**, *96*, 368–374. [CrossRef]
9. Kwasny, T.; Dobernic, K.; Riefler, P. Towards Reduced Meat Consumption: A Systematic Literature Review of Intervention Effectiveness, 2001–2019. *Appetite* **2022**, *168*, 105379. [CrossRef]
10. Kim, S.E.; Lee, S.M.; Kim, K.O. Consumer Acceptability of Coffee as Affected by Situational Conditions and Involvement. *Food Qual. Prefer.* **2016**, *52*, 124–132. [CrossRef]
11. Hersleth, M.; Monteleone, E.; Segtnan, A.; Næs, T. Effects of Evoked Meal Contexts on Consumers' Responses to Intrinsic and Extrinsic Product Attributes in Dry-Cured Ham. *Food Qual. Prefer.* **2015**, *40*, 191–198. [CrossRef]
12. Pennanen, K.; Närväinen, J.; Vanhatalo, S.; Raisamo, R.; Sozer, N. Effect of Virtual Eating Environment on Consumers' Evaluations of Healthy and Unhealthy Snacks. *Food Qual. Prefer.* **2020**, *82*, 103871. [CrossRef]
13. Bagozzi, R.P.; Gopinath, M.; Nyer, P.U. The Role of Emotions in Marketing. *J. Acad. Mark. Sci.* **1999**, *27*, 184–206. [CrossRef]
14. Rychalski, A.; Hudson, S. Asymmetric Effects of Customer Emotions on Satisfaction and Loyalty in a Utilitarian Service Context. *J. Bus. Res.* **2017**, *71*, 84–91. [CrossRef]
15. Lefebvre, S.; Hasford, J.; Wang, Z. The Effects of Guilt and Sadness on Sugar Consumption. *J. Bus. Res.* **2019**, *100*, 130–138. [CrossRef]
16. Tyrväinen, L.; Ojala, A.; Korpela, K.; Lanki, T.; Tsunetsugu, Y.; Kagawa, T. The Influence of Urban Green Environments on Stress Relief Measures: A Field Experiment. *J. Environ. Psychol.* **2014**, *38*, 1–9. [CrossRef]
17. Olafsdottir, G.; Cloke, P.; Schulz, A.; van Dyck, Z.; Eysteinsson, T.; Thorleifsdottir, B.; Vögele, C. Health Benefits of Walking in Nature: A Randomized Controlled Study Under Conditions of Real-Life Stress. *Environ. Behav.* **2020**, *52*, 248–274. [CrossRef]
18. Lu, J.; Huet, C.; Dubé, L. Emotional Reinforcement as a Protective Factor for Healthy Eating in Home Settings. *Am. J. Clin. Nutr.* **2011**, *94*, 254–261. [CrossRef] [PubMed]
19. Zelenski, J.M.; Dopko, R.L.; Capaldi, C.A. Cooperation is in Our Nature: Nature Exposure May Promote Cooperative and Environmentally Sustainable Behavior. *J. Environ. Psychol.* **2015**, *42*, 24–31. [CrossRef]
20. White, K.; Habib, R.; Hardisty, D.J. How to SHIFT Consumer Behaviors to be More Sustainable: A Literature Review and Guiding Framework. *J. Mark.* **2019**, *83*, 22–49. [CrossRef]
21. Russell, S.V.; Young, C.W.; Unsworth, K.L.; Robinson, C. Bringing Habits and Emotions into Food Waste Behaviour. *Resour. Conserv. Recycl.* **2017**, *125*, 107–114. [CrossRef]
22. Statista. 2015. Available online: https://www.statista.com/statistics/615827/locations-europeans-eat-their-lunch/ (accessed on 4 May 2021).
23. Lin, I.Y.; Mattila, A.S. Restaurant Servicescape, Service Encounter, and Perceived Congruency on Customers' Emotions and Satisfaction. *J. Hosp. Mark. Manag.* **2010**, *19*, 819–841. [CrossRef]
24. Stroebele, N.; De Castro, J.M. Effect of Ambience on Food Intake and Food Choice. *Nutrition* **2004**, *20*, 821–838. [CrossRef]

25. Hoppu, U.; Puputti, S.; Mattila, S.; Puurtinen, M.; Sandell, M. Food Consumption and Emotions at a Salad Lunch Buffet in a Multisensory Environment. *Foods* **2020**, *9*, 1349. [CrossRef] [PubMed]
26. Peng-Li, D.; Mathiesen, S.L.; Chan, R.C.K.; Byrne, D.V.; Wang, Q.J. Sounds Healthy: Modelling Sound-Evoked Consumer Food Choice Through Visual Attention. *Appetite* **2021**, *164*, 105264. [CrossRef] [PubMed]
27. Barton, J.; Pretty, J. What Is the Best Dose of Nature and Green Exercise for Improving Mental Health? A Multi-Study Analysis. *Environ. Sci. Technol.* **2010**, *44*, 3947–3955. [CrossRef] [PubMed]
28. Bratman, G.N.; Hamilton, P.J.; Daily, G.C. The Impact of Nature Experience on Human Cognitive Function and Mental Health. *Ann. N. Y. Acad. Sci.* **2012**, *1249*, 118–136. [CrossRef]
29. Elliot, A.J. Color and Psychological Functioning: A Review of Theoretical and Empirical Work. *Front. Psychol.* **2015**, *6*, 1–8. [CrossRef]
30. Dillman Carpentier, F.R.; Potter, R.F. Effects of Music on Physiological Arousal: Explorations into Tempo and Genre. *Media Psychol.* **2007**, *10*, 339–363. [CrossRef]
31. Alvarsson, J.J.; Wiens, S.; Nilsson, M.E. Stress Recovery During Exposure to Nature Sound and Environmental Noise. *Int. J. Environ. Res. Public Health* **2010**, *7*, 1036–1046. [CrossRef]
32. Richins, M.L. Measuring Emotions in the Consumption Experience. *J. Consum. Res.* **1997**, *24*, 127–146. [CrossRef]
33. Järvelin-Pasanen, S.; Sinikallio, S.; Tarvainen, M.P. Heart Rate Variability and Occupational Stress—Systematic Review. *Ind. Health* **2019**, *56*, 500–511. [CrossRef]
34. Dorado, R.; Chaya, C.; Tarrega, A.; Hort, J. The Impact of Using a Written Scenario When Measuring Emotional Response to Beer. *Food Qual. Prefer.* **2016**, *50*, 38–47. [CrossRef]
35. Piqueras-Fiszman, B.; Jaeger, S.R. Emotion Responses Under Evoked Consumption Contexts: A Focus on the Consumers' Frequency of Product Consumption and the Stability of Responses. *Food Qual. Prefer.* **2014**, *35*, 24–31. [CrossRef]
36. Van Baak, M.A. Meal-Induced Activation of the Sympathetic Nervous System and Its Cardiovascular and Thermogenic Effects in Man. *Physiol. Behav.* **2008**, *94*, 178–186. [CrossRef] [PubMed]
37. Corti, R.; Binggeli, C.; Sudano, I.; Spieker, L.; Hänseler, E.; Ruschitzka, F.; Chaplin, W.F.; Lüscher, T.F.; Noll, G. Coffee Acutely Increases Sympathetic Nerve Activity and Blood Pressure Independently of Caffeine Content Role of Habitual Versus Nonhabitual Drinking. *Circulation* **2002**, *106*, 2935–2940. [CrossRef] [PubMed]
38. Friis, R.; Skov, L.R.; Olsen, A.; Appleton, K.M.; Saulais, L.; Dinnella, C.; Hartwell, H.; Depezay, L.; Monteleone, E.; Giboreau, A.; et al. Comparison of Three Nudge Interventions (Priming, Default Option, and Perceived Variety) to Promote Vegetable Consumption in a Self-Service Buffet Setting. *PLoS ONE* **2017**, *12*, 1–16. [CrossRef]
39. Vermote, M.; Versele, V.; Stok, M.; Mullie, P.; D'Hondt, E.; Deforche, B.; Clarys, P.; Deliens, T. The Effect of a Portion Size Intervention on French Fries Consumption, Plate Waste, Satiety and Compensatory Caloric Intake: An On-Campus Restaurant Experiment. *Nutr. J.* **2018**, *17*, 43. [CrossRef]
40. Von Massow, M.; McAdams, B. Table Scraps: An Evaluation of Plate Waste in Restaurants. *J. Foodserv. Bus. Res.* **2015**, *18*, 437–453. [CrossRef]
41. Kuo, C.F.; Shih, Y. Gender Differences in the Effects of Education and Coercion on Reducing Buffet Plate Waste. *J. Foodserv. Bus. Res.* **2016**, *19*, 223–235. [CrossRef]
42. Attiq, S.; Habib, M.D.; Kaur, P.; Hasni, M.J.S.; Dhir, A. Drivers of Food Waste Reduction Behaviour in the Household Context. *Food Qual. Prefer.* **2021**, *94*, 104300. [CrossRef]
43. Wang, F.; Basso, F. "Animals Are Friends, Not Food": Anthropomorphism Leads to Less Favorable Attitudes Toward Meat Consumption by Inducing Feelings of Anticipatory Guilt. *Appetite* **2019**, *138*, 153–173. [CrossRef]
44. Menghini, L.; Gianfranchi, E.; Cellini, N.; Patron, E.; Tagliabue, M.; Sarlo, M. Stressing the Accuracy: Wrist-Worn Wearable Sensor Validation Over Different Conditions. *Psychophysiology* **2019**, *56*, e13441. [CrossRef]
45. Nelson, B.W.; Low, C.A.; Jacobson, N.; Areán, P.; Torous, J.; Allen, N.B. Guidelines for Wrist-Worn Consumer Wearable Assessment of Heart Rate in Biobehavioral Research. *NPJ Digit. Med.* **2020**, *3*, 1–9. [CrossRef] [PubMed]

foods

Article

Inulin as a Fat-Reduction Ingredient in Pork and Chicken Meatballs: Its Effects on Physicochemical Characteristics and Consumer Perceptions

Leidy Montoya, Natalia Quintero, Stevens Ortiz, Juan Lopera, Patricia Millán and Aída Rodríguez-Stouvenel *

School of Food Engineering, Universidad del Valle, Cali 76001, Colombia;
leidy.montoya@correounivalle.edu.co (L.M.); quintero.natalia@correounivalle.edu.co (N.Q.);
brayan.ortiz@correounivalle.edu.co (S.O.); juan.lopera@correounivalle.edu.co (J.L.);
patricia.millan@correounivalle.edu.co (P.M.)
* Correspondence: ayda.rodriguez@correounivalle.edu.co

Abstract: Fat reduction in meat products represents a technological challenge, as it affects the physicochemical and sensory properties of foods. The objective of the present investigation was to develop reduced-fat pork and chicken meatballs. In the initial stage, a survey was performed on 387 individuals, in order to determine the consumer perception of the meaning of a healthy meatball and the likelihood that they would consume such a product. In the second stage, four pork and chicken meatball formulations were developed: control meatballs (AC), meatballs with inulin (AI), meatballs with fructo-oligosaccharides (AF), and meatballs with inulin and fructo-oligosaccharides (AM). In the third stage, physicochemical properties were evaluated (water activity, humidity, fat, protein, ash, weight loss, pH, color, and texture) and a sensorial profile was created with semi-trained panelists for the four meatball formulations. In the fourth stage, AI was selected as the meatball with sensorial and physicochemical characteristics most similar to AC. An analysis of nutritional characteristics and a home test (84 consumers) were performed. The present study established that the inclusion of inulin as a fat substitute in the preparation of pork and chicken meatballs, in the amount of 3.5 g of fiber/100 g of the mixture, imitates the technological properties characteristic of fat and showed acceptance by consumers.

Keywords: chicken; consumer test; FOS; inulin; meatball; penalty analysis; pork; product development; reduced-fat

Citation: Montoya, L.; Quintero, N.; Ortiz, S.; Lopera, J.; Millán, P.; Rodríguez-Stouvenel, A. Inulin as a Fat-Reduction Ingredient in Pork and Chicken Meatballs: Its Effects on Physicochemical Characteristics and Consumer Perceptions. *Foods* **2022**, *11*, 1066. https://doi.org/10.3390/foods11081066

Academic Editor: Witoon Prinyawiwatkul

Received: 16 November 2021
Accepted: 10 March 2022
Published: 7 April 2022

1. Introduction

International nutrition policies are encouraging the reformulation of meat products either by eliminating or reducing some ingredients or incorporating bioactive compounds. The World Health Organization (WHO) recommends that fats should not exceed 30% of calories in the diet. Of this percentage, saturated fats should only encompass 10% of calories, and cholesterol consumption should be limited to 300 mg/day. In addition, the elimination of trans-fatty acid from the food supply are suggested by WHO [1]. In accordance with Resolution 810 of 2021 of the Colombian Ministry for Social Protection [2], which establishes nutritional and food labelling regulations, the Colombian food industry will begin to implement warning statements for foods high in saturated fat, sodium, and added sugar.

Researchers and processed meat producers seek healthier products by way of reducing nutritional deficiencies, without affecting product sensorial qualities or healthiness [3]. The options to the reformulation of meat products include the addition of antioxidants [4–7], sodium reduction [8,9], addition of dietary fiber [10–13], incorporation of monounsaturated and polyunsaturated fats [14–17] and fat reduction [18–24].

Fat replacement in meat products is achieved by the use of lean meat and the addition of water, in higher levels than in traditional products, as well as the addition of other ingredients with little or no caloric content [25]. Research in the development of reduced-fat meat products includes the use of cereal and legume fibers, β-glucans, starches, gums, pectin, aloe vera, and prebiotics. The development of processed meat faces a technological challenge, in terms of fat reduction, as this significantly alters the sensorial characteristics thereof, as fat serves important functions in the determination of three principal sensorial characteristics: appearance (color and surface uniformity), texture (viscosity, elasticity, and firmness), and flavor intensity. [13,25–28].

Fiber has been employed recently for the development of reduced-fat meat products, given the textural and organoleptic characteristics that it contributes, as well as the reduced caloric value and nutritional effects [29–31]. The fat substitution by canola and olive oils has caused greater hardness in low-fat hamburger meat; however, the incorporation of inulin (3.1%) and β-glucan (2.2%) could reduce this undesirable effect on the sensorial properties of hamburgers [14]. In addition, adding prebiotics to meat may improve the technological and sensorial characteristics of hamburgers [26]. The oatmeal β-glucans (OG) and their hydrolysates (OGH) as fat substitutes has been used in the production of pork meatballs, obtaining improved overall acceptance between samples [18]. Rye and pea fiber affect the sensorial qualities of meatballs resulting in gritty texture, less juiciness and crumbly texture for pea fiber meatballs [32].

Even though palatability in meat products has changed over time, tenderness, juiciness, and flavor are still considered important aspects for acceptability and preference for cooked meat. This is not only determined by the meat, in and of itself, but also by demographic aspects [33]. The development of healthier meat products represents a technological challenge that are finally evaluated by consumers, as when they are purchased, characteristics such as aroma, appearance, texture, and flavor become decisive factors. It is important to identify the way in which consumers perceive the sensorial attributes of food products, as demonstrated by their growing participation in the characterization of foods. [34,35]. The objective of the present study was to develop reduced-fat pork and chicken meatballs and to apply sensorial analysis methodologies to measure consumer perceptions of this type of product.

2. Materials and Methods

2.1. Consumer Sensorial Perceptions during Meatball Consumption

In order to evaluate the stimuli with descriptions of meatballs of different types, the Polizer et al. [36] methodology was employed. This was carried out with traditional meatballs (Stimulus A), one that was reduced in fat (Stimulus B), one with fiber (Stimulus C), and one without preservatives (Stimulus D). By way of a virtual format, the stimuli were presented to participants following the monadic presentation, and they were instructed to evaluate their perceptions, as related to both health and probability of consumption. One structured scale was utilized for each of the two evaluations, in accordance with Table 1.

Table 1. Scales used for the evaluation meatball consumption perceptions.

Perception of Healthiness		Probability of Consumption		Product Opinion	
Unhealthy 1	Extremely healthy 7	Very unlikely to try 1	Very likely to try 7	I don't like them at all 1	I like them very much 7

Data were collected from 387 individuals. They were recruited in Colombia, by way of social networks, email, and telephone. These individuals were chosen in accordance with the following characteristics: they were all over 18 years of age, consumers of pork and chicken, consumed processed meat products at least biweekly, and were from social division by strata on three, four, and five. Participants consumed meat products at least once per month, 84.5% of whom were in the age range of 18–45 years, 61.5% were women, 38.5% were men, 78% had undergraduate and/or postgraduate degrees, and 72.61% resided in the Valle del Cauca, Colombia.

2.2. Raw Materials for Meatball Preparation

Meats were purchased at a local supermarket. The chicken breast, lean pork loin, and bacon used complied with the quality requirements, in accordance with National Colombian Standards and Regulations [37–41]. High-purity refined salt (Refisal), parboiled P.A.N. white corn flour (Polar Inc.), paprika (Stravaganza), powdered oregano (El cocinerito), powdered bay leaf (El Rey), powdered garlic, pure ground thyme, powdered basil, powdered ginger, ground pepper, inulin, oligofructose, and Carl natural color (Tecnas-Ingredient Company at Colombia) were used.

2.3. Meatball Preparation

Meatballs are defined as processed meat products, whether cooked or raw, that are not packed as sausage, and are prepared with meat, the addition of substances whose use is permitted [42]. They were prepared in the Experimental Kitchen area at the Unitary Food Engineering Operations at the Universidad del Valle (Colombia). Previous studies with focal groups determined the base formulation for meatball preparation as well as concentrations of added fiber. The meats were received and stored at −18 °C, until their use. The remaining ingredients were stored in a fresh, dry area, at room temperature. Raw materials were weighed, in accordance with this formulation, for the creation of 320 meatballs in 1 kg batches (48 meatballs). The chicken breast, lean pork loin, and bacon fat were ground, using a meat grinder (ESSEN), and were added to a mixer (Kitchen Aid), consecutively adding the spices, salt, color, water, and corn flour over the course of five minutes. The meat mixture was stored in the freezer until it reached 4 ± 1 °C, in order to permit uniform shapes. Pieces of the meat mixture were weighed in amounts of 20.5 ± 0.5, using an Adventurer (OHASUS) scale. Meatballs were formed manually into spheres, and were boiled in water at 85 ± 2 °C, until their internal temperature rose to 80 ± 1 °C. They were then extracted from the water, drained for 30 s, and subjected to a thermal shock until their internal temperatures dropped to 35 ± 1 °C. Meatballs were packed on PET/WINPACK easy-open, transparent, thermoformed plastic trays measuring 615 µm and a PET/Coextruded. HBA (PE-EVOH) sheet measuring 80 µm, using a modified atmosphere of 40% CO_2 and 60% N_2.

Table 2 shows the formulations which correspond AC control meatball samples (neither reduced in fat, nor with added fiber), AI inulin meatballs (with reduced fat and added inulin dissolved in water), AF meatballs with fructo-oligosaccharides (with reduced fat and added FOS dissolved in water), and AM mixed meatballs (with reduced fat and added inulin and FOS dissolved in water). In the AI, AF, and AM samples, fat was replaced with the addition of 3.5 g of fiber (inulin and/or FOS) dissolved in water in a 1:2 proportion (fiber: water).

Table 2. Formulation of pork and chicken meatballs with added inulin and/or fructo-oligosaccharides.

Ingredients	Amount (g/100 g of Raw Meatballs)			
	AC	AF	AI	AM
Chicken breast	30.0	30.0	30.0	30.0
Lean pork loin	39.7	39.7	39.7	39.7
Pork fat	16.5	6.0	6.0	6.0
Corn flour	3.5	3.5	3.5	3.5
Water	6.0	6.0	6.0	6.0
Fiber water	0.0	7.0	7.0	7.0
FOS	0.0	3.5	0.0	1.8
Inulin	0.0	0.0	3.5	1.8
Salt	1.0	1.0	1.0	1.0
Ground garlic	0.7	0.7	0.7	0.7
Powdered onion	0.5	0.5	0.5	0.5
Ground oregano	0.2	0.2	0.2	0.2
Ground thyme	0.1	0.1	0.1	0.1
Ground bay leaf	0.1	0.1	0.1	0.1
Ground basil	0.1	0.1	0.1	0.1
Ginger	0.1	0.1	0.1	0.1
Ground black pepper	0.1	0.1	0.1	0.1
Carl natural color	1.5	1.5	1.5	1.5

2.4. Measurement of Physicochemical and Microbiological Parameters

Water activity. Water activity measurement occurred in accordance with the Association of Official Agricultural Chemists AOAC 978.19 method [43], using an AquaLab Vapor Sorption Analyzer Hygrometer VSA (METER Group, Washington, DC, USA). Slices of meatball centers were obtained and placed in the sample cup. Measurements were taken in triplicate.

Meatball pH. This measurement was taken in accordance with the AOAC 981.12 method [44], in which 10 g of meatballs were weighed, placed in a blender, 90 mL of distilled water was added, and these were blended for one minute. This was filtered, and a meat suspension was obtained, from which the pH measurement was taken, using an Orion Star A215 pHmeter (Thermo Scientific, Madrid, Spain). Measurements were taken in triplicate.

Humidity content. The preparation of the meat sample was performed in accordance with those methods established in the AOAC 983.18 [45], beginning with 200 g of ground meatball sample. The humidity determination occurred via an air stove drying method according to National Colombian Standard NTC 1663 [46], using a Binder stove, and placing 5 ± 0.01 g of sample at 125 ± 1 °C for five hours. Measurements were taken in triplicate.

Color. Chromatic properties were evaluated in septuplicate, in the crust and internal areas of each meatball, in accordance with those criteria described in International Organization for Standardization ISO 7724-2 [47] using a CM-5 Chroma Meter spectrophotometer (Konica Minolta Business Technologies, Tokyo, Japan), and as a reference, illuminant D 65 and 10° observer.

Texture measurement. The texture measurement occurred using texture analysis equipment (EZ-SX, Shimadzu Corp., Kyoto, Japan) with a load cell of 500 N and a five-blade Kramer cell. Measurements were taken in septuplicate, and samples were placed at 25 °C. Four meatballs were placed into the cell, the blade displacement velocity was 50 mm/min.

Physicochemical analysis. Total protein—ISO 1871:2009 [48], total fat—ether extract AOAC 920.39C [49], ash—AOAC 920.153 [50], fiber—AOAC 999.03 [51], fatty acid profile AOAC 969.33 [52], cholesterol—AOAC 976.26 [53], iron and sodium—AOAC 984.27/ICP—OES [54]. All measurements were taken in triplicate.

Microbiological analysis. Recount of aerobic mesophiles—AOAC 966.23 [55], NMP of total coliforms—ICMSF:2000 [56], recount of *Staphylococcus* positive coagulase—UNE EN ISO 6888 [57], *Salmonella* detection in 25 g—UNE EN ISO 6579 [58], recount of *Clostridium* spores reducing sulphite—ICMSF:2000 [56], detection of *Listeria monocytogenes*—UNE EN ISO 11290 [59] and detection of *Escherichia coli*—ISO 4832 [60] were all performed in triplicate.

2.5. Meatball Sensorial Profile

The descriptive analysis of the meatballs was performed in 60 sessions of 1–3 h, in accordance with the methodology established by ISO [61], NTC [62–65], and Kemp et al. [66]. A total of 8 candidates were selected based on: (i) normal acuity for the different senses include non-existence of physical or physiological deficiencies; (ii) discriminative ability, evaluated through of triangular and ranking test; and (iii) ability to describe their sensory perceptions. In order to establish the evaluation scale, a consensus was reached between panelists for descriptors. The intensities of the descriptors were determined using a non-structured 10 cm long scale next to each descriptor that was anchored on the left side, for the terms "slight", "none", and "light", and on the right side, "strong", "a lot", "dark", and "high" (Table 3). The methodology indicated by Saldaña et al. [67] and Quadros et al. [68] was used to define both the scale and references.

Table 3. Definitions and references of descriptive terms for meatballs, which correspond to the ends of the measurement scales.

Attribute	Definition	Reference
Surface homogeneity	Visual perception of homogeneity on the surface of the meatball.	Slight: Cracked meatball. Meatball formulation without corn flour High: Well-formed meatball. Meatball formulation with corn flour (8.5%)
Meat aroma	Intensity of the aroma characteristic of chicken and pork.	None: Distilled water Strong: Chicken breast and/or pork loin cooked in water
Spiced aroma	Intensity of the aroma characteristic of the spices (oregano, thyme, bay leaf, basil, pepper, garlic, onion).	None: Spiceless meatball Strong: Meatball with 3.5% spices
Fatty aroma	Intensity of the aroma of fat.	None: Chicken breast cooked in water Strong: Fried bacon
Meat flavor	Intensity of the flavors characteristic of chicken and pork.	None: Distilled water Strong: Chicken breast and/or pork loin boiled in water
Spiced flavor	Intensity of the flavors characteristic of the spices (oregano, thyme, bay leaf, basil, garlic, onion).	None: Spiceless meatball Strong: Meatball with 3.5% spices
Fatty flavor	Intensity of the flavor of fat.	None: Chicken breast boiled in water Strong: Fried bacon
Salty flavor	Intensity of the salty flavor, associated with the presence of salt.	None: Saltless meatball Strong: Meatball with 2.5% salt
Sweet flavor	Intensity of the sweet flavor.	None: Sugarless meatball Strong: Meatball with 14% sugar
Firmness on cutting	Degree of meatball firmness when cut. This is closely related to cohesiveness. It is rated firm and cohesive when the meatball is cut with a knife without any loss of structure or crumbling.	Slight: Commercial meatball High: Meatball formulation with corn flour (8.5%)
Juiciness	Perception of water absorbed or freed from the meatball during chewing. A succulent meatball frees a great deal of liquid as the product is chewed, and one that is not frees very little liquid, producing the sensation of a dry product.	Slight: Meatball formulation with corn flour (8.5%) High: Commercial meatball
Cohesiveness	Degree to which the meatball stays together or compact.	Slight: Commercial meatball High: Meatball formulation with corn flour (8.5%)
Fatty sensation	Amount of fat perceived in the mouth, especially on the palate and lips.	None: Chicken breast boiled in water High: Fried bacon

The objective of the training consisted of the memorization of references at the extremes of each attribute. The panel's performance and final evaluator selection occurred considering three criteria: discrimination, reproducibility, and consensus. The final evaluation occurred in accordance with the vocabulary developed by the panel, using three meatballs per session. The panel consisted of eight judges, who evaluated 13 sensorial attributes related to appearance, aroma, texture, and flavor.

2.6. Consumer Sensorial Evaluation of Meatballs

The sensorial evaluation of reduced-fat meatballs occurred by way of a home test. A total of 84 consumers were recruited at Cali (City of Colombia), by way of social networks, email, and telephone. These individuals were chosen in accordance with the following characteristics: they were all over 18 years of age, consumers of pork and chicken, consumed processed meat products at least biweekly, were not allergic to any of the product ingredients, and were from strata three, four, and five. The descriptive statistics of consumers are shown in the Figure 1. Samples were distributed the day after their preparation, and meatballs were guaranteed to be kept in portable refrigerators during delivery to each home. One tray of meatballs, packed in modified atmospheres, and one 100 g packet of ranchera sauce were provided. Preference studies performed previously, with 100 consumers, had established that this type of sauce was considered adequate as an accompaniment to pork and chicken meatballs. Each individual signed informed consent, and received a product evaluation survey with the instructions to be followed. On the form, it indicated that the meatball was already cooked and ready for consumption, and that it only required heating. In the tasting process, the meatballs could be heated together with a moderate amount of ranchera sauce, or could be prepared in the way in which the individual wished to consume these (with pasta, Bolognese sauce, without sauce, etc.).

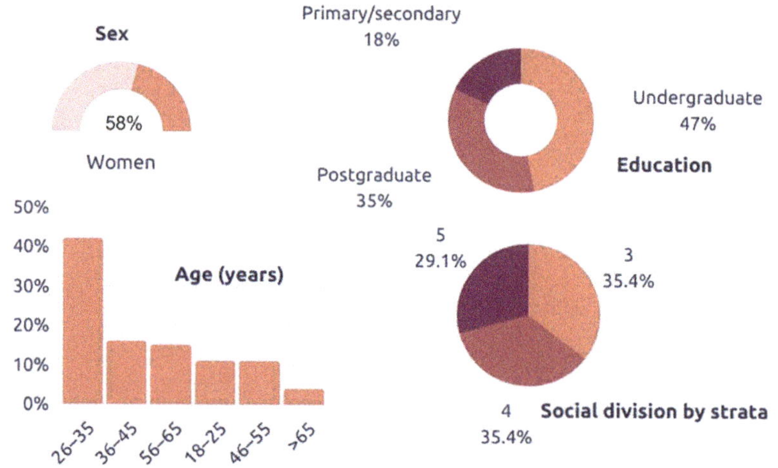

Figure 1. Descriptive statistics of consumers who resided in the Valle del Cauca, Colombia. (n = 84).

The scale structure was designed considering NTC 5328 [65]. All panelists evaluated the samples in terms of overall acceptance, appearance, aroma, texture, and meatball flavor, on a nine-point hedonic scale (9 = I like them very much, 1 = I strongly dislike them). In addition, a three-point "JAR Just About Right" scale was used to evaluate color, firmness, juiciness, saltiness, the intensity of the spiced flavor, and fat level (3 = more than I like, 2 = just as I like, 1 = less than I like), and an open question was prepared, such that panelists could describe the reasons for which they did or did not like the samples evaluated. At the end, the consumer was asked to indicate how healthy they considered the meatballs to be (on a scale of 1–7, where 1 was not at all healthy, and 7 was very healthy) and the probability that they would purchase them (on a scale of 1–7, where 1 was not at all probable, and 7 was very probable). The Penalty Analysis was performed in accordance with the methodology indicated by Concha-Meyer et al. [69], Davis et al. [70], Naes et al. [71], and Lawless et al. [72].

2.7. Statistical Analysis

A completely randomized, unifactorial design, with fixed effects, and two replicates was employed. The factor which corresponded to fiber and the four levels which corresponded to the different types of meatball formulations with fiber were as follows: inulin, FOS, mixture (inulin-FOS), and control (fiberless meatball). All results were subjected to Kolmogorov–Smirnov and Shapiro–Wilk normality tests occurred by way of SPSS statistical software (IBM SPSS v25.0, Chicago, IL, USA).

The physico-chemical results were analyzed using ANOVA and the Tukey test (parametric statistics). For the data of probability of consumption and perception of healthiness for meatballs a paired comparison test (nonparametric statistics) was used. The data during the training of the panelists were analyzed by means of binomial test, adjusted chi-square, normal distribution test and Walds analysis (parametric statistics). Sensory profile data of the 4 meatball formulations evaluated by trained panelists were analyzed by ANOVA. The hedonic data from the consumers were analyzed using nonparametric statistics (Kruskall–Wallis test) because they did not have a normal distribution. The penalty analysis for the sensorial analysis occurred by way of Xlstat software (Addinsoft, Paris, France), with which "Just About Right" scale (JAR) results permitted the identification of those attributes which influenced overall acceptance scores for meatballs with inulin. Variables were considered significant with $p < 0.05$, and differences were determined based on the Tukey HSD test.

3. Results

3.1. Consumer Sensorial Comparison of Meatball Consumption

Consumers considered the reduced-fat, preservative free, and added-fiber meatballs to be equally healthy. However, they perceived these meatballs to be significantly healthier (median = 6.0) than traditional meatballs (median = 4.0) (Figure 2 and Table 4).

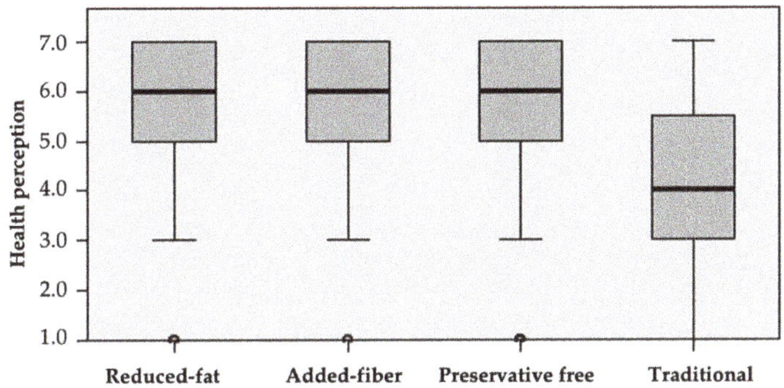

Figure 2. Median test for health perception of meatballs.

Table 4. Paired comparison test for meatballs perception of healthiness.

Sample 1–Sample 2	Test Statistic	Std. Error	Std. Test Statistic	Sig.	Adj. Sig.
Traditional-Added fiber	314.540	28.965	10.859	0.000	0.000
Traditional-Reduce fat	338.440	28.965	11.684	0.000	0.000
Traditional-Preservative free	366.763	28.955	12.662	0.000	0.000
Added fiber-Reduce fat	23.900	28.955	0.825	0.409	1.000
Added fiber-Preservative-free	−52.222	28.955	−1.803	0.071	0.428
Reduce fat-Preservative free	−28.322	28.955	−0.978	0.328	1.000

Figure 3 identifies that there is a significantly higher probability of meatball consumption when they have added fiber, are reduced in fat, or are preservative free (median = 6.0), than a traditional meatball (median = 5.0). There is a greater tendency to consume preservative-free meatballs than those with added fiber, whereas the probability of consumption is equal between added-fiber and reduced-fat meatballs, as well as for reduced-fat and preservative-free meatballs (Table 5).

In the "Voice of the Industry: Food & Nutrition" survey, performed by Euromonitor in 2021, it was found that the functionality of foods as key components in well-being has become an important element for consumers. Over 75% of industry professionals hope that individuals focus much more on health and well-being in the next five years, given the impact of COVID-19 on the way in which consumers eat [73]. The results obtained in the present study indicate that new global consumer tendencies are also reflected on a national level, where reduced-fat meatballs are considered healthier than traditional meatballs, and as such, these are more likely to be consumed.

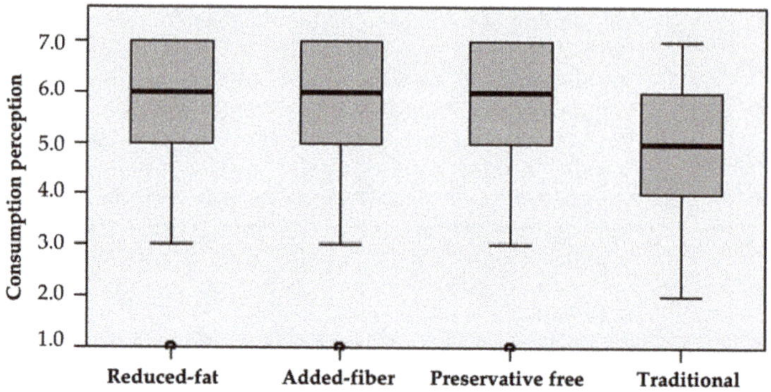

Figure 3. Median test for probability of meatball consumption.

Table 5. Paired comparison test of meatball probability of consumption.

Sample 1–Sample 2	Test Statistic	Std. Error	Std. Test Statistic	Sig.	Adj. Sig.
Traditional-Added fiber	120.381	28.781	4.183	0.000	0.000
Traditional-Reduce fat	166.140	28.738	5.781	0.000	0.000
Traditional-Preservative free	221.446	28.803	7.688	0.000	0.000
Added fiber-Reduce fat	45.759	28.781	1.590	0.112	0.671
Added fiber-Preservative-free	−101.065	28.846	−3.504	0.000	0.003
Reduce fat-Preservative free	−55.306	28.803	−1.920	0.055	0.329

The preservative free term is perhaps the most important for consumers. The preservative free meatballs were predominantly perceived as healthy relative to other associations, and as such, their probability of consumption is also greater. In accordance with the information reported by Mintel, the main meat product launches in 2020 were food apt for microwaves (28.1%), that were additive/preservative free (20%), low in/without allergens (18.2%), easy to use/prepare (17.6%), and gluten free (16.8%) [74].

3.2. Physicochemical, Microbiological Characteristics and Sensorial Profiles (Trained Panelists) of Four Meatball Formulations (AC, AI, AF, AM)

Table 6 shows the results of the physicochemical characteristics of pork and chicken meatballs. Protein and ash contents did not present significant differences between samples, while differences in fat content were directly related to the amount of added fat in the AI, AF, and AM formulations. Color parameters (L, a, b, C, H) in the internal and external parts of the meatballs did not present significant differences between treatments, except for luminosity of the internal part of samples AI, AF, and AM, which were higher than in AC.

The fat reduction and addition of fiber-water significantly affected ($p < 0.05$) meatball humidity content, and neither affected the water activity thereof, nor the pH. The loss of weight was lower in AC than in AI or AF, and was highest in AM. Kramer maximum force was significantly higher for AC meatballs than added-fiber meatballs. This could be due to the fact that fiber prevented the rigidity characteristic of reduced-fat meat products.

Certain soluble fibers are used as fat imitators, as is the case of inulin, a prebiotic dietary fiber with broad use as a fat replacement in various food matrices, it improves viscosity, forms gels, increases water-retention capacity, and improves mouth texture [30]. In the present study, the addition of water, together with fiber, may explain many of significant differences in physicochemical parameters and between treatments. It caused less toughness, greater humidity, and loss of weight, compared to the AC control sample (Table 6).

Observe that, in Table 7, the pork and chicken meatballs comply with the microbiological requirements of Technical Colombian Regulation 1325 [42], in terms of the recount of mesophilic aerobes (CFU), Coliforms (CFU), *Staphylococcus aureus* (CFU), *Salmonella* sp./25 g of sample, *C. perfringens reducing sulphite* (CFU/g), *Listeria monocytogenes*/25 g of sample, and *E. coli*, indicating that it is an innocuous product, and as such, is apt for consumption.

Table 6. Physicochemical pork and chicken meatball parameters.

Physicochemical Parameter	Treatment			
	AC	AI	AF	AM
Humidity (%wb)	59.9 ± 0.000 [a]	65.9 ± 0.000 [b]	65.6 ± 0.002 [b]	65.8 ± 0.001 [b]
pH	6.008 ± 0.010 [a]	6.032 ± 0.004 [a]	6.028 ± 0.002 [a]	6.016 ± 0.007 [a]
Water activity	0.987 ± 0.000 [a]	0.986 ± 0.001 [a]	0.989 ± 0.001 [a]	0.988 ± 0.001 [a]
Kramer maximum force (N)	248.275 ± 8.049 [a]	220.689 ± 6.956 [b]	217.305 ± 4.651 [b]	219.618 ± 5.414 [b]
Weight loss (%)	11.96 ± 0.44 [a]	17.16 ± 4.99 [b]	17.59 ± 2.87 [b]	19.50 ± 0.98 [c]
Color				
Crust				
L*	59.420 ± 0.183 [a]	60.171 ± 0.037 [a]	60.081 ± 0.037 [a]	59.523 ± 0.085 [a]
a*	6.342 ± 0.009 [a]	6.522 ± 0.0497 [a]	6.411 ± 0.012 [a]	6.513 ± 0.011 [a]
b*	13.136 ± 0.007 [a]	12.793 ± 0.026 [a]	12.428 ± 0.137 [a]	12.723 ± 0.174 [a]
C*	14.591 ± 0.012 [a]	14.369 ± 0.016 [a]	13.985 ± 0.129 [a]	14.297 ± 0.163 [a]
h*	64.167 ± 0.001 [a]	62.950 ± 0.281 [a]	62.699 ± 0.163 [a]	62.893 ± 0.204 [a]
Center				
L*	58.471 ± 0.014 [a]	59.455 ± 0.172 [b]	59.222 ± 0.007 [b]	59.194 ± 0.021 [b]
a*	6.620 ± 0.007 [a]	6.821 ± 0.057 [a]	6.826 ± 0.064 [a]	6.818 ± 0.016 [a]
b*	14.280 ± 0.158 [a]	14.435 ± 0.040 [a]	14.489 ± 0.064 [a]	14.124 ± 0.131 [a]
C*	15.741 ± 0.166 [a]	15.969 ± 0.003 [a]	16.020 ± 0.093 [a]	15.685 ± 0.126 [a]
h*	65.116 ± 0.0253 [a]	64.709 ± 0.054 [a]	64.770 ± 0.034	64.235 ± 0.046 [a]
Proximate composition				
Fat (g/100 g)	15.715 ± 0.488 [a]	6.65 ± 0.424 [b]	6.145 ± 0.615 [b]	7.095 ± 0.898 [b]
Protein (g/100 g)	18.205± 2.751 [a]	17.940 ± 0.424 [a]	18.815 ± 0.615 [a]	18.320 ± 0.898 [a]
Ash (g/100 g)	1.455 ± 0.035 [a]	1.450 ± 0.042 [a]	1.45 ± 0.000 [a]	1.51 ± 0.014 [a]

Data averages shown ± standard deviation. Different superscript letters in the same row indicate different significance ($p < 0.05$). Lightness (L*), redness+/greenness (a*), yellowness+/blueness (b*), Chroma (C*) and Hue angle (h*).

Table 7. Microbiological meatball analysis.

Microbiological Analysis	NTC 1325	Result
Mesophilic aerobes, CFU/g	<100,000	86.67
Coliforms, CFU/g	100–500	13
S. aureus positive coagulase, CFU/g	<100	<100
Salmonella detection, 25 g	None	None
C. perfringens reducing sulphite spores, CFU/g	<10–100	<10
Listeria monocytogenes, 25 g	None	None
E. coli, g	<10	None

Figure 4 shows the sensorial profile of pork and chicken meatballs accomplished by eight semi-trained panelists. The AC meatballs had the greatest homogeneity, fatty aroma, fatty sensation, juiciness, and meat flavor. The AI, AF, and AM meatballs had greater sweet flavor intensities than AC, as sweetness is characteristic of inulin and oligo-fructose. The challenge in the development of reduced-fat meat products is to obtain the texture and flavor which is as close as possible to those foods with high fat content. Despite the differences between meatballs with fiber and control meatballs, AI presented the sensorial characteristics most closely related to those of AC; for this reason, it was selected for the performance of hedonic sensorial tests and the determination of final consumer acceptance.

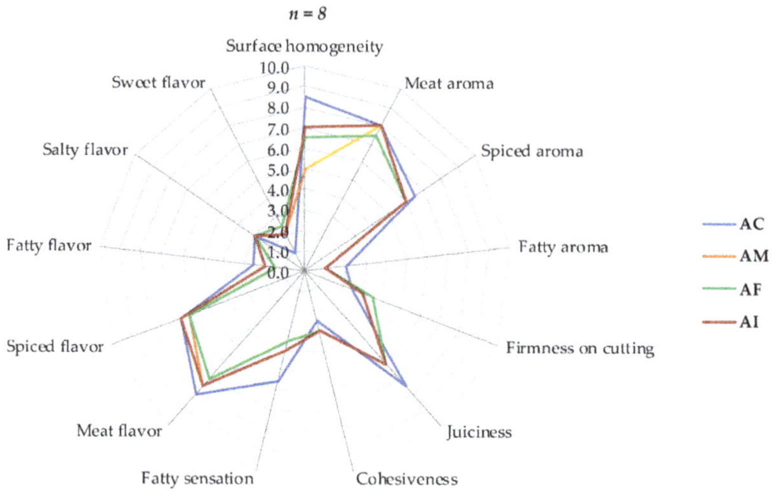

Figure 4. Sensorial profile of pork and chicken meatballs performed by semi-trained panelists (n = 8). Control AC meatballs (full fat, no fiber added), meatballs with inulin AI (reduced fat and added inulin dissolved in water), meatballs with fructo- oligosaccharides AF (reduced fat and with added FOS dissolved in water), and meatballs with the AM mixture (reduced fat and added inulin mixture and FOS dissolved in water).

3.3. Selected Meatball (AI) Sensorial Evaluation with Consumers (Home-Based Test)

The objective of the present investigation with consumers was to establish the acceptability of reduced-fat pork and chicken meatballs with added inulin. On a nine-point hedonic scale (1 = I strongly dislike them, 9 = I like them very much), consumers indicated an overall product acceptance of 7.4, appearance of 7.0, smell of 7.5, texture of 7.4, and flavor of 7.6 (Figure 5), indicating high consumer product acceptance.

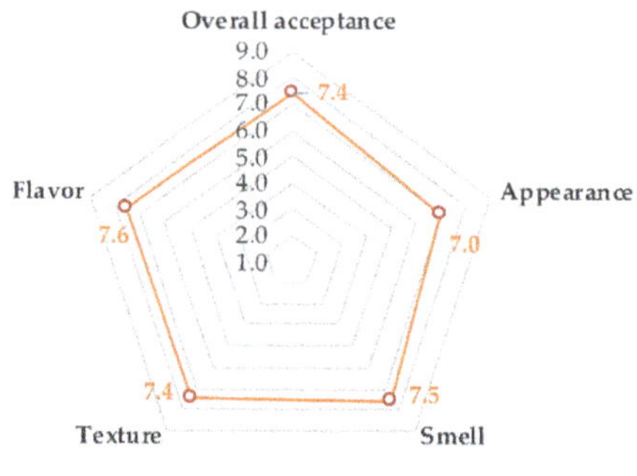

Figure 5. Radial graph of the sensorial characteristics of reduced-fat, preservative-free (AI) pork and chicken meatballs performed by consumers (n = 84) via a hedonic test.

The evaluated JAR scale identified that the toughness and salt levels were correct, provided that the spice and fat content were also at the right levels for the majority of the population. Exactly 48% of the consumers evaluated indicated that the juiciness was adequate, while 26% considered it to be very high, and 27% very low. Further, 61% of consumers considered the meatball to have a very light color, while 37% considered it adequate (Figure 6). This parameter is very dependent on the characteristics of raw materials and meatball color associations with beef, which is darker. Despite the addition of natural coloring to the formulation, in order to improve the color intensity, without being too near to the color of beef meatballs. This study determined that a more intense color is preferred.

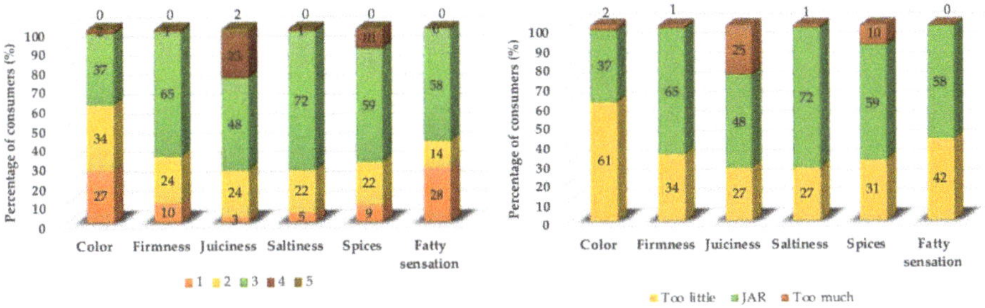

Figure 6. Results of the Just About Right (JAR) scale for reduced fat, preservative-free meatballs (AI) performed by consumers (n = 84).

The penalty analysis was performed to determine the influence of color, toughness, juiciness, saltiness, spice intensity, and fatty flavor on the overall acceptance of the meatballs, in order to guide the development of pork and chicken meatballs. Table 8 shows the penalty analysis. Observe that there was statistical significance ($p < 0.05$) in the penalty for the very light color, low spice levels, and low fatty flavor, which generated an impact on consumer acceptance. If one or several of these parameters were improved, the overall product acceptance (7.4) could increase. Specifically, it is recommended that the product color be improved, with an increased proportion of the natural color used.

Table 8. Penalty analysis for reduced-fat pork and chicken meatballs without preservatives (AI).

Variable	Level	%	Mean Drops	p-Value for the Endpoint	Penalties	p-Value for the Attribute
Color	Very clear	60.76	0.654	0.012	0.619	0.017
	JAR	36.71				
	Very dark	2.53	−0.241			
Firmness	Very soft	34.18	0.399	0.139	0.403	0.128
	JAR	64.56				
	Very firm	1.27	0.510			
Juiciness	Not juicy	26.58	1.197		0.865	0.000
	JAR	48.10				
	Very juicy	25.32	0.516			
Saltiness	Saltless	26.58	0.947		0.887	0.001
	JAR	72.15				
	Very salty	1.27	−0.386			
Spiced flavor	Very little	30.38	0.784	0.006	0.617	0.016
	JAR	59.49				
	Very intense	10.13	0.117			
Fat sensation	Very low	41.77	0.630	0.013		
	JAR	58.23				
	Very high	0.00				

3.4. Nutritional Composition of Selected Meatballs (AI)

The nutritional composition of pork and chicken meatballs, with the inclusion of inulin, is shown in Table 9. The initial formulation was created, considering Technical Colombian Regulation NTC 1325 [42], for premium products. In the table, the characteristics corresponding to protein above 14% and fat below 28% are corroborated. This characteristic is attributed to the fact that the pork loin and chicken breast employed in this formulation are lean meats with high protein content.

Table 9. The nutritional composition of reduced fat preservative-free pork and chicken meatballs (AI).

Nutrient	Amount
Protein (g/100 g)	17.94 ± 0.0
Fat (g/100 g)	6.65 ± 0.30
Saturated fat (g/100 g)	2.325 ± 0.18
Monounsaturated (g/100 g)	2.865 ± 0.03
Polyunsaturated (g/100 g)	1.46 ± 0.10
Trans isomers (g/100 g)	0.035 ± 0.02
Ash (g/100 g)	1.44 ± 0.01
Iron (mg/100 g)	1.615 ± 0.11
Cholesterol (mg/100 g)	5.253 ± 3.50
Sodium (mg/100 g)	340.75 ± 34.21
Fiber (g/100 g)	0.56 ± 0.01

The combination of two protein types (pork loin and chicken breast) permitted the meatballs obtained in the present study to be considered an excellent source of protein, with 17.94 g of protein in 100 g of product. Authors such as Aripin et al. [75] reported the preparation of meatballs with chicken and duck meat in various concentrations, to obtain low protein content (from 5.88–10.45%). Yeung and Huang [76] prepared pork meatballs with the incorporation of different additives, to obtain elevated protein levels (from 15.74–16.04%).

The total and saturated fat in meatballs with inulin represent just 10% of daily intake (based on a 2000 kcal diet), as compared to commercial meatballs, which provide over 17% of the daily intake value and saturated fat total. The sodium content of the prepared meatballs was low, as per 100 g of product, 340.75 mg is ingested, representing 17.03% of the intake established by the World Health Organization, which is 2000 mg of sodium per day, per person. In commercial meatballs, however, one ingests 500 mg of sodium per 68 g of product, or 25% of the recommended intake.

The result of fiber content in the inulin meatballs was unexpected, as 3.5% of inulin was added during preparation, and only 0.56 g of inulin is reported in 100 g of product. There may be inulin losses throughout the cooking process which oscillate within 3–5%, and this yields content of greater than 3 g of inulin in 100 g of product. Angiolillo et al. [26] reported similar results on the inclusion of inulin, FOS, and oat bran to hamburger meat, where the fiber analysis came out null for meat with FOS and inulin, and at 0.21 g per 100 g of meat with the inulin and oats.

4. Discussion

The consumer sensorial perception survey for meatball consumption permitted confirmation that claims are an important part of the development and labelling of a food. Meatballs with additional claims, such as reduced-fat, added fiber, or preservative free have better consumer perceptions than those without claims on their label. In the case of preservatives, it was shown that, despite the fact that claims of reduced-fat and extra fiber are considered the same for healthy components, this is not part of the consumption perception, as there is a higher probability that an individual consumes meatballs labeled preservative free rather than added fiber.

In the preparation of the four meatball formulations (AC, AI, AF, and AM), significant differences were found in terms of certain physicochemical and sensorial parameters measured. AC was the sample with the lowest humidity, because in the reduced fat formulation, the replacement of fat was accompanied by the addition of fiber dissolved in water. As such, the amount of water present in AC was lower than in the other samples. Despite the differences in the amount of added water, the available water is equal in all formulations, as fiber has a high-water retention capacity, generating an interaction with the water added to these formulations. In the present study, neither water activity nor pH changed owing to fat reduction, similar to that which occurred with the addition of short-chain fructo-oligosaccharides in low-fat sausage [77].

The samples' losses of weight during cooking in AI, AF, and AM indicate that, despite the fibers' water retention capacity [78], there is a portion of water that was not immobilized by the interactions created in the gel network, permitting their liberation to the external cooking media. Some researchers have reported this characteristic as a technological disadvantage of inulin [30], and others have found that inulin improves cooking performance and presents advantages in emulsion stabilization [28,79,80]. In certain cases, it has been observed that soluble, insoluble, and starch fiber mixtures result in a reduction in the weight loss of hamburgers and meat emulsions [26,81].

In terms of color, the luminosity of the internal part of meatballs AI, AF, and AM was greater than in AC. This is likely due to the brightness provided by the fiber gels in the meat matrix. Huang et al. [82] reported that there were no color changes in sausages with added fiber: inulin, wheat fiber, and oat fiber, while Menegas et al. [83] observed a tendency toward lighter and reddish coloring in fermented chicken sausages with added inulin and corn oil. However, Cáceres et al. [77] reported decreases in luminosity values in reduced-fat sausages with added fructo-oligosaccharides. Paglarini et al. [84] found that the addition of a high-fiber emulsion gel containing inulin in Bologna increased L* (lightness) values and reduced a* (redness/greenness) values comparing to control treatment.

The texture, measured as the firmness in meatballs (Kramer) was significantly higher in control meatballs than in those with added fiber. The firmness of cut and cohesiveness measurements of the panelists did not present significant differences between the AI, AF, AM, and AC treatments, while juiciness was highest in AC. Huang et al. [82] reported an increase in firmness in sausages with added wheat and oat fiber, while those with added inulin did not present texture differences. Keenan et al. [79] found that firmness values increased with the concentration of inulin, and panelists also rated products with inulin to be less tender. Other authors establish that the addition of FOS to meat products permitted slightly more tender products than those without FOS or fat reduction [77].

Inulin is used as a fat replacement, as in the presence of water, it develops particulate gels which improve the product texture, with a mouth feel similar to that produced by fat [85]. In the present study, it was observed that, depending on the added fiber and water content, as well as the fat replaced, lower fat content may reduce juiciness, causing greater toughness. However, fiber may mimic this characteristic, to a certain point, and provide tenderness. It may also generate a mass that is excessively tender, so as to produce unpleasant sensorial characteristics that may significantly affect product homogeneity. For this reason, control meatballs had a more defined form than those with added fiber.

The addition of fiber to meat matrices induces the formation of stable gel networks [25,85], which permits the improvements of meatball rheology and texture characteristics, in this case. However, many of these characteristics are dependent on the remaining food matrix components, emulsion formation, meat grind, product cooking, and especially the amount of water added in preparation. Considering the physicochemical characterization, results with trained panelists permitted the selection of the formulation with sensorial characteristics most like the control meatballs, which was that with inulin (AI), which was then put to consumer sensorial analysis.

The results of the reduced-fat pork and chicken meatball (AI) sensorial analysis performed by consumers showed high levels of product acceptability in terms of overall acceptance characteristics, appearance, smell, texture, and flavor. The flavor of spices perceived by consumers was found to be at the correct level.

For consumers, the salt level was also found to be at the correct level (JAR) Meat products are characterized by high levels of sodium chloride (table salt), as this may be used, industrially, at concentrations of 1–2.5% in cooked products, and from 2–6% in raw, cured products [86]. In the formulation, 1% salt (1 g/100 g of mixture prior to cooking), reflecting a sodium content in the final product that is not high. This responds to public health policies that have established regulations in different countries to monitor sodium consumption, thus creating a need in the food industry to reduce sodium in their products.

The fatty flavor was adequate (JAR) for consumers and reduced-fat meatball acceptance was good. Juiciness is another important characteristic in the development of reduced-fat foods. The results were inconclusive for consumer perceptions thereof, as 48% indicated that the product's juiciness was adequate, and 27% reported it to be very low, and 26% reported it to be quite high. However, the population rated the texture parameter as 7.4 on a hedonic scale of 1–9, which indicates good acceptability.

The penalty analysis indicated that the light color, minimal spice flavor, and minimal fat flavor characteristics are penalized highly by the consumer, when providing overall product acceptance ratings. Less than 20% of consumers identified a "minimal spice flavor" or "minimal fat flavor", for which reason improvements to these parameters are not proposed. Most consumers indicated that the meatballs presented a "very light color", which statistically penalized the overall acceptance of the product. As such, a formulation with greater natural color content could be proposed.

This product is reduced in fat, and an excellent source of protein. Saturated fat and sodium were both below those limits established for food labelling regulations. It was expected that 90% of the added fiber would be retained, since theoretically only a maximum of 5% of soluble fiber is lost in cooking processes. Although the fiber generated a positive technological effect, the fiber content at the end of cooking was very low. It would be important to set the modifications to the fiber quantification methodologies after meat cooking processes.

The COVID-19 pandemic situation generated restrictions for access to laboratories and forced us to change the methodology of sensory analysis with consumers from a room location test to a home test. Initially, it was planned to evaluate the 4 formulations of meatballs with consumers, but finally, the evaluation was done with trained panelists. Only one formulation was selected to evaluation with consumers.

In future research it would be important to determine the effect of inulin in other types of products such as hamburgers, sausages and nuggets made from pork and chicken. This would allow identifying if it is possible that inulin, in addition to the technological effect on fat reduction, also allows obtaining a food that is a good source of fiber.

The influence of an ingredient in the development of reduced-fat foods should be accompanied by a shelf-life study of this food, especially when there are no preservatives in its formulation. It is recommended to always perform stability studies to complement the effect of these ingredients. In this case, shelf-life studies with the selected formulation were carried out and will be published later.

5. Conclusions

The declarations preservative free, reduced fat, and added fiber are of great importance for consumers when they consider meatballs as health food. However, there is greater probability of meatball consumption when they claim to be "preservative-free", as compared to others that claim to be "reduced-fat" or a "good source of fiber". Definitively, a meatball with no nutritional claim presents minimal perceptions of a healthy food, and has reduced probability of consumption.

The acceptance of the appearance of pork and chicken meatballs is predetermined by the association with the color of beef meatballs. Even though this does not represent a significant parameter on evaluation of their general sensorial acceptance and the purchase decision, this may increase appearance acceptance, if higher concentrations of natural color are used in the formulation, so as to obtain a food product with a darker brown color.

The inclusion of inulin as a fat substitute in the preparation of pork and chicken meatballs (3.5 g of fiber/100 g of mixture) permits the imitation of the technological properties of fat, without significantly affecting the sensorial characteristics of the food.

Author Contributions: Conceptualization, L.M., A.R.-S., P.M.; Methodology, L.M., N.Q., J.L., S.O., P.M.; Software and data analysis L.M., N.Q., J.L., S.O.; Writing and preparation of the first draft, L.M.; Writing, review, and editing, L.M., A.R.-S.; Supervision, L.M., A.R.-S., P.M. All authors have read and agreed to the published version of the manuscript.

Funding: The present article was developed within the framework of a project entitled, "Fortalecimiento de las capacidades científicas y tecnológicas para impulsar la innovación y competitividad de la cadena productiva de proteína blanca Valle del cauca," which was financed by Colciencias, through the Science, Technology, and Innovation Fund-FCTeI of the General Royalties System-SGR. The funding number is BPIN 2017000100075 code. Eleven companies from the White Protein Cluster made economic contributions for the development of this project (Agrícola Colombiana S.A., Avícola Santa Rita S.A.S., Avidesa De Occidente S.A., Carnes Frías Enriko Ltda, Cerdos Del Valle S.A, Incubadora Santander S.A., Nutrientes Avícolas S.A., Pollos El Bucanero S.A., Productora Nacional, Avícola S.A., Santa Anita Nápoles S.A., Súper Pollos Del Galpón S.A.S.), and the FENAVI association.

Institutional Review Board Statement: The study was conducted in accordance with the Declaration of Helsinki, and approved by the Ethics Committee of the Faculty of Engineering (CEIFI) at UNIVERSIDAD DEL VALLE (protocol number: 012021).

Informed Consent Statement: Informed consent was obtained from all subjects involved in the study.

Data Availability Statement: Data is contained within the article.

Acknowledgments: We thank the Cali Chamber of Commerce for acting as coordinators of the White Protein Cluster Initiative with the Government of the Valle del Cauca, as a liaison in the development of projects for the FCTeI-SGR and the Universidad del Valle as an implementing entity.

Conflicts of Interest: The authors declare no conflict of interest. Neither the Cali Chamber of Commerce nor the White Protein Business Cluster had any role in the design, collection, analysis, or interpretation of data, the writing of the document, or decision to publish results.

References

1. World Health Organization. Guidelines: Saturated Fatty Acid and Trans-Fatty Acid Intake for Adults and Children. 2018. Available online: https://www.who.int/docs/default-source/healthy-diet/healthy-diet-fact-sheet-394.pdf?sfvrsn=69f1f9a1_2&download=true (accessed on 22 June 2021).
2. Ministry of Health and Social Protection (Colombia). Resolution 810 of 2021. Available online: https://www.minsalud.gov.co/Normatividad_Nuevo/Resoluci%C3%B3n%20No.810de%202021.pdf (accessed on 22 October 2021).
3. Ursachi, C.Ș.; Perța-Crișan, S.; Munteanu, F.D. Strategies to improve meat products' quality. *Foods* **2020**, *9*, 1883. [CrossRef] [PubMed]
4. Ding, Y.; Wang, S.Y.; Yang, D.J.; Chang, M.H.; Chen, Y.C. Alleviative effects of litchi (Litchi chinensis Sonn.) flower on lipid peroxidation and protein degradation in emulsified pork meatballs. *J. Food Drug Anal.* **2015**, *23*, 501–508. [CrossRef] [PubMed]
5. Turgut, S.S.; Işıkçı, F.; Soyer, A. Antioxidant activity of pomegranate peel extract on lipid and protein oxidation in beef meatballs during frozen storage. *Meat Sci.* **2017**, *129*, 111–119. [CrossRef]
6. Ribeiro, J.S.; Santos, M.J.M.C.; Silva, L.K.R.; Pereira, L.C.L.; Santos, I.A.; da Silva Lannes, S.C.; da Silva, M.V. Natural antioxidants used in meat products: A brief review. *Meat Sci.* **2019**, *148*, 181–188. [CrossRef] [PubMed]
7. Cunha, L.C.M.; Monteiro, M.L.G.; Lorenzo, J.M.; Munekata, P.E.S.; Muchenje, V.; de Carvalho, F.A.L.; Conte-Junior, C.A. Natural antioxidants in processing and storage stability of sheep and goat meat products. *Food Res. Int.* **2018**, *111*, 379–390. [CrossRef]
8. Lilić, S.; Nikolić, D.; Pejkovski, Z.; Velebit, B.; Lakićević, B.; Koričanac, V.; Vranić, D. Sodium and potassium content and their ratio in meatballs in tomato sauce produced with lower amounts of sodium. *IOP Conf. Ser. Earth Environ. Sci.* **2017**, *85*. [CrossRef]
9. de Sousa, A.M.B.; de Araujo Alves, R.; Madeira, D.S.S.; Santos, R.M.; Pereira, A.L.F.; de Oliveira Lemos, T.; Abreu, V.K.G. Storage of beef burgers containing fructooligosaccharides as fat replacer and potassium chloride as replacing sodium chloride. *J. Food Sci. Technol.* **2020**, *57*, 3232–3243. [CrossRef]
10. Gupta, R.; Thind, S.S.; Kaur, A.; Bhise, S. Development of Chicken Meat Cutlets Incorporating Carrots and Oats as Functional Ingredients. *Int. J. Food Ferment. Technol.* **2017**, *7*, 41. [CrossRef]
11. Hartmann, G.; Teixeira, F.; Soares, J.M.; da Silva, K.A.; Schwarz, K.; Schiessel, D.L.; Novello, D. Effect of fat replacement by fructooligosaccharide in hamburger: Physicochemical, technological and sensorial analysis. *Int. J. Innov. Educ. Res.* **2020**, *8*, 15–25. [CrossRef]
12. Glisic, M.; Baltic, M.; Glisic, M.; Trbovic, D.; Jokanovic, M.; Parunovic, N.; Dimitrijevic, M.; Suvajdzic, B.; Boskovic, M.; Vasilev, D. Inulin-based emulsion-filled gel as a fat replacer in prebiotic- and PUFA-enriched dry fermented sausages. *Int. J. Food Sci. Technol.* **2019**, *54*, 787–797. [CrossRef]
13. Berizi, E.; Shekarforoush, S.S.; Mohammadinezhad, S.; Hosseinzadeh, S.; Farahnaki, A. The use of inulin as fat replacer and its effect on texture and sensory properties of emulsion type sausages. *Iran. J. Vet. Res.* **2017**, *18*, 253–257. [CrossRef] [PubMed]

14. Afshari, R.; Hosseini, H.; Mousavi Khaneghah, A.; Khaksar, R. Physico-chemical properties of functional low-fat beef burgers: Fatty acid profile modification. *LWT Food Sci. Technol.* **2017**, *78*, 325–331. [CrossRef]
15. Delgado-Pando, G.; Cofrades, S.; Ruiz-Capillas, C.; Jiménez-Colmenero, F. Healthier lipid combination as functional ingredient influencing sensory and technological properties of low-fat frankfurters. *Eur. J. Lipid Sci. Technol.* **2010**, *112*, 859–870. [CrossRef]
16. Pérez-Palacios, T.; Ruiz-Carrascal, J.; Jiménez-Martín, E.; Solomando, J.C.; Antequera, T. Improving the lipid profile of ready-to-cook meat products by addition of omega-3 microcapsules: Effect on oxidation and sensory analysis. *J. Sci. Food Agric.* **2018**, *98*, 5302–5312. [CrossRef] [PubMed]
17. Pérez-Palacios, T.; Ruiz-Carrascal, J.; Solomando, J.C.; Antequera, T. Strategies for Enrichment in ω-3 Fatty Acids Aiming for Healthier Meat Products. *Food Rev. Int.* **2019**, *35*, 485–503. [CrossRef]
18. Liu, R.; Wang, N.; Li, Q.; Zhang, M. Comparative studies on physicochemical properties of raw and hydrolyzed oat β-glucan and their application in low-fat meatballs. *Food Hydrocoll.* **2015**, *51*, 424–431. [CrossRef]
19. Schmiele, M.; Nucci Mascarenhas, M.C.C.; da Silva Barretto, A.C.; Rodrigues Pollonio, M.A. Dietary fiber as fat substitute in emulsified and cooked meat model system. *LWT Food Sci. Technol.* **2015**, *61*, 105–111. [CrossRef]
20. Abbasi, E.; Amini Sarteshnizi, R.; Ahmadi Gavlighi, H.; Nikoo, M.; Azizi, M.H.; Sadeghinejad, N. Effect of partial replacement of fat with added water and tragacanth gum (Astragalus gossypinus and Astragalus compactus) on the physicochemical, texture, oxidative stability, and sensory property of reduced fat emulsion type sausage. *Meat Sci.* **2019**, *147*, 135–143. [CrossRef]
21. Carvalho, L.T.; Pires, M.A.; Baldin, J.C.; Munekata, P.E.S.; de Carvalho, F.A.L.; Rodrigues, I.; Polizer, Y.J.; de Mello, J.L.M.; Lapa-Guimarães, J.; Trindade, M.A. Partial replacement of meat and fat with hydrated wheat fiber in beef burgers decreases caloric value without reducing the feeling of satiety after consumption. *Meat Sci.* **2019**, *147*, 53–59. [CrossRef]
22. Soltanizadeh, N.; Ghiasi-Esfahani, H. Qualitative improvement of low meat beef burger using Aloe vera. *Meat Sci.* **2014**, *99*, 75–80. [CrossRef]
23. Alaei, F.; Hojjatoleslamy, M.; Hashemi Dehkordi, S.M. The effect of inulin as a fat substitute on the physicochemical and sensory properties of chicken sausages. *Food Sci. Nutr.* **2018**, *6*, 512–519. [CrossRef] [PubMed]
24. Menegas, L.Z.; Pimentel, T.C.; Garcia, S.; Prudencio, S.H. Effect of adding inulin as a partial substitute for corn oil on the physicochemical and microbiological characteristics during processing of dry-fermented chicken sausage. *J. Food Process. Preserv.* **2017**, *41*, 1–7. [CrossRef]
25. Balestra, F.; Bianchi, M.; Petracci, M. Applications in meat products. In *Dietary Fiber: Properties, Recovery, and Applications*; Galanakis, C., Ed.; Academic Press: London, UK, 2019; ISBN 9780128164952.
26. Angiolillo, L.; Conte, A.; Del Nobile, M.A. Technological strategies to produce functional meat burgers. *LWT Food Sci. Technol.* **2015**, *62*, 697–703. [CrossRef]
27. Rodriguez Furlán, L.T.; Padilla, A.P.; Campderrós, M.E. Development of reduced fat minced meats using inulin and bovine plasma proteins as fat replacers. *Meat Sci.* **2014**, *96*, 762–768. [CrossRef]
28. Yousefi, M.; Khorshidian, N.; Hosseini, H. An overview of the functionality of inulin in meat and poultry products. *Nutr. Food Sci.* **2018**, *48*, 819–835. [CrossRef]
29. Han, M.; Bertram, H.C. Designing healthier comminuted meat products: Effect of dietary fibers on water distribution and texture of a fat-reduced meat model system. *Meat Sci.* **2017**, *133*, 159–165. [CrossRef]
30. Öztürk, B.; Serdaroğlu, M. A Rising Star Prebiotic Dietary Fiber: Inulin and Recent Applications in Meat Products. *J. Food Health Sci.* **2016**, *3*, 12–20. [CrossRef]
31. Ponnampalam, E.N.; Bekhit, A.; Holman, B.; Jayasundera, M.; Lewandowski, P.; Dunshea, F.R.; Hopkins, D.L.; Gill, H. Use of plant materials to enhance the nutritional appeal of processed meat products. In *Advances in Meat Processing Technology*; Alaa El-Din, A.B., Ed.; CRC Press: Boca Raton, FL, USA, 2017; p. 604; ISBN 1498700489,978-1-4987-0048-1.
32. Kehlet, U.; Pagter, M.; Aaslyng, M.D.; Raben, A. Meatballs with 3% and 6% dietary fibre from rye bran or pea fibre—Effects on sensory quality and subjective appetite sensations. *Meat Sci.* **2017**, *125*, 66–75. [CrossRef]
33. Garmyin, A. Consumer Preferences and Acceptance of Meat Products. *Kirk-Othmer Encycl. Chem. Technol.* **2020**, *9*, 708. [CrossRef]
34. Schouteten, J.J.; De Steur, H.; De Pelsmaeker, S.; Lagast, S.; De Bourdeaudhuij, I.; Gellynck, X. An integrated method for the emotional conceptualization and sensory characterization of food products: The EmoSensory®Wheel. *Food Res. Int.* **2015**, *78*, 96–107. [CrossRef]
35. Ventanas, S.; Gonzalez, A.; Estevez, M.; Carvalho, L. Innovation in sensory assessment of meat and meat products. In *Meet Quality Analysis*; Biswas, A., Mandal, P., Eds.; Academic Press: London, UK, 2020; pp. 393–418. [CrossRef]
36. Polizer Rocha, Y.J.; de Noronha, R.L.F.; Trindade, M.A. Understanding the consumer's perception of traditional frankfurters and frankfurters with healthy attributes through sorting task and hard laddering techniques. *Meat Sci.* **2019**, *149*, 70–78. [CrossRef] [PubMed]
37. Colombian Institute of Technical Standards and Certification. Food Industries. Chicken Carcasses and Chicken Cuts. Requirements. NTC 3644-2:2018. Available online: https://tienda.icontec.org/gp-industrias-alimentarias-pollo-en-canal-y-sus-cortes-requisitos-ntc3644-2-2018.html (accessed on 10 August 2021).

38. Colombian Institute of Technical Standards and Certification. Guide for Evaluation and Prevention of Microorganisms in Food Industry. Listeria Monocytogenes. GTC 155:2007. Available online: https://tienda.icontec.org/gp-guia-para-la-evaluacion-y-prevencion-de-microorganismos-en-plantas-de-alimentos-listeria-monocytogenes-gtc155-2007.html (accessed on 1 August 2021).
39. Ministry of Health and Social Protection (Colombia). Resolution 242 of 2013. Available online: https://www.minsalud.gov.co/sites/rid/Lists/BibliotecaDigital/RIDE/DE/DIJ/resolucion-0242-de-2013.pdf (accessed on 28 September 2021).
40. Ministry of Health and Social Protection (Colombia). Decree 1500 of 2007. Available online: https://corponarino.gov.co/expedientes/juridica/2007decreto1500.pdf (accessed on 28 September 2021).
41. Ministry of Health and Social Protection (Colombia). Resolution 4287 of 2007. Available online: https://www.minsalud.gov.co/Normatividad_Nuevo/RESOLUCI%C3%93N%204287%20DE%202007.pdf (accessed on 18 August 2021).
42. Colombian Institute of Technical Standards and Certification. Food Industries. Processed Meat Products Non Canned. NTC 1325:2008. Available online: https://tienda.icontec.org/gp-industrias-alimentarias-productos-carnicos-procesados-no-enlatados-ntc1325-2008.html (accessed on 1 January 2021).
43. Association of Official Agricultural Chemists. *Official Methods of Analysis International, Agricultural Chemicals, Contaminants, Drug. AOAC 978.19*, 17th ed.; Association of Official Agricultural Chemists: Rockville, MD, USA, 2000.
44. Association of Official Agricultural Chemists. PH of Acidified Foods. AOAC 981.12-1982. Available online: http://www.aoacofficialmethod.org/index.php?main_page=product_info&cPath=1&products_id=1159 (accessed on 15 November 2021).
45. Association of Official Agricultural Chemists. Meat and Meat Products—Preparation of Test Sample Procedure. AOAC 983.18-1983. Available online: http://www.aoacofficialmethod.org/index.php?main_page=product_info&cPath=1&products_id=558 (accessed on 4 March 2021).
46. Colombian Institute of Technical Standards and Certification. Meat and Meat Products. Determination of Moisture Content. (Reference Methods and Rutine Method). NTC 1663:2009. Available online: https://tienda.icontec.org/gp-carne-y-productos-carnicos-metodos-de-determinacion-del-contenido-de-humedad-metodo-de-referencia-y-metodo-de-rutina-ntc1663-2009.html (accessed on 4 March 2021).
47. International Organization for Standardization. Paints and Varnishes—Colorimetry—Part 2: Colour Measurement. ISO 7724-2:1984. Available online: https://www.iso.org/standard/14558.html (accessed on 1 September 2021).
48. International Organization for Standardization. Food and Feed Products—General Guidelines for the Determination of Nitrogen by the Kjeldahl Method (ISO 1871:2009). Available online: https://www.iso.org/standard/41320.html (accessed on 1 August 2021).
49. Association of Official Agricultural Chemists. Fat (Crude) or Ether Extract in Animal Feed. AOAC 920.39-1920. Available online: http://www.aoacofficialmethod.org/index.php?main_page=product_info&cPath=1&products_id=1088 (accessed on 1 August 2021).
50. Association of Official Agricultural Chemists. Ash of Meat. AOAC 920.153-1920. Available online: http://www.aoacofficialmethod.org/index.php?main_page=product_info&cPath=1&products_id=1694 (accessed on 1 August 2021).
51. Association of Official Agricultural Chemists. Measurement of Fructan in Foods. Enzymatic/spe. AOAC 999.03-2005. Available online: http://www.aoacofficialmethod.org/index.php?main_page=product_info&cPath=1&products_id=2385 (accessed on 20 March 2021).
52. Association of Official Agricultural Chemists. Fatty Acids in Oils and Fats. Preparatio. AOAC 969.33-1969(1997). Available online: http://www.aoacofficialmethod.org/index.php?main_page=product_info&cPath=1&products_id=1733 (accessed on 1 August 2021).
53. Association of Official Agricultural Chemists. Cholesterol in Multicomponent Foods. Gas Chromatographic Method. AOAC 976.26-1977(2010). Available online: http://www.aoacofficialmethod.org/index.php?main_page=product_info&products_id=2321 (accessed on 1 August 2021).
54. Association of Official Agricultural Chemists. Calcium, Copper, Iron, Magnesium, Manganese, Hosphorus, Potassium, Sodium and Zinc. AOAC 984.27-1986. Available online: http://www.aoacofficialmethod.org/index.php?main_page=product_info&cPath=1&products_id=2126 (accessed on 7 September 2021).
55. Association of Official Agricultural Chemists. Microbiological Methods. AOAC 966.23-1989. Available online: http://www.aoacofficialmethod.org/index.php?main_page=product_info&cPath=1&products_id=1922 (accessed on 15 June 2021).
56. International Commission on Microbiological Specifications for Foods. *Microorganisms in Foods 7: Microbiological Testing in Food Safety Management*, 2nd ed.; Springer: Cham, Switzerland, 2018; ISBN 9783319684604.
57. International Organization for Standardization. Microbiology of Food and Animal Feeding Stuffs—Horizontal Method for the Enumeration of Coagulase-Positive Staphylococci (Staphylococcus aureus and other Species). UNE-EN ISO 6888-1:2000/A2:2019. Available online: https://www.une.org/encuentra-tu-norma/busca-tu-norma/norma?c=N0061754 (accessed on 15 June 2021).
58. International Organization for Standardization. Microbiology of the Food Chain—Horizontal Method for the Detection, Enumeration and Serotyping of Salmonella. UNE-EN ISO 6579-1:2017. Available online: https://www.une.org/encuentra-tu-norma/busca-tu-norma/norma/?c=N0058760 (accessed on 15 June 2021).
59. International Organization for Standardization. Microbiology of the Food Chain—Horizontal Method for the Detection and Enumeration of Listeria Monocytogenes and of Listeria spp.—Part 1: Detection Method. UNE-EN ISO 11290-1:2018. Available online: https://www.une.org/encuentra-tu-norma/busca-tu-norma/norma/?c=N0059546 (accessed on 15 June 2021).

60. International Organization for Standardization. Microbiology of Food and Animal Feeding Stuffs—Horizontal Method for the Enumeration of Coliforms—Colony-Count Technique. ISO 4832:2006. Available online: https://www.iso.org/standard/38282.html (accessed on 15 June 2021).
61. International Organization for Standardization. Sensory Analysis—General Guidelines for the Selection, Training and Monitoring of Selected Assessors and Expert Sensory Assessors (ISO 8586:2012). Available online: https://www.iso.org/standard/45352.html (accessed on 23 October 2021).
62. Colombian Institute of Technical Standards and Certification. Sensory Analysis. Methodology. Method of Investigating Sensitivity of Taste. NTC 3915:2012. Available online: https://tienda.icontec.org/gp-analisis-sensorial-metodologia-metodo-para-investigar-la-sensibilidad-del-gusto-ntc3915-2012.html (accessed on 23 October 2021).
63. Colombian Institute of Technical Standards and Certification. Sensory Analysis. Methodology. Initiation and Training of Assessors in the Detection and Recognition of Odours. NTC 4503:2011. Available online: https://tienda.icontec.org/gp-analisis-sensorial-metodologia-iniciacion-y-entrenamiento-de-evaluadores-en-la-deteccion-y-reconocimiento-de-olores-ntc4503-2011.html (accessed on 23 October 2021).
64. Colombian Institute of Technical Standards and Certification. Sensory Analysis. Methodology. Ranking. NTC 3930:2015. Available online: https://tienda.icontec.org/gp-analisis-sensorial-metodologia-ordenamiento-de-acuerdo-con-un-criterio-especifico-ranking-ntc3930-2015.html (accessed on 23 October 2021).
65. Colombian Institute of Technical Standards and Certification. Sensory Analysis—Guidelines for Use of Quantitative Response Scale. NTC 5328:2004. Available online: https://tienda.icontec.org/gp-analisis-sensorial-directrices-para-el-uso-de-escalas-de-respuesta-cuantitativas-ntc5328-2004.html (accessed on 23 October 2021).
66. Kemp, S.; Hort, J.; Hollowood, T. *Descriptive Analysis in Sensory Evaluation*; Willey: West Sussex, UK, 2018; ISBN 9781118991671.
67. Saldaña, E.; de Oliveira Garcia, A.; Selani, M.M.; Haguiwara, M.M.H.; de Almeida, M.A.; Siche, R.; Contreras-Castillo, C.J. A sensometric approach to the development of mortadella with healthier fats. *Meat Sci.* **2018**, *137*, 176–190. [CrossRef] [PubMed]
68. de Quadros, D.A.; Rocha, I.F.d.O.; Ferreira, S.M.R.; Bolini, H.M.A. Low-sodium fish burgers: Sensory profile and drivers of liking. *LWT Food Sci. Technol.* **2015**, *63*, 236–242. [CrossRef]
69. Concha-Meyer, A.A.; Durham, C.A.; Colonna, A.E.; Hasenbeck, A.; Sáez, B.; Adams, M.R. Consumer Response to Tomato Pomace Powder as an Ingredient in Bread: Impact of Sensory Liking and Benefit Information on Purchase Intent. *J. Food Sci.* **2019**, *84*, 3774–3783. [CrossRef] [PubMed]
70. Davis, L.; Jung, J.; Colonna, A.; Hasenbeck, A.; Gouw, V.; Zhao, Y. Quality and Consumer Acceptance of Berry Fruit Pomace–Fortified Specialty Mustard. *J. Food Sci.* **2018**, *83*, 1921–1932. [CrossRef]
71. Naes, T.; Varela, P.; Berget, U. *Individual Differences in Sensory and Consumer Science*; Elsevier: Kidlington, UK, 2018; ISBN 9780081003947.
72. Lawless, H.T.; Heymann, H. *Sensory Evaluation of Food: Principles and Practices*, 2nd ed.; Springer: New York, NY, USA, 2010; ISBN 9781441964878.
73. Euromonitor International Voice of the Industry: Food & Nutrition. 2021. Available online: https://www.euromonitor.com/voice-of-the-industry-food-and-nutrition/report (accessed on 28 October 2021).
74. Mintel Global Consumer Trends. 2021. Available online: https://www.mintel.com/global-consumer-trends?utm_term=%2Bmintel%2Btrends&utm_campaign=SCH_(CP-LeadGen)_(G-LAR)_%7BBrand%7D_(D-All)_B_&utm_source=adwords&utm_medium=ppc&hsa_acc=9603040692&hsa_cam=10445632723&hsa_grp=104537149340&hsa_ad=445726075827&hsa_s (accessed on 28 October 2021).
75. Aripin, N.; Huda, N. Quality Characteristics of Meatball Prepared from Different Ratios of Chicken and Duck Meat. *Asia Pac. J. Sustain. Agric. Food Energy* **2018**, *6*, 6–9. [CrossRef]
76. Yeung, C.K.; Huang, S.C. Effects of Food Proteins on Sensory and Physico-Chemical Properties of Emulsified Pork Meatballs. *J. Food Nutr. Res.* **2018**, *6*, 8–12. [CrossRef]
77. Cáceres, E.; García, M.L.; Toro, J.; Selgas, M.D. The effect of fructooligosaccharides on the sensory characteristics of cooked sausages. *Meat Sci.* **2004**, *68*, 87–96. [CrossRef]
78. Franck, A. Technological functionality of inulin and oligofructose. *Br. J. Nutr.* **2002**, *87*, S287–S291. [CrossRef]
79. Keenan, D.F.; Resconi, V.C.; Kerry, J.P.; Hamill, R.M. Modelling the influence of inulin as a fat substitute in comminuted meat products on their physico-chemical characteristics and eating quality using a mixture design approach. *Meat Sci.* **2014**, *96*, 1384–1394. [CrossRef]
80. Álvarez, D.; Barbut, S. Effect of inulin, β-Glucan and their mixtures on emulsion stability, color and textural parameters of cooked meat batters. *Meat Sci.* **2013**, *94*, 320–327. [CrossRef]
81. Felisberto, M.H.F.; Galv o, M.T.E.L.; Picone, C.S.F.; Cunha, R.L.; Pollonio, M.A.R. Effect of prebiotic ingredients on the rheological properties and microstructure of reduced-sodium and low-fat meat emulsions. *LWT Food Sci. Technol.* **2015**, *60*, 148–155. [CrossRef]
82. Huang, S.C.; Tsai, Y.F.; Chen, C.M. Effects of wheat fiber, oat fiber, and inulin on sensory and physico-chemical properties of Chinese-style sausages. *Asian-Australas. J. Anim. Sci.* **2011**, *24*, 875–880. [CrossRef]
83. Menegas, L.Z.; Pimentel, T.C.; Garcia, S.; Prudencio, S.H. Dry-fermented chicken sausage produced with inulin and corn oil: Physicochemical, microbiological, and textural characteristics and acceptability during storage. *Meat Sci.* **2013**, *93*, 501–506. [CrossRef] [PubMed]

84. Paglarini, C.; Vidal, V.A.S.; Ribeiro, W.; Badan Ribeiro, A.P.; Bernardinelli, O.D.; Herrero, A.M.; Ruiz-Capillas, C.; Sabadini, E.; Rodrigues Pollonio, M.A. Using inulin-based emulsion gels as fat substitute in salt reduced Bologna sausage. *J. Sci. Food Agric.* **2021**, *101*, 505–517. [CrossRef] [PubMed]
85. Shoaib, M.; Shehzad, A.; Omar, M.; Rakha, A.; Raza, H.; Sharif, H.R.; Shakeel, A.; Ansari, A.; Niazi, S. Inulin: Properties, health benefits and food applications. *Carbohydr. Polym.* **2016**, *147*, 444–454. [CrossRef] [PubMed]
86. Grossi, A.; Søltoft-Jensen, J.; Knudsen, J.C.; Christensen, M.; Orlien, V. Reduction of salt in pork sausages by the addition of carrot fibre or potato starch and high pressure treatment. *Meat Sci.* **2012**, *92*, 481–489. [CrossRef] [PubMed]

foods

MDPI

Article

The Consumption of Insects in Switzerland: University-Based Perspectives of Entomophagy

Aline Oliveira Penedo [1], Sophie Bucher Della Torre [1], Franziska Götze [2], Thomas A. Brunner [2] and Wolfram Manuel Brück [3,*]

1 Geneva School of Health Sciences, HES-SO University of Applied Sciences and Arts Western Switzerland, Rue des Caroubiers 25, 1227 Carouge, Switzerland
2 Food Science & Management, School of Agricultural, Forest and Food Sciences (HAFL), Bern University of Applied Sciences, Länggasse 85, 3052 Zollikofen, Switzerland
3 Institute of Life Technologies, School of Engineering, HES-SO University of Applied Sciences and Arts Western Switzerland Valais-Wallis, 1950 Sion, Switzerland
* Correspondence: wolfram.bruck@hevs.ch; Tel.: +41-58-606-86-64

Abstract: Although insects have long been part of the human diet in many countries, they are poorly received and accepted in European and North American countries. Therefore, this cross-sectional observational study, based on a structured questionnaire, aimed to evaluate the level of acceptability of entomophagy among young adults in a Swiss university context. The variable "acceptability of consuming insects" (ACI) was calculated according to the perception of entomophagy of each participant. The ACI was related to various socio-demographic and behavioral aspects. A total of 290 responses were validated and analyzed. The mean ACI score was 3.7 out of 6.0 (SD 1.1). Most participants responded that the most likely reason for eating insect foods was curiosity. The most common reason for not eating such foods was disgust. None of the socio-demographic variables showed a significant association with ACI. Generally, participants in this study showed a potential interest in entomophagy—on a theoretical level, as measured here by the ACI. In practice, however, there are still barriers, including disgust, which contribute to the low consumption of these foods, at least in Switzerland.

Keywords: insects; entomophagy; consumer acceptance; sustainable foods

Citation: Penedo, A.O.; Bucher Della Torre, S.; Götze, F.; Brunner, T.A.; Brück, W.M. The Consumption of Insects in Switzerland: University-Based Perspectives of Entomophagy. *Foods* **2022**, *11*, 2771. https://doi.org/10.3390/foods11182771

Academic Editor: Witoon Prinyawiwatkul

Received: 8 August 2022
Accepted: 5 September 2022
Published: 8 September 2022

1. Introduction

Entomophagy is defined as the practice of eating insects [1]. In many cultures, particularly in the sub-Saharan Africa, south and southeast Asia, and Latin America, insects have been consumed for centuries [2,3]. As such, more than 1900 species of insects have already been used as food with insects being part of the daily diet of at least two billion people [1,4–6]. Among the vast number of edible insect species are beetles, caterpillars, bees, wasps, ants, grasshoppers, locusts, and crickets [5].

Edible insects are generally valued for their high energy content and the quality of their protein, thanks to an amino acid profile that can match human nutritional requirements. For some species, the proportion of protein can exceed 20 g per 100 g portion, which is equivalent or even higher than chicken breast. Despite the long history and nutritional benefits of entomophagy, it was only in 2013, following a major publication by the Food and Agriculture Organization of the United Nations (FAO), that discussions about this practice and its possible role in global food security expanded, especially in Europe and North America [1,7]. As a result, insects are now seen as one of the future pillars of a sustainable human diet.

Compared to other alternative protein sources, the advantage of insects is their low soil requirement for rearing, which allows for an intensive production in small areas [8]. Indeed, vertical farming of insects may greatly optimize the space used for their rearing [9].

Insect farming also requires much less water, compared to the rearing of other animals for food [2]. Therefore, the production of insects as foods has a better feed conversion ratio and generates less greenhouse gases [10]. For example, to produce 1 kg of protein, mealworms need only 10% of the land required for beef production [9]. Crickets require only 1.7 kg of food for every 1 kg of their body weight to grow [11]. Moreover, insect food production carries a low risk of zoonotic disease transmission—a strong argument in the wake of the COVID-19 pandemic [4,11–13].

However, the current global scale of insect production is insufficient to be considered a sustainable source of nutrients. Thus, to establish entomophagy and its potential role in reducing the environmental impact of the food industry, the European Union has adopted the Novel Foods Regulation in 2015 (Regulation (EC) No 2015/2283), allowing companies to market innovative foods and ensuring that these same companies meet advanced food safety and quality standards for all European consumers [14]. Similarly, in 2017, the Swiss Federal Office for Food Safety and Veterinary Affairs (FSVO) published technical guidelines on official controls in primary insect production, as well as an information sheet containing a description of insects authorized as foodstuffs, the conditions for their production and release [15]. Three species of insects have been approved as foodstuffs in Switzerland, namely: (i) *Tenebrio molitor* in the larval stage (yellow mealworm); (ii) *Acheta domesticus*, adult form (house cricket); and (iii) *Locusta migratoria*, adult form (migratory locust).

Although legislation is being passed, consumer acceptance of insect-based foods is being investigated internationally. Studies generally show that most Western consumers are not yet ready to consume insects [16]. Indeed, a recent study found that there is the need for caution regarding the prospects of edible insects as the "food of the future" [17]. The consumers surveyed responded with a natural aversion to insects as foods [10,18,19]. This food neophobia is, therefore, an important factor to consider in the creation of new, insect-based foods [16]. Consumer food neophobia, together with perception of disgust, has been shown to significantly influence the individual decision to try insect-based foods. Thus, these are important predictors that contribute to the acceptability of new food products [4,20,21].

From a socio-cultural point of view, one of the beliefs associated with insect consumption is that only economically disadvantaged groups adhere to it, especially in conditions of food scarcity [22]. Indeed, preconceived notions about insect foods, as well as lack of knowledge about their benefits, are factors associated with low consumer acceptability of eating insect foods [2]. In a survey with 542 participants, Schlup and Brunner observed that only a small segment of the Swiss population (9.3%) was in favor of eating insects [23]. Swiss consumers who were open to entomophagy had a high level of education and scored low on the food neophobia scale. However, only one in three people reported having ever eaten insects. These people showed high nutritional knowledge and had read about entomophagy in the media. These results show that in Switzerland this type of food was still considered a modern and innovative product in 2015. Another study found that, in Switzerland, people who consumed insect products (e.g., insect burgers) were perceived as more health-conscious and environmentally friendly, amongst other qualities, than meat consumers [24]. In view of this positive image, it is possible that group dynamics influence the eating behavior of the Swiss population.

In Switzerland, there is a need to better understand the aspects that determine the willingness to consume insect-based foods so that they can, in fact, be integrated into an overall sustainable food strategy. There is a lack of studies on the acceptability of entomophagy following recent changes in the regulatory framework, which have potentially impacted the supply of products [15]. On this basis, the aim of this study was to better understand the viability of insect-based food products in Switzerland from the perspectives of the younger generation of consumers. In this sense, the study proposed to evaluate the level of acceptability of entomophagy among young adults in a Swiss university context, considering different socio-demographic and behavioral aspects. This study aimed to answer the following research questions:

- Question 1: What proportion of the target population is potentially interested in eating insect-based foods, based on environmental, nutritional, economic, gastronomic, and cultural arguments?
- Question 2: What are the facilitating and inhibiting factors for the consumption of insect-based foods in the studied population?
- Question 3: What socio-demographic and behavioral aspects are associated with the acceptability of consuming insect-based foods?

2. Materials and Methods

The present work is a cross-sectional observational study aimed at assessing the acceptability of eating insect-based foods among young adults enrolled at a university or college in Switzerland, using a structured questionnaire. In the study, "acceptability of consuming insects" was used as a dependent variable, which was then related to socio-demographic and behavioral aspects (independent variables).

2.1. Definition of Acceptability (Dependent Variable)

For the definition of the dependent variable, this work was based on a similar study conducted in Switzerland by Schlup and Brunner [23]. In their study, the authors used the term willingness to consume to refer to the acceptability of participants to eat insects. Here, we will use the term 'acceptability of consuming insects', henceforth referred to as 'ACI'. Like Schlup and Brunner, the ACI was measured based on 5 questions (see questions Q1 to Q5 in the questionnaire, Table S1). These questions assessed the level of agreement or disagreement of the participants with arguments in favor of entomophagy, covering 5 themes (sustainability, health, price, taste, and trend), according to a Likert scale ranging from 1 to 6 (1 = strongly disagree; 6 = strongly agree). This six-point scale is identical to that of Schlup and Brunner and was chosen because it ensures that an opinion is obtained on each question, avoiding "neutral" responses [23]. Using the 5 questions mentioned above, it was possible to assign each participant a score in the continuous range of 1.0 to 6.0 corresponding to their ACI. An ACI score of 3.5 (mean value of range between 1 and 6) or higher (an individual score of either 4, 5, or 6) was interpreted as potentially indicating an interest in eating insect-based foods (and vice versa for values lower than 3.5). The calculation of the ACI score is explained in the statistical analysis section.

2.2. Target Population and Inclusion Criteria

The study focused on the population of students enrolled at Swiss-based universities and universities of applied sciences. This choice of target audience was made based on the study's access to the participating population. This was facilitated by the availability of emails from people studying and working at the universities, as recruitment for participation in the questionnaire was done by sending emails and invitations to participate via the universities' social media.

Younger adults tend to be more sensitive to health issues and more sustainable in their food choices than older people, who are more resistant to changes such as accepting new foods [24]. Furthermore, compared to adolescents, young adults are at a point in their lives where they are beginning to have the opportunity to make more conscious and independent food choices, with more opportunities to explore new foods [16]. Most importantly, today's youth are tomorrow's consumers, which makes understanding their food preferences particularly interesting for assessing the viability of insect-based foods.

2.3. Recruitment and Exclusion Criteria

Participants were recruited by an e-mail invitation sent to the French-, German-, and Italian-speaking regions in Switzerland. Different databases containing the e-mail addresses of students enrolled at a university or university of applied sciences in Switzerland were used for recruitment. The email contained a direct link to the survey, which was developed on the Google Forms platform. An e-mail address was provided in case participants wished to send a comment or make a remark about the study.

Data collection took place over three months, from July to September 2021. The questionnaire was first sent out in July 2021, and then a reminder was sent out at the start of the academic year in September 2021—thus reaching the minimum number of participants for the study. In total, the recruitment email was sent to over 1500 people.

It is possible that the email was informally forwarded by the recipients themselves, and that more people received the invitation without necessarily being part of the target audience. In addition, the databases used did not exclusively contain students, but anyone affiliated with a university or college in Switzerland. In view of these two aspects, it was necessary to sort the participants to exclude those who were not part of the target audience, i.e., those not being students in Switzerland.

An exclusion criterion was therefore set up, based on age: all participants older than 36 years were excluded. This threshold was determined by considering the age of a person at the end of their PhD. In addition to the age, location (Switzerland) and presence at a university of university of applied sciences, no other exclusion criteria such as health status or diet were established.

2.4. Questionnaire

The questionnaire used in this study is an adaptation of the one developed and validated by Schlup and Brunner [23]. A digital version of the questionnaire was constructed using the Google Forms platform. To avoid incomplete questionnaires being returned, all questions were mandatory so that each participant was required to answer all questions to submit their questionnaire. For the sake of inclusiveness, the questionnaire was distributed in French, German, and Italian (three of the Swiss national languages). Before being distributed, the adapted questionnaire was tested with a convenience sample. The English version of the questionnaire is provided in Supplement 1.

The questionnaire was structured as follows:

- Section 1: arguments in favor of insects (5 questions, Q1 to Q5 of the questionnaire);
- Section 2: eating behavior (4 questions, Q6 to Q9 of the questionnaire);
- Section 3: socio-demographic data (1 question, Q10 of the questionnaire, broken down into 6 points).

As mentioned above, the first section addressed the arguments in favor of consuming insect-based foods, through five themes: sustainability, health, price, taste, and trend. The answers to these five initial questions were used to calculate each participant's ACI score. The second section looked at cultural and behavioral issues, such as participants' values and beliefs around entomophagy, meat consumption, and reaction to new types of food. This section also looked at the presentation of insect-based foods to understand how this may impact their acceptability to participants. This considers recent studies that show a correlation between the acceptability of consuming insect-based foods if they are visible as a whole or non-visible (processed) on the plate [23]. The third section dealt with socio-demographic aspects considered relevant in the literature, such as gender, age, country of origin, or linguistic region [25–27].

Except for Section 3 (socio-demographic data), for all questions participants were asked to respond on a Likert scale. The scale ranged from 1 ("strongly disagree" or "very unlikely") to 6 ("strongly agree" or "very likely"). The intermediate values were anchored in numerical form only [4,20,28].

2.5. Statistical Analysis

Statistical analyses were carried out using STATA (version 16.1) and RStudio (version 2021.09.0) software. First, a purely descriptive analysis was carried out to obtain the profile of the participants, based on the following variables: gender, age category, linguistic region, current education level, main field of study, and nationality. This allowed for the identification and exclusion of all participants over the age of 36.

Then, to answer research question 1, each participant's ACI score was calculated by averaging the answers to the first 5 questions (Q1 to Q5 of the questionnaire, see Supplement 1), resulting in a continuous numerical value between 1.0 and 6.0. As mentioned above, participants with an ACI score of 3.5 or higher were considered potentially interested in eating insect-based foods. This threshold (3.5) corresponds to the middle value of the interval given by the Likert scale. It should be noted that, as in the study by Schlup and Brunner, the ACI is in fact a theoretical acceptability, defined according to the affinity of the participants with the arguments presented in favor of entomophagy—thus being measured independently of whether the participants had previously consumed insect-based foods [23]. To check the relevance of the ACI score in our questionnaire, it was compared with the participants' previous experiences, who were asked to specify whether they had ever eaten such foods.

To answer research question 2, the analysis focused on the answers obtained in the second section of the questionnaire (eating behavior). It was the answers to question 7 (broken down into 3 points, i.e., Q7.1, Q7.2, and Q7.3) that were relevant at this stage of the analysis. Thus, to highlight the facilitating or inhibiting factors to the consumption of insect-based foods among the participants, an analysis of the frequency of responses to questions 7.2 and 7.3 was carried out. This made it possible to identify: (i) what was the most frequent reason for not eating insect-based foods among those who have never eaten them, and (ii) what was the most frequent reason for eating insect-based foods among those who have eaten them.

To answer research question 3, the ACI score was related to the behavioral and socio-demographic variables identified in Sections 2 and 3 of the questionnaire. This was done through Chi-square and Fischer tests, where the ACI score was used as the dependent variable, and the behavioral and socio-demographic variables as independent variables. This time, the ACI was treated as a discrete variable: for each participant, its value was rounded to an integer between 1 and 6 (thus generating 6 categories). This made it possible to compare the ACI with the various socio-demographic categories. Here, it should be noted that, given the low response from the Swiss Italian population, it was decided to group Italian and French speakers into a single group (Latin Swiss). The behavioral variables, on the other hand, required a similar pre-processing to the ACI. First, for each participant, the averages of the responses to different groups of questions (each group being linked to a specific behavioral aspect) were calculated, giving continuous numerical values between 1.0 and 6.0. Like the ACI, these values were rounded to discrete variables (integer 1–6). In this way, the behavioral aspects could be inserted as categorical variables in the Chi-square and Fischer models to test their association with the ACI. The accepted level of significance for testing the null hypothesis was set at $p < 0.05$.

2.6. Internal Consistency and Cronbach's Alpha

As mentioned above in Section 2 (questions about eating behavior in general), different responses were aggregated to form behavioral variables. All these responses followed a scale from 1 to 6; however, the direction of the scale was not constant, being sometimes reversed: for example, in one question the value 6 corresponded to "I constantly taste different or new foods"; while, in another question, 6 corresponds to "I do not trust new foods". As a result, some questions had to be reverse coded before being translated into a discrete variable (as described above in the statistical analysis). The use of reversed statements allows for testing the overall consistency of each participant's responses.

To assess the internal consistency of the questionnaire, Cronbach's alpha (α) was calculated for each variable derived from the aggregation of responses to 2 or more questions—i.e., the ACI score and the behavioral variables. This allowed for a check of the overall consistency of the responses composing each variable according to the Likert scale. Based on other studies, the threshold for Cronbach's alpha consistency was set at $\alpha \geq 0.70$ [29].

2.7. Ethical Considerations

In this study, no data on the health of the participants were collected. Therefore, it does not fall within the scope of the Federal Act on Research Involving Human Beings and was, therefore, not submitted to an ethics committee for approval.

At the beginning of the questionnaire, an introductory section informed the participants about their contribution to this research, as well as about the confidentiality of their answers. Participation in the study was voluntary and responses were treated as confidential. All data were anonymized during processing (through aggregation processes), so that no answers could be linked to a specific person participating in the study. Finally, participants did not receive any compensation.

3. Results

3.1. General Information about the Participants

A total of 405 people responded to the questionnaire. Of these participants, 115 did not meet the inclusion criteria (people over 36 years old). This resulted in a total of 290 valid responses which met the quantitative requirements of the study.

3.2. Characteristics of the Participants

A summary of the respondents' gender, age group, linguistic region, level of education, main field of study, and nationality is given in Table 1. Overall, 57% of participants were female, with the largest proportion of participants being in the field of Health Sciences (32%); followed by Natural, Computer and Life Sciences (24%); and Engineering and Architecture (14%). Respondents aged between 18 and 24 were the most numerous, representing 49% of the sample. Most of the study population were currently in Masters' degree programs (37%), followed by Bachelor degree students (33%). In terms of language region, most participants (73%) reported living in Latin Switzerland (Suisse Romande or Ticino). Overall, 80% of the participants were of Swiss nationality.

Table 1. Socio-demographic characteristics of the sample.

Features	$N = 290$
Type	
Female	165 (57%)
Male	121 (42%)
I prefer not to answer	4 (1.4%)
Age Category	
18–24	142 (49%)
25–30	106 (37%)
31–36	42 (14%)
Language Region	
Latin Switzerland	211 (73%)
German-speaking Switzerland	79 (27%)
Current Level of Education	
High School Diploma	49 (17%)
Bachelor	97 (33%)
Master	108 (37%)
PhD	11 (3.8%)
Other	25 (8.6%)
Main Field of Study	
Health Sciences	94 (32%)
Natural sciences, computer sciences, and life sciences	71 (24%)
Engineering and architecture	40 (14%)
Economics, trade, management, and services	26 (9.0%)
Other	25 (8.6%)
Humanities and social sciences	24 (8.3%)
Pedagogy and educational sciences	6 (2.1%)
Law	4 (1.4%)
Nationality	
Switzerland	231 (80%)
France	23 (7.9%)
Brazil	8 (2.8%)
Portugal	7 (2.4%)
Germany	5 (1.7%)
Spain	5 (1.7%)
Other	11 (3.5%)

3.3. Acceptability of Eating Insect-Based Foods

To meet the first objective (determine the proportion of the study population that would be willing to eat insect-based foods), we calculated the ACI score. Figure 1 and Table 2 show that the mean ACI score (represented by the red line) was 3.7 (with a standard deviation of 1.1) and the median was 3.8 (orange line). The ACI appears to follow a normal distribution with the number of responses decreasing as the score approaches the extremes (1 and 6). The largest proportion of responses is concentrated in the range between 3.5 and 3.9, with 22.1% of participants having an ACI score in this range.

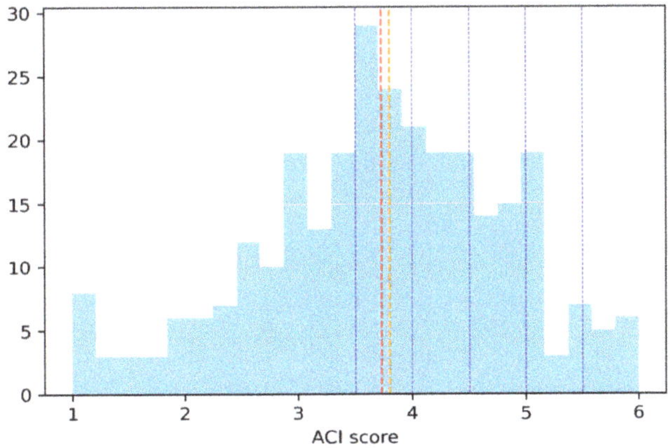

Figure 1. Distribution of ACI scores, by intervals. Red line: the mean of ACI score was 3.7; Orange line: the median was 3.8; Blue lines mark the ACI score intervals are shown in Table 2.

Table 2. Distribution of ACI scores, by intervals.

ACI Score	Proportion (%) of Participants with ACI Score within Given Interval
3.5–4.0	22.1
4.1–4.5	16.6
4.6–5.0	14.1
5.1–5.5	5.9
5.6–6.0	3.8

In total, 62.5% of the participants had an ACI score of 3.5 or higher. Most participants were, therefore, potentially interested in entomophagy. As mentioned above, the ACI was reported in a theoretical dimension, namely the participants' level of agreement with different arguments in favor of eating insect-based foods. To check the relevance of the ACI score, it was compared with the participants' previous experiences with eating this type of food (Table 3).

Table 3. Previous experiences with insect food consumption and respective ACI scores.

Have You Eaten Insect-Based Foods?	*N*	%	Average ACI
No, never	130	44.8%	3.4 (SD = 1.1)
Yes, once	87	30.0%	3.8 (SD = 1.0)
Yes, several times	73	25.2%	4.4 (SD = 0.9)
TOTAL	290	100.0%	3.7 (SD = 1.1)

More than half of the respondents had already consumed insects (55%). However, of these, most had done so only once (87 out of 160). It is noted that the average ACI scores increase with the reported frequency of insect food consumption (average ACI of 3.4 among those who have never eaten insects compared to an average ACI score of 4.4 among those who have eaten them several times).

Table 4 shows the results in more detail and highlights the differences in responses for men and women. Of those who had never eaten insects, women made up the vast majority (n = 90). A total of 50 women reported having eaten insect food once, compared to 37 men. For the most frequent consumption, only 25 women responded that they had eaten insect food several times, compared to 46 men.

Table 4. Previous experiences with insect food consumption, by gender.

Have You Eaten Insect Food (M/F)	Type	N	%
No, never	Female	90	31%
	Male	38	13%
Yes, once	Female	50	17%
	Male	37	13%
Yes, several times	Female	25	9%
	Male	46	16%
Total, excluding gender not declared (n = 4)		286	100%

3.4. Facilitating and Inhibiting Factors for the Consumption of Insect-Based Foods

To meet the second objective of the study (identify the facilitating and inhibiting factors for the consumption of insect-based foods among the target population) the analysis focused on the participants' previous consumption. These analyses were performed according to two subgroups: (1) people who had never eaten insects and (2) people who had already eaten insects (once or several times).

First, to better understand what reasons would encourage people who have never eaten insect food to do so, response options were given on different aspects that could be considered (Figure 2). Of the 128 respondents who had never eaten insect food, half (n = 64) stated that the most likely reason for them to eat such food would be curiosity. Of these, 45 (70%) were women. The category 'other reasons' ranked second (n = 35), followed by 'because eating insects is sustainable' (n = 9).

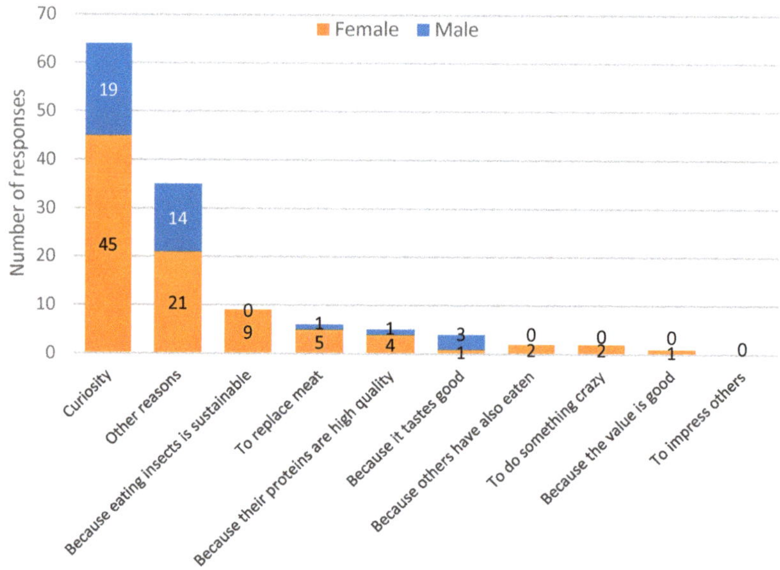

Figure 2. Most likely reason for participants who have never eaten insect-based foods to try such foods.

The same question was analyzed among the second group, i.e., those 158 people who had already eaten insect-based food once or several times (Figure 3). This made it possible to identify the most common reason for eating insect food. Here, as well, curiosity was the most frequently mentioned reason, being present in most responses (n = 135, or 85% of the total).

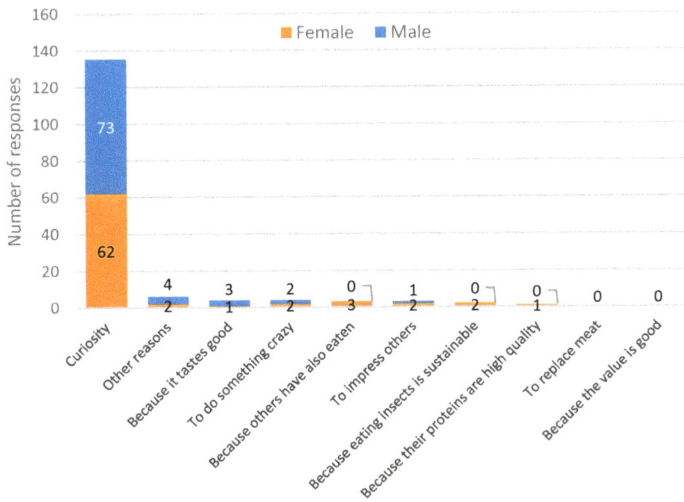

Figure 3. Most likely reason for eating insect food among those who have ever eaten it.

Answers to the question: "Why have you never eaten insects before, or why don't you eat them more often?" are shown in Figure 4. To this question, almost 40% of the participants answered that eating insects is disgusting. To this question, most responses were given by female participants (73%). In addition, 27% said they had not had the opportunity to eat or had not found these foods, while 8% said they preferred "real" meat.

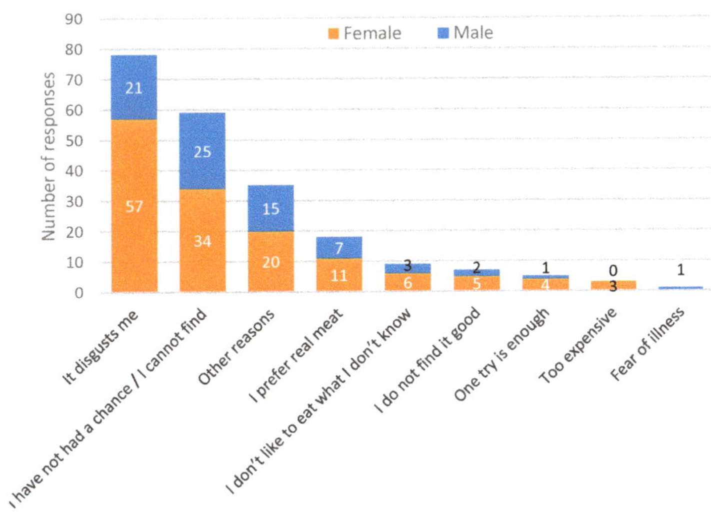

Figure 4. Reasons for never having eaten insect food or having eaten only once.

3.5. Acceptability by Socio-Demographic and Behavioral Variables

As shown in Table 5, none of the socio-demographic variables were significantly associated with the ACI. In Table 6, the following variables concerning eating habits were significantly (positively or negatively) associated with the ACI: "Foods with recognizable insects", "Foods with non-recognizable insects", "Eating meat is synonymous with health", "Food neophobia", and "Neophobia to new food technologies".

Table 5. Acceptability of consuming insect-based foods by socio-demographic variables.

		Not at All in Agreement				Totally Agree		
Socio-Demographic Variable	Total	1, N = 11 [1]	2, N = 29 [1]	3, N = 69 [1]	4, N = 112 [1]	5, N = 53 [1]	6, N = 16 [1]	p-Value [2]
Type								
Total	286	10 (3.5%)	29 (10%)	67 (23%)	112 (39%)	53 (19%)	15 (5.2%)	0.075
Female	165	5 (3.0%)	22 (13%)	43 (26%)	63 (38%)	23 (14%)	9 (5.5%)	
Male	121	5 (4.1%)	7 (5.8%)	24 (20%)	49 (40%)	30 (25%)	6 (5.0%)	
Age Category								
Total	290	11 (3.8%)	29 (10%)	69 (24%)	112 (39%)	53 (18%)	16 (5.5%)	0.079
18–24	142	5 (3.5%)	20 (14%)	35 (25%)	50 (35%)	25 (18%)	7 (4.9%)	
25–30	106	3 (2.8%)	7 (6.6%)	24 (23%)	40 (38%)	26 (25%)	6 (5.7%)	
31–36	42	3 (7.1%)	2 (4.8%)	10 (24%)	22 (52%)	2 (4.8%)	3 (7.1%)	
Language Region								
Total	290	11 (3.8%)	29 (10%)	69 (24%)	112 (39%)	53 (18%)	16 (5.5%)	0.6
German-speaking Switzerland	79	2 (2.5%)	8 (10%)	17 (22%)	37 (47%)	11 (14%)	4 (5.1%)	
Latin Switzerland	211	9 (4.3%)	21 (10%)	52 (25%)	75 (36%)	42 (20%)	12 (5.7%)	
Level of Training								
Total	290	11 (3.8%)	29 (10%)	69 (24%)	112 (39%)	53 (18%)	16 (5.5%)	0.1
High School	49	2 (4.1%)	9 (18%)	10 (20%)	22 (45%)	4 (8.2%)	2 (4.1%)	
Bachelor	97	3 (3.1%)	8 (8.2%)	31 (32%)	33 (34%)	17 (18%)	5 (5.2%)	
Master	108	3 (2.8%)	9 (8.3%)	19 (18%)	43 (40%)	25 (23%)	9 (8.3%)	
PhD	11	1 (9.1%)	0 (0%)	3 (27%)	7 (64%)	0 (0%)	0 (0%)	
Other	25	2 (8.0%)	3 (12%)	6 (24%)	7 (28%)	7 (28%)	0 (0%)	
Field of Study								
Total	290	11 (3.8%)	29 (10%)	69 (24%)	112 (39%)	53 (18%)	16 (5.5%)	0.073
Law	4	0 (0%)	1 (25%)	2 (50%)	0 (0%)	0 (0%)	1 (25%)	
Economy, Trade, Management and Services	26	0 (0%)	3 (12%)	9 (35%)	7 (27%)	2 (7.7%)	5 (19%)	
Engineering, Architecture	40	2 (5.0%)	2 (5.0%)	10 (25%)	14 (35%)	10 (25%)	2 (5.0%)	
Pedagogy or Education	6	0 (0%)	0 (0%)	3 (50%)	3 (50%)	0 (0%)	0 (0%)	
Nat. Sci., Life Sci., and Computer Sci.	71	4 (5.6%)	8 (11%)	19 (27%)	29 (41%)	10 (14%)	1 (1.4%)	
Health Sciences	94	1 (1.1%)	11 (12%)	19 (20%)	38 (40%)	19 (20%)	6 (6.4%)	
Humanities and Social Sciences	24	2 (8.3%)	1 (4.2%)	4 (17%)	13 (54%)	3 (12%)	1 (4.2%)	
Other	25	2 (8.0%)	3 (12%)	3 (12%)	8 (32%)	9 (36%)	0 (0%)	
Nationality								
Total	290	11 (3.8%)	29 (10%)	69 (24%)	112 (39%)	53 (18%)	16 (5.5%)	0.13
Germany	5	0 (0%)	1 (20%)	0 (0%)	4 (80%)	0 (0%)	0 (0%)	
England	1	0 (0%)	0 (0%)	1 (100%)	0 (0%)	0 (0%)	0 (0%)	
Brazil	8	1 (12%)	2 (25%)	2 (25%)	3 (38%)	0 (0%)	0 (0%)	
Cameroon	1	0 (0%)	0 (0%)	0 (0%)	1 (100%)	0 (0%)	0 (0%)	
Colombia	1	0 (0%)	0 (0%)	0 (0%)	1 (100%)	0 (0%)	0 (0%)	
Spain	5	0 (0%)	2 (40%)	0 (0%)	2 (40%)	0 (0%)	1 (20%)	
France	23	0 (0%)	0 (0%)	5 (22%)	5 (22%)	11 (48%)	2 (8.7%)	
Greece	1	0 (0%)	0 (0%)	1 (100%)	0 (0%)	0 (0%)	0 (0%)	
Iran	1	0 (0%)	0 (0%)	1 (100%)	0 (0%)	0 (0%)	0 (0%)	
Italy	2	0 (0%)	0 (0%)	1 (50%)	1 (50%)	0 (0%)	0 (0%)	
Liechtenstein	1	0 (0%)	0 (0%)	1 (100%)	0 (0%)	0 (0%)	0 (0%)	
Philippines	1	0 (0%)	0 (0%)	0 (0%)	1 (100%)	0 (0%)	0 (0%)	
Portugal	7	0 (0%)	0 (0%)	2 (29%)	5 (71%)	0 (0%)	0 (0%)	
DRC [3]	1	0 (0%)	0 (0%)	0 (0%)	1 (100%)	0 (0%)	0 (0%)	
Switzerland	231	9 (3.9%)	24 (10%)	55 (24%)	88 (38%)	42 (18%)	13 (5.6%)	
Vietnam	1	1 (100%)	0 (0%)	0 (0%)	0 (0%)	0 (0%)	0 (0%)	

[1] n (%); [2] Chi-square/Fisher test (depending on n); [3] Democratic Republic of the Congo.

Table 6. Acceptability of consuming insect-based foods by behavioral variables.

Behavioral Variables	Total	Not at All in Agreement			Totally Agree			*p*-Value [2]
		1	2	3	4	5	6	
		N = 11 [1]	*N* = 29 [1]	*N* = 69 [1]	*N* = 112 [1]	*N* = 53 [1]	*N* = 16 [1]	
Visible Insects								**<0.001**
1	104 (36%)	11 (100%)	28 (97%)	33 (48%)	25 (22%)	5 (9.4%)	2 (12%)	
2	70 (24%)	0 (0%)	1 (3.4%)	25 (36%)	33 (29%)	9 (17%)	2 (12%)	
3	60 (21%)	0 (0%)	0 (0%)	10 (14%)	35 (31%)	14 (26%)	1 (6.2%)	
4	35 (12%)	0 (0%)	0 (0%)	1 (1.4%)	17 (15%)	16 (30%)	1 (6.2%)	
5	14 (4,8%)	0 (0%)	0 (0%)	0 (0%)	1 (0.9%)	8 (15%)	5 (31%)	
6	7 (2.4%)	0 (0%)	0 (0%)	0 (0%)	1 (0.9%)	1 (1.9%)	5 (31%)	
Insects Not Visible								**<0.001**
1	52 (18%)	10 (91%)	22 (76%)	14 (20%)	6 (5.4%)	0 (0%)	0 (0%)	
2	59 (20%)	0 (0%)	7 (24%)	21 (30%)	27 (24%)	4 (7.5%)	0 (0%)	
3	90 (31%)	1 (9.1%)	0 (0%)	26 (38%)	45 (40%)	15 (28%)	3 (19%)	
4	62 (21%)	0 (0%)	0 (0%)	7 (10%)	28 (25%)	23 (43%)	4 (25%)	
5	20 (6.9%)	0 (0%)	0 (0%)	1 (1.4%)	5 (4.5%)	9 (17%)	5 (31%)	
6	7 (2.4%)	0 (0%)	0 (0%)	0 (0%)	1 (0.9%)	2 (3.8%)	4 (25%)	
Meat = Health								**<0.001**
1	25 (8.6%)	4 (36%)	6 (21%)	4 (5.8%)	5 (4.5%)	4 (7.5%)	2 (12%)	
2	61 (21%)	1 (9.1%)	9 (31%)	11 (16%)	22 (20%)	14 (26%)	4 (25%)	
3	58 (20%)	1 (9.1%)	2 (6.9%)	10 (14%)	29 (26%)	14 (26%)	2 (12%)	
4	94 (32%)	0 (0%)	5 (17%)	27 (39%)	40 (36%)	16 (30%)	6 (38%)	
5	35 (12%)	1 (9.1%)	3 (10%)	14 (20%)	11 (9.8%)	5 (9.4%)	1 (6.2%)	
6	17 (5.9%)	4 (36%)	4 (14%)	3 (4.3%)	5 (4.5%)	0 (0%)	1 (6.2%)	
Meat Lovers								0.057
1	25 (8.6%)	5 (45%)	4 (14%)	7 (10%)	7 (6.2%)	1 (1.9%)	1 (6.2%)	
2	30 (10%)	1 (9.1%)	5 (17%)	10 (14%)	10 (8.9%)	2 (3.8%)	2 (12%)	
3	56 (19%)	0 (0%)	7 (24%)	9 (13%)	27 (24%)	12 (23%)	1 (6.2%)	
4	95 (33%)	4 (36%)	8 (28%)	22 (32%)	35 (31%)	21 (40%)	5 (31%)	
5	60 (21%)	0 (0%)	3 (10%)	16 (23%)	24 (21%)	12 (23%)	5 (31%)	
6	24 (8.3%)	1 (9.1%)	2 (6.9%)	5 (7.2%)	9 (8.0%)	5 (9.4%)	2 (12%)	
Food Neophobia								**0.004**
2	4 (1.4%)	0 (0%)	0 (0%)	3 (4.3%)	1 (0.9%)	0 (0%)	0 (0%)	
3	23 (7.9%)	2 (18%)	5 (17%)	7 (10%)	6 (5.4%)	2 (3.8%)	1 (6.2%)	
4	74 (26%)	5 (45%)	5 (17%)	20 (29%)	33 (29%)	8 (15%)	3 (19%)	
5	106 (37%)	2 (18%)	13 (45%)	30 (43%)	39 (35%)	16 (30%)	6 (38%)	
6	83 (29%)	2 (18%)	6 (21%)	9 (13%)	33 (29%)	27 (51%)	6 (38%)	
Neophobia of New Technology								**<0.001**
1	6 (2.1%)	2 (18%)	0 (0%)	0 (0%)	1 (0.9%)	0 (0%)	3 (19%)	
2	31 (11%)	4 (36%)	1 (3.4%)	10 (14%)	11 (9.8%)	2 (3.8%)	3 (19%)	
3	56 (19%)	1 (9.1%)	12 (41%)	15 (22%)	14 (12%)	14 (26%)	0 (0%)	
4	118 (41%)	2 (18%)	9 (31%)	27 (39%)	52 (46%)	21 (40%)	7 (44%)	
5	57 (20%)	2 (18%)	6 (21%)	15 (22%)	23 (21%)	10 (19%)	1 (6.2%)	
6	22 (7.6%)	0 (0%)	1 (3.4%)	2 (2.9%)	11 (9.8%)	6 (11%)	2 (12%)	
Interest in Health								0.5
1	3 (1.0%)	0 (0%)	2 (6.9%)	1 (1.4%)	0 (0%)	0 (0%)	0 (0%)	
2	21 (7.2%)	0 (0%)	2 (6.9%)	10 (14%)	6 (5.4%)	3 (5.7%)	0 (0%)	
3	15 (5.2%)	0 (0%)	1 (3.4%)	2 (2.9%)	7 (6,2%)	4 (7.5%)	1 (6.2%)	
4	126 (43%)	8 (73%)	14 (48%)	24 (35%)	48 (43%)	22 (42%)	10 (62%)	
5	51 (18%)	1 (9.1%)	3 (10%)	16 (23%)	20 (18%)	9 (17%)	2 (12%)	
6	74 (26%)	2 (18%)	7 (24%)	16 (23%)	31 (28%)	15 (28%)	3 (19%)	
Practicality								0.5
2	1 (0.3%)	0 (0%)	0 (0%)	1 (1.4%)	0 (0%)	0 (0%)	0 (0%)	
3	39 (13%)	2 (18%)	5 (17%)	7 (10%)	12 (11%)	9 (17%)	4 (25%)	
4	194 (67%)	6 (55%)	20 (69%)	46 (67%)	77 (69%)	35 (66%)	10 (62%)	
5	54 (19%)	2 (18%)	4 (14%)	15 (22%)	23 (21%)	8 (15%)	2 (12%)	
6	2 (0.7%)	1 (9.1%)	0 (0%)	0 (0%)	0 (0%)	1 (1.9%)	0 (0%)	
Value for Money								0.2
1	3 (1.0%)	1 (9.1%)	1 (3.4%)	1 (1.4%)	0 (0%)	0 (0%)	0 (0%)	
2	40 (14%)	0 (0%)	4 (14%)	12 (17%)	17 (15%)	7 (13%)	0 (0%)	
3	80 (28%)	4 (36%)	8 (28%)	16 (23%)	34 (30%)	13 (25%)	5 (31%)	
4	103 (36%)	4 (36%)	11 (38%)	27 (39%)	40 (36%)	17 (32%)	4 (25%)	
5	54 (19%)	2 (18%)	4 (14%)	12 (17%)	20 (18%)	12 (23%)	4 (25%)	
6	10 (3.4%)	0 (0%)	1 (3.4%)	1 (1.4%)	1 (0.9%)	4 (7.5%)	3 (19%)	

[1] n (%); [2] Fisher test. Bold in this case indicated that these *p*-values were significant.

4. Discussion

4.1. Acceptability of Eating Insect-Based Foods

To find out what the proportion of the study population that was interested in ACI was, the threshold of acceptability proposed for the analysis was the central value of the interval corresponding to the Likert scale (1–6), i.e., 3.5. This means that 62.5% of the participants would be potentially interested in entomophagy. A more restrictive threshold, where it would be even more certain that participants would accept entomophagy, could also have been used; for example, a threshold of 4.0 (40.4%) or even 5.0 (9.7%). Indeed, it is difficult to define a threshold in the spectrum of acceptability, based on arguments and not on practice (because of the intention-behavior gap). In order to contextualize these results, the scores of the variables composing the ACI were compared with those obtained in the previous study of Schlup and Brunner, who used exactly the same five arguments to calculate a score comparable to the ACI (also ranging from 1 to 6), which they named "Willingness To Consume" (WTC) [23]. This comparison is shown in Table 7.

Table 7. Comparison of variables composing the acceptance score (ACI) obtained in the present study and those obtained by Schlup and Brunner, based on the participants' responses [23].

Argument		Mean Score (This Study)	SD	Mean Score (Schlup and Brunner)	SD (Schlup and Brunner)	T-Statistics	Degrees of Freedom	p-Value
Sustainability	Extent of agreement	4.8	1.4	4	1.7	6.67	663.38	**<0.001**
	Perceived relevance	3.7	1.7	2.8	1.6	6.96	602.07	**<0.001**
Health	Extent of agreement	4.4	1.3	3.9	1.6	4.46	664.58	**<0.001**
	Perceived relevance	3.4	1.7	2.9	1.6	3.87	602.07	**<0.001**
Price	Extent of agreement	4.4	1.4	3.7	1.6	6.02	655.07	**<0.001**
	Perceived relevance	3.3	1.7	2.5	1.5	6.34	578.84	**<0.001**
Taste	Extent of agreement	4	1.6	3.6	1.6	3.2	622.03	**<0.001**
	Perceived relevance	3.4	1.7	2.8	1.6	4.64	602.07	**<0.001**
Trend	Extent of agreement	3.2	1.5	3.3	1.5	−0.85	622.03	0.39
	Perceived relevance	2.8	1.6	2.6	1.5	1.65	600.7	0.1

Bold in this case indicated that these p-values were significant.

Comparing the results obtained in this study with those of Schlup and Brunner, the difference between the values of the means relating to the first four arguments (sustainability, health, price, and taste) were highly significant ($p < 0.001$) [23]. These same arguments scored higher in the present study than in that of Schlup and Brunner, suggesting a possible evolution of the customs of recent years regarding the acceptability of eating insect-based foods [23]. More specifically, the "level of agreement" for the first four arguments had means above 4 for this study, being systematically higher than in the reference study. In terms of trend, however, no significant differences were found in the responses.

Considering that Schlup and Brunner's study was conducted in 2015, just before the changes in the regulatory framework in Switzerland on the sale and marketing of insect products, people's acceptability of entomophagy has evidently changed over the last 6 years. As mentioned above, these legislative changes may have had a positive impact on the supply and demand for this type of product. It probably encouraged consumers to try this new food that was appearing on the shelves of mainstream supermarkets. However, there is also a possible population bias, as in this study we focused on the young adult population who are probably more open-minded towards novel foods such as insects than the rest of the Swiss population. This contrasts with the study by Schlup and Brunner where age had a marginally negative effect on WTC in the general Swiss population [23].

Despite the differences between the means, the relative importance of the arguments did not change between the two studies. Sustainability remains the most accepted argument, while the questions related to taste and tendency to consume insects showed the most disagreement. In both studies, taste received a score close to the central value of the scale (of 3.5), which means that these arguments for consuming insects are still relatively weak or that respondents had not formed a strong opinion on these issues.

As presented in the article by Schlup and Brunner, there is an apparent disconnect between the perceived relevance and the actual influence of the argument on the participants' decision to consume insects [23]. It is perceived that positive arguments allow participants to become aware of the benefits of insect consumption. However, certain obstacles to the effective consumption of this type of food remain. This is illustrated by the proportion of participants with a potential interest in entomophagy (ACI $\geq$ 3.5, i.e., 62.5%), and the proportion of participants who have consumed insect-based foods more than once (25.2%).

4.2. Factors Motivating or Inhibiting Entomophagy

In this study, curiosity was the main reason why people might consume insect-based foods. Curiosity is indeed a factor that motivates people to try new types of food [2]. Curiosity about new foods is often aroused in a group dynamic. Studies have shown that people are more willing to try new foods such as insect-based foods in a relaxed setting with friends, for example, as an aperitif, given the variety of insect-based chips and burgers [20]. One can also consider "peer pressure" as a factor among people to at least accept to taste this type of food [30,31]. However, peer pressure can also have the opposite effect, triggering reactions ranging from disgust to food neophobia—depending on the group dynamic [32]. Sustainability was of rather marginal importance as a reason for participants to eat insect-based foods. Indeed, while participants showed a high level of agreement with the environmental arguments presented at the beginning of the questionnaire, the same participants only rarely mentioned sustainability as the main reason for eating insects. Interestingly, sustainability was more frequently cited as a reason among those who had never eaten insects than among those who had. This suggests that the awareness for climate change may be changing eating habits.

4.3. Association between ACI and Socio-Demographic Variables

To investigate whether acceptability was associated with socio-demographic variables, this study conducted Chi-square and Fisher tests. Notably, none of the selected socio-demographic variables (gender, age category, language region, current level of education, main field of study, and nationality) were significantly associated with ACI. These results contrast with previous studies where it was shown that gender shows a significant association with acceptability (females/males are more/less likely to ...), while other studies show that age and education level have significant associations (explanation ...) [20,27,33].

According to recent studies, gender seems to impact the acceptability of alternative proteins. In the case of insects, they are generally better accepted by men [16,34–37]. Looking at the distribution of the ACI score, this association was not found in the present study. However, this result can be qualified by the data concerning previous insect consumption (which was not considered in the ACI). Of the 25% of participants who had consumed insects more than once, men were in the majority (46 out of 71, or 65%), whereas women were in the majority among all participants (57.7% of the total). It is true that women are more attentive to healthy eating and are, therefore, more open to new foods with a positive impact on health [38]. However, studies suggest that this same audience has a strong sense of disgust and high levels of food neophobia towards insect-based foods [32]. This was also observed in the present study, as disgust was one of the main reasons that inhibits insect consumption (Figure 4). At the same time, men were less reluctant and more open to eating foods with a perceived "disgusting" appearance. Indeed, studies show that their threshold for food neophobia is generally higher than that of women [39].

Another result that contrasts with the existing literature is the lack of a significant association between age and ACI. This may be explained by the small differences in the age ranges of the participants, who were all relatively young. In a study in Germany, young people were open to entomophagy than older individuals [35]. Another study in Belgium found that a 10-year age difference in age was associated with a 27% decrease in the likelihood of participants being willing to adopt insects as food [39].

For the other variables, limitations related to the representativeness of the sample certainly biased the results. The level of education lacked contrast, being high for all participants (university or university of applied sciences). The nationalities of the participants were varied, but the non-Swiss remained few. Finally, the linguistic regions were not well represented, considering the prevalence of French speakers and the low number of Italian speakers.

4.4. Association between ACI and Behavioral Variables

In the present study, the majority (5 out of 9) of the behavioral variables, i.e., (i) food with visible insects, (ii) foods with non-visible insects, (iii) meat is synonymous with health, (iv) food neophobia, and (v) neophobia to new food technologies, were significantly ($p < 0.05$) associated with ACI. These results are consistent with the literature.

Regarding the first two variables (insect visibility), recent studies have shown increased acceptability when the insect is processed, e.g., as flour, and an ingredient of, for example, baked goods, pasta, brownies, crisps, or protein bars [40]. A study conducted in Switzerland in 2015 aimed at understanding the emotional experiences of insect-based snacks confirmed the impact of processing on consumer perception, showing that the more processed the insect-based food is, the more positive the emotional evaluation of the product is [41]. Furthermore, the authors considered that repeated exposure is necessary to accustom consumers to these new foods. They thus reveal the role of marketing and the supply of these foods in the market: with the introduction of new products, closer to traditional products, emotional barriers can be overcome and the willingness to consume insects increases. This impact of the product offer on the perception of the increase in the acceptance of insect-based foods can be seen by comparing the results of the present study with those of Schlup and Brunner: the number of people who reported having eaten insects (at least once) increased from 18.9% in 2015 to 55% in 2021 (present study) [23]. The new Swiss regulatory framework (established in 2017) and the evolution of the product offer taking place during these six years have likely contributed to increase the acceptability of these products among the Swiss population.

Regarding the variables linked to meat consumption, according to a qualitative study carried out in Switzerland, the increase in the supply of various sources of protein leads to an equally varied consumption of food products [42]. However, meat consumption seems to be very present among the participants of the present study: 15% of the participants declared that they did not consume meat, 7% consumed it less than once a month and a little more than half consumed it 2 to 4 times a week. It is likely that these people are aware of the need to reduce meat consumption and take advantage of new food technologies to discover meat substitutes, such as insects [43].

It should be noted that the variable "meat lovers" was not significantly positively or negatively associated with ACI, whereas the variable related to neophobia of food technologies (techneo) was. This result shows that liking meat is not associated with greater acceptability (or rejection) of insect food consumption. Alexander et al. (2017) argue that gradual (non-disruptive) changes in consumer eating habits should be induced, facilitating a transition to more sustainable diets [44]. Indeed, meat and insect consumption are not mutually exclusive. Insects can gradually be inserted (in different forms) onto our plates as food technologies advance.

Although meat consumption was not clearly associated with ACI, another diet-related factor was relevant to understanding the refusal of entomophagy. Some people who received the recruitment email refused to participate, stating that they were vegans or vegetarians and did not agree to participate in this type of research.

4.5. Limitations of the Study

This study has limitations related to the convenience sampling method, which is not random and may have selection biases. For example, the percentage of women (57.7%) and men (42.3%) differed slightly from the proportions observed in Switzerland in 2020, which were 50.4% and 49.6%, respectively [45]. This discrepancy can certainly be explained by the age range and field of study of the target population. For example, according to the Swiss Federal Statistical Office, in 2018, women were in the majority among young doctors, accounting for 62% of doctors aged 35–39 years [46]. This may explain the high prevalence of women among the participants, as the questionnaire reached more people whose main field of study is health sciences.

Moreover, in studies with voluntary sampling (such as this one), it is possible to produce a bias that could overestimate the real consumption because the participants who agree to answer the questionnaire are more concerned or who identify with the topic being studied.

In addition, considering that the authors had greater access so the universities in the French-speaking part of Switzerland, there was an increased probability that an above-average proportion of the participants would come from this linguistic region. For the same practical reasons, most of the participants were Swiss. This makes it difficult to extrapolate the results of this study beyond Western Switzerland. In a future study, the questionnaire could be offered in more languages, for example. In addition, the recruitment process could be expanded, for example, by recruiting participants from the various local universities on site.

5. Conclusions

In a context of climate emergency and increasing limitations in terms of arable land availability, it is essential to reassess the nutritional potential and possible role of insects in global food security. The first step is to assess the acceptability of insect-based foods to consumers, particularly in the West where there is no tradition and habit toward them yet.

This study investigated the perception of entomophagy among young adults in a Swiss university context. Overall, the theoretical arguments in favor of entomophagy were well accepted. In practice, however, the study participants were less inclined to consume insect-based foods. Nevertheless, the overall acceptability of these foods, as well as the number of people who have eaten insects at least once, appears to have increased between 2015 and 2021. This has probably been influenced by the regulatory framework put in place in Switzerland since 2017, which has helped to expose consumers to this new product and, thus, to attract their attention.

Significantly, among the participants in this study, curiosity was by far the most mentioned motivating factor for eating insect-based foods. Whereas no socio-demographic variables were found to be significant in predicting theoretical acceptability (ACI), five behavioral variables were significantly related to ACI: (i) food with visible insects, (ii) food with non-visible insects, (iii) meat is synonymous with health, (iv) food neophobia, and (v) neophobia to new food technologies. These results suggest that there is still resistance to these products due to factors, such as presentation (level of processing) and food neophobia.

Given the trends discussed in this study, it is reasonable to assume that in the coming years entomophagy will expand in Switzerland. However, entomophagy still needs to be better considered by different sectors: research, health, food industry, associations, and governments (through the regulatory framework of these foods). There is also a need for further research into the benefits of insects as a food and as a potential substitute for meat, so that they can be used as arguments in marketing to consumers.

Supplementary Materials: The following supporting information can be downloaded at: https://www.mdpi.com/article/10.3390/foods11182771/s1, Table S1: Survey: Consumption of insect-based foods in Switzerland.

Author Contributions: Conceptualization, A.O.P., W.M.B. and S.B.D.T.; methodology, A.O.P.; software, A.O.P.; validation, A.O.P., S.B.D.T., F.G. and T.A.B.; formal analysis, A.O.P.; investigation, A.O.P.; data curation, A.O.P.; writing—original draft preparation, A.O.P. and W.M.B.; writing—review and editing, A.O.P., S.B.D.T., F.G. and W.M.B.; visualization, A.O.P.; supervision, W.M.B.; project administration, W.M.B. and S.B.D.T. All authors have read and agreed to the published version of the manuscript.

Funding: This research received no external funding.

Informed Consent Statement: Informed consent was obtained from all subjects involved in the study.

Data Availability Statement: The data presented in this study are available on request from the corresponding author. The data are not publicly available due to due privacy and data protections issues.

Acknowledgments: The authors would like to thank the participants of this study. This article is based on the results of a master thesis conducted within the joint Master of Science (MSc) in Health Sciences of HES-SO (University of Applied Sciences and Arts Western Switzerland) and the University of Lausanne (UNIL), within the orientation of Nutrition and Dietetics, at HES-SO Master.

Conflicts of Interest: The authors declare no conflict of interest.

References

1. Van Huis, A. Potential of Insects as Food and Feed in Assuring Food Security. *Annu. Rev. Entomol.* **2013**, *58*, 563–583. [CrossRef] [PubMed]
2. Stone, H.; FitzGibbon, L.; Millan, E.; Murayama, K. Curious to eat insects? Curiosity as a Key Predictor of Willingness to try novel food. *Appetite* **2022**, *168*, 105790. [CrossRef]
3. Durst, P.B.; Hanboonsong, Y. Small-scale production of edible insects for enhanced food security and rural livelihoods: Experience from Thailand and Lao People's Democratic Republic. *J. Insects Food Feed* **2015**, *1*, 25–31. [CrossRef]
4. Tso, R.; Lim, A.J.; Forde, C.G. A Critical Appraisal of the Evidence Supporting Consumer Motivations for Alternative Proteins. *Foods* **2020**, *10*, 24. [CrossRef] [PubMed]
5. Van Huis, A. *Edible Insects: Future Prospects for Food and Feed Security*; FAO: Rome, Italy, 2013; p. 201. Available online: https://www.fao.org/3/i3253e/i3253e.pdf (accessed on 4 June 2022).
6. Kim, T.K.; Yong, H.I.; Kim, Y.B.; Kim, H.W.; Choi, Y.S. Edible Insects as a Protein Source: A Review of Public Perception, Processing Technology, and Research Trends. *Food Sci. Anim. Resour.* **2019**, *39*, 521–540. [CrossRef] [PubMed]
7. Raheem, D.; Carrascosa, C.; Oluwole, O.B.; Nieuwland, M.; Saraiva, A.; Millán, R.; Raposo, A. Traditional consumption of and rearing edible insects in Africa, Asia and Europe. *Crit. Rev. Food Sci. Nutr.* **2019**, *59*, 2169–2188. [CrossRef] [PubMed]
8. Oonincx, D.G.A.B. Environmental impact of insect production. In *Insects as Food and Feed: From Production to Consumption*; van Huis, A., Tomberlin, J.K., Eds.; Wageningen Academic Publishers: Wageningen, The Netherlands, 2017; pp. 79–93.
9. Baiano, A. Edible insects: An overview on nutritional characteristics, safety, farming, production technologies, regulatory framework, and socio-economic and ethical implications. *Trends Food Sci. Technol.* **2020**, *100*, 35–50. [CrossRef]
10. Bisconsin-Júnior, A.; Rodrigues, H.; Behrens, J.H.; da Silva, M.A.A.P.; Mariutti, L.R.B. Food made with edible insects: Exploring the social representation of entomophagy where it is unfamiliar. *Appetite* **2022**, *173*, 106001. [CrossRef]
11. Doi, H.; Gałęcki, R.; Mulia, R.N. The merits of entomophagy in the post COVID-19 world. *Trends Food Sci. Technol.* **2021**, *110*, 849–854. [CrossRef]
12. Van Huis, A. Nutrition and health of edible insects. *Curr. Opin. Clin. Nutr. Metab. Care* **2020**, *23*, 228–231. [CrossRef]
13. Kosečková, P.; Zvěřina, O.; Pěchová, M.; Krulíková, M.; Duborská, E.; Borkovcová, M. Mineral profile of cricket powders, some edible insect species and their implication for gastronomy. *J. Food Comp. Anal.* **2022**, *107*, 104340. [CrossRef]
14. European Union. Regulation (EU) 2015/2283 of the European Parliament and of the Council of 25 November 2015 on Novel Foods, Amending Regulation (EU) No 1169/2011 of the European Parliament and of the Council and Repealing Regulation (EC) No 258/97 of the European Parliament and of the Council and Commission Regulation (EC) No 1852/2001. 2015/2283 2015. Available online: https://eur-lex.europa.eu/legal-content/EN/TXT/HTML/?uri=CELEX:32015R2283 (accessed on 4 June 2022).
15. Insects as foods [Insekten als Lebensmittel] Federal Food Safety and Veterinary Office FSVO. Available online: https://www.blv.admin.ch/blv/de/home/lebensmittel-und-ernaehrung/lebensmittelsicherheit/einzelne-lebensmittel/insekten.html (accessed on 4 June 2022).
16. Sogari, G.; Menozzi, D.; Mora, C. Exploring young foodies' knowledge and attitude regarding entomophagy: A qualitative study in Italy. *Int. J. Gastron. Food Sci.* **2017**, *7*, 16–19. [CrossRef]
17. Dagevos, H. A literature review of consumer research on edible insects: Recent evidence and new vistas from 2019 studies. *J. Insects Food Feed* **2021**, *7*, 249–259. [CrossRef]
18. Sogari, G.; Menozzi, D.; Mora, C. The food neophobia scale and young adults' intention to eat insect products. *Int. J. Consum. Stud.* **2019**, *43*, 68–76. [CrossRef]
19. Van Huis, A. Edible insects: Challenges and prospects. *Entomol. Res.* **2022**, *52*, 161–177. [CrossRef]
20. Dupont, J.; Fiebelkorn, F. Attitudes and acceptance of young people toward the consumption of insects and cultured meat in Germany. *Food Qual. Prefer.* **2020**, *85*, 103983. [CrossRef]
21. Videbæk, P.N.; Grunert, K.G. Disgusting or delicious? Examining attitudinal ambivalence towards entomophagy among Danish consumers. *Food Qual. Pref.* **2020**, *83*, 103913. [CrossRef]
22. Looking at Edible Insects from a Food Safety Perspective. *Challenges and Opportunities for the Sector*; FAO: Rome, Italy, 2021; p. 108. Available online: http://www.fao.org/documents/card/en/c/cb4094en (accessed on 19 January 2022).
23. Schlup, Y.; Brunner, T. Prospects for insects as food in Switzerland: A tobit regression. *Food Qual. Prefer.* **2018**, *64*, 37–46. [CrossRef]

24. Hartmann, C.; Ruby, M.B.; Schmidt, P.; Siegrist, M. Brave, health-conscious, and environmentally friendly: Positive impressions of insect food product consumers. *Food Qual. Prefer.* **2018**, *68*, 64–71. [CrossRef]
25. Myers, G.; Pettigrew, S. A qualitative exploration of the factors underlying seniors' receptiveness to entomophagy. *Food Res. Int.* **2018**, *103*, 163–169. [CrossRef]
26. Skotnicka, M.; Karwowska, K.; Kłobukowski, F.; Borkowska, A.; Pieszko, M. Possibilities of the Development of Edible Insect-Based Foods in Europe. *Foods* **2021**, *10*, 766. [CrossRef] [PubMed]
27. Onwezen, M.C.; Bouwman, E.P.; Reinders, M.J.; Dagevos, H. A systematic review on consumer acceptance of alternative proteins: Pulses, algae, insects, plant-based meat alternatives, and cultured meat. *Appetite* **2021**, *159*, 105058. [CrossRef] [PubMed]
28. La Barbera, F.; Amato, M.; Fasanelli, R.; Verneau, F. Perceived Risk of Insect-Based Foods: An Assessment of the Entomophagy Attitude Questionnaire Predictive Validity. *Insects* **2021**, *12*, 403. [CrossRef] [PubMed]
29. Taber, K.S. The Use of Cronbach's Alpha When Developing and Reporting Research Instruments in Science Education. *Res. Sci. Educ.* **2018**, *48*, 1273–1296. [CrossRef]
30. Barberis, N.; Gugliandolo, M.C.; Costa, S.; Cannavò, M. Healthy and binge eating behaviours: The motivational processes underlying peer pressure. *Psychol. Health Med.* **2022**, *27*, 1144–1153. [CrossRef] [PubMed]
31. Berger, S.; Christandl, F.; Bitterlin, D.; Wyss, A.M. The social insectivore: Peer and expert influence affect consumer evaluations of insects as food. *Appetite* **2019**, *141*, 104338. [CrossRef]
32. Poortvliet, P.M.; Van der Pas, L.; Mulder, B.C.; Fogliano, V. Healthy, but Disgusting: An Investigation Into Consumers' Willingness to Try Insect Meat. *J. Econ. Entomol.* **2019**, *112*, 1005–1010. [CrossRef]
33. Piha, S.; Pohjanheimo, T.; Lähteenmäki-Uutela, A.; Křečková, Z.; Otterbring, T. The effects of consumer knowledge on the willingness to buy insect food: An exploratory cross-regional study in Northern and Central Europe. *Food Qual. Prefer.* **2018**, *70*, 1–10. [CrossRef]
34. Orkusz, A. Edible Insects versus Meat—Nutritional Comparison: Knowledge of Their Composition Is the Key to Good Health. *Nutrients* **2021**, *13*, 1207. [CrossRef]
35. Lammers, P.; Ullmann, L.M.; Fiebelkorn, F. Acceptance of insects as food in Germany: Is it about sensation seeking, sustainability consciousness, or food disgust? *Food Qual. Prefer.* **2019**, *77*, 78–88. [CrossRef]
36. Menozzi, D.; Sogari, G.; Veneziani, M.; Simoni, E.; Mora, C. Eating novel foods: An application of the Theory of Planned Behaviour to predict the consumption of an insect-based product. *Food Qual. Prefer.* **2017**, *59*, 27–34. [CrossRef]
37. Woolf, E.; Zhu, Y.; Emory, K.; Zhao, J.; Liu, C. Willingness to consume insect-containing foods: A survey in the United States. *LWT* **2019**, *102*, 100–105. [CrossRef]
38. Caparros Megido, R.; Gierts, C.; Blecker, C.; Brostaux, Y.; Haubruge, É.; Alabi, T.; Francis, F. Consumer acceptance of insect-based alternative meat products in Western countries. *Food Qual. Prefer.* **2016**, *52*, 237–243. [CrossRef]
39. Verbeke, W. Profiling consumers who are ready to adopt insects as a meat substitute in a Western society. *Food Qual Prefer.* **2015**, *39*, 147–155. [CrossRef]
40. Mishyna, M.; Chen, J.; Benjamin, O. Sensory attributes of edible insects and insect-based foods—Future outlooks for enhancing consumer appeal. *Trends Food Sci. Technol.* **2020**, *95*, 141–148. [CrossRef]
41. Gmuer, A.; Nuessli Guth, J.; Hartmann, C.; Siegrist, M. Effects of the degree of processing of insect ingredients in snacks on expected emotional experiences and willingness to eat. *Food Qual. Prefer.* **2016**, *54*, 117–127. [CrossRef]
42. Sahakian, M.; Godin, L.; Courtin, I. Promoting 'pro', 'low', and 'no' meat consumption in Switzerland: The role of emotions in practices. *Appetite* **2020**, *150*, 104637. [CrossRef]
43. Roma, R.; Ottomano Palmisano, G.; De Boni, A. Insects as Novel Food: A Consumer Attitude Analysis through the Dominance-Based Rough Set Approach. *Foods* **2020**, *9*, 387. [CrossRef]
44. Alexander, P.; Brown, C.; Arneth, A.; Dias, C.; Finnigan, J.; Moran, D.; Rounsevell, M.D.A. Could consumption of insects, cultured meat or imitation meat reduce global agricultural land use? *Glob. Food Secur.* **2017**, *15*, 22–32. [CrossRef]
45. Demographic portrait of Switzerland [Demografisches Porträt der Schweiz] Federal Statistical Office. Available online: https://www.bfs.admin.ch/bfs/en/home/statistics/catalogues-databases.assetdetail.21764558.html (accessed on 19 January 2022).
46. Medical Practices and Ambulatory Centres 2018. Federal Statistical Office. Available online: https://www.bfs.admin.ch/bfs/en/home/statistics/catalogues-databases.assetdetail.14667071.html (accessed on 19 January 2022).

foods

MDPI

Article

Consumer Preference Segments for Plant-Based Foods: The Role of Product Category

Armand V. Cardello [1], Fabien Llobell [2], Davide Giacalone [3], Sok L. Chheang [4] and Sara R. Jaeger [4,*]

1 A.V. Cardello Consulting and Editing Services, Framingham, MA 01701, USA
2 Addinsoft, XLSTAT, 75018 Paris, France
3 SDU Innovation & Design Engineering, Department of Technology and Innovation, University of Southern Denmark, 5230 Odense, Denmark
4 The New Zealand Institute for Plant and Food Research Limited, Mt Albert Research Centre, Private Bag 92169, Auckland 1142, New Zealand
* Correspondence: sara.jaeger@plantandfood.co.nz

Abstract: A survey of willingness to consume (WTC) 5 types of plant-based (PB) food was conducted in USA, Australia, Singapore and India (n = 2494). In addition to WTC, emotional, conceptual and situational use characterizations were obtained. Results showed a number of distinct clusters of consumers with different patterns of WTC for PB foods within different food categories. A large group of consumers did not discriminate among PB foods across the various food categories. Six smaller, but distinct clusters of consumers had specific patterns of WTC across the examined food categories. In general, *PB Milk* and, to a much lesser extent, *PB Cheese* had highest WTC ratings. *PB Fish* had the lowest WTC, and two PB meat products had intermediate WTC. Emotional, conceptual and situational use characterizations exerted significant lifts/penalties on WTC. No penalty or lifts were imparted on WTC by the situational use of 'moving my diet in a sustainable direction', whereas uses related to 'when I want something I like' and 'when I want something healthy' generally imparted WTC lifts across clusters and food categories. The importance of this research for the study of PB foods is its demonstration that consumers are not monolithic in their willingness to consume these foods and that WTC is often a function of the food category of the PB food.

Keywords: plant-based foods; consumer research; preference segments; product category; milk; cheese; fish; meat

Citation: Cardello, A.V.; Llobell, F.; Giacalone, D.; Chheang, S.L.; Jaeger, S.R. Consumer Preference Segments for Plant-Based Foods: The Role of Product Category. *Foods* **2022**, *11*, 3059. https://doi.org/10.3390/foods11193059

Academic Editor: Witoon Prinyawiwatkul

Received: 2 September 2022
Accepted: 28 September 2022
Published: 1 October 2022

1. Introduction

Over the past decades, it has become increasingly clear that the consumption of animal products has had unsustainable effects on the environment through high demand on land, water, feed, housing and the production of greenhouse gases [1–10]. In addition, excess consumption of animal products is known to have harmful effects on human health, including cancer, cardiovascular disease and obesity [11–15]. In response to these known detrimental effects of animal protein consumption on the environment and human health, global food policy has shifted to place greater emphasis on more sustainable farming practices and protein sources [1,16,17]. Leading this trend has been the emphasis on plant-based protein consumption [18–20]. This trend has resulted in a dramatic increase in the number of plant-based foods, beverages, product extenders and/or meat alternatives now available globally. It is estimated that the plant-based food market will further grow from approximately $US 30 billion in 2020 to $US 160 billion by 2030 [21]. In addition to plant-based proteins, a variety of other alternative proteins are now being investigated for their potential to replace animal protein in foods, e.g., algae, insects, and cultured meat [22–28].

1.1. Factors Influencing Acceptance of Plant-Based and Other Alternative Proteins

With the rapid growth in the alternative protein food market, research into the factors that drive consumer choice, purchase, acceptance and consumption of foods containing these proteins has grown correspondingly. Among the many factors that influence consumer acceptance and choice behavior toward these products are their sensory properties, e.g., taste, odor, texture, appearance [29–38], their familiarity [23,25,26,39–44], price/brand [37,45–47] and how appropriate the alternative protein is within the meal or its context of use [37,48–51].

Other important factors relate more specifically to the individual. These include the health concerns of individuals and the perceived benefits attributed to the alternative protein [52–60], the consumers' attitudes [36,44,55,61,62], their values and cultural norms [3,63–68] and their neophobic tendencies [25,28,32,38,40,44,69–73].

Due to the important impact of individual factors in alternative protein acceptance, several studies have searched for distinct segments of consumers with different preferences for alternative proteins. Thus, de Boer et al. [29] examined consumer involvement with food and its influence on the acceptance of alternative proteins. These researchers identified 'trendsetters' who sought more authentic proteins, like lentils and seaweeds, but eschewed hybrid products like meat extended with plant proteins [29]. Van der Zanden et al. [74] examined preference segments among elderly consumers and found that the type of carrier (food type) for the protein was an important differentiator among different clusters of consumers, as was the meal type (meal component vs. snack). Health-orientation was another important variable differentiating a majority of consumer segments in a study of seaweed protein consumption by Palmieri and Forleo [75], as well as in a study by Possidónio et al. [76] who identified 3 segments of alternative protein consumers; (1) hedonically motivated meat eaters uninterested in meat substitutes, (2) health-oriented meat eaters open to some meat substitutes, and (3) ethically conscious meat avoiders positive toward protein alternatives to meat. Finally, in a study by Aschemann-Witzel and Peschel [34], individuals who previously purchased or consumed vegetarian products had more differentiated associations to different protein types, suggesting that different segments of consumers may respond differentially to plant proteins depending on the category of food.

Despite the environmental and health reasons that support consumption of alternative protein foods, research has shown that consumer acceptance and choice behavior toward these products is lower than for their animal counterparts [77–80]. Despite the need to identify alternative proteins with the greatest potential for acceptance by consumers, most research has examined consumer attitudes and acceptance within the context of a single protein source, e.g., plants, pulses, algae, seaweed, insects, etc., and relatively fewer have studied multiple alternative proteins [80]. Among studies comparing multiple proteins, results have generally shown that plant-based proteins are more acceptable to consumers than other protein sources, e.g., seaweed and cultured meat, and that insect protein is the least preferred [47,53,76,81–84]. Nevertheless, a recent review of the literature on alternative proteins found that the vast majority of published research studies focused on insect protein and far fewer examined plant proteins [80].

1.2. Role of Food Product and Food Category

Although several studies have examined different alternative proteins or protein mixes within the same product, e.g., hamburger, milk alternatives, lasagna [47,48,82,85,86], relatively few have examined differences in acceptance/willingness to consume as a function of product type, in spite of the fact that in their review of the area, Hartmann and Siegrist [77] found that acceptance of alternative proteins varies by product and that "it is of little value to ask consumers about abstract concepts (e.g., are you willing to eat insects?"). Still fewer studies have compared consumer reactions to alternative proteins based on differences in the target food group or category (e.g., milk, cheese, meat, fish, etc.). This, too, in spite of the fact that the food category to which a product belongs has been shown to have a strong influence on consumer responses to alternative proteins.

In one series of studies, Elzerman et al. [49,50,87] showed that alternative proteins differ in acceptance depending upon the food (or food category) in which they are incorporated. For example, consumers were found to have a greater willingness to accept meat substitutes when served with spaghetti than as part of a soup. Similarly, Michel et al. [37] observed significant interactions between protein type (meat vs. non-meat) and product groups in the cognitive associations to different test products (e.g., nuggets, sausages). In still another study on alternative plant proteins, Aschemann-Witzel and Peschel [34] found differences in consumer perceptions of these products depending on food category. In their study, consumers were shown ingredient lists that accompanied sketches of two different products identified as being in the food categories of "protein drink" or "sorbets." The ingredient list for each product category was varied to include either the term 'protein', 'vegetable protein', 'soy protein', 'pea protein' or 'potato protein'. Results showed that the product category evoked specific associations for the protein(s), e.g., 'nutrition' for protein drinks and 'functional' for sorbets, and that more positive associations accrued to the protein drinks and more negative associations to the sorbets. Further, the nature of the protein evoked different associations depending on the food category in which it appeared. Functional roles, e.g., serving as a cohesive ingredient, were evoked for the potato and pea proteins, but only within the sorbet category.

The food category to which a specific food belongs is an important factor influencing consumer behavior toward it, because it serves a variety of purposes and needs for the consumer [34]. Primary among these is that food categories establish the context within which the product is conceptualized by the consumer. For example, 'salmon', as a product, is conceptually a member of the food category 'fish' and, thereby, accrues a variety of cognitive, emotional and situational use associations and expectations that are shared by all types of fish, e.g., specific sensory (odor, texture) expectations, expectations related to preparation, cooking, etc. Similarly, for insect proteins, although, ants, beetles, and mealworms all have different sensory attributes, consumers react to these products in a similar way, simply because they are all members of the 'insect food' category, which evokes both generalized neophobic and disgust responses in many individuals [88–93]. Even in routine CLT tests on familiar products, it has been shown that over 75% of consumers, when asked after the test whether they rated their liking of that food within the context of 'this kind of food' or 'all foods', reported that their ratings were made within the context of 'this *kind* of food' [48].

To our knowledge, it has not previously been shown across a range of possible food categories that consumers hold distinct PB food category preferences or that (for example) a high WTC for one PB food category is an unreliable predictor of WTC (and preferences) for other PB food categories. However, if true, this finding would have implications for the promotion of PB diets, since it could imply that promotional campaigns, if category generic, may resonate with fewer people than a more category-targeted campaign. Thus, one goal of the present research is to explore the degree to which WTC for PB foods is dependent on the food category to which the PB food belongs.

1.3. Objectives and Overview of the Empirical Approach

Building on the above research threads, the specific objectives of the present research were: (1) to examine consumers' willingness to consume plant-based protein within a number of distinct food categories, (2) to determine if there are different segments of consumers who have different patterns of their willingness to consume plant-based proteins, and to (3) assess differences in the identified segments of consumers in terms of their emotional, cognitive or situational use characterizations of the different plant-based categories of food and the impact of these variables on their judgments of willingness to consume these products. The latter objective is aimed at filling the literature gap identified by Onwezen et al. [94] in which it was noted that, with minor exceptions [37,82,94] few studies have examined either the role of emotions or other affective variables on acceptance of alternative proteins or the physical or social environment in which consuming such proteins is most appropriate or contextually acceptable.

2. Materials and Methods

2.1. Participants

Consumer insights for a global challenge like sustainable food production are more comprehensive when research is conducted in multiple countries. For this reason, participants in the present research came from the United States of America (US, n = 629), the Commonwealth of Australia (AU, n = 623), the Republic of Singapore (SG, n = 627), and the Republic of India (IN, n = 615). These countries differed on several dimensions including geographical location, population size, national cultures [95], importance of F&B sectors in national economies [96], sustainable energy [97], proportion of people following a vegetarian/vegan diet [98–101], growth rates of plant-based foods in retail [102], as well as the regulatory and legal policies regarding the labeling of PB foods. Country selection was also informed by a desire to lessen the dominance of Sensory-Consumer research taking place in Western, Educated, Industrialized, Rich, and Democratic (WEIRD) and especially Anglocentric countries [103].

Participants had self-registered on a database managed by a web panel provider with ISO 20252:2019 accreditation (ISO: International Organization for Standardization [104]). A quota sampling strategy was imposed by country with interlocking quota for men (50%) and women (50%) across two age groups (18–45 y.o. (50%), 46–69 y.o. (50%)). The samples were not nationally representative, but diverse across a range of characteristics such as living location, educational attainment, marital status, etc. (Table 1) (Part S1 of Supplementary Materials has country specific details). High proficiency in English and regular participation household grocery shopping and food preparation (more than once a week) were imposed as eligibility criteria.

Table 1. Participant characteristics for aggregate sample and by consumer segments based on willingness to consume (WTC) for plant-based (PB) food categories.

	Aggregate	Cluster 1	Cluster 2	Cluster 3	Cluster 4	Cluster 5	Cluster 6	Cluster 7
N	2494	1382	127	214	296	150	143	182
Country (%)								
Australia	25	27.6	11.8	18.7	13.5	34.7	22.3	34
India	24.7	17.2	63.8	26.2	44.6	30	25.9	14.3
Singapore	25.1	28.4	14.2	25.7	25.7	15.3	18.9	19.8
United States	25.2	26.8	10.2	29.4	16.2	20	32.9	31.9
Biological sex (%)								
Female	50	49.1	52.8	54.2	46.6	59.3	59.4	40.7
Male	50	50.9	47.2	45.8	53.4	40.7	40.6	59.3
Age group (%)								
18–45 y.o.	49.5	47.8	47.2	54.2	51	45.3	55.2	55.5
46–69 y.o.	50.5	52.2	52.8	45.8	49	54.7	45.8	44.5
Education (%)								
High school, vocational or short graduate	36.8	41.2	19.7	32.7	29.7	37.3	25.2	40.1
University education (Bachelor or higher)	62.8	58.4	80.3	66.8	70.3	62	74.8	58.8
Prefer to not answer	0.4	0.4	0	0.5	0	0.7	0	1.1
Dietary preference (%)								
Flexitarian	33.9	29.2	35.4	35.5	45.9	38	48.3	32.4
Omnivore	59.7	68	35.4	60.3	41.9	52	35.7	66.5
Vegetarian	6.5	2.8	29.1	4.2	12.2	10	16.1	1.1

Table 1. *Cont.*

	Aggregate	Cluster 1	Cluster 2	Cluster 3	Cluster 4	Cluster 5	Cluster 6	Cluster 7
Food Neophobia	36.8	37.4	37.8	35.1	37.2	35.1	35.4	35.7
[M (SD)]	(10.2)	(10.6)	(8.7)	(9.1)	(9.0)	(11.9)	(9.4)	(10.5)
Food Technology Neophobia	56.3	57.5	55.6	54.4	53.7	57.6	53.2	55.1
[M (SD)]	(11.6)	(11.9)	(11.2)	(10.7)	(11.1)	(11.0)	(11.0)	(10.0)
Environmental concern	62.4	60	67.1	64.3	64.6	64.3	66.1	62.4
[M (SD)]	(12.4)	(11.9)	(11.7)	(11.0)	(10.7)	(11.0)	(12.6)	(11.1)

Food Neophobia (FN) scores could range between 10 and 70, with higher scores reflecting higher levels of FN. Values are mean (M) and standard deviation (SD). Food Technology Neophobia (FTN) scores could range between 13 and 91, with higher scores reflecting higher levels of FTN. Values are mean (M) and standard deviation (SD). Environmental concern (ENV) scores could range between 12 and 84, with higher scores reflecting more positive attitude toward the environment. Values are mean (M) and standard deviation (SD).

Human Ethics Statement

The study was covered by a general approval for sensory and consumer research from the Human Ethics Committee at the New Zealand Institute for Plant and Food Research Limited. Participants gave voluntary consent and were assured that their responses would remain confidential. They were informed that they could end their participation at any time. As compensation, participants received reward points which could be redeemed for online purchases.

2.2. Plant-Based Food Stimuli

The research focused on 5 plant-based (PB) foods (abbreviations for figures and tables in brackets):

- Milk—from 100% plant-based ingredients (*PB Milk*)
- Cheese—from 100% plant-based ingredients (*PB Cheese*)
- Meat—blend containing 33% plant-based ingredients (*PB Meat 33%)*
- Meat—from 100% plant-based ingredients (*PB Meat*)
- Fish—from 100% plant-based ingredients (*PB Fish*)

Milk and cheese were chosen because they are both familiar but quite different subcategories within the overall dairy category and have quite different levels of availability and consumer acceptance (milk—readily available, well accepted; cheese—less readily available, lower acceptance). Meat (100% and 33% plant-based) were chosen because meat protein is the most highly targeted protein for replacement, but percentage replacement can influence acceptance and willingness to consume. Meat hybrids like *PB Meat 33%* offer a more sustainable food alternative for consumers who consider meat to be an essential and integral element of their daily diet [105]. Lastly, fish was chosen as it is a more novel food category for plant-based products and identified as a niche for PB food innovations by Alcorta et al. [106].

To provide a common frame of reference, participants read a short text (written by authors SRJ and DG) before answering questions about the focal foods. It read: "Our current approaches to food production and our consumption levels are global problems because they drive climate change and environmental degradation. Animal farming is particularly damaging for the environment, and to reduce environmental impacts we must change the way we eat—increasing our consumption of plant foods while substantially limiting our intake from animal sources. To support this transition, interest has centered on those plant foods that are good sources of proteins—soybeans, nuts, peas, and some grains—and can provide a nutritionally sound alternative to foods from animals, while also being suitable for the growing number of people who do not, or cannot, eat certain animal-sourced foods (e.g., vegetarians/vegans, people who have a dairy allergy or are lactose intolerant). In the past decade, more and more of these 100% plant-based foods have become available in the [United States/Australia/Singapore/India]. PB "milks" made from, for example, soy, almonds, oats, and even cashew nuts are fairly familiar now, and plant-based "yoghurt" and "cheese" are no longer uncommon. PB "meat", "fish", and "seafood" are much more novel. Besides replacing foods from animals with plant-based foods, or creating blends of animal-plant foods, there is also considerable interest in developing alternative sources of animal-derived proteins. These include insects."

The text continued to describe other novel types of foods/food technologies (Part S2 of Supplementary Materials has the text in full). The reason was to provide background information for other foods/food technologies also included in the survey but not relevant for the present research (Part S2 of Supplementary Materials lists these other foods).

2.3. *Empirical Procedures*

2.3.1. Stimulus Evaluation

Stimulus evaluation proceeded in two parts. First, the plant-based food names were assessed using two check-all-that-apply (CATA) question [107], pertaining to (i) emotional and cognitive product perceptions, and (ii) situational use product perceptions. This evaluation proceeded sequentially and both CATA questions were completed before the next food name appeared. Randomization of stimuli and terms were used within CATA questions and across participants.

Drawing on a general vocabulary [108], the CATA question relating to emotional and cognitive product conceptualizations included 16 terms: 'adventurous', 'boring', 'classy', 'comforting', 'dissatisfied', 'easygoing', 'energetic', 'enthusiastic', 'feminine', 'happy', 'inspiring', 'nervous', 'passive', 'powerful', 'pretentious', 'sophisticated', 'tense', 'uninspired', 'unique', and 'youthful'. The CATA question relating to situational use included 8 terms: 'When I want something I like', 'When I feel like trying something new', 'To move my diet in a more sustainable direction', 'When I want something healthy', 'As part of meals that I post on social media', 'To set a good example to those around me', 'As a regular part of my diet', and 'As part of easy and convenient meals'. These were developed to address aspects of pleasure, health, environmental concern, social status, fit to diet and convenience, and drew in part on extant literature [109]. The suitability of CATA questions for measuring perceived situational appropriateness was previously demonstrated, for example, by Jaeger et al. [110]. When answering each of these CATA questions, participants were instructed to "Please think about this [stimulus name]. Select all the words that apply."

The second part of the stimulus evaluation obtained stated willingness to consume, which was measured using the question: "How often would you consume the following foods and beverages?" and a fully labelled 9-pt category scale with anchors: 1 = 'Never or less than once yearly', 2 = '2–3 times a year', 3 = 'Every 2–3 months', 4 = 'Once every month', 5 = '1–3 times per month', 6 = 'Once every week', 7 = '2–4 times per week', 8 = '5–6 times per week', and 9 = 'Once daily or more often' [109,111]. Stimulus presentation order was randomized.

2.3.2. Psychographic Variables and Dietary Habit

Responses to psychographic variables were obtained after stimulus evaluation. The scales included in all countries were Food Neophobia (FN) [112], Food Technology Neophobia (FTN) [113,114] and environmental concern (ENV) [115]. The 10-item food neophobia scale is widely used to capture consumers' stable propensity to avoid novel and unfamiliar foods (e.g., 'I don't trust new foods'). Across 13 items, FTN measured consumers' fears of novel food technologies (e.g., 'new food technologies are something I am uncertain about' and 'it can be risky to switch to new food technologies too quickly'). Based on a recent review of scales to measure concern for the environment, a composite scale was created combining four items from each of three existing scales [116–118] seeking to mitigate criticisms of existing scales. The constructed scale included 12 items (e.g., 'If things continue on their present course, we will soon experience a major ecological catastrophe', 'I am worried about future children's chance of living in a clean environment', and 'We shouldn't worry about environmental problems because science and technology will solve them before very long), which are given in full in Part S3 of Supplementary Materials.

All responses were obtained on fully labelled 7-point Likert scales with anchors: 'Disagree strongly' (1), 'Disagree moderately' (2), 'Disagree slightly' (3), 'Neither agree nor disagree' (4), 'Agree slightly' (5), 'Agree moderately' (6), and 'Agree strongly' (7). Participants were instructed to indicate their degree of agreement or disagreement with each of the statements. Within the psychographic scales, statement order was randomized across participants.

Dietary habit was categorized using a question from De Backer and Hudders [119] with nine available options. Participants were classified as Omnivores (no limitation on consumption of meat and fish), Flexitarians (consciously limits quantity of either all types or specific types of meat) or Vegetarians (who completely avoid the consumption of meat and fish).

2.3.3. Data Collection

The survey was conducted in English and was appropriate given the status of this language as *lingua franca* in all four countries. High proficiency in English as an eligibility criterion further ensured that participants had the necessary language skills to complete the survey.

Demographic and socio-economic information was obtained either at the start of the survey (for quota sampling purposes) or at the end of the survey.

Data collection took place in December 2021 and January 2022, following careful revision of test links and evaluation of responses from ~10% of the total sample in each country to ensure that the survey performed as expected.

The data were obtained as part of a survey that also included other questions, which are not described further due to lack of relevance for the present research. Participants completed the task from a location of their own choosing, using a desktop or laptop computer.

Drawing on Jaeger and Cardello [120] who identify factors affecting data quality in online questionnaires, data-driven exclusion criteria were implemented relating to completion time and straight-line responding (see Part S4 of Supplementary Materials for details).

2.4. Data Analysis

All analyses were performed in XLSTAT [121], using a 5% significance level for inference tests.

2.4.1. Willingness to Consume

As directed by Obj. 1 and the exploration of consumers' WTC for PB foods, an ANOVA followed by Tukey's HSD multiple comparisons was performed with WTC ratings as the dependent variable and PB food category as the explanatory variable. In a second step, violin plots [122] were drawn to show the heterogeneity of the distributions of WTC ratings between PB food categories.

2.4.2. Emotional, Conceptual and Situational Use Terms

Extending the analyses linked to Obj. 1, consumer-derived profiling of the PB food categories was explored. Upon confirming that each CATA term was discriminant via Cochran's Q tests, Correspondence Analysis was applied on the PB category x term contingency tables [123].

To determine the effect of each term on the WTC ratings, penalty/lift analysis was performed [123]. The purpose is to determine for each of the PB food categories which terms positively or negatively affect WTC. The change in WTC for each PB food is calculated and a student's test between the average WTC when the term is checked and when the term is unchecked to establish if this difference is significantly different from zero.

2.4.3. Consumer Segmentation

To perform consumer segmentation (Obj. 2), an agglomerative hierarchical clustering based on the WTC scores across the 5 PB food categories was used. This cluster analysis was computed with the Euclidean distance and Ward's criterion [124]. To build clusters that discriminated between the PB food categories, a centering of the WTC scores by subject was carried out beforehand. The number of clusters to retain was determined by visual inspection of the dendrogram, where a significant change of within-cluster variation highlighted a merge of two heterogeneous clusters.

To visualize differences in the WTC patterns between the clusters, the matrix of clusters x PB foods centers of gravity was calculated and submitted to a PCA based on the covariance matrix. Cluster confidence ellipses (95%) were computed by bootstrapping [125].

An ANOVA measuring the PB food category effect on the WTC scores was performed for each cluster separately. As the sample size differed greatly between clusters, effect sizes (η^2) were used in addition to p-values as indicators of degree of product discrimination within each cluster. ANOVAs to measure the cluster effect on each of the PB food categories separately, as well as on the category means were also computed and followed by Tukey's multiple comparisons.

Per Obj. 3, the final step was to perform penalty/lift analysis within each of the retained clusters to determine the impact on average WTC scores on CATA term selection for each of the 5 PB food categories. Within clusters, the analysis was performed in the same manner as described in Section 2.4.2.

2.4.4. Psychographic and Socio-Demographic Variables

For trait scales (FN, FTN) and ENV concern, summed scores for each participant were calculated across all scale items (following reverse coding as needed). Cronbach's alpha values exceeded the 0.7 threshold for internal reliability [126].

To investigate whether the clusters were influenced by psychographic and/or socio-demographic characteristics, an analysis of the proportions of presence of different variables was performed: country, age, gender, education, and dietary preferences.

3. Results

3.1. Aggregate Level Findings

3.1.1. Willingness to Consume

The first objective of the present research was to examine consumers' willingness to consume plant-based protein within a number of distinct food categories (Obj. 1). Across all countries and PB food categories, slightly more than one-third of all willingness to consume (WTC) ratings (37.3%) were for the response option 'never or less than once yearly' (see also Figure 1A), while 9.6% were for the two response options indicating willingness to consume most frequently ('5–6 times per week' and 'once daily or more often'). Figure 1B shows the mean WTC ratings by PB food category for all participants across countries. The *PB Milk* category had the highest WTC (between 'once every month' and '1–3 times per month') followed, in order, by *PB Cheese*, *PB Meat* and *PB Meat 33%*. The food category for which there was, on average, the lowest WTC was *PB Fish* (all between 'every 2–3 months' and 'once every month'). Part S5 of Supplementary Materials shows that similar preference ordering was obtained when summarizing and analyzing the data non-parametrically.

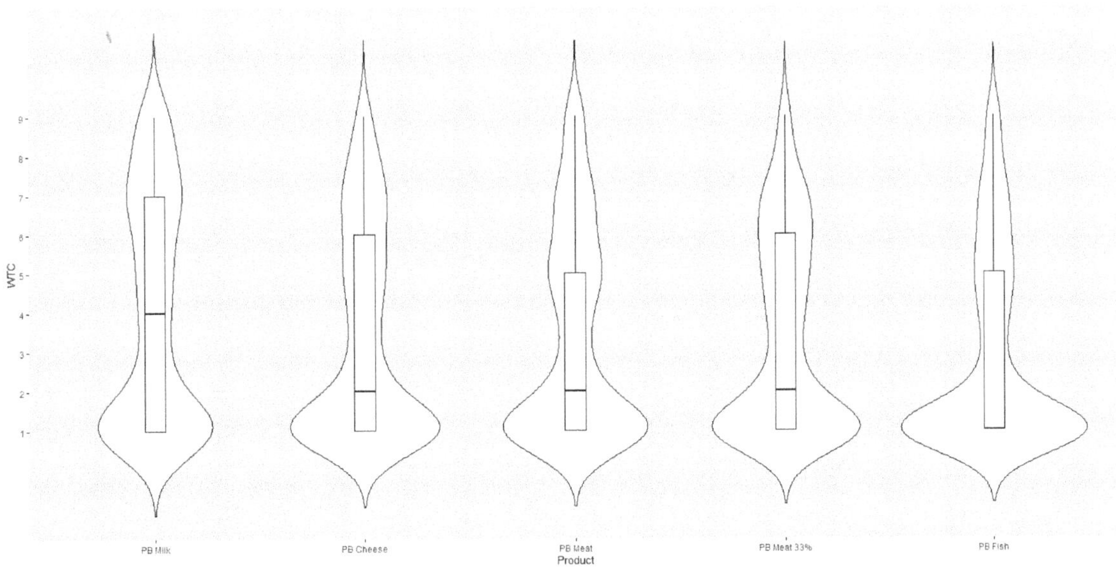

(A)

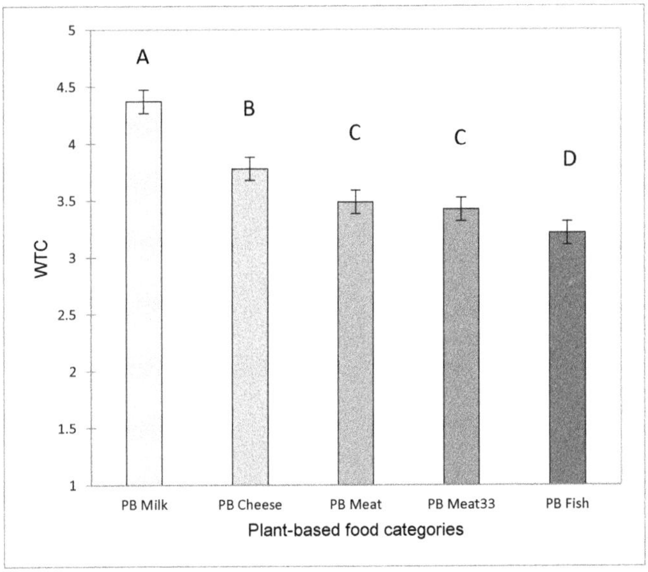

(B)

Figure 1. Mean willingness to consume (WTC) ratings for aggregate sample across 2494 consumers in USA, Australia, Singapore and India shown for the five plant-based (PB) food categories included in the research. (**A**) Violin plot of WTC ratings for PB food categories, with median and interquartile range; (**B**) Mean WTC ratings, with standard error. WTC was measured on a 9-point scale (1 = 'Never or less than once yearly', 5 = '1–3 times per month', 9 = 'Once daily or more'). In (**B**), PB food categories with different letters following Tukey's post hoc test are significantly different at the 5% level of significance.

Heterogeneity in WTC ratings for PB food categories was revealed in Figure 1A which shows violin plots for each of the PB food categories. It was particularly obvious that WTC ratings with notably higher than the mean and median values were provided by a modest proportion of participants. This pointed to heterogeneity in consumers' WTC ratings and paved the way for the second objective of the present research (Obj. 2).

3.1.2. Emotional and Conceptual Product Associations

According to Cochran's Q tests, the five PB food categories were significantly differentiated ($p < 0.05$) on all emotional and conceptual terms except 'passive' ($p = 0.21$). Overall, the most frequently used terms were 'sophisticated' (22%) and 'happy' (20%), while 'tense' (6%) and 'energetic' (7%) were least frequently used (Part S6 of Supplementary Materials has full details). Figure 2A shows a biplot of the two first dimensions after CA, with a dominating first dimension (87.7%) (Part S7 of Supplementary Materials shows the average stimulus positions with 95% confidence intervals). *PB Milk* was separated from the other PB food categories on the first dimension and most strongly associated with 'energetic' and 'powerful', and least strongly associated with 'sophisticated'. The second dimension (8.7%) separated *PB Cheese* from *PB Fish,* with the former being significantly more frequently associated with 'feminine' and 'classy', while negative emotions – 'dissatisfied', 'tense' and 'nervous' were dominant for *PB Fish*. The associations for *PM Meat* and *PM Meat 33%* were very similar.

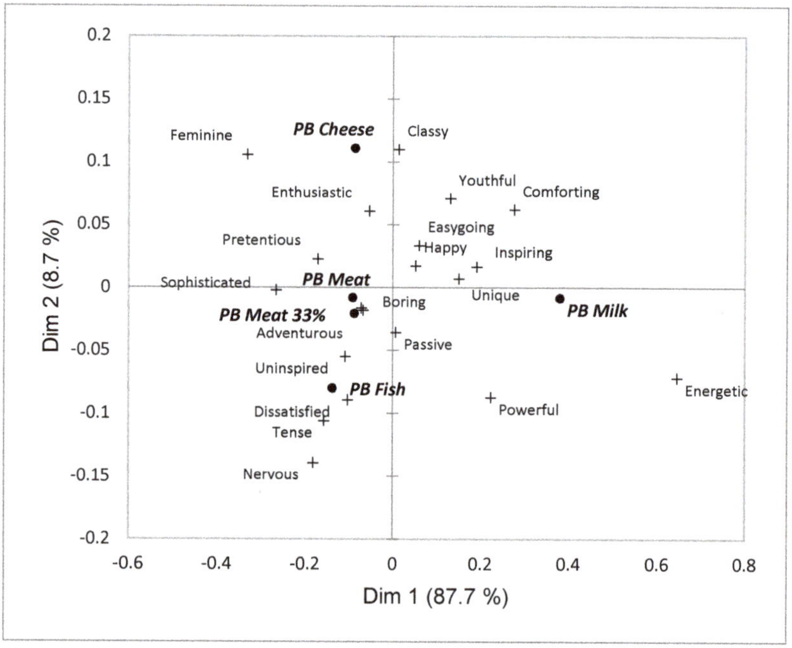

(A)

Figure 2. *Cont.*

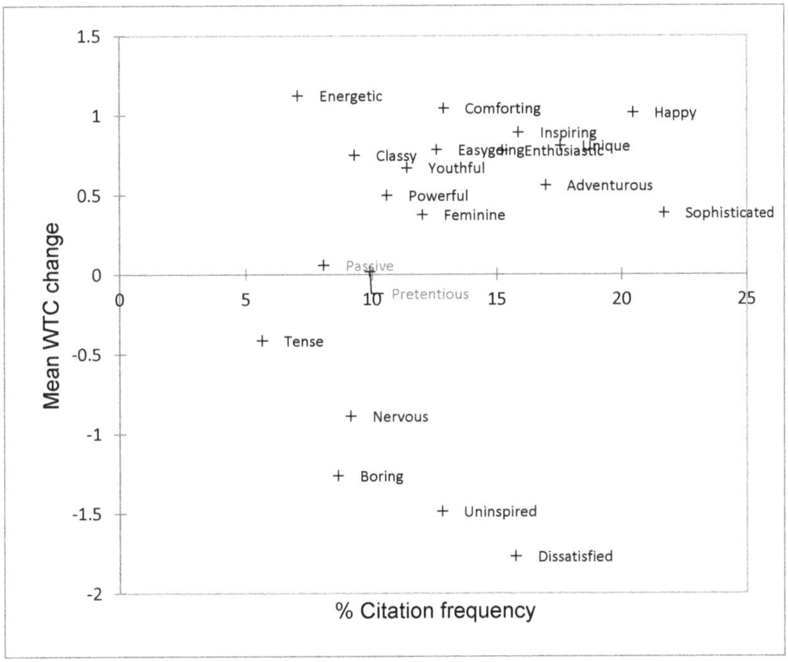

(B)

Figure 2. Results linked to emotional and conceptual product associations for aggregate sample across 2494 consumers in USA, Australia, Singapore and India shown for aggregate sample the five plant-based (PB) food categories included in the research. (**A**) Plot of the first two dimensions following Correspondence Analysis; (**B**) Impact on average willingness to consume (WTC) based on Penalty/Lift analysis. In (**A**), PB food categories are shown in bold and italic font. In (**B**), grey font used for terms where change in WTC is not significantly different from zero at the 5% level of significance.

Penalty/Lift analysis was used to identify the relationships between stimulus characterization and WTC, with Figure 2B showing that selection of 'happy, 'comforting' and 'energetic' was associated with a positive change in WTC of more than one scale point, and that selection of 'boring', 'uninspired' and 'dissatisfied' was associated with a negative change in WTC of more than one scale point. Smaller positive WTC changes (0.5 to 1 scale point) were observed for many positive and conceptual terms, while negative WTC change of a similar magnitude was associated with 'nervous'.

3.1.3. Situational Use Product Associations

For situational use associations, the PB food categories were significantly differentiated ($p < 0.05$) according to Cochran's Q tests (except for 'as part of meals that I post on social media', $p = 0.18$). Across all stimuli, the most frequent situational use associations were 'when I feel like trying something new' (43%) and When I want something healthy' (33%), while the least frequent association was 'as part of meals that I post on social media' (12%). Figure 3A shows a biplot of the two first dimensions after CA, with a dominating first dimension (82.5%) (Part S7 of Supplementary Materials shows the average stimulus positions with 95% confidence intervals). *PB Milk* was separated from the other four PB food categories on Dimension 1 and was the PB category most strongly associated with 'as a regular part of my diet' and least frequently associated with 'when I feel like trying something new'. The second dimension (12.7%) separated *PB Meat* and *PB Cheese*, and the former was significantly more frequently associated with 'to set a good example to those around me' and significantly less associated with 'when I want something I like'.

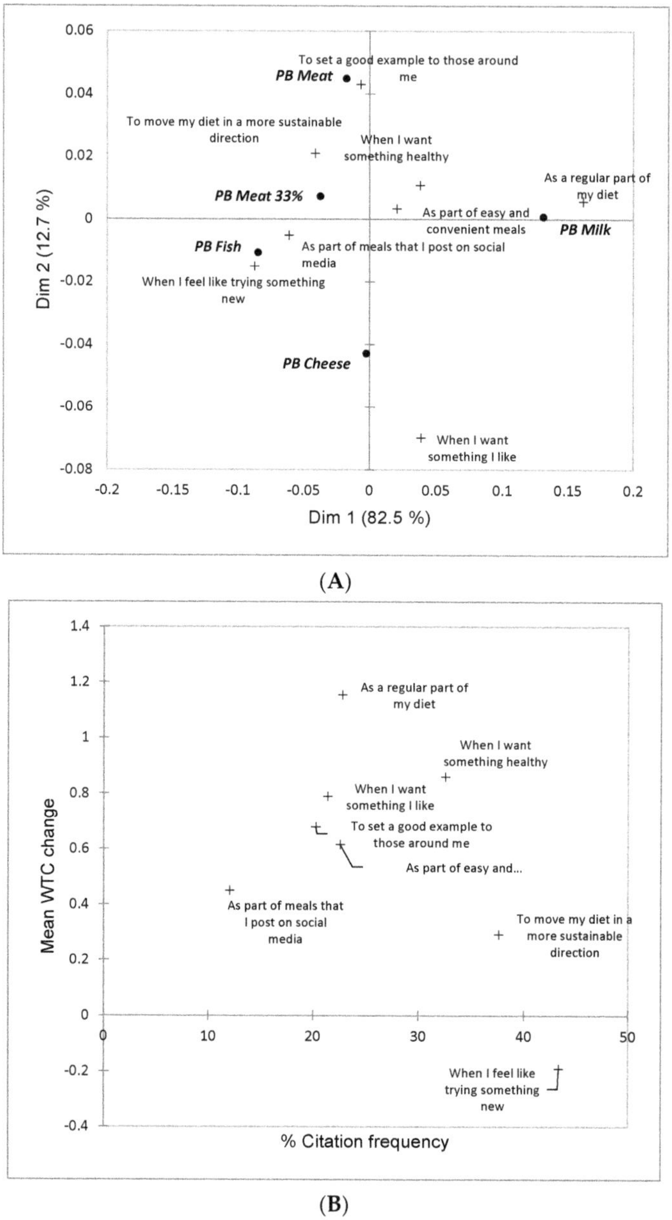

Figure 3. Results linked to situational use product associations for aggregate sample across 2494 consumers in USA, Australia, Singapore and India shown for the five plant-based (PB) food categories included in the research. (**A**) Plot of the first two dimensions following Correspondence Analysis; (**B**) Impact on average willingness to consume (WTC) based on Penalty/Lift analysis. In (**A**), PB food categories are shown in bold and italic font. In (**B**), change in WTC is significantly different from zero ($p < 0.05$) for all terms.

The Penalty/Lift analysis (Figure 3B) revealed that a positive WTC change greater than one scale point was only found for 'As a regular part of my diet'. Smaller positive, but still significant ($p < 0.05$) changes in WTC were found for all other situational use situations. The exception was 'When I feel like trying something new'. On average, selection of this use situation reduced WTC by about 0.2 WTC scale points.

3.2. Consumer Segmentation Based on Willingness to Consume

Directed by Objective 2, hierarchical cluster analysis of WTC ratings was performed on the total sample of 2494 people across the four countries. Based on the dendrogram (Part S8 of Supplementary Materials), a 7-cluster solution was retained. With the way of constructing the clusters being ascending, the dendrogram showed that the first notable "jump" in within-cluster variation took place between a 7-cluster and a 6-cluster solution. Moreover, the number of participants in each of the retained clusters was greater than 100, which ensured that the clusters were both homogeneous and large.

Among the 7 retained clusters there was one large cluster (1382 people; 55.4% of total sample) which could be described as PB category non-discriminators. Additionally, there was six smaller clusters (127 to 296 people per cluster; 44.6% of total sample) with distinct patterns of WTC ratings for the different PB food categories. Because the individual clusters had nuanced and complex WTC profiles for PB food category, arbitrary cluster naming (Cluster 1 to Cluster 7) was used to simplify the presentation of results and retain focus on the existence of multiple minor clusters rather than their specific individual WTC profiles.

The demographic profiles of the 7 clusters are summarized in Table 1. There were few between-cluster differences in relation to biological sex, age group or level of education. Participants from India were notably overrepresented in Cluster 2 (63.8%), and to a lesser degree in Cluster 4 (44.6%). Other country differences among clusters were minor, with the percentage distributions ranging between 10% and 34% (relative to 25% for even distributions by country). When considering self-declared dietary preferences, omnivores were overrepresented in Clusters 1, 3 and 7 (60–68%). Flexitarians were more evenly distributed, but with a tendency to higher representation in Clusters 4 and 6 (46–48%). Compared to the overall sample, vegans were strongly overrepresented in Cluster 2 (29%), which is likely attributable to the high percentage of consumers from India falling into Cluster 2. Cluster differences for FN, FTN and environmental concern were minor and not clearly related to WTC for PB food categories.

3.2.1. PB Category Non-Discriminators: Cluster 1

In the retained solution, there was one large group of consumers (55.4%, $n = 1382$) whose WTC responses revealed these people to be PB category non-discriminators. While significantly different, the average WTC ratings for the five PB food categories in this cluster were very similar and the effect size was 'nil' (Cluster 1, Table 2).

Table 2. Mean willingness to consume (WTC) ratings by plant-based (PB) food category for the 7-cluster solution based on the aggregate sample ($n = 2494$).

Cluster	PB Foods (Average)	PB Milk	PB Cheese	PB Fish	PB Meat	PB Meat 33%	*p*-Value (by Cluster)	Effect Size (by Cluster)
1 (55.4%)	3.2 C	3.1 B	3.0 B	3.2 AB	3.4 A	3.2 AB	0.006	0.002
2 (5.1%)	4.1 B	7.3 A	6.4 B	2.2 CD	1.9 D	2.6 C	<0.0001	0.68
3 (8.6%)	4.7 AB	6.6 A	3.4 D	4.3 C	4.1 C	4.9 B	<0.0001	0.19
4 (11.9%)	4.4 AB	5.1 B	5.8 A	3.9 C	3.7 C	3.6 C	<0.0001	0.14
5 (6.0%)	3.0 C	7.6 A	2.3 B	1.7 C	1.7 C	1.9 BC	<0.0001	0.72
6 (5.7%)	4.9 A	6.9 A	5.1 C	3.9 D	6.1 B	2.5 E	<0.0001	0.37
7 (7.3%)	4.1 B	3.9 B	4.9 A	2.4 C	3.9 B	5.5 A	<0.0001	0.20

WTC was measured on a 9-point scale (1 = 'Never or less than once yearly', 5 = '1–3 times per month', 9 = 'Once daily or more'). Cluster sizes are given as % of the total sample ($n = 2494$). Standard errors around WTC means were between 0.1 and 0.2 for all clusters and all PB food categories). The last two columns show *p*-value and effect size (η^2) following analysis of variance within clusters. Results from Tukey's post hoc tests are shown, and within rows, PB food categories with same capital letters are not significant at the 5% level of significance.

A follow-up cluster analysis identified three sub-groups of consumers within Cluster 1 (Part S9 of Supplementary Materials includes a dendrogram and a plot of WTC means). Although each of these three sub-groups were non-discriminating among food categories, they differed significantly in the magnitude of their stated WTC. The largest group (n = 865, 34.7% of total sample) gave low WTC ratings to all five PB food categories (between '2–3 times a year' and 'never or less than once a year'). The smallest group (n = 203, 8.1% of total sample) gave higher WTC ratings across food categories (between 'once every month' and 'every 2–3 months'), while the third group (n = 314, 12.6% of total sample) gave high WTC ratings to all PB food categories (around '2–4 times a week').

Part S10 of Supplementary Materials has the demographic profiles for the Cluster 1 sub-groups and it fits that Group 2 which had the lowest average WTC across the five PB food categories (between 'never or less than once yearly' and '2–3 times a year') comprised more older people (60%), more people with lower educational attainment (46.5%), was dominated by self-declared omnivores (73.4%) and was most food neophobic (38.4) and food technology neophobic (60.2). This was contrasted with Group 3 where average WTC was highest. In this group, people from the younger age group (18 to 45 years old) were in the majority (65.3%), as were those who had higher educational attainment (70.7%). Consumers from India were also relatively overrepresented (39.2%) in this latter group.

3.2.2. PB Category Discriminators: Clusters 2 to 7

Each of the 6 smaller segments had distinct WTC patterns by PB food category, and based on effect size, the PB category differences were largest in Cluster 5 (η^2 = 0.72) and Cluster 2 (η^2 = 0.68) (Table 2). In Cluster 5 (n = 150, 6.0% of total sample), average WTC ratings were high for *PB Milk* (between '2–4 times per week' and '5–6 times per week') and much lower for all other PB food categories. Among these consumers, *PB Cheese* had the highest WTC, followed by *PB Meat 33%* and then *100% PB Meat* and *PB Fish*. In Cluster 2 (n = 127, 5.1% of total sample), the high average WTC for *PB Milk* was similar to Cluster 5. Furthermore, in this cluster WTC for *PB Cheese* was also high (between 'once every week' and '2–4 times every week'). In a further parallel to Cluster 5, *PB Meat 33%* was the PB food category with the third highest average WTC rating (between 'every 2–3 months' and '2–3 times a year'). *PB Meat* and *PB Fish* had the lowest average WTC (around '2–3 times a year'). These distinct WTC profiles are seen in Figure 4 which presents two-dimensional biplots following Principal Components Analysis of mean WTC ratings by PB product category across the six category discriminating clusters (Figure 4A plots PC1 vs. PC2; Figure 4B plots PC1 vs. PC3). Briefly, PC1 and PC2 separated, respectively, *PB Milk* and *PB Cheese* from the other PB food categories, while PC3 separated the two variants of *PB Meat*.

Cluster 6 (n = 143, 5.7% of total sample) resembled Clusters 2 and 5 by having high average WTC ratings for *PB Milk* and *PB Cheese* (Table 1), as well as high WTC for *PB Meat* (between '2–4 times per week' and '5–6 times per week') (Table 2). Conversely, in this cluster, WTC was lowest for *PB Meat 33%* (between 'every 2–3 months' and '2–3 times a year'), although still higher than in Clusters 2 and 5. Overall, Cluster 6 comprised the people who gave the highest average WTC ratings to PB foods.

In Cluster 3 (n = 214, 8.6% of total sample), WTC for *PB Milk* was again high (Table 2), and Cluster 3 resembled Cluster 5 in having a much lower average WTC rating for *PB Cheese* than *PB Milk* (Figure 4A). Differentiating its WTC profile from the other clusters was a higher WTC for 33% *PB Meat* compared to *100% PB Meat* (Figure 4B). Finally, Cluster 3 resembled Cluster 6 in having an overall high WTC for PB foods (Table 1).

Cluster 7 (n = 182, 7.3% of total population) had the lowest WTC for *PB Milk* among the six category discriminating clusters (Part S11 of Supplementary Materials has results from ANOVA for each of the five PB food categories separately). Consumers in this cluster rated, on average, WTC for *PB Meat 33%* the highest (between 'once every month' and 'every 2–3 months'). *PB Meat* had the same average rating as *PB Milk* (Table 2). The distinct WTC profile for Cluster 7 was clearly seen in Figure 4.

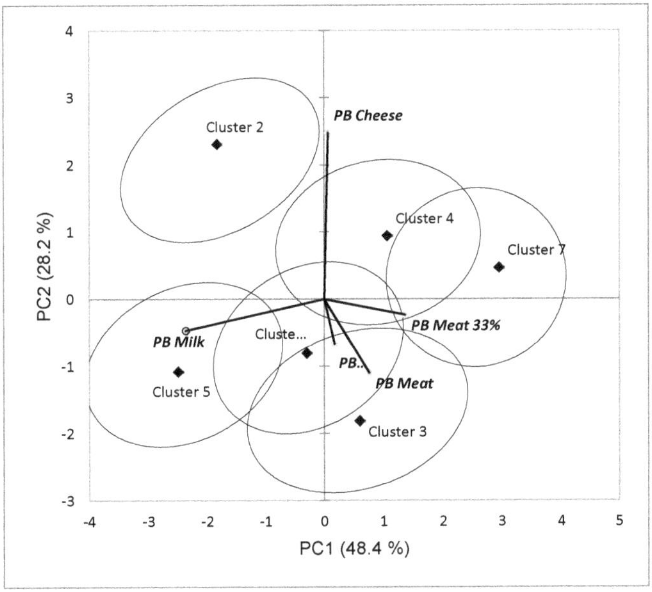

(A)

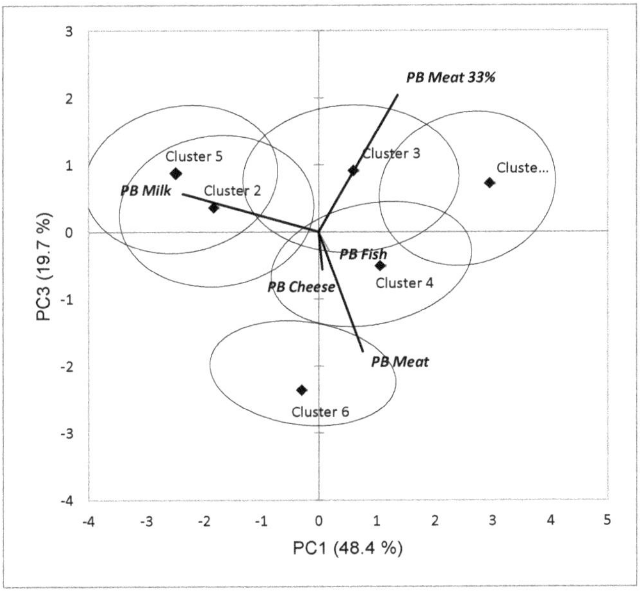

(B)

Figure 4. Two-dimensional biplots following Principal Components Analysis of mean willingness to consume (WTC) ratings for PB food categories in the six PB category discriminating clusters (Cluster 2 to Cluster 7). Confidence ellipses (95%) around average cluster positions were obtained by bootstrapping. (**A**) PC1 vs. PC2; (**B**) PC1 vs. PC3. (**A**,**B**), PB food categories are shown in bold and italic font.

Cluster 4 was the largest of the six category discriminating clusters (n = 296, 11.9% of total sample) and, compared to the other five clusters, differences in mean WTC for the

PB food categories were smallest ($\eta^2 = 0.14$) (Table 2), which could also be seen from its position closer to the origins of the plots in Figure 4. The WTC profile in Cluster 4 most strongly resembled that from the aggregate sample (Figure 1), with highest WTC for *PB Milk* followed by *PB Cheese*, and then lower ratings for the other three PB food categories (Table 2).

3.3. Emotional, Conceptual and Situational Use Drivers of Willingness to Consume in PB Category Discriminating Consumer Segments

Directed by the third objective of the research (Obj. 3), Penalty/Lift analysis was performed to identify emotional, conceptual and situational use drivers of WTC in each of the PB category discriminating consumer segments (Cluster 2 to Cluster 7). Supported by the findings from Penalty/Lift analysis, where none of the explanatory CATA terms strongly impacted WTC (change in WTC was with one exception always less than 0.5 scale points), no further consideration was given to the non-discriminating PB category cluster (Cluster 1).

3.3.1. Emotional and Conceptual Drivers of WTC by Consumer Segment

The results following Penalty/Lift analysis for PB food categories within each of the six discriminating clusters are shown in Table 3 for the 20 emotional and conceptual terms. When term selection was associated with a change in WTC that was significantly different from zero ($p < 0.05$), the five terms with negative valence—'boring', 'dissatisfied', 'nervous', 'tense', and 'uninspired'—overwhelmingly reduced WTC. Table 3 shows this by the red shading of cells and a negative value up to one WTC scale points (blank cells in Table 3 indicate that significance testing of change in WTC was not performed due to infrequent term use (<5%) for a specific PB food category; tables of citation frequencies for terms by PB food category in individual clusters are given in Part S12 of Supplementary Materials). The two exceptions to this pattern of WTC penalties were Cluster 2 and Cluster 3, where an association between 'nervous' and *PB Meat 33%* resulted in a significant WTC increase (Table 3).

Differences between the PB food categories regarding the penalty on WTC of negative emotional associations were seen in Table 3, with reduced WTC for *PB Milk* often associated with the terms 'boring' and 'uninspired'. Conversely, 'dissatisfied' was most often associated with reduced WTC for *PB Cheese*. Table 3 also showed that many negative emotional associations reduced WTC for *PB Milk* in Cluster 7 and the same applied to *PB Cheese* in Cluster 6. The average negative impact on WTC was highest for *PB Milk* in Cluster 7, which could tentatively be explained by the low average WTC for this food category in this cluster (Table 1). The largest WTC penalty (1 scale point) was observed for 'uninspired' in response to *33% PB Meat* in Cluster 6 and is likely explained by the much lower WTC for this hybrid PB category than the other PB food categories in this cluster. Just as it was an infrequent emotion at the aggregate level (Section 3.1.2), 'tense' was infrequently used for individual food categories within clusters and was often <5% (blank cells in Table 3). The only instance where selection of this term was associated with a significant WTC penalty was for *PB Fish* in Cluster 4.

When citation of terms with positive valence ('enthusiastic', 'comforting', 'easygoing', 'energetic', 'happy', and 'inspiring') significantly impacted WTC, the change was always positive, as seen from the green shading in Table 3. The results aligned well with the WTC changes for terms with negative valence. For example, WTC significantly increased when 'inspiring' was used to characterize *PB Meat 33%* in Cluster 6, which fitted with the significant WTC decrease for 'uninspired' for this product-by-cluster combination. For *PB Cheese* in Cluster 5, significant WTC increases for 'happy' and 'enthusiastic' fitted with significant WTC decreases for, respectively, 'dissatisfied' and 'uninspired'. Another example was *PB Meat* in Cluster 7, where a WTC increase for 'comforting' fitted with a WTC decrease for 'tense'. In the two instances where 'easygoing' was significantly associated with a WTC increase, it was for *PB Fish* (Cluster 3 and Cluster 6).

Table 3. Results following Penalty/Lift analysis for PB food categories in each of the six plant-based (PB) category discriminating clusters (Cluster 2 to Cluster 7), showing emotional and conceptual drivers of willingness to consume (WTC).

Prodx clus	Product	Cluster	PB category	Boring	Dissatisfied	Nervous	Tense	Uninspired	Enthusiastic	Comforting	Easygoing	Energetic	Happy	Inspiring	Adventurous	Classy	Feminine	Passive	Powerful	Pretentious	Sophisticated	Unique	Youthful
Milk-2	1. Milk	2	PB Milk		−0.2				0.3	−0.2	−0.4	−0.3	−0.1	−0.2	−0.3	−0.1		0.4	0.1		−0.1	0.0	0.3
Cheese-2	2. Cheese	2	PB Cheese		0.1	0.1	0.0	0.0	0.4	−0.2	−0.2		0.1	0.0	0.1	0.5	0.3	−0.2		0.4	0.0	0.4	0.1
Fish-2	3. Fish	2	PB Fish	0.0	−0.2	0.1	−0.3	−0.2	−0.2	0.0	0.4		0.2	−0.1	0.6	0.3	0.5	0.2	0.0	0.0	0.0	−0.1	−0.2
Meat-2	4. Meat	2	PB Meat	−0.2	−0.2	−0.2	−0.1	0.0	0.0	0.1	0.0		0.1	−0.1	0.3	−0.1	0.1	−0.2	0.1	0.1	0.0	0.3	0.0
Meat33-2	5. Meat33	2	PB Meat 33%		−0.3	0.8	0.4	0.5	−0.3	0.0	0.0		0.0	−0.5	−0.5	0.1	−0.5	−0.2	0.3	−0.4	0.0	−0.1	0.5
Milk-3	1. Milk	3	PB Milk	−0.6	−0.4		−0.3	−0.2	0.0	0.2	0.0	−0.1	0.3	0.0	0.0	−0.6		−0.3	0.0	−0.3	0.3	−0.1	−0.4
Cheese-3	2. Cheese	3	PB Cheese	−0.3	−0.9	0.1	−0.4	−0.6	−0.3	−0.3	−0.1		0.1	0.1	0.1	0.0	0.0	0.0	0.0	−0.1	0.2	−0.1	−0.2
Fish-3	3. Fish	3	PB Fish	−0.2	−0.5	−0.3		−0.6	0.3	0.3	0.6	0.4	0.5	0.0	0.5	0.5	0.5	0.8	0.2	0.3	0.5	−0.1	0.6
Meat-3	4. Meat	3	PB Meat	0.0	−0.3	−0.5	−0.5	0.1	−0.3	−0.4	0.4	0.4	0.1	0.6	0.6	0.3	0.2	0.1	0.5	−0.2	0.1	0.2	0.8
Meat33-3	5. Meat33	3	PB Meat 33%	−0.6	−0.3	0.7		−0.3	−0.1	−0.1	−0.4	0.0	0.0	0.2	0.0	0.2	0.0	0.2	0.9	−0.1	−0.2	0.3	−0.1
Milk-4	1. Milk	4	PB Milk	−0.2	−0.1		−0.3	0.2	−0.1	0.2	0.0	−0.2	−0.1	0.2	0.0	0.0	0.0	0.1	0.1	0.0	0.2	−0.2	0.1
Cheese-4	2. Cheese	4	PB Cheese	−0.5	0.0	0.2		−0.1	−0.3	−0.1	−0.2		−0.2	−0.3	0.0	−0.2	−0.1	0.0	−0.1	0.0	−0.1	0.0	0.0
Fish-4	3. Fish	4	PB Fish	0.0	−0.2	−0.3	−0.6	−0.2	0.1	0.1	0.2	0.3	−0.1	0.0	0.0	0.2	−0.3	0.0	−0.2	0.0	0.1	0.0	−0.1
Meat-4	4. Meat	4	PB Meat	−0.1	0.1	0.2	0.2	−0.1	0.0	0.1	0.1		0.1	−0.1	−0.2	0.2	0.1	0.0	0.2	−0.2	−0.1	0.2	0.0
Meat33-4	5. Meat33	4	PB Meat 33%	−0.4	0.1	0.2	0.2	0.1	0.1	0.2	0.0		0.1	0.1	−0.1	0.1	0.1	0.2	−0.2	−0.1	0.0	0.1	0.0
Milk-5	1. Milk	5	PB Milk		0.0			−0.6	0.2	0.0	0.2	0.1	0.1	0.0	−0.1	−0.4		0.4	0.1		0.0	−0.3	−0.4
Cheese-5	2. Cheese	5	PB Cheese	−0.2	−0.5	−0.2		−0.6	0.9	0.2	0.1		0.8	0.4	0.0	1.2	0.6	0.4	0.2	0.2	0.4	0.1	0.0
Fish-5	3. Fish	5	PB Fish	−0.1	−0.1	0.1		0.1	−0.2	0.2	−0.1		−0.1	−0.1	0.1	0.2	0.1	0.1	0.2	−0.1	−0.1	0.0	−0.1
Meat-5	4. Meat	5	PB Meat	−0.2	−0.2	−0.2	0.3	−0.2	0.1	0.3	0.1		0.1	0.3	0.1	0.1	0.1	0.4	0.1	−0.1	−0.1	0.4	0.3
Meat33-5	5. Meat33	5	PB Meat 33%	−0.2	−0.1	−0.3	−0.1	−0.4	0.2	−0.2	0.0	0.3	0.1	0.2	0.0	−0.1	−0.3	0.2		0.2	−0.2	0.0	−0.1
Milk-6	1. Milk	6	PB Milk	−0.7				−0.2	0.2	0.4	0.3	0.2	0.2	0.4	0.4	0.0		−0.4	−0.1	−0.3	0.1	−0.4	0.0
Cheese-6	2. Cheese	6	PB Cheese		−0.6	−0.6		−0.6	0.4	0.3	0.4		0.7	0.2	0.2	−0.1	0.2	0.2	0.1	0.8	0.0	0.6	0.3
Fish-6	3. Fish	6	PB Fish	0.4	−0.5	−0.7		−0.8	0.5	0.5	0.7		0.3	0.1	−0.2	0.5	0.7		0.2	0.6	0.4	0.6	0.3
Meat-6	4. Meat	6	PB Meat	−0.4	0.2	0.2			0.3	0.1	0.1	0.1	0.1	0.1	−0.1	−0.4	−0.6	−0.4	0.5	−0.5	−0.3	−0.3	−0.5
Meat33-6	5. Meat33	6	PB Meat 33%	−0.4	−0.2	−0.1		−1.0	0.0	−0.2	0.4		−0.1	0.8	0.4	−0.2	−0.5	0.1	0.7	−0.3	0.3	0.1	0.4
Milk-7	1. Milk	7	PB Milk	−0.9	−0.9	−0.9		−0.8	0.2	0.4	0.2	0.3	−0.1	0.1	0.6	0.4		−0.5	−0.1	−0.7	0.5	0.5	0.1
Cheese-7	2. Cheese	7	PB Cheese	−0.5	−0.6	−0.2	−0.2	−0.5	0.0	0.3	−0.1		0.1	0.1	−0.1	0.0	0.1	0.1	−0.1	0.1	0.0	−0.3	0.1
Fish-7	3. Fish	7	PB Fish	0.1	0.2	0.1	0.1	0.2	0.3	0.1	0.4	0.6	0.5	0.2	0.2	0.4	0.4	−0.1	−0.3	0.5	0.3	0.0	0.6
Meat-7	4. Meat	7	PB Meat	0.0	−0.2	−0.2	−0.6	−0.2	0.3	0.8	−0.2	0.5	0.1	0.3	0.0	0.2	0.0	0.1	−0.1	−0.4	0.3	0.6	−0.4
Meat33-7	5. Meat33	7	PB Meat 33%	0.3	0.0	−0.1			0.2	0.3	−0.1		−0.1	0.1	−0.1	0.2	−0.1		−0.3	0.1	−0.2	−0.3	0.1

WTC was measured on a 9-point scale (1 = 'Never or less than once yearly', 5 = '1–3 times per month', 9 = 'Once daily or more'). Cell values indicate average WTC change, where red shading used for emotional/conceptual terms with average WTC impact significantly less than zero ($p < 0.05$) and green shading used for terms with average WTC impact significantly less than zero ($p < 0.05$). Blank cells indicate that significance testing of change in WTC was not performed due to infrequent term use (<5%) for a focal PB food category. The presentation order for terms is based on grouping as 'negative valence', 'positive valence' and 'conceptual', shown in three blocks from left to right in the figure.

Contrary to the systematic WTC impact for terms with negative or positive valence (respectively, WTC penalty and WTC lift), selection of conceptual terms ('adventurous', 'classy', 'feminine', 'passive', 'powerful', 'pretentious', 'sophisticated', 'unique', 'youthful') largely resulted in WTC increases, although some occasional WTC decreases were noted (a mix of green and red shading in Table 3). In the case of 'classy', WTC significantly increased for *PB cheese* in Cluster 2 and Cluster 5. However, in Cluster 3, WTC was penalized when 'classy' was associated with *PB Milk*, while it increased when this term was associated with *PB Fish*. The term 'feminine', was associated with a WTC increase for *PB Fish* in Cluster 6, but a WTC decrease for *PB Meat*. Furthermore, in Cluster 6, mixed WTC impacts were seen for 'pretentious', which was negative for *PB Meat* but positive for *PB Fish* and *PB Cheese*.

For 'adventurous', 'powerful' and 'unique', WTC changes were positive (green shading), although never for the same clusters and PB food categories. However, in Cluster 7, where *PB Milk* had the lowest within-cluster WTC rating, selection of 'adventurous' increased WTC. When 'powerful' was significantly associated with WTC increase it was only for *PB Meat* or *PB Meat 33%*. Perceiving PB food categories as 'sophisticated' never resulted in a significant WTC change. Least significant impacts on WTC for emotional and conceptual terms were observed for Cluster 4, which was the largest of the six PB category discriminating clusters (11.9% of total sample). There were only two instances (Table 3), which both revealed a WTC penalty – *PB Cheese* (highest WTC in cluster) when associated with 'boring', *PB Fish* (lowest WTC in cluster) when associated with 'tense'.

3.3.2. Situational Use Drivers of WTC by Consumer Segment

Table 4 pertains to situational use terms but is otherwise the same as Table 3. It, shows the impact on WTC for PB food categories in each of the six plant-based (PB) category discriminating clusters (Cluster 2 to Cluster 7) as determined by Penalty/Lift analysis (Part S13 of Supplementary Materials has citation frequency for situational use CATA terms by cluster for the five plant-based (PB) food categories). Compared with the WTC changes observed for emotional and conceptual terms, the average impacts were smaller for situation use situations.

Selection of 'when I want something I like' was associated with significant WTC increases in Clusters 3, 4, 6 and 7, but never pertained to *PB Meat 33%*. For the situational use 'as part of easy and convenient meals', the significant WTC increases only pertained to *PB Fish* (Clusters 3 and 6). The greater familiarity of *PB Milk* and *PB Cheese* compared with the other PB food categories likely explained why selection of 'when I feel like trying something new' was only significant in connection with these two PB food categories and negatively impacted WTC. Selection of 'when I want something healthy' resulted in eight significant WTC changes (6 were increases and 2 were decreases), both across and within clusters and across and within PB food categories. In the case of *PB Cheese*, the WTC impact was positive in Cluster 2, 3 and 5 and negative in Cluster 4. In Cluster 7, there was a significant WTC lift for *PB Meat*, but a significant WTC penalty for *PB Meat 33%*. There was also a mix of significant WTC increases and decreases for 'as a regular part of my diet' across clusters and PB food categories. There were mostly positive WTC changes associated with selection of 'to set a good example to those around me', except for Cluster 4 where a significant lift was found for *PB Meat* but a significant penalty for *PB Cheese*.

Two situational uses – 'to move my diet in a more sustainable direction' and 'to set a good example to those around me' – were never associated with significant WTC changes.

Table 4. Results following Penalty/Lift analysis for PB food categories in each of the six plant-based (PB) category discriminating clusters (Cluster 2 to Cluster 7), showing situational use drivers of willingness to consume (WTC).

Product	Cluster	Product	When I want something I like	When I feel like trying something new	To move my diet in a more sustainable direction	When I want something healthy	As part of meals that I post on social media	To set a good example to those around me	As a regular part of my diet	As part of easy and convenient meals
1. Milk	2	PB Milk	−0.2	−0.1	0.0	0.1	0.0	−0.1	0.3	0.5
2. Cheese	2	PB Cheese	−0.1	0.1	0.3	0.4	0.2	0.0	0.0	0.3
3. Fish	2	PB Fish	0.1	0.1	−0.3	−0.1	−0.1	0.1	−0.2	0.1
4. Meat	2	PB Meat	0.2	−0.1	0.2	0.1	0.0	0.0	0.1	0.0
5. Meat33	2	PB Meat 33%	−0.2	0.4	0.1	−0.3	−0.2	−0.2	−0.4	0.2
1. Milk	3	PB Milk	0.3	−0.4	0.1	0.1	−0.2	0.0	0.2	−0.1
2. Cheese	3	PB Cheese	0.2	−0.4	0.0	0.5	0.4	0.2	0.3	0.2
3. Fish	3	PB Fish	0.5	−0.1	0.2	0.2	0.4	0.8	0.5	0.5
4. Meat	3	PB Meat	−0.1	−0.1	0.2	0.1	−0.4	0.1	0.7	0.4
5. Meat33	3	PB Meat 33%	−0.1	0.2	0.2	0.2	0.1	0.3	−0.1	0.0
1. Milk	4	PB Milk	0.1	0.0	0.1	0.0	0.1	−0.2	0.0	−0.1
2. Cheese	4	PB Cheese	0.1	−0.1	−0.2	−0.3	−0.3	−0.3	−0.3	0.0
3. Fish	4	PB Fish	0.0	0.1	0.0	−0.2	−0.2	−0.2	0.0	0.1
4. Meat	4	PB Meat	0.3	0.0	0.1	0.1	0.0	0.4	−0.1	0.0
5. Meat33	4	PB Meat 33%	0.1	−0.1	0.1	0.1	0.2	0.2	0.3	0.1
1. Milk	5	PB Milk	−0.1	−0.1	−0.2	−0.1	0.1	0.0	0.2	0.0
2. Cheese	5	PB Cheese	0.3	0.0	0.0	0.5	−0.2	0.4	0.2	0.1
3. Fish	5	PB Fish	−0.1	0.0	−0.1	0.0	0.1	−0.1	0.2	−0.1
4. Meat	5	PB Meat	0.0	0.0	0.1	0.2	0.2	0.4	0.3	0.1
5. Meat33	5	PB Meat 33%	−0.1	0.0	0.1	−0.1	−0.1	−0.3	−0.1	0.0
1. Milk	6	PB Milk	0.1	−0.5	0.0	0.2	−0.2	−0.2	0.5	0.2
2. Cheese	6	PB Cheese	0.5	−0.1	0.0	0.4	0.3	0.8	0.6	0.2
3. Fish	6	PB Fish	0.4	0.3	0.2	0.9	0.0	0.0	0.5	0.6
4. Meat	6	PB Meat	0.1	−0.2	−0.1	−0.2	−0.1	0.3	0.3	−0.3
5. Meat33	6	PB Meat 33%	−0.1	0.2	0.1	0.4	−0.3	0.3	0.2	0.4
1. Milk	7	PB Milk	0.6	−0.1	−0.2	0.7	0.4	−0.1	0.7	0.2
2. Cheese	7	PB Cheese	0.3	−0.1	0.0	0.2	0.3	−0.2	0.0	0.1
3. Fish	7	PB Fish	0.2	−0.1	0.1	0.0	0.3	0.0	−0.2	−0.1
4. Meat	7	PB Meat	0.2	0.2	0.3	0.3	0.3	−0.1	0.2	0.2
5. Meat33	7	PB Meat 33%	−0.3	−0.1	0.0	−0.3	−0.2	−0.1	−0.1	−0.1

WTC was measured on a 9-point scale (1 = 'Never or less than once yearly', 5 = '1–3 times per month', 9 = 'Once daily or more'). Cell values indicate average WTC change, where red shading used for emotional/conceptual terms with average WTC impact significantly less than zero ($p < 0.05$) and green shading used for terms with average WTC impact significantly less than zero ($p < 0.05$).

4. Discussion

4.1. Willingness to Consume, Emotional, Conceptual, and Situational Use Characterizations for PB Food Categories (Aggregate Level)

About 2500 consumers in four countries took part in the present research, and rated WTC for five categories of PB foods – *PB Milk, PB Cheese, PB Meat, PB Meat 33%* and *PB Fish*. Across all countries and participants, WTC declined in the order as listed above, which separated the "dairy" categories from the "meat" and "fish" categories (Figure 1A). This finding was consistent with initial expectations that the food category to which a product belongs has an important influence on consumer responses to alternative proteins, and the contraposition of PB dairy foods with PB fish was consistent with projective mapping data that show PB dairy products on a perceptual dipole with PB fish products [127].

More generally, the ordering of the PB food categories according to WTC was consistent with the established market for PB milk being the most developed (Good Food Institute, 2022) and, therefore, most familiar to consumers. In contrast, existing challenges with product quality for PB cheese [128,129] and PB meat alternatives [130,131] has hampered migration of these products into the mainstream of consumer purchases. Lastly, PB fish is still extremely novel and not generally available to consumers. The category of fish also suffers from a relatively low level of product interest due to generally lower acceptance of fish products due to sensory, preparation and cost barriers [132–135].

The average WTC ratings for all PB food categories were low and between '1–3 times per week' and 'every 2–3 month' (Figure 1A), which corresponded with low citation frequencies of positive emotional associations (<25% on average) (Part S6 of Supplementary Materials). These findings were consistent with the observed low uptake of PB food categories in general [109] but are concern for the needed transition to more sustainable food systems and consumption patterns [14]. As previously noted by, for example, [51,136], findings ways to help consumers to incorporate PB food categories into their diet is paramount. The present results confirmed this by showing that the situational use 'as a regular part of my diet' was associated with the strongest positive impact on WTC in the Penalty/Lift analysis (Figure 3B), followed by associations with specific benefits related to health ('when I want something healthy') and taste ('when I want something I like').

With this in mind (also known as, helping consumers to find ways to make PB foods a regular part of their diets), it was encouraging that emotional associations to the PB food categories were more positive than negative (Part S6 of Supplementary Materials). Thus, negative overall attitude to PB foods may not be the primary consumption barrier. This may instead lie with product and category specific factors which is an encouraging finding, since such factors can be addressed through the combined efforts of product development, information/education efforts and marketing initiatives [137–139]. The conceptual associations supported the notion that marketing and branding efforts can play a role in increasing uptake since significant impacts on WTC were seen for 'sophisticated', 'feminine', 'powerful', 'classy', 'youthful, 'comforting' (Figure 3B). Shaping perceptions of PB food categories in positive ways through such conceptual associations can help to overcome negative stereotypes related to early market introductions of poorer quality plant-based products, as described by Cardello et al. [48] in the case of PB milk.

Associations to 'adventurous' and 'unique' were, on average, also found to positively impact WTC, and as above, branding and product positioning can capitalize on these associations, although care should be taken to ensure that such market positioning does not evoke conceptualizations of extreme novelty that may evoke neophobic responses. This is the case, because consumers with higher levels of food neophobia are known to be laggards in the uptake of PB foods [69,140–142], and further associating such foods to novelty might slow their uptake among this group of consumers. More generally, harnessing the novelty of PB food categories as a pathway to greater consumption also seems questionable given the finding that WTC was slightly, but significantly reduced (~0.2 scale points) with the selection of the situational use 'when I feel like trying something new'. This suggests an expectation that unfamiliar PB foods may not provide enjoyable sensory experiences.

Such negative expectations have been raised previously with regard to PB foods [48,143], although as the development of PB foods progresses and products of higher sensory and hedonic quality appear in the market, such negative expectations are likely to diminish.

4.2. Consumer Segmentation Based on Willingness to Consume Different PB Food Categories

4.2.1. Consumer Segments Based on WTC for PB Food Categories

The average WTC ratings for the PB food categories masked significant heterogeneity in ratings from individual consumers (Figure 1B) and a 7-cluster solution was derived comprising one large PB non-category discriminating cluster (~55% of participants) and six smaller PB category discriminating clusters (~45% of participants). It was a key finding of the present research that about half of the participants viewed all PB food categories similarly be it relatively positively (with higher WTC ratings) or quite negatively (with low/lower WTC ratings)—while the remaining participants displayed varying preferences (as proxied by WTC) for the different PB food categories. To our knowledge, it has not previously been shown within the context of a wide range of food categories that consumers hold highly specific PB food category preferences, and that, for these people, a high WTC for one PB food category is an unreliable predictor of WTC (and preferences) for other PB food categories. This insight has implications for the promotion of PB diets, since it implies that generic marketing efforts to improve PB consumption may not have their desired effect, because consumers are more or less open to PB foods within one or more food categories, but not others. Future research is needed to explore this possibility and better understand whether consumers are receptive for persuasive arguments regarding increasing PB food consumption for categories where they have low WTC.

While the six PB discriminating clusters (Clusters 2 to 7 in Table 2) had distinct WTC profiles (Figure 2), they also showed consistency, in that *PB Milk* always had the highest/second highest WTC rating, while *PB Fish* always had the lowest/second lowest WTC rating. This corresponded with the aggregate sample WTC results discussed in Section 4.1. The greatest between-cluster differences in WTC were seen for *PB Cheese, PB Meat* and *PB Meat 33%,* which in individual clusters varied between having the highest and/or lowest WTC ratings. The PB category discriminating clusters also differed in overall WTC for PB foods, as seen in the average WTC ratings in the second column of Table 2. Clusters 4 and 6 had similar high overall WTC, but this was underpinned by different WTC for 4 of the 5 individual PB food categories (except *PB Fish*) (Part S8 of Supplementary Material has post hoc cluster comparisons of WTC within PB food categories). In Clusters 4 and 6, the greatest share of people who self-declared as flexitarians (46–48%) were found (Table 1). This was in sharp contrast to Cluster 2 which had the lowest overall WTC and "rejected" PB food categories other than *PB Milk.* The highest proportion of self-declared vegans/vegetarians were found in this cluster (29%), confirming previous reports that many PB foods lack appeal for such consumers [37]. The overall WTC in Clusters 3 and 7 was equally low (about 'once a month' on average), but this was underpinned by significant differences in WTC for 3 of the 5 PB food categories (*PB Milk, PB Cheese* and *PB Fish*; Part S8 of Supplementary Material). The low overall WTC fitted with the self-declared dietary preferences, which was dominated by omnivores (60–65%) and the near absence of veg*ns (1–4%) (Table 1).

The drivers of the uncovered cluster specific PB food category preferences is likely multifactorial, and also an interactive function of personal cognitive, experiential and value-based associations with PB foods and the specific food categories used in this study. For example, *PB Milk* had highest WTC among most clusters, but it was also perceived as healthier and tastier than most of the other PB food categories according to the situational appropriateness data (Part S10B of Supplementary Materials). While consumers may believe that all PB foods are equally sustainable, these environmental benefits may not lead to an increased WTC, unless other benefits (sensory, health) are also associated with some of the less familiar food categories, like *PB Fish.* Similarly, previous disappointing experiences with PB foods within a specific category, e.g., PB cheeses, can disrupt any generalized halo effect across categories. Likewise, it is possible to envisage consumers making the inference that that if they dislike, e.g., fish, they will also likely dislike PB fish, and in the case of *PB Meat 33%*, it is possible to imagine that people may eschew such a product with the reasoning that they prefer to just eat less meat than eating a PB blend.

By gaining a better understanding of the preference drivers within each cluster, it becomes possible to understand the unique profiles more fully and how to achieve greater PB food uptake. This could take many forms, including a focus on increasing consumption for the PB food category with lowest WTC or focusing on achieving a complete substitution to PB alternatives in the food category where WTC was highest, or combinations hereof. Suggestions like these resonate with previous authors including Lang [136] who note that understanding consumer context and motivations for adopting PB foods has important consequences for marketing positioning, messaging and promotion.

4.2.2. Emotional, Cognitive and Situational Use Drivers of WTC in PB Food Category Discriminating Clusters

Systematically, across all PB food categories and clusters, WTC decreased significantly when associations were made to negative emotions. Similarly, WTC increased significantly when associations were made to positive emotions. However, based on number of significant WTC changes (Table 3), an overall positive effect on WTC was less certain. This could suggest that care must be taken to avoid negative emotions for PB product consumption experiences and understanding that, if they occur, their impact may be greater, in accordance with prospect theory where "losses loom larger" [144,145].

An association between 'nervous' and *PB Meat 33%* resulted in a significant WTC increase in Clusters 2 and 3 (Table 3. At a first glance this result was counter intuitive in the sense that a negative emotion would be associated with greater WTC. However, it is possible that consumers in these two clusters were positive about such meat hybrids being available in the marketplace, while at the same time experiencing some nervousness and unease about them because they do not understand what they are. For example, the stimulus description used in this study (*Meat—blend containing 33% plant-based ingredients*) lacked specificity about the source of the non-PB part of the blend. In spite of years of development efforts on hybrid meat products, e.g., PB-extended hamburger products [146], sensory quality issues remain. Grasso and Jaworska [147] who summarized recent commercial developments in the UK in the hybrid meat category noted mixed results in relation to attempts to bring products to market. The authors reported that several early launches had not maintained a place in the market and suggested that hybrids have been received with confusion and are not understood by consumers, leaving them as a minor alternative for those most attached to meat [86]. Desire for ingredient transparency and clean labels by consumers in relation to PB foods [34,148] is also consistent with the need to reduce this continuing confusion with hybrid products.

Health and environmental concern are often regarded as the two key motivators for PB food consumption [109]. In this regard it was interesting that the health-related situational use situation ('when I want something healthy') was significantly associated with WTC change in the PB category discriminating clusters, while the environment-related use situation ('when I want to move my diet in a more sustainable direction') never was (Table 4). Tentatively both factors are important for motivating PB food uptake, but only personal health concerns are effective at regulating the frequency of consumption. A slightly different interpretation would be that consumers feel that they are actively contributing to the needed sustainability transition if they sometimes eat PB foods, and that as long as they do so sometimes, environmental and biodiversity challenges are not sufficient to motivate increases in PB food consumption. In a study on meat hybrids, Profeta et al. [105] found that choice probability increased with perceived product healthiness, and that this self-centered motive seemed to outperform altruistic motives like animal welfare or environmental concerns when it came to choices within this emerging product category. In a similar vein, Grasso and Jaworska [147] noted that the most recent meat hybrid product launches did not mention flexitarianism and instead stressed flavor, healthiness and convenience, including messages such as "5-a-day", the convenience of having vegetables already in minced meat and the use of vegetables as flavor enhancers.

Of further interest regarding the role of healthiness in driving WTC for PB foods, was the result in Cluster 7. For the situational use 'when I want something healthy' there was a significant WTC lift for *PB Meat*, but a significant WTC penalty for *PB Meat 33%*. Interestingly, the latter had the highest within-cluster WTC, suggesting good intention to consume, but not because of a positive health perception of this hybrid product. None of the other situational uses were associated with significant WTC increases for this cluster-product combination suggesting that other drivers likely exist which explain reasons for consumption or lack thereof. A worthy extension of the present research would be to include more PB category consumption drivers and to disregard those not having relevance. The use 'as part of meal that I post on social media about' did not provide useful insights, suggesting that "bragging" about PB food consumption is not something to which consumers in this sample aspired. Had the sample comprised more young people, then social media use may have been cited more frequently, because use of social media is known to decline with age [149].

The result from Cluster 4 further supported the notion that the situational uses included in the present research were not fully encompassing of uses and motivations for consumption. Notably, *PB Cheese* had the highest within-cluster WTC (around 'once every week'), but significant WTC decreases were associated with selection of 'when I want something healthy', 'as a regular part of my diet', and 'to set a good example to those around me'. This suggested that these were not reasons that motivated consumption for consumers in Cluster 4.

4.3. Limitations and Suggestions for Future Research

It should be noted that the short text presented to all participants prior to the study, although designed to establish a common frame of reference for all participants, may also have had an influence on participant responses. However, any influence of this text on responses would be a random effect, unlikely to differentially influence responses among consumer segments or food categories. In addition, it should be noted that PB food categories extend far beyond those included in the present research. Some of these categories would be interesting to consider further in relation to category vs. product-specific preferences and would add desirable diversity in terms encompassing ready-to-eat foods (e.g., PB pizza), menu dishes (e.g., PB lasagna), condiments (e.g., PB mayonnaise), and desserts (e.g., PB ice-cream). Whether an ice-cream is PB or not may not matter greatly to consumers considering that ice-cream is an indulgence product and sensory quality and enjoyment is paramount [150,151]. However, it is likely that PB meat topping on a pizza is perceived differently from the patty of a PB meat burger or from the PB meat category in general. Previously, in support of the notion that "gastronomic context" matters, Elzerman et al. [49,50] found that meat substitutes are perceived differently depending on their shape (e.g., pieces vs. mince) and the type of meals they are in (e.g., pasta, soup, and salad). Future research should also look to emerging PB categories, notably PB eggs, which is a new growth category in the PB space [152]. The latter products are interesting because of the important nutritional properties of eggs, their multi-functionality in cooking, and the fact that they may have an important impact on overall PB food consumption and dietary health.

There is often considerable interest in the effect of demographic and socio-economic variables on consumer responses. Where testable hypotheses relating to these can be developed we encourage their testing as part of future research.

5. Conclusions

It is clear from the present study that the food category to which a plant-based food belongs can be an important factor in the willingness to consume PB alternatives to animal products. In particular, a large number of distinct clusters of consumers with different WTC for PB foods falling into different food categories was observed in the present research. Among these clusters, there also exists a large cluster of consumers who are less differentiating in their WTC PB products, although they differ significantly in the magnitude of their WTC for these products. In general, WTC is greatest among consumers for *PB Milk* vs. *PB Cheese* or *PB Meat* (100% or 33%). *PB Fish* had generally low WTC for all consumer groups. Emotional, conceptual and situational use characterizations of the PB foods by food category, combined with penalty/lift analysis of the effect of these characterizations on WTC, showed significant effects on WTC. Positive and negative emotional associations effected corresponding positive and negative effects on WTC. Specific conceptualizations, like 'adventurous', 'classy', 'feminine', and 'pretentious' had important contributions to WTC for specific PB food categories and within specific consumer clusters. Similarly, most situational use characterizations drove WTC positively, e.g., 'when I want something healthy' or 'when I want something I like', or negatively, e.g., 'when I feel like trying something new', within PB food discriminating clusters, although specific PB food category-cluster combinations produced alternative effects. In sum, the data urge caution when interpreting data on plant-based food preferences/WTC without taking consideration of the specific food category to which the PB food belongs. Findings regarding PB food preferences within one food category should not be generalized to other food categories.

Supplementary Materials: The following supporting information can be downloaded at: https://www.mdpi.com/article/10.3390/foods11193059/s1: (1) Summary of participant characteristics by country; (2) List of 12 food technologies and background text given to participants about each of these; (3) Items included in the environmental concern scale; (4) Information regarding online data quality by country; (5) Descriptive statistics and non-parametric tests for the 5 PB food categories (aggregate sample); (6) Citation frequency for emotion, conceptual and situational use terms by for the five plant-based (PB) food categories; (7) Stimuli plots with 95% confidence intervals following Correspondence Analysis (aggregate sample) for emotional/conceptual and situational use responses; (8) Results following hierarchical cluster analysis on willingness to consume (WTC) ratings (7-cluster solution for aggregate sample); (9) Results following hierarchical cluster analysis on willingness to consume (WTC) ratings (3 sub-groups in C1 from aggregate sample); (10) Summary of participant characteristics in the 3 sub-groups of the PB food non-discriminating cluster (Cluster 1); (11) Results of ANOVA of WTC across clusters by PB food category; (12) Citation frequency for emotional and conceptual CATA terms by cluster for the five plant-based (PB) food categories; (13) Citation frequency for situational use CATA terms by cluster for the five plant-based (PB) food categories [153,154].

Author Contributions: A.V.C.: Conceptualization, Writing—Original draft, Writing—Review and Editing. F.L.: Formal analysis, Visualization, Writing—Review and Editing. D.G.: Conceptualization, Methodology, Writing—Review and Editing. S.L.C.: Investigation, Data curation. S.R.J.: Conceptualization, Methodology, Visualization, Writing—Original draft, Writing—Review and Editing. All authors have read and agreed to the published version of the manuscript.

Funding: Financial support was received from two sources: (1) The New Zealand Institute for Plant and Food Research Limited, and (2) The New Zealand Ministry for Business, Innovation & Employment.

Institutional Review Board Statement: The study was conducted in accordance with the Declaration of Helsinki. It was covered by a general approval for sensory and consumer research from the Human Ethics Committee at the New Zealand Institute for Plant and Food Research Limited (205-2021, 649-2021).

Informed Consent Statement: Informed consent was obtained from all subjects involved in the study.

Data Availability Statement: Data that support the findings of this study are available upon request to the authors.

Acknowledgments: Staff at the Sensory & Consumer Science Team at PFR are thanked for help with data collection and management.

Conflicts of Interest: The authors declare no conflict of interest. The funders had no role in the design of the study; in the collection, analyses, or interpretation of data; in the writing of the manuscript; or in the decision to publish the results. The companies did not affect the authenticity and objectivity of the experimental results of this work.

References

1. Aiking, H. Protein production: Planet, profit, plus people? *Am. J. Clin. Nutr.* **2014**, *100*, 483S–489S. [CrossRef] [PubMed]
2. Aleksandrowicz, L.; Green, R.; Joy, E.J.; Smith, P.; Haines, A. The impacts of dietary change on greenhouse gas emissions, land use, water use, and health: A systematic review. *PLoS ONE* **2016**, *11*, e0165797. [CrossRef] [PubMed]
3. Gallen, C.; Pantin-Sohier, G.; Peyrat-Guillard, D. Cognitive acceptance mechanisms of discontinuous food innovations: The case of insects in France. *Rech. Appl. Mark.* **2019**, *34*, 48–73. [CrossRef]
4. Godfray, H.C.J.; Aveyard, P.; Garnett, T.; Hall, J.W.; Key, T.J.; Lorimer, J.; Pierrehumbert, R.T.; Scarborough, P.; Springmann, M.; Jebb, S.A. Meat consumption, health, and the environment. *Science* **2018**, *361*, eaam5324. [CrossRef]
5. National Institutes of Health. Eating Red Meat Daily Triples Heart Disease-Related Chemical. Available online: https://www.nih.gov/news-events/nih-research-matters/eating-red-meat-daily-triples-heart-disease-related-chemical (accessed on 1 August 2022).
6. Nijdam, D.; Rood, T.; Westhoek, H. The price of protein: Review of land use and carbon footprints from life cycle assessments of animal food products and their substitutes. *Food Policy* **2012**, *37*, 760–770. [CrossRef]
7. Poore, J.; Nemecek, T. Reducing food's environmental impacts through producers and consumers. *Science* **2018**, *360*, 987–992. [CrossRef]
8. Rojas-Downing, M.M.; Nejadhashemi, A.P.; Harrigan, T.; Woznicki, S.A. Climate change and livestock: Impacts, adaptation, and mitigation. *Clim. Risk Manag.* **2017**, *16*, 145–163. [CrossRef]
9. Röös, E.; Sundberg, C.; Tidåker, P.; Strid, I.; Hansson, P.-A. Can carbon footprint serve as an indicator of the environmental impact of meat production? *Ecol. Indicators* **2013**, *24*, 573–581. [CrossRef]
10. Campbell, B. *Climate Change Impacts and Adaptation Options in the Agrifood System—A Summary of Recent IPCC Sixth Assessment Report Findings*; FAO: Rome, Italy, 2022.
11. Aiking, H. Future protein supply. *Trends Food Sci. Technol.* **2011**, *22*, 112–120. [CrossRef]
12. Ocké, M.; Larranaga, N.; Grioni, S.; Van Den Berg, S.; Ferrari, P.; Salvini, S.; Benetou, V.; Linseisen, J.; Wirfält, E.; Rinaldi, S. Energy intake and sources of energy intake in the European Prospective Investigation into Cancer and Nutrition. *Eur. J. Clin. Nutr.* **2009**, *63*, S3–S15. [CrossRef]
13. Tilman, D.; Clark, M. Global diets link environmental sustainability and human health. *Nature* **2014**, *515*, 518–522. [CrossRef] [PubMed]
14. Willett, W.; Rockström, J.; Loken, B.; Springmann, M.; Lang, T.; Vermeulen, S.; Garnett, T.; Tilman, D.; DeClerck, F.; Wood, A. Food in the Anthropocene: The EAT–Lancet Commission on healthy diets from sustainable food systems. *Lancet* **2019**, *393*, 447–492. [CrossRef]
15. Lippi, G.; Mattiuzzi, C.; Cervellin, G. Meat consumption and cancer risk: A critical review of published meta-analyses. *Crit. Rev. Oncol. Hematol.* **2016**, *97*, 1–14. [CrossRef] [PubMed]
16. Westhoek, H.; Lesschen, J.P.; Rood, T.; Wagner, S.; De Marco, A.; Murphy-Bokern, D.; Leip, A.; van Grinsven, H.; Sutton, M.A.; Oenema, O. Food choices, health and environment: Effects of cutting Europe's meat and dairy intake. *Global Environ. Chang.* **2014**, *26*, 196–205. [CrossRef]
17. Lamb, A.; Green, R.; Bateman, I.; Broadmeadow, M.; Bruce, T.; Burney, J.; Carey, P.; Chadwick, D.; Crane, E.; Field, R. The potential for land sparing to offset greenhouse gas emissions from agriculture. *Nat. Clim. Chang.* **2016**, *6*, 488–492. [CrossRef]
18. Aiking, H.; de Boer, J. The next protein transition. *Trends Food Sci. Technol.* **2020**, *105*, 515–522. [CrossRef]
19. Graça, J.; Godinho, C.A.; Truninger, M. Reducing meat consumption and following plant-based diets: Current evidence and future directions to inform integrated transitions. *Trends Food Sci. Technol.* **2019**, *91*, 380–390. [CrossRef]
20. Van der Weele, C.; Feindt, P.; van der Goot, A.J.; van Mierlo, B.; van Boekel, M. Meat alternatives: An integrative comparison. *Trends Food Sci. Technol.* **2019**, *88*, 505–512. [CrossRef]
21. Bloomberg. Press Release: Plant-based Foods Market to Hit $162 Billion in Next Decade, Projects Bloomberg Intelligence. 11 August 2021. 2021. Available online: https://www.foodmarket.com/News/A/1205792/Plant-based-Foods-Market-to-Hit-162-Billion-in-Next-Decade-Projects-Bloomberg-Intelligence (accessed on 1 August 2022).
22. Ali, A.E. A semiotic approach to entomophagy: The language, localization, and reimagining of insects as foodstuffs in America. *Perspect. Glob. Dev. Technol.* **2016**, *15*, 391–405. [CrossRef]
23. Balzan, S.; Fasolato, L.; Maniero, S.; Novelli, E. Edible insects and young adults in a north-east Italian city an exploratory study. *Br. Food J.* **2016**, *118*, 318–326. [CrossRef]
24. Bekker, G.A.; Fischer, A.R.; Tobi, H.; van Trijp, H.C. Explicit and implicit attitude toward an emerging food technology: The case of cultured meat. *Appetite* **2017**, *108*, 245–254. [CrossRef]
25. Birch, D.; Skallerud, K.; Paul, N. Who eats seaweed? An Australian perspective. *J. Int. Food Agribus. Mark.* **2019**, *31*, 329–351. [CrossRef]

26. Brayden, W.C.; Noblet, C.L.; Evans, K.S.; Rickard, L. Consumer preferences for seafood attributes of wild-harvested and farm-raised products. *Aquacult. Econ. Manag.* **2018**, *22*, 362–382. [CrossRef]
27. Siegrist, M.; Sütterlin, B.; Hartmann, C. Perceived naturalness and evoked disgust influence acceptance of cultured meat. *Meat Sci.* **2018**, *139*, 213–219. [CrossRef]
28. Wilks, M.; Phillips, C.J.; Fielding, K.; Hornsey, M.J. Testing potential psychological predictors of attitudes towards cultured meat. *Appetite* **2019**, *136*, 137–145. [CrossRef]
29. de Boer, J.; Schösler, H.; Boersema, J.J. Motivational differences in food orientation and the choice of snacks made from lentils, locusts, seaweed or "hybrid" meat. *Food Qual. Prefer.* **2013**, *28*, 32–35. [CrossRef]
30. Lea, E.; Worsley, A.; Crawford, D. Australian adult consumers' beliefs about plant foods: A qualitative study. *Health Educ. Behav.* **2005**, *32*, 795–808. [CrossRef]
31. Tucker, C.A. The significance of sensory appeal for reduced meat consumption. *Appetite* **2014**, *81*, 168–179. [CrossRef]
32. Hartmann, C.; Shi, J.; Giusto, A.; Siegrist, M. The psychology of eating insects: A cross-cultural comparison between Germany and China. *Food Qual. Prefer.* **2015**, *44*, 148–156. [CrossRef]
33. Vainio, A.; Niva, M.; Jallinoja, P.; Latvala, T. From beef to beans: Eating motives and the replacement of animal proteins with plant proteins among Finnish consumers. *Appetite* **2016**, *106*, 92–100. [CrossRef]
34. Aschemann-Witzel, J.; Peschel, A.O. Consumer perception of plant-based proteins: The value of source transparency for alternative protein ingredients. *Food Hydrocoll.* **2019**, *96*, 20–28. [CrossRef]
35. McCarthy, K.; Parker, M.; Ameerally, A.; Drake, S.; Drake, M. Drivers of choice for fluid milk versus plant-based alternatives: What are consumer perceptions of fluid milk? *J. Dairy Sci.* **2017**, *100*, 6125–6138. [CrossRef] [PubMed]
36. Menozzi, D.; Sogari, G.; Veneziani, M.; Simoni, E.; Mora, C. Eating novel foods: An application of the Theory of Planned Behaviour to predict the consumption of an insect-based product. *Food Qual. Prefer.* **2017**, *59*, 27–34. [CrossRef]
37. Michel, F.; Hartmann, C.; Siegrist, M. Consumers' associations, perceptions and acceptance of meat and plant-based meat alternatives. *Food Qual. Prefer.* **2021**, *87*, 104063. [CrossRef]
38. Powell, P.A.; Jones, C.R.; Consedine, N.S. It's not queasy being green: The role of disgust in willingness-to-pay for more sustainable product alternatives. *Food Qual. Prefer.* **2019**, *78*, 103737. [CrossRef]
39. Hartmann, C.; Siegrist, M. Becoming an insectivore: Results of an experiment. *Food Qual. Prefer.* **2016**, *51*, 118–122. [CrossRef]
40. Verbeke, W. Profiling consumers who are ready to adopt insects as a meat substitute in a Western society. *Food Qual. Prefer.* **2015**, *39*, 147–155. [CrossRef]
41. Mancini, M.C.; Antonioli, F. Exploring consumers' attitude towards cultured meat in Italy. *Meat Sci.* **2019**, *150*, 101–110. [CrossRef]
42. Tan, H.S.G.; Tibboel, C.J.; Stieger, M. Why do unusual novel foods like insects lack sensory appeal? Investigating the underlying sensory perceptions. *Food Qual. Prefer.* **2017**, *60*, 48–58. [CrossRef]
43. Woolf, E.; Zhu, Y.; Emory, K.; Zhao, J.; Liu, C. Willingness to consume insect-containing foods: A survey in the United States. *Lwt* **2019**, *102*, 100–105. [CrossRef]
44. Hoek, A.C.; Luning, P.A.; Weijzen, P.; Engels, W.; Kok, F.J.; De Graaf, C. Replacement of meat by meat substitutes. A survey on person-and product-related factors in consumer acceptance. *Appetite* **2011**, *56*, 662–673. [CrossRef]
45. Elzerman, J.E.; Van Boekel, M.A.; Luning, P.A. Exploring meat substitutes: Consumer experiences and contextual factors. *Br. Food J.* **2013**, *115*, 700–710. [CrossRef]
46. Gómez-Luciano, C.A.; Vriesekoop, F.; Urbano, B. Towards food security of alternative dietary proteins: A comparison between Spain and the Dominican Republic. *Amfiteatru Econ.* **2019**, *21*, 393–407. [CrossRef]
47. Van Loo, E.J.; Caputo, V.; Lusk, J.L. Consumer preferences for farm-raised meat, lab-grown meat, and plant-based meat alternatives: Does information or brand matter? *Food Policy* **2020**, *95*, 101931. [CrossRef]
48. Cardello, A.V.; Llobell, F.; Giacalone, D.; Roigard, C.M.; Jaeger, S.R. Plant-based alternatives vs. dairy milk: Consumer segments and their sensory, emotional, cognitive and situational use responses to tasted products. *Food Qual. Prefer.* **2022**, *100*, 104599. [CrossRef]
49. Elzerman, J.E.; Hoek, A.C.; Van Boekel, M.A.; Luning, P.A. Consumer acceptance and appropriateness of meat substitutes in a meal context. *Food Qual. Prefer.* **2011**, *22*, 233–240. [CrossRef]
50. Elzerman, J.E.; Hoek, A.C.; van Boekel, M.J.; Luning, P.A. Appropriateness, acceptance and sensory preferences based on visual information: A web-based survey on meat substitutes in a meal context. *Food Qual. Prefer.* **2015**, *42*, 56–65. [CrossRef]
51. Jaeger, S.R.; Giacalone, D. Barriers to consumption of plant-based beverages: A comparison of product users and non-users on emotional, conceptual, situational, conative and psychographic variables. *Food Res. Int.* **2021**, *144*, 110363. [CrossRef]
52. Adámek, M.; Adámková, A.; Mlček, J.; Borkovcová, M.; Bednářová, M. Acceptability and sensory evaluation of energy bars and protein bars enriched with edible insect. *Potr. S. J. F. Sci.* **2018**, *12*, 431–437. [CrossRef]
53. Grasso, A.C.; Hung, Y.; Olthof, M.R.; Verbeke, W.; Brouwer, I.A. Older consumers' readiness to accept alternative, more sustainable protein sources in the European Union. *Nutrients* **2019**, *11*, 1904. [CrossRef]
54. Kornher, L.; Schellhorn, M.; Vetter, S. Disgusting or innovative-consumer willingness to pay for insect based burger patties in Germany. *Sustainability* **2019**, *11*, 1878. [CrossRef]
55. Lemken, D.; Knigge, M.; Meyerding, S.; Spiller, A. The value of environmental and health claims on new legume products: A non-hypothetical online auction. *Sustainability* **2017**, *9*, 1340. [CrossRef]

56. Siegrist, M.; Hartmann, C. Impact of sustainability perception on consumption of organic meat and meat substitutes. *Appetite* **2019**, *132*, 196–202. [CrossRef] [PubMed]
57. Van Loo, E.J.; Hoefkens, C.; Verbeke, W. Healthy, sustainable and plant-based eating: Perceived (mis) match and involvement-based consumer segments as targets for future policy. *Food Policy* **2017**, *69*, 46–57. [CrossRef]
58. Weinrich, R.; Elshiewy, O. Preference and willingness to pay for meat substitutes based on micro-algae. *Appetite* **2019**, *142*, 104353. [CrossRef]
59. Sogari, G.; Amato, M.; Biasato, I.; Chiesa, S.; Gasco, L. The potential role of insects as feed: A multi-perspective review. *Animals* **2019**, *9*, 119. [CrossRef]
60. Marcus, N.; Klink-Lehmann, J.; Hartmann, M. Exploring factors determining German consumers' intention to eat meat alternatives. *Food Qual. Prefer.* **2022**, *100*, 104610. [CrossRef]
61. Lombardi, A.; Vecchio, R.; Borrello, M.; Caracciolo, F.; Cembalo, L. Willingness to pay for insect-based food: The role of information and carrier. *Food Qual. Prefer.* **2019**, *72*, 177–187. [CrossRef]
62. Lucas, S.; Gouin, S.; Lesueur, M. Seaweed consumption and label preferences in France. *Mar. Resour. Econ.* **2019**, *34*, 143–162. [CrossRef]
63. De Boer, J.; Hoogland, C.T.; Boersema, J.J. Towards more sustainable food choices: Value priorities and motivational orientations. *Food Qual. Prefer.* **2007**, *18*, 985–996. [CrossRef]
64. Hayley, A.; Zinkiewicz, L.; Hardiman, K. Values, attitudes, and frequency of meat consumption. Predicting meat-reduced diet in Australians. *Appetite* **2015**, *84*, 98–106. [CrossRef]
65. Myers, G.; Pettigrew, S. A qualitative exploration of the factors underlying seniors' receptiveness to entomophagy. *Food Res. Int.* **2018**, *103*, 163–169. [CrossRef]
66. McCarthy, S.N.; O'Rourke, D.; Kearney, J.; McCarthy, M.; Henchion, M.; Hyland, J. Excessive Food Consumption in Irish Adults: Implications for Climatic Sustainability and Public Health. In Proceedings of the European Association of Agricultural Economists (EAAE), 166th Seminar, Galway, Ireland, 30–31 August 2018.
67. de Graaf, S.; Van Loo, E.J.; Bijttebier, J.; Vanhonacker, F.; Lauwers, L.; Tuyttens, F.A.; Verbeke, W. Determinants of consumer intention to purchase animal-friendly milk. *J. Dairy Sci.* **2016**, *99*, 8304–8313. [CrossRef]
68. Haas, R.; Schnepps, A.; Pichler, A.; Meixner, O. Cow milk versus plant-based milk substitutes: A comparison of product image and motivational structure of consumption. *Sustainability* **2019**, *11*, 5046. [CrossRef]
69. Bryant, C.; Szejda, K.; Parekh, N.; Deshpande, V.; Tse, B. A survey of consumer perceptions of plant-based and clean meat in the USA, India, and China. *Front. Sustain. Food Syst.* **2019**, *3*, 11. [CrossRef]
70. Laureati, M.; Cattaneo, C.; Bergamaschi, V.; Proserpio, C.; Pagliarini, E. School children preferences for fish formulations: The impact of child and parental food neophobia. *J. Sens. Stud.* **2016**, *31*, 408–415. [CrossRef]
71. Moons, I.; Barbarossa, C.; De Pelsmacker, P. The determinants of the adoption intention of eco-friendly functional food in different market segments. *Ecol. Econ.* **2018**, *151*, 151–161. [CrossRef]
72. Piha, S.; Pohjanheimo, T.; Lähteenmäki-Uutela, A.; Křečková, Z.; Otterbring, T. The effects of consumer knowledge on the willingness to buy insect food: An exploratory cross-regional study in Northern and Central Europe. *Food Qual. Prefer.* **2018**, *70*, 1–10. [CrossRef]
73. Schäufele, I.; Albores, E.B.; Hamm, U. The role of species for the acceptance of edible insects: Evidence from a consumer survey. *Br. Food J.* **2019**, *121*, 2190–2204. [CrossRef]
74. Van der Zanden, L.D.; van Kleef, E.; de Wijk, R.A.; van Trijp, H.C. Examining heterogeneity in elderly consumers' acceptance of carriers for protein-enriched food: A segmentation study. *Food Qual. Prefer.* **2015**, *42*, 130–138. [CrossRef]
75. Palmieri, N.; Forleo, M.B. The potential of edible seaweed within the western diet. A segmentation of Italian consumers. *Int. J. Gastron. Food Sci.* **2020**, *20*, 100202. [CrossRef]
76. Possidónio, C.; Prada, M.; Graça, J.; Piazza, J. Consumer perceptions of conventional and alternative protein sources: A mixed-methods approach with meal and product framing. *Appetite* **2021**, *156*, 104860. [CrossRef]
77. Hartmann, C.; Siegrist, M. Consumer perception and behaviour regarding sustainable protein consumption: A systematic review. *Trends Food Sci. Technol.* **2017**, *61*, 11–25. [CrossRef]
78. La Barbera, F.; Verneau, F.; Amato, M.; Grunert, K. Understanding Westerners' disgust for the eating of insects: The role of food neophobia and implicit associations. *Food Qual. Prefer.* **2018**, *64*, 120–125. [CrossRef]
79. Neff, R.A.; Edwards, D.; Palmer, A.; Ramsing, R.; Righter, A.; Wolfson, J. Reducing meat consumption in the USA: A nationally representative survey of attitudes and behaviours. *Public Health Nutr.* **2018**, *21*, 1835–1844. [CrossRef]
80. Onwezen, M.C.; Bouwman, E.P.; Reinders, M.J.; Dagevos, H. A systematic review on consumer acceptance of alternative proteins: Pulses, algae, insects, plant-based meat alternatives, and cultured meat. *Appetite* **2021**, *159*, 105058. [CrossRef]
81. Iannuzzi, E.; Sisto, R.; Nigro, C. The willingness to consume insect-based food: An empirical research on Italian consumers. *Agric. Econ.* **2019**, *65*, 454–462. [CrossRef]
82. Schouteten, J.J.; De Steur, H.; De Pelsmaeker, S.; Lagast, S.; Juvinal, J.G.; De Bourdeaudhuij, I.; Verbeke, W.; Gellynck, X. Emotional and sensory profiling of insect-, plant-and meat-based burgers under blind, expected and informed conditions. *Food Qual. Prefer.* **2016**, *52*, 27–31. [CrossRef]
83. Slade, P. If you build it, will they eat it? Consumer preferences for plant-based and cultured meat burgers. *Appetite* **2018**, *125*, 428–437. [CrossRef] [PubMed]

84. Circus, V.E.; Robison, R. Exploring perceptions of sustainable proteins and meat attachment. *Br. Food J.* **2018**, *121*, 533–545. [CrossRef]
85. Cordelle, S.; Redl, A.; Schlich, P. Sensory acceptability of new plant protein meat substitutes. *Food Qual. Prefer.* **2022**, *98*, 104508. [CrossRef]
86. Tarrega, A.; Rizo, A.; Murciano, A.; Laguna, L.; Fiszman, S. Are mixed meat and vegetable protein products good alternatives for reducing meat consumption? A case study with burgers. *Curr. Res. Food Sci.* **2020**, *3*, 30–40. [CrossRef]
87. Elzerman, J.E.; Keulemans, L.; Sap, R.; Luning, P.A. Situational appropriateness of meat products, meat substitutes and meat alternatives as perceived by Dutch consumers. *Food Qual. Prefer.* **2021**, *88*, 104108. [CrossRef]
88. Fischer, A.R.; Steenbekkers, L.B. All insects are equal, but some insects are more equal than others. *Br. Food J.* **2018**, *120*, 852–863. [CrossRef]
89. La Barbera, F.; Verneau, F.; Videbæk, P.N.; Amato, M.; Grunert, K.G. A self-report measure of attitudes toward the eating of insects: Construction and validation of the Entomophagy Attitude Questionnaire. *Food Qual. Prefer.* **2020**, *79*, 103757. [CrossRef]
90. Lammers, P.; Ullmann, L.M.; Fiebelkorn, F. Acceptance of insects as food in Germany: Is it about sensation seeking, sustainability consciousness, or food disgust? *Food Qual. Prefer.* **2019**, *77*, 78–88. [CrossRef]
91. Ribeiro, J.C.; Gonçalves, A.T.S.; Moura, A.P.; Varela, P.; Cunha, L.M. Insects as food and feed in Portugal and Norway–cross-cultural comparison of determinants of acceptance. *Food Qual. Prefer.* **2022**, *102*, 104650. [CrossRef]
92. Videbæk, P.N.; Grunert, K.G. Disgusting or delicious? Examining attitudinal ambivalence towards entomophagy among Danish consumers. *Food Qual. Prefer.* **2020**, *83*, 103913. [CrossRef]
93. Kröger, T.; Dupont, J.; Büsing, L.; Fiebelkorn, F. Acceptance of Insect-Based Food Products in Western Societies: A Systematic Review. *Front. Nutr.* **2021**, *8*, 759885. [CrossRef]
94. Onwezen, M.C.; Verain, M.C.; Dagevos, H. Positive emotions explain increased intention to consume five types of alternative proteins. *Food Qual. Prefer.* **2022**, *96*, 104446. [CrossRef]
95. Hofstede Insights. Country Comparison. Available online: https://www.hofstede-insights.com/country-comparison/australia,india,singapore,the-usa/ (accessed on 30 August 2022).
96. World Bank. World Bank National Accounts Data, and OECD National Accounts Data Files. Agriculture, Forestry, and Fishing, % of GDP (Constant 2015 US$). Available online: https://data.worldbank.org/indicator/NV.AGR.TOTL.ZS?end=2021&name_desc=false&start=1960&view=chart (accessed on 30 August 2022).
97. International Energy Agency. World Energy Outlook 2018, IEA, Paris. Available online: https://www.iea.org/reports/world-energy-outlook-2018 (accessed on 30 August 2022).
98. Green, R.; Milner, J.; Joy, E.J.; Agrawal, S.; Dangour, A.D. Dietary patterns in India: A systematic review. *Br. J. Nutr.* **2016**, *116*, 142–148. [CrossRef]
99. Khara, T.; Riedy, C.; Ruby, M.B. The evolution of urban Australian meat-eating practices. *Front. Sustain. Food Syst.* **2021**, *5*, 624288. [CrossRef]
100. Statista. Share of People Who Are Vegan or Vegetarian Australia. 2019. Available online: https://www.statista.com/statistics/1267834/australia-vegan-or-vegetarian-population-share/ (accessed on 7 August 2022).
101. Statista. Share of Vegetarians in Select Countries Worldwide in 2021. Available online: https://www.statista.com/statistics/1280079/global-country-ranking-vegetarian-share/ (accessed on 7 August 2022).
102. Statista. Growth Rate of Meat Substitutes Worldwide in 2020, by Region. Available online: https://www.statista.com/statistics/1278503/growth-rate-of-meat-substitutes-worldwide/ (accessed on 7 August 2022).
103. Meiselman, H.; Jaeger, S.; Carr, B.; Churchill, A. Approaching 100 years of sensory and consumer science: Developments and ongoing issues. *Food Qual. Prefer.* **2022**, *100*, 104614. [CrossRef]
104. *ISO 20252*; Market, Opinion and Social Research, Including Insights and Data Analytics—Vocabulary and Service Requirements. International Organization for Standardization: Geneva, Switzerland, 2015. Available online: https://www.iso.org/obp/ui/#iso:std:iso:20252:ed-3:v1:en (accessed on 30 March 2022).
105. Profeta, A.; Baune, M.-C.; Smetana, S.; Bornkessel, S.; Broucke, K.; Van Royen, G.; Enneking, U.; Weiss, J.; Heinz, V.; Hieke, S. Preferences of german consumers for meat products blended with plant-based proteins. *Sustainability* **2021**, *13*, 650. [CrossRef]
106. Alcorta, A.; Porta, A.; Tárrega, A.; Alvarez, M.D.; Vaquero, M.P. Foods for plant-based diets: Challenges and innovations. *Foods* **2021**, *10*, 293. [CrossRef]
107. Ares, G.; Jaeger, S. Check-all-that-apply (CATA) questions with consumers in practice: Experimental considerations and impact on outcome. In *Rapid Sensory Profiling Techniques*; Delarue, J., Lawlor, J.B., Rogeaux, M., Eds.; Elsevier: Cambridge, UK, 2015; pp. 227–245.
108. Thomson, D.M. Conceptual profiling. In *Emotion Measurement*; Meiselman, H.L., Ed.; Elsevier: Cambridge, UK, 2016; pp. 239–272.
109. Giacalone, D.; Llobell, F.; Jaeger, S.R. "Beyond liking" measures in food-related consumer research supplement hedonic responses and improve ability to predict consumption. *Food Qual. Prefer.* **2022**, *97*, 104459. [CrossRef]
110. Jaeger, S.R.; Lee, P.-Y.; Jin, D.; Chheang, S.L.; Rojas-Rivas, E.; Ares, G. The item-by-use (IBU) method for measuring perceived situational appropriateness: A methodological characterisation using CATA questions. *Food Qual. Prefer.* **2019**, *78*, 103724. [CrossRef]
111. Jaeger, S.R.; Prescott, J.; Worch, T. Food neophobia modulates importance of food choice motives: Replication, extension, and behavioural validation. *Food Qual. Prefer.* **2022**, *97*, 104439. [CrossRef]

112. Pliner, P.; Hobden, K. Development of a scale to measure the trait of food neophobia in humans. *Appetite* **1992**, *19*, 105–120. [CrossRef]
113. Cox, D.; Evans, G. Construction and validation of a psychometric scale to measure consumers' fears of novel food technologies: The food technology neophobia scale. *Food Qual. Prefer.* **2008**, *19*, 704–710. [CrossRef]
114. Evans, G.; Kermarrec, C.; Sable, T.; Cox, D. Reliability and predictive validity of the Food Technology Neophobia Scale. *Appetite* **2010**, *54*, 390–393. [CrossRef]
115. Cruz, S.M.; Manata, B. Measurement of environmental concern: A review and analysis. *Front. Psychol.* **2020**, *11*, 363. [CrossRef]
116. Dunlap, R.E.; Van Liere, K.D.; Mertig, A.G.; Jones, R.E. New trends in measuring environmental attitudes: Measuring endorsement of the new ecological paradigm: A revised NEP scale. *J. Soc. Issues* **2000**, *56*, 425–442. [CrossRef]
117. Lounsbury, J.W.; Tornatzky, L.G. A scale for assessing attitudes toward environmental quality. *J. Soc. Psychol.* **1977**, *101*, 299–305. [CrossRef]
118. Maloney, M.P.; Ward, M.P.; Braucht, G.N. A revised scale for the measurement of ecological attitudes and knowledge. *Am. Psychol.* **1975**, *30*, 787–790. [CrossRef]
119. De Backer, C.J.; Hudders, L. Meat morals: Relationship between meat consumption consumer attitudes towards human and animal welfare and moral behavior. *Meat Sci.* **2015**, *99*, 68–74. [CrossRef]
120. Jaeger, S.R.; Cardello, A.V. Factors affecting data quality of online questionnaires: Issues and metrics for sensory and consumer research. *Food Qual. Prefer.* **2022**, *102*, 104676. [CrossRef]
121. Addinsoft. *XLSTAT Statistical and Data Analysis Solution*; Addinsoft: New York, NY, USA, 2022.
122. Hintze, J.L.; Nelson, R.D. Violin plots: A box plot-density trace synergism. *Am. Stat.* **1998**, *52*, 181–184.
123. Meyners, M.; Castura, J.C.; Carr, B.T. Existing and new approaches for the analysis of CATA data. *Food Qual. Prefer.* **2013**, *30*, 309–319. [CrossRef]
124. Arabie, P.; Hubert, L.; De Soete, G. *Clustering and Classification*; World Scientific: River Edge, NJ, USA, 1996.
125. Lebart, L. Which bootstrap for principal axes methods. In *Selected Contributions in Data Analysis and Classification*; Brito, P., Cucumel, G., Bertrand, P., de Carvalho, F., Eds.; Springer: Berlin/Heidelberg, Germany, 2007; pp. 581–588.
126. Bland, J.M.; Altman, D.G. Statistics notes: Cronbach's alpha. *BMJ* **1997**, *314*, 572. [CrossRef]
127. Varela, P.; Arvisenet, G.; Gonera, A.; Myhrer, K.S.; Fifi, V.; Valentin, D. Meat replacer? No thanks! The clash between naturalness and processing: An explorative study of the perception of plant-based foods. *Appetite* **2022**, *169*, 105793. [CrossRef] [PubMed]
128. Short, E.C.; Kinchla, A.J.; Nolden, A.A. Plant-based cheeses: A systematic review of sensory evaluation studies and strategies to increase consumer acceptance. *Foods* **2021**, *10*, 725. [CrossRef] [PubMed]
129. Falkeisen, A.; Gorman, M.; Knowles, S.; Barker, S.; Moss, R.; McSweeney, M.B. Consumer perception and emotional responses to plant-based cheeses. *Food Res. Int.* **2022**, *158*, 111513. [CrossRef] [PubMed]
130. Fiorentini, M.; Kinchla, A.J.; Nolden, A.A. Role of sensory evaluation in consumer acceptance of plant-based meat analogs and meat extenders: A scoping review. *Foods* **2020**, *9*, 1334. [CrossRef]
131. Wang, Y.; Tuccillo, F.; Lampi, A.M.; Knaapila, A.; Pulkkinen, M.; Kariluoto, S.; Coda, R.; Edelmann, M.; Jouppila, K.; Sandell, M. Flavor challenges in extruded plant-based meat alternatives: A review. *Compr. Rev. Food Sci. Food Safe.* **2022**, *21*, 2898–2929. [CrossRef]
132. Pérez-Rodrigo, C.; Ribas, L.; Serra-Majem, L.; Aranceta, J. Food preferences of Spanish children and young people: The enKid study. *Eur. J. Clin. Nutr.* **2003**, *57*, S45–S48. [CrossRef]
133. Sawyer, F.; Cardello, A.; Prell, P. Consumer evaluation of the sensory properties of fish. *J. Food Sci.* **1988**, *53*, 12–18. [CrossRef]
134. Carlucci, D.; Nocella, G.; De Devitiis, B.; Viscecchia, R.; Bimbo, F.; Nardone, G. Consumer purchasing behaviour towards fish and seafood products. Patterns and insights from a sample of international studies. *Appetite* **2015**, *84*, 212–227. [CrossRef]
135. Birch, D.; Lawley, M. Buying seafood: Understanding barriers to purchase across consumption segments. *Food Qual. Prefer.* **2012**, *26*, 12–21. [CrossRef]
136. Lang, M. Consumer acceptance of blending plant-based ingredients into traditional meat-based foods: Evidence from the meat-mushroom blend. *Food Qual. Prefer.* **2020**, *79*, 103758. [CrossRef]
137. Thomson, D.M.; Crocker, C. Development and evaluation of measurement tools for conceptual profiling of unbranded products. *Food Qual. Prefer.* **2014**, *33*, 1–13. [CrossRef]
138. Thomson, D.M.; Crocker, C. Application of conceptual profiling in brand, packaging and product development. *Food Qual. Prefer.* **2015**, *40*, 343–353. [CrossRef]
139. Thomson, D.M. Predicting purchase and consumption of new products. In *Handbook of Eating and Drinking: Interdisciplinary Perspectives*; Meiselman, H.L., Ed.; Springer: Cham, Switzerland, 2020; pp. 843–872.
140. Pagliarini, E.; Spinelli, S.; Proserpio, C.; Monteleone, E.; Fia, G.; Laureati, M.; Gallina Toschi, T.; Dinnella, C. Sensory perception and food neophobia drive liking of functional plant-based food enriched with winemaking by-products. *J. Sens. Stud.* **2022**, *37*, e12710. [CrossRef]
141. Pasqualone, A. Balancing Innovation and Neophobia in the Production of Food for Plant-Based Diets. *Foods* **2022**, *11*, 1702. [CrossRef]
142. Tuorila, H.; Hartmann, C. Consumer responses to novel and unfamiliar foods. *Curr. Opin. Food Sci.* **2020**, *33*, 1–8. [CrossRef]
143. He, J.; Evans, N.M.; Liu, H.; Shao, S. A review of research on plant-based meat alternatives: Driving forces, history, manufacturing, and consumer attitudes. *Compr. Rev. Food Sci. Food Safe.* **2020**, *19*, 2639–2656. [CrossRef]

144. Ariely, D.; Huber, J.; Wertenbroch, K. When do losses loom larger than gains? *J. Mark. Res.* **2005**, *42*, 134–138. [CrossRef]
145. Barberis, N.C. Thirty years of prospect theory in economics: A review and assessment. *J. Econ. Perspect.* **2013**, *27*, 173–196. [CrossRef]
146. Cardello, A.V.; Secrist, J.; Smith, J. Effects of soy particle size and color on the sensory properties of ground beef patties. *J. Food Qual.* **1983**, *6*, 139–151. [CrossRef]
147. Grasso, S.; Jaworska, S. Part meat and part plant: Are hybrid meat products fad or future? *Foods* **2020**, *9*, 1888. [CrossRef]
148. Aschemann-Witzel, J.; Gantriis, R.F.; Fraga, P.; Perez-Cueto, F.J. Plant-based food and protein trend from a business perspective: Markets, consumers, and the challenges and opportunities in the future. *Crit. Rev. Food Sci. Nutr.* **2021**, *61*, 3119–3128. [CrossRef]
149. Hruska, J.; Maresova, P. Use of social media platforms among adults in the United States—behavior on social media. *Societies* **2020**, *10*, 27. [CrossRef]
150. Bullock, K.; Lahne, J.; Pope, L. Investigating the role of health halos and reactance in ice cream choice. *Food Qual. Prefer.* **2020**, *80*, 103826. [CrossRef]
151. Oh, G.-E.G.; van der Lans, R.; Mukhopadhyay, A. Choice Architecture Effects on Indulgent Consumption: Evidence from Combinations of Nudges at an Ice-Cream Store. *J. Assoc. Consum. Res.* **2022**, *7*, 1410–1427. [CrossRef]
152. Boukid, F.; Gagaoua, M. Vegan egg: A future-proof food ingredient? *Foods* **2022**, *11*, 161. [CrossRef]
153. Goldammer, P.; Annen, H.; Stöckli, P.L.; Jonas, K. Careless responding in questionnaire measures: Detection, impact, and remedies. *Lead. Q.* **2020**, *31*, 101384. [CrossRef]
154. Schonlau, M.; Toepoel, V. Straightlining in Web survey panels over time. *Surv. Res. Methods* **2015**, *9*, 125–137.

foods

Review

Behavioural Nudges, Physico-Chemical Solutions, and Sensory Strategies to Reduce People's Salt Consumption

Charles Spence

Department of Experimental Psychology, New Radcliffe House, University of Oxford, Oxford OX2 6BW, UK; charles.spence@psy.ox.ac.uk

Abstract: This narrative historical review examines the wide range of approaches that has been trialled/suggested in order to reduce the consumption of salt. While sodium is an essential micronutrient, there is widespread evidence that high levels of consumption are leading to various negative health outcomes. This review summarises the evidence relating to the various approaches that have been put forward to date to help reduce salt consumption over the years, while also highlighting a number of important questions that remains for future research. Solutions to reducing salt consumption include everything from the gradual reduction in salt in foods through to the reduction in the number/size of holes in saltshakers (what one might consider a behavioural nudge). Physico-chemical solutions have included salt replacers, such as monosodium glutamate (MSG) through to the asymmetric distribution of salt in processed (e.g., layered) foods. A wide range of sensory approaches to modulating expected and perceived saltiness have also been suggested, including the use of salty aromas, as well as suggesting the use of colour cues, sonic seasoning, and even textural primes. It is currently unclear whether different salty aromas can be combined to increase odour-induced taste enhancement (OITE) effectiveness. In the years ahead, it will be interesting to assess how long such solutions remain effective, as well as whether different solutions can be combined to help reduce salt consumption without having to compromise on taste/flavour

Keywords: salt-reduction strategies; gastrophysics; aroma; flavour; multisensory; congruency

Citation: Spence, C. Behavioural Nudges, Physico-Chemical Solutions, and Sensory Strategies to Reduce People's Salt Consumption. *Foods* **2022**, *11*, 3092. https://doi.org/10.3390/foods11193092

Academic Editor: Witoon Prinyawiwatkul

Received: 29 August 2022
Accepted: 1 October 2022
Published: 5 October 2022

1. Introduction

Salt is an essential mineral in the human diet and has long been used to help season and, thus, add flavour to food [1]. Since pre-history, salt has also played an important preserving function (e.g., in the case of salting fish and meat) [2,3]. It has been suggested that salt has occupied a place of culinary dominance across cultures for centuries [4]. According to Rachel Herz [5]: "We like the taste of salt innately because salt is a signal of protein in nature" (quoted in [6]). Others have argued, perhaps more persuasively, that the allure of salt taste is that it signals the presence of essential minerals instead [7]. Interestingly, however, research from van Dongen and colleagues has revealed that in a range of 50 commonly consumed foods, the presence of both salty and savoury tastes was independently correlated with the presence of both sodium and protein [8]. Humans need a small amount of salt to regulate fluid balance and help the nerves and muscles to function properly. The preference/need for salt is, however, not uniquely human. Famously, the monkeys at Koshima Inlet on an island off the coast of Japan salt sweet potatoes by washing them in sea water prior to consuming them [9,10]. Amazonian parrots flock to certain cliffs (mineral licks) to get the sodium and other nutrients they need, which is otherwise hard to come by in their diet [11–13]. It is commonly believed that providing salt licks for deer can give rise to improved antler quality [14]. Dogs, meanwhile, are fairly insensitive to the taste of salt, the argument being that meat, which would have featured heavily in their ancestral carnivorous diet, would naturally have been high in sodium [15,16].

Salt is both a very successful flavour enhancer and can also mask the taste of bitterness [17,18]. While salt is by no means the only basic tastant to perform such a function—umami, for example, has also been reported to suppress other tastes, while, at the same time enhancing the tastes of sweet and salt [19,20] and, therefore, it does a better job than any other tastant. However, while salt is undoubtedly an essential nutrient in the human diet, there is mounting evidence of the negative health consequences that are associated with the overconsumption of sodium. This, in turn, has led to growing calls for innovative salt-reduction strategies to help improve public health outcomes [21], strategies that go beyond merely taxing high-sodium foods as has been trialled in some countries [22].

Over the years, several different approaches have been proposed, from physico-structural solutions in the case of processed foods, through to various kinds of nudges, such as reducing the hole sizes in salt dispensers. Furthermore, everything from aromas to colours and from certain sounds to particular tactile/oral-somatosensory attributes have been shown to modulate expected, and often also perceived, saltiness, at least under carefully controlled laboratory conditions. The key question to be addressed in this review is how effective these various approaches to salt reduction might be in the long term in the real world and whether different approaches to salt reduction can be combined effectively. Finding long-term solutions to reduce salt intake is especially important given that salt warning and nutritional labels tend to have little effect [23,24]. In this narrative historical review, I summarize the evidence relating to the various approaches that have been put forward to date to help reduce salt consumption, while also highlighting a number of important questions that remain for future research.

2. Negative Health Consequences of the Overconsumption of Salt

Despite the fact that small amounts of salt are essential for healthy nutrition, a growing body of robust scientific evidence has demonstrated that the overconsumption of salt in our diets is leading to a number of negative health consequences, including hypertension (i.e., high blood pressure), which increases the risk of heart disease and stroke [25], currently two of the leading causes of mortality. At the outset, though, it is important to note how the salt we consume comes not only from the foods that we choose to eat, but also from the sodium that is added at the table, either in the form of table salt (see [26,27] on the origins of table salt) or as a condiment, such as soy sauce, fish sauce, or Maggi-type seasoning [28]. Given the ready availability of table salt and salty condiments at the dining table, the danger is that simply reducing the salt levels in popular food products might well result in consumers adding salt (or salty condiments) to enhance the taste at the table [29]. Here it is intriguing to note that there have been some legislative efforts to remove salt from the dining tables of schools and kindergartens [30].

Bear in mind here that just a single tablespoon of mayonnaise contains around 11 g of fat, 100 cal, and 85 mg of sodium [31]. Some of the most shocking epidemiological evidence to have emerged recently suggests that those who regularly add salt to their food lower their life expectancy at 50 years of age by an average of 1.5 years for women and 2.3 years for men, relative to those who do not salt their food [25]. This was the result of a study of more than half a million participants from the UK biobank who were followed-up for a median of 9 years. While the current recommended daily intake of salt is 5–6 g [32,33], typical consumption figures are closer to double that, depending on the culture/country [34–36]. This contrasts markedly with the diets of our pre-agricultural ancestors, which may well have been low in salt, with estimated daily sodium intake of 768 mg, equivalent to a little under 2 g of salt [37].

While the rhetoric in the popular science press in recent years has been to blame the food industry for loading processed foods with sugar, salt, and fat [38,39], it should be noted that saltiness is not always a desirable characteristic. Indeed, food companies have been actively trying to suppress the salty taste of isotonic drinks to help make their beverages more palatable. At the same time, however, given the flavour-enhancing qualities of salt, reducing its presence in processed foods often results in an unacceptable loss in taste quality, leading to a backlash from consumers when the salt level of their favourite brands is reduced suddenly [40–43]. What is more, there are certain traditional foods, such as bread, where salt plays an essential structural function. In the latter case, the research shows that the addition of salt influences dough elasticity, improves taste, and also enhances the browning of the crust [44–46]. Salt is also a key ingredient in cheese making [47]. At the same time, however, there is a growing awareness of the exceptionally high levels of salt in some cheeses, such as Roquefort [48], where a 30 g serving contains more salt (1.06 g) than a small bag of crisps/potato chips. It has been estimated that more than 20–30% of dietary salt comes from meat and meat products [49] and that around 80% of salt consumed is hidden in processed, canteen, restaurant, and fast food [50]. Salt also provides an important role in helping to preserve meat products. Consequently, any attempt to reduce the salt levels may give rise to increased issues with food safety [51].

What Determines the Preferred Salt Level?

There is some debate as to whether our salt preferences are hard-wired or learnt. For instance, one large global study (of more than 10,000 men and women over a 24 h period) found very similar levels of salt intake across a wide range of countries/cultures [52], thus, suggesting an innate preference. There is widespread evidence for an innate hunger for salt across many species, with salt hunger presumably helping to ensure a sufficient intake of sodium [53,54]. At the same time, however, other evidence shows that those on restricted sodium diets soon adapt to the reduced saltiness in their food [55]. In particular, Gary Beauchamp and colleagues conducted a study in which people put on a controlled, low-sodium diet showed taste adaptation. Specifically, within a period of one to two months, the amount of salt that the participants found optimal in soup or crackers declined by 40 or 50 percent. Put the other way, the more salt people eat, the more they crave it. The way in which people taste their food has also been shown to influence perceived saltiness [56]. Separately, there has been an ongoing debate in the food science literature about how best to measure salt thresholds [57,58]. Bear in mind here only how in perception threshold studies, participants are normally presented with water-based pure salt solutions, whereas most food products represent complex matrices, which may result in multi-factor perceptual interactions. One of the challenges here in a real-world context is that the way in which a food is labelled (or described) can also modulate rated saltiness as a function of expectations or even disconfirmed expectations [59,60].

Interestingly, infants only start to respond to the taste of salt at around six months of age [61,62]. At the other end of the age spectrum, the declining chemosensory abilities that have been extensively documented in the elderly can all too easily lead to unhealthy eating habits [63,64], as the latter increase their intake of salt and sugar to make up for their inability to taste these ingredients in food at lower concentrations [65,66]. According to Stevens and colleagues [67], older individuals may need to add as much as two- or three-times more salt to perceive the same intensity in a tomato soup as those who are younger. Worryingly, this figure was found to increase to twelve-times as much salt added for those older individuals who were on five or more medications, which turns out to be the majority of them [68]. Given the negative health consequences of the overconsumption of salt (e.g., hypertension), this likely represents a very serious issue and one that needs to be tackled by those hoping to optimise food delivery amongst the elderly. Note here that according to observational data from the Framingham study in North America, the lifetime risk of developing hypertension in those who are 55–65 years of age is 90% [69].

3. Behavioural Nudging to Modify Salt-Related Food Behaviours

In recent years, there has been growing interest in the potential use of a range of sensory nudges to influence consumer behaviour in the context of food and drink [70,71]. "Nudging" refers to the use of more-or-less subtle sensory contextual cues that are designed to encourage people to behave in ways that are likely to help in improving the health, social support welfare, sustainability, and/or happiness of the individual and/or the society in which they live. One strategy that has worked effectively to reduce salt levels in processed foods has involved what has been referred to as a 'health-by-stealth' approach [72]. Cereal companies have been gradually decreasing the amount of salt in their breakfast cereals. Overall, they have managed to reduce the salt content by up to 47% (between 1992 and 2015), without any noticeable customer backlash, simply by doing it so gradually that customers never became aware of any difference in the formulation [73]. Consequently, each successive reduction was essentially imperceptible to the consumer and, over the long term, salt levels have dropped substantially cf. [51,74–77].

It is important to note that behavioural nudges operate at different timescales, from the slow health-by-stealth approaches to reducing the salt content of, e.g., breakfast cereals, over the decades, through to the suggestion that the number and diameter of holes on saltshakers should be reduced [78–80]. The latter solution apparently significantly reduces the amount of salt that people add to their food, at least in the short term. Note that a similar approach has been suggested more recently in terms of reducing the use of fish sauce in university canteens serving noodles in Thailand. In particular, the results of one study indicated that a simple change in how fish sauce was served can reduce consumption. Specifically, serving fish sauce in a bowl with a spoon reduced the amount used per noodle bowl by 0.25 g, as compared to the typical situation in which fish sauce is served from a bottle [29]. Meanwhile, using a specially designed spoon with a hole induced a larger reduction of 0.58 g of fish sauce used per bowl. By contrast, the reduction in fish sauce usage associated with cognitive-/affectively oriented nudges failed to reach statistical significance. Similar solutions would presumably also help with soya sauce dispensers in Japan, given that it represents a significant source of dietary sodium for those living there [81]. According to Fukutome [82], soy sauce and other sauces containing soybean paste account for 15% of the amount of table sauce consumed in Thailand.

One of the key challenges, though, is that no other element plays such a flavour-enhancing function as salt. What is more, and as mentioned earlier, salt also plays a functional/structural role in many foods, such as bread and cheese/dairy, thus, making its effective replacement that much more challenging. Researchers have also considered the use of a range of other salt-reduction strategies. In the next section, I will review the evidence concerning various physico-chemical attempts to find salt replacers that are not derived from sodium and potential solutions associated with modifying the structural properties in processed foods.

4. Physico-Chemical Solutions of Salt Reduction While Maintaining Taste

There has long been interest in the possibility of finding stimuli that taste salty without relying on sodium. Molecular salt replacers, such as potassium chloride (KCl), iodized salt, etc., have been on the market for decades, often marketed as 'low-sodium' [83]. However, none of the other metal salts give rise to anything like as salty a taste as sodium. Furthermore, potassium chloride imparts a bitter taste if present in too high a proportion in foods [48,84]. More worryingly, an excess of potassium may cause hyperkalaemia (the name given to an abnormally high level of potassium in the blood leading to cardiac arrest), which is why many authorities do not recommend the use of potassium chloride as a salt substitute [85]. Others have been championing the replacement of sodium chloride with monosodium glutamate as an effective route to maintain taste/flavour, while reducing the amount of sodium consumed [86]. Note that MSG also gives rise to a somewhat salty taste [87]. Despite containing only one-third the amount of sodium (12.3 g/100 g) found in salt (39.3 g/100 g), it has been suggested that this makes MSG a particularly promising salt alternative in sodium-reduction strategies, though it is important to stress the independence of salty and umami as taste qualities, as stressed in ISO 3972. However, as will be mentioned later, some consumers show an aversion to 'artificial' flavour enhancers, as MSG is often labelled.

Separate to work on the development of salt replacers, a number of innovative physico-chemical food structural solutions to salt reduction have also emerged in recent years.

Therefore, for example, it has been demonstrated that it is possible to maintain the perceived flavour of layered processed foods, such as lasagne, by asymmetrically distributing the salt through the food matrix rather than distributing it evenly [88,89], loading the salt taste in the first mouthful of other foods [90]. The evidence suggests that asymmetrically distributing the salt can give rise to enhanced (or maintained) taste perception while reducing the absolute quantity of salt. There may also be grounds for trying to target the salt to those parts of the oral cavity where taste receptors are more densely distributed [91,92].

Another promising approach to salt reduction involves the use of non-gustatory sensory stimuli in order to modulate the expected, and hopefully also the experienced, saltiness of foods. Therefore, for example, colours, aromas, textures, and even sonic stimuli have been shown to be associated with saltiness and can, under laboratory conditions at least, influence perceived saltiness.

5. Retronasal Enhancement of Saltiness Perception

One of the most popular/promising approaches to reducing salt, while, at the same time, maintaining taste, involves targeting the key role played by so-called salty aromas (see [93] for a recent review). The evidence from a growing body of empirical research demonstrates that those aromas that commonly co-occur with a salty taste in food (aromas that can be considered as congruent with a salty taste [94]) tend to take on the association with the co-occurring taste [95] and become more perceptually similar to the taste with which they commonly co-occur [96]. Therefore, for example, the odour of soy sauce and dried ham is classed as salty odour by those who are familiar with them. What is more, having become associated with a salty taste, these aromas, when added to foods give rise to odour-induced taste enhancement (OITE). This is where perceived saltiness is enhanced by the addition of odorants that are congruent (i.e., which commonly co-occur) with salty-tasting foods. At the same time, however, the presence of a specific taste (e.g., salty, umami, or sweet), has also been documented to enhance the associated food odour, no matter whether experienced orthonasally or retronasally [97–99].

Relevant to the theme of the present review, it has been widely suggested that OITE could be used as an effective strategy in order to enhance the perceived saltiness of reduced-salt food products. As highlighted in Table 1, many salty aromas that are clearly capable of giving rise to an OITE effect have been identified, there are several fundamental questions concerning OITE effects that remain to be addressed.

Table 1. Chronological summary of published studies that have specifically studied the salt/umami taste-enhancing properties of volatile aromas. Adapted and updated from [93].

Study	Participants	Volatile Aroma	Taste Quality	Result	Comments
Murphy & Cain [100]	20 experienced p's	Citral (citrus)	Sweet Salt	Sig. Sig.	Aroma increased both congruent and incongruent taste intensity
Frank & Byram [101]	E1: 20; E2: 20; E3: 18 untrained p's	E1: Strawberry; E2: Peanut butter; E3: Strawberry	E1, E2: Sweet E3: Salt	Sig. n.s.	Only congruent strawberry aroma led to OITE in whipped cream base
Djordjevic et al. [102]	E1 (40 untrained p's) E2 (same 40 p's)	Strawberry & soy sauce	Sweet Salt	E1: Sig. E2: Sig.	Both actual and imagined odours enhanced congruent taste quality
Lawrence et al. [103]	59 untrained p's	Range of salty food aromas including bacon and sardine	Salt	Sig.	7 salty aromas gave rise to significant enhancement in salt perception
Batenburg & van der Velden [104]	10 p's & 2 trained panels of 10–12; Consumer panels	Chicken flavouring (inc. individual salty volatiles; e.g., sotolon)	Salt	Sig.	Untrained panel exhibited greater odour-induced salt enhancement in bouillons
Lawrence et al. [105]	27 untrained consumers	Comté cheese, sardine, & carrot	Salt	Sig.	Comté cheese & sardine aroma increased saltiness of model cheese
Nasri et al. [106]	64 untrained p's	Sardine	Salt	Sig.	Sig. effect at low/medium salt levels, but no effect at high salt intensity
Nasri et al. [107]	61 untrained p's	Sardine	Salt	Sig.	No effect of odour intensity on OITE
Seo et al. [108]	E1 (25 p's); E2 (25 p's)	Salty bacon & sweet strawberry	Salt Sweet	Sig. Sig.	Psychophysical & neuroimaging study
Manabe et al. [109]	70 p's	Dried bonito stock	Umami	Sig.	Increased palatability of saltiness
Niimi et al. [110]	10 trained panellists	Cheese aroma mixture containing 10 volatiles	Umami Bitter	Sig. Sig.	Increasing OITE with increasing aroma intensity
Chokumnoyporn et al. [111]	10 panellists	Soy sauce	Salt	Sig.	Sig. effect on salty solutions above & below salt threshold
Emorine et al. [112]	Consumer panel (82)	Ham	Salt	Sig.	Sig. effect on salt perception
Syarifuddin et al. [113]	31 panellists	Sardine or butter	Salt/Fat	Sig.	Congruent aroma enhanced salt or fat
Kakutani et al. [114]	12 p's	Soy sauce	Salt	Sig.	Retronasal odour after drinking, but not orthonasal odour before drinking, significantly increased salty taste
Onuma et al. [115]	E1 (12 p's); E2 (20 p's) E3 (12 p's)	Soy sauce	Salt	Sig.	Saltiness enhancement demonstrated psychophysically & using neuroimaging
Manabe et al. [116]	E1 (75 p's) E2 (75 p's)	Soy sauce & 3-methyl-1-butanol	Salt	Sig.	Saltiness enhancement demonstrated & role of key compound identified
Sinding et al. [117]	13 untrained p's	Beef stock	Salt	Sig.	Saltiness of reduced salt green pea soup

p's—participants; Sig.—A significant enhancement on rated salty taste of the volatile was reported; n.s.—non-significant; OITE—odour-induced taste enhancement; inc.—including.

Therefore, for example, one might wonder whether certain salty odorants are more effective than others at enhancing the perceived salty taste in food. It is currently also rather unclear whether different salty odorants can be combined to enhance their effectiveness. Recent findings also question whether heterogenous salt distribution can be combined effectively with OITE in realistic food substrates, such as hot flan [118]. The role of cultural differences in the OITE effects also constitutes an intriguing area for further study. To what extent does the aroma of specific local salty condiments come to take on enhanced OITE effects? Think here of Thai fish sauce in Thailand [29], soy sauce in Japan [81], or Maggi-type seasoning in a number of other countries [28]. Intriguingly, researchers have already started to probe the different representations of the qualities of soy sauces amongst consumers and chefs as a function of their sensory qualities cf. [119–121]. Another intriguing issue that has yet to be resolved is the relative importance of congruency to OITE effects (i.e., as distinct from the effects of perceptual similarity, see [94,96,99,122] on this subtle/intriguing distinction). Nevertheless, despite these outstanding issues, it is becoming increasingly clear that OITE represents a promising sensory approach to the maintenance of an acceptable taste/flavour profile in reduced-salt processed food products.

6. Visual Modulation of Saltiness Expectations and Perception

Beyond the olfactory contribution to salt perception, there has been much interest in the use of food colour to enhance perceived saltiness (e.g., in chicken stock/broth/bouillon [123–125]). In fact, a large body of research conducted over the last 90 years or so has demonstrated that colour cues in food and drink often exert a significant effect on taste thresholds [126], suprathreshold intensity ratings, and flavour-identification responses [127,128]. In an influential early psychophysical study, Maga found that adding colour (red, green, or yellow) to otherwise colourless salt solution had no impact over the salt taste detection threshold, despite the significant influence of adding colour on the threshold for sweet, bitter, and sour [126]. Other researchers though have reported that increasing the intensity of orange colour from a lighter to a darker colour increased saltiness expectation (based on visual observation) in the context of mayonnaise-dipping sauce [129,130]. Wongthahan and colleagues conducted a study of soy sauces (light, medium, and dark) using both regular users (consumers) and culinary chefs. Their results highlighted an association between brown colour intensity with saltiness expectation/perception [119].

One problem, though, when it comes to the use of colour to modulate salt perception is that salty foods come in all manner of colours (this is the argument put forward by Maga [126] to explain the null results on salt detection thresholds he reports) and, hence, it has been suggested that colour may be a less effective cue to saltiness perception that in the case of these colours associated with other tastes (e.g., consider the association between pinkish-red and sweetness; [131,132]; see also [133]). While early research tended to focus on the association between orangey-brown colour of savoury stocks and broth [29,123,124,134,135] and brown colour of condiments/seasonings, such as fish sauce and soy sauce [119], contemporary research on the cross-modal correspondences that exist between colour patches and taste have highlighted a robust association between the colours white and blue (either when presented individually or else together) and a salty taste [136–139]. While the link between salt and white may well relate to the typically white colour of salt crystals, the blue association may well be the result of packaging colour conventions in the marketplace. It should also be noted that the colours orange/brown were often not available to participants in the laboratory studies of colour-taste correspondences, hence, perhaps explaining why this pairing did not appear in the results that have been reported [140].

Nevertheless, taken together, the research that has been published to date demonstrates that despite the fact that multiple colours are associated with saltiness, there may be the opportunity to enhance perceived saltiness through the use of colour cues that help to set expectations of saltiness in the consumer (cf. [141]). Further, going beyond the use of colour, there may be fruitful opportunities to combine salty colours with salty aromas in order to enhance their combined effectiveness in conveying an impression of saltiness in reduced-salt food products (cf. [131,142]).

7. Sonic Seasoning to Modify Saltiness Perception

Background noise has been shown to suppress the perception of salt (and sweetness) in both food products and pure taste solutions [143–145]. At the same time, however, the presence of food consumption sounds that are typically associated with a salty taste (such as the sound of someone noisily eating potato chips) has been shown to increase perceived saltiness [146], though note the danger in triggering misophonia in some proportion of consumers on hearing such sounds [72]. Intriguingly, sonic seasoning is where music and soundscapes with particular sonic properties are found to accentuate the corresponding taste property [147–151]. According to research from Wang and colleagues, the sonic qualities that are most strongly associated with saltiness were a long decay time, high auditory roughness, and a regular rhythm. Meanwhile, in terms of emotional associations, saltiness was matched with negative valence, high arousal, and minor mode [151]. A few years ago, one Beijing café even went so far as to introduce sweet sonic seasoning so that they could reduce the sugar content in their drinks without having to compromise on taste [152]. In the future, it would not seem beyond the realms of possibility to use sonic seasoning to modulate perceived saltiness in food and drink.

8. Tactile/Haptic Modulation of Saltiness Perception

In terms of tactile/oral-somatosensory influences on saltiness perception, early research from Christensen demonstrated an impact of the viscosity in solutions on perceived saltiness and sweetness [153], though, in this case, the mechanism behind underlying the effect might be a cross-modal correspondence between viscosity and taste or perhaps a physicochemical impact of reduced OITE due to increased viscosity. Elsewhere, Van Rompay and Groothedde [154] demonstrated that perceived saltiness of potato chips could be enhanced through the use of rough surface texture design of a serving bowl (cf. [155–159]). Importantly, however, these saltiness-enhancement effects were only documented for medium and full-salt crisps but not for no-salt crisps.

9. Challenges to Salt Reduction: The Knowledge–Behaviour Gap and Self-Efficacy

Self-efficacy refers to the perceived ability of an individual to exert personal control [160]. The first of the four main constructs in the Knowledge Behaviour Gap model is knowledge, followed by acceptance, intention, and behaviour [161]. Self-efficacy has been shown to play a major role in the maintenance of health behaviours across a variety of health domains [162]. Most chronic diseases are rooted in lifestyle factors and enhancing knowledge is useful to the extent that it subsequently leads to the modification of people's behaviour [163–165]. Intriguingly, dietary modification through enhancement of self-efficacy and knowledge was a principal area of focus of the FCP (Stanford Five-City Project) campaign with some success [166]. That said, a recent study among people living in Hong Kong highlighted a worrying disengagement with salt-reduction behaviour, such as rarely/never checking the sodium or salt content listed on the food label and rarely/never purchasing food labelled as containing low-salt or no-salt [167]. Meanwhile, nearly 90% of the participants in one Greek study did not know what the recommended daily salt intake was [168]. While 90% of the participants in another study conducted in the Australian state of Victoria were aware that excessive salt intake can cause health damage, more than 80% still acknowledged that they were eating "far too much" (i.e., more than the recommended daily intake), with less than half of the participants actively attempting to reduce their salt intake [169]. Importantly, for fundamental knowledge regarding the recommended daily intake, the primary food sources of high salt content and the differences between salt and sodium continues to be lacking, even amongst those living in high-income countries [170]. According to the results of a Chinese study, promoting better public knowledge, enhancing the public's awareness of salt reduction, and encouraging more active salt-reduction behaviour can help to suppress the transition from normal blood pressure to hypertension [171]. Ultimately, researchers believe that the population's knowledge, attitudes, and behaviours will affect their salt consumption and are thought to be adjustable and controllable intermediate factors over the short term [172].

10. Conclusions

The negative health consequences of the overconsumption of salt are becoming increasingly apparent [173–175]. This is despite the discrepancy between the WHO recommendations that people should keep their salt intake below 5 g per day (2000 mg/day of sodium) and their potassium intake above 3500 mg/day, and the dietary reference intakes from the National Academy of Sciences, Engineering, and Medicine in the USA that the adequate intake in adults is 3.75 g/day of salt (1500 mg/day of sodium), while the adequate potassium intake is 3400 mg/day in men and 2600 mg/day in women [176]. The dangers have, for example, been highlighted by the latest epidemiological research, showing that people who salt their food have a 1.5–2.3-year reduction in average life expectancy when compared to those who do not [25]. At the same time, however, the potential health benefits of reducing salt intake (e.g., in terms of lowering blood pressure) are also becoming increasingly well-established [177,178]. In fact, effective salt-reduction strategies have been rolled out in several countries in recent years [179] and these have provided learnings for those wanting to introduce effective public health strategies around salt reduction [180]. That said, while the major sources of dietary salt have now been identified [49–51,181,182], one of the challenges is that unless salt reduction occurs very gradually, it can lead to a negative consumer response in terms of impaired taste/flavour perception [41,183–186]. Numerous different solutions have been proposed, including behavioural nudges, such as reducing the number/size of holes in saltshakers [78–80]. In recent years, researchers have increasingly been attempting to take such suggested solutions from the laboratory to increasingly realistic contexts [187–189]. In one recent series of four controlled laboratory studies, five-holed saltshakers were shown to deliver around 34% of the salt of 17-holed saltshakers [186]. The question of which solutions are likely to have the greatest long-lasting effect, and separately whether the various factors can be combined, is, though, an important current area for research ([118,190]).

It is currently unclear the extent to which sensory modulations of perceived saltiness, such as via OITE effects, or the use of colour cues or sonic seasoning are capable of influencing neural activity in primary taste areas [191]. Relevant here, verbal labels/descriptions have been documented to modulate primary taste areas [192,193]. That said, a colour's effect on consumer responses can sometimes trigger a response bias [128], while even visually presented shapes can at least under certain conditions influence taste thresholds, arguing that at least some of the cross-modal effects may have genuine perceptual consequences [194]. It would seem likely that such a cross-modal modulation of salt perception is likely to be more acceptable in the long term than the electric cutlery that has been invented by Japanese researchers [195]. Indeed, it is worth noting that the various salt-reduction strategies that have been studied over the years clearly differ markedly in their ease/practicality of implementation.

According to Liem and colleagues [185], a sodium reduction of up to 30% may be acceptable in processed foods if introduced gradually (i.e., over a period of 3 years) and that the reduction can be up to 50% as long as it is in parallel with the addition of a flavour-boosting ingredient, such as soy sauce or dried bonito. The addition of chilli can also be used to help spice-up foods, as can black pepper, etc., without the negative health consequences associated with the overconsumption of salt [196]. Others have proposed that reductions in salt (and sodium) can be mitigated with monosodium glutamate (MSG) in what they have termed the 'salt flip' [83,197,198]. MSG is the sodium salt of L-glutamic acid. It is the most abundant amino acid in nature, constituting up to 8% to 10% of most dietary proteins, either as free glutamate or bound to other amino acids. At the same time, however, consumer concerns with the consumption of MSG (often labelled as an 'artificial' taste or flavour enhancer), although misplaced, continue to limit the applicability of such an approach [199–206]. The lack of familiarity of many Western consumers with umami may not help either [207,208]. Finally, here, one might consider Nico Ladenis, the Greek-born restaurateur who was known for kicking customers out of his London restaurant should they be so foolish as to ask for the salt [209] (p. 215). Given the appeal of salt, which can all too easily turn to salt hunger [210], one might wonder whether such extreme responses will ultimately be needed in order to deal with the very serious problem of the overconsumption of salt, though given that it has been estimated that more than 70% of our salt consumption comes from food consumption outside the home (i.e., from ready meals and eating out) [211,212], it is clear that, ultimately, various other strategies are also going to be needed (see also [213]).

Funding: This research received no external funding.

Data Availability Statement: No data is contained within the article.

Conflicts of Interest: The author declares no conflict of interest.

References

1. Kaushik, S.; Kumar, R.; Kain, P. Salt an Essential Nutrient: Advances in Understanding Salt Taste Detection Using Drosophila as a Model System. *J. Exp. Neurosci.* **2018**, *12*, 1179069518806894. [CrossRef] [PubMed]
2. Roberts, J. *Salted and Cured: Savoring the Culture, Heritage, and Flavor of America's Preserved Meats*; Chelsea Green Publishing: Chelsea, VT, USA, 2017.
3. Weller, O.; Dumitroaia, G. The earliest salt production in the world: An early Neolithic exploitation in Poiana Slatinei-Lunca, Romania. *Antiquity* **2005**, 79.
4. Kurlansky, M. *Salt: A World History*; Penguin Books: London, UK, 2003.
5. Herz, R. *Why You Eat What You Eat: The Science Behind Our Relationship with Food*; Norton & Co.: New York, NY, USA, 2017.
6. Jacewicz, N. How Did Salt and Pepper Become the Mainstays of Western Cuisine. Available online: https://www.npr.org/sections/thesalt/2018/02/02/582477785/how-did-salt-and-pepper-become-the-soulmates-of-western-cuisine (accessed on 28 August 2022).
7. Schulkin, J. The allure of salt. *Psychobiology* **1991**, *19*, 116–121. [CrossRef]
8. Van Dongen, M.V.; van der Berg, M.C.; Vink, N.; Kok, F.J.; De Graaf, C. Taste–nutrient relationships in commonly consumed foods. *Br. J. Nutr.* **2012**, *108*, 140–147. [CrossRef] [PubMed]
9. Kawai, M. Newly-acquired pre-cultural behavior of the natural troop of Japanese monkeys on Koshima islet. *Primates* **1965**, *6*, 1–30. [CrossRef]
10. Matsuzawa, T. Sweet-potato washing revisited: 50th anniversary of the Primates article. *Primates* **2015**, *56*, 285–287. [CrossRef]
11. Emmons, L.H.; Stark, N.M. Elemental Composition of a Natural Mineral Lick in Amazonia. *Biotropica* **1979**, *11*, 311. [CrossRef]
12. Blair-West, J.R.; Coghlan, J.P.; Denton, D.A.; Nelson, J.F.; Orchard, E.; Scoggins, B.A.; Wright, R.D.; Myers, K.; Junqueira, C.L. Physiological, morphological and behavioural adaptation to a sodium deficient environment by wild native Australian and introduced species of animals. *Nature* **1968**, *217*, 922–928. [CrossRef]
13. Wiener, J.G. Nutrient cycles, nutrient limitation and vertebrate populations. *Biologist* **1975**, *57*, 104–124.
14. Schultz, S.R.; Johnson, M.K. Effects of supplemental mineral licks on white-tailed deer. *Wildl. Soc. Bull.* **1992**, *20*, 303–308.
15. Coile, C. Accounting for Taste: What Do Dogs Find Most Delicious? Available online: https://www.akc.org/expert-advice/nutrition/accounting-taste-probing-mysteries-dogs-find-delicious/ (accessed on 28 August 2022).
16. Kitchell, R.L. Taste perception and discrimination by the dog. *Adv. Vet. Sci. Comp. Med.* **1978**, *22*, 287–314. [PubMed]
17. Breslin, P.A.S.; Beauchamp, G.K. Salt enhances flavor by suppressing bitterness. *Nature* **1997**, *387*, 563. [CrossRef] [PubMed]
18. Breslin, P.A.S.; Beauchamp, G.K. Suppression of bitterness by sodium: Variation among bitter taste stimuli. *Chem. Senses* **1995**, *20*, 609–623. [CrossRef]
19. Suwankanit, C.; Dermiki, M.; Kennedy, O.B.; Methven, L. Umami: Suppressed by all other tastes but itself an enhancer of salty and sweet perception. In Proceedings of the 10th Pangborn Sensory Science Symposium, Rio de Janeiro, Brazil, 11–15 August 2013.
20. Baines, D.; Brown, M. Flavor enhancers: Characteristics and uses. In *Encyclopedia of Food and Health*; Elsevier: Amsterdam, The Netherlands, 2016; pp. 716–723. [CrossRef]
21. Webster, J.; Dunford, E.; Huxley, R.; Li, N.; Nowson, C.A.; Neal, B. The development of a national salt reduction strategy for Australia. *Asia Pac. J. Clin. Nutr.* **2009**, *18*, 303–309. [PubMed]
22. Dodd, R.; Santos, J.A.; Tan, M.; Campbell, N.R.C.; Ni Mhurchu, C.; Cobb, L.; Jacobson, M.F.; He, F.J.; Trieu, K.; Osornprasop, S.; et al. Effectiveness and feasibility of taxing salt and foods high in sodium: A systematic review of the evidence. *Adv. Nutr. Int. Rev. J.* **2020**, *11*, 1616–1630. [CrossRef] [PubMed]

23. Rojas-Rivas, E.; Antúnez, L.; Cuffia, F.; Otterbring, T.; Aschemann-Witzel, J.; Giménez, A.; Ares, G. Time orientation and risk perception moderate the influence of sodium warnings on food choice: Implications for the design of communication campaigns. *Appetite* **2020**, *147*, 104562. [CrossRef]
24. Pavia, W. New York peppers menus with salt warnings. *The Times*, 1 December 2015.
25. Ma, H.; Xue, Q.; Wang, X.; Li, X.; Franco, O.H.; Li, Y.; Heianza, Y.; Manson, J.E.; Qi, L. Adding salt to foods and hazard of premature mortality. *Eur. Heart J.* **2022**, *43*, 2878–2888. [CrossRef]
26. Lander, N. *The Art of the Restaurateur*; Phaidon Press: London, UK, 2012.
27. Time. A Brief History of Salt. *Time*, 15 March 1982.
28. Spence, C. The psychology of condiments: A review. *Int. J. Gastron. Food Sci.* **2018**, *11*, 41–48. [CrossRef]
29. andenbroucke, J.P.; von Elm, E.; Altman, D.G.; Gøtzsche, P.C.; Mulrow, C.D.; Pocock, S.J.; Poole, C.; Schlesselman, J.J.; Egger, M.; STROBE Initiative. Strengthening the Reporting of Observational Studies in Epidemiology (STROBE): Explanation and elaboration. *PLoS ONE* **2007**, *4*, e297. [CrossRef]
30. UK Government. Department of Education. Guidance: School Food Standards Practical Guide (Updated 11 May 2022). Available online: https://www.gov.uk/government/publications/school-food-standards-resources-for-schools/school-food-standards-practical-guide (accessed on 28 August 2022).
31. Patel, A. Bad condiments: 15 unhealthy condiments include ketchup, BBQ sauce and mayonnaise. *Huffington Post*, 21 February 2013.
32. World Health Organization. *Diet, Nutrition and the Prevention of Chronic Diseases*; World Health Organization: Geneva, Switzerland, 2003; Volume 916.
33. Food Safety Authority of Ireland. *Salt and Health: Review on the Scientific Evidence and Recommendations for Public Policy in Ireland*; Food Safety Authority of Ireland: Dublin, Ireland, 2005.
34. Durack, E.; Alonso-Gomez, M.; Wilkinson, M.G. Salt: A review of its role in food science and public health. *Curr. Nutr. Food Sci.* **2008**, *4*, 290–297. [CrossRef]
35. Beer-Borst, S.; Costanza, M.C.; Pechère-Bertschi, A.; Morabia, A. Twelve-year trends and correlates of dietary salt intakes for the general adult population of Geneva, Switzerland. *Eur. J. Clin. Nutr.* **2009**, *63*, 155–164. [CrossRef] [PubMed]
36. Brown, I.J.; Tzoulaki, I.; Candeias, V.; Elliott, P. Salt intakes around the world: Implications for public health. *Int. J. Epidemiol* **2009**, *38*, 791–813. [CrossRef] [PubMed]
37. Eaton, S.B.; Konner, M.J. Paleolithic nutrition revisited: A twelve-year retrospective on its nature and implications. *Eur. J. Clin. Nutr.* **1997**, *51*, 207. [CrossRef] [PubMed]
38. Moss, M. *Salt, Sugar, Fat: How the Food Giants Hooked Us*; Random House: New York, NY, USA, 2013.
39. Moss, M. The Extraordinary Science of Addictive Junk Food. *The New York Times*, 20 February 2013.
40. Tozer, J. The great HP Sauce revolt: Online fury over salt cut that 'ruins the taste'. *Daily Mail Online*, 12 September 2011.
41. Charles, D. The Paradox and Mystery of Our Taste for Salt. *NPR*, 20 December 2012.
42. Kilcast, D.; den-Ridder, C. Sensory issues in reducing salt in food products. In *Reducing Salt in Foods: Practical Strategies*; Kilcast, D., Angus, F., Eds.; Woodhead Publishing: New York, NY, USA, 2007; pp. 201–220.
43. Ainsworth, P.; Plunkett, A. Reducing salt in snack products. In *Reducing Salt in Foods: Practical Strategies*; Kilcast, D., Angus, F., Eds.; Woodhead Publishing: New York, NY, USA, 2007; pp. 296–315.
44. Nahar, N.; Madzuki, I.; Izzah, N.; Ab Karim, M.; Ghazali, H.; Karim, R. Bakery science of bread and the effect of salt reduction on quality: A review. *Borneo J. Sci. Technol.* **2019**, *1*, 9–14.
45. Belz, M.C.E.; Ryan, L.A.M.; Arendt, E.K. The impact of salt reduction in bread: A review. *Crit. Rev. Food Sci. Nutr.* **2012**, *52*, 514–524. [CrossRef]
46. Carcea, M.; Narducci, V.; Turfani, V.; Mellara, F.A. Comprehensive study on the influence of sodium chloride on the technological quality parameters of soft wheat dough. *Foods* **2020**, *9*, 952. [CrossRef]
47. Goy, D.; Häni, J.-P.; Piccinali, P.; Wehrmüller, K.; Jakob, E.; Fröhlich-Wyder, M.-T. *Salt and Its Significance in Cheese Making*; Agroscope Liebefeld-Posieux Research Station ALP: Berne, Switzerland, 2012.
48. Lever, A.-M. 'Unnecessary' High Salt Levels in Cheese, Health Group Warns. *BBC News*, 29 November 2012.
49. Wirth, F. Reducing the fat and sodium content of meat products. What possibilities are there? *Fleischwirtsch* **1991**, *71*, 294–297.
50. He, F.J.; MacGregor, G.A. Salt in food. *Lancet* **2005**, *365*, 844–845. [CrossRef]
51. Inguglia, E.S.; Zhang, Z.; Tiwari, B.K.; Kerry, J.P.; Burgess, C. Salt reduction strategies in processed meat products—A review. *Trends Food Sci. Technol.* **2017**, *59*, 70–78. [CrossRef]
52. Intersalt Cooperative Research Group. Intersalt: An international study of electrolyte excretion and blood pressure. Results for 24 hour urinary sodium and potassium excretion. Intersalt Cooperative Research Group. *BMJ* **1988**, *297*, 319–328. [CrossRef]
53. Wolf, G. Innate mechanisms for regulation of sodium intake. In *Olfaction and Taste*; Pfaffman, C., Ed.; Rockefeller University Press: New York, NY, USA, 1969; pp. 548–553.
54. Denton, D. *The Hunger for Salt*; Springer: New York, NY, USA, 1982.
55. Bertino, M.; Beauchamp, G.K.; Engelman, K. Long-term reduction in dietary sodium alters the taste of salt. *Am. J. Clin. Nutr.* **1982**, *36*, 1134–1144. [CrossRef] [PubMed]
56. Jeon, S.-Y.; Lee, E.-K.; Kim, K.-O. The perceived saltiness of soup affected by tasting protocols. *Food Qual. Prefer.* **2014**, *35*, 98–103. [CrossRef]

57. O'Mahoney, M.; Heintz, C. Direct magnitude estimation of salt taste intensity with continuous correction for salivary adaptation. *Chem. Senses* **1981**, *6*, 101–112. [CrossRef]
58. O'Mahony, M.; Kingsley, L.; Harji, A.; Davies, M. What sensation signals the salt taste threshold? *Chem. Senses Flavor* **1976**, *2*, 177–188. [CrossRef]
59. Barrós-Loscertales, A.; González, J.; Pulvermüller, F.; Ventura-Campos, N.; Bustamante, J.C.; Costumero, V.; Ávila, C. Reading salt activates gustatory brain regions: fMRI evidence for semantic grounding in a novel sensory modality. *Cereb. Cortex* **2012**, *22*, 2554–2563. [CrossRef]
60. Yeomans, M.R.; Chambers, L.; Blumenthal, H.; Blake, A. The role of expectancy in sensory and hedonic evaluation: The case of smoked salmon ice-cream. *Food Qual. Prefer.* **2008**, *19*, 565–573. [CrossRef]
61. Desor, J.A.; Maller, O.; Andrews, K. Ingestive responses of human newborns to salty, sour, and bitter stimuli. *J. Comp. Physiol. Psychol.* **1975**, *89*, 966–970. [CrossRef]
62. Desor, J.A.; Greene, L.S.; Maller, O. Preferences for sweet and salty in 9- to 15-year-old and adult humans. *Science* **1975**, *190*, 686–687. [CrossRef]
63. Davis, M.A.; Randall, E.; Forthofer, R.N.; Lee, E.S.; Margen, S. Living arrangements and dietary patterns of older adults in the United States. *J. Gerontol.* **1985**, *40*, 434–442. [CrossRef]
64. Hughes, G.; Bennett, K.M.; Hetherington, M.M. Old and alone: Barriers to healthy eating in older men living on their own. *Appetite* **2004**, *43*, 269–276. [CrossRef]
65. Schiffman, S.; Graham, B. Taste and smell perception affect appetite and immunity in the elderly. *Eur. J. Clin. Nutr.* **2000**, *54*, S54–S63. [CrossRef]
66. Ferris, A.M.; Duffy, V.B. Effect of olfactory deficits on nutritional status: Does age predict persons at risk? *Ann. N. Y. Acad. Sci.* **1989**, *561*, 113–123. [CrossRef] [PubMed]
67. Stevens, J.C.; Cain, W.S.; Demarque, A.; Ruthruff, A.M. On the discrimination of missing ingredients: Aging and salt flavour. *Appetite* **1991**, *16*, 129–140. [CrossRef]
68. Frank, M.E.; Hettinger, T.P.; Mott, A.E. The sense of taste: Neurobiology, aging, and medication effects. *Crit. Rev. Oral Biol. Med.* **1992**, *3*, 371–393. [CrossRef] [PubMed]
69. Vasan, R.S.; Beiser, A.; Seshadri, S.; Larson, M.G.; Kannel, W.B.; D'Agostino, R.B.; Levy, D. Residual lifetime risk for developing hypertension in middle-aged women and men: The Framingham Heart Study. *JAMA* **2002**, *287*, 1003–1010. [CrossRef]
70. Seo, H.-S. Sensory Nudges: The Influences of Environmental Contexts on Consumers' Sensory Perception, Emotional Responses, and Behaviors toward Foods and Beverages. *Foods* **2020**, *9*, 509. [CrossRef]
71. Thomas-Danguin, T.; Guichard, E.; Salles, C. Cross-modal interactions as a strategy to enhance salty taste and to maintain liking of low-salt food: A review. *Food Funct.* **2019**, *10*, 5269–5281. [CrossRef]
72. Spence, C. *Gastrophysics: The New Science of Eating*; Viking Penguin: London, UK, 2017.
73. Stuckey, B. *Taste What You're Missing: The Passionate Eater's Guide to Why Good Food Tastes Good*; Free Press: London, UK, 2012.
74. BBC News. Kraft Cuts Dairylea Fat and Salt. Available online: http://news.bbc.co.uk/2/hi/business/6345205.stm (accessed on 28 August 2022).
75. Pombo-Rodrigues, S.; Hashem, K.M.; He, F.J.; MacGregor, G.A. Salt and sugars content of breakfast cereals in the UK from 1992 to 2015. *Public Health Nutr.* **2017**, *20*, 1500–1512. [CrossRef]
76. Ma, Y.; He, F.J.; Yin, Y.; Hashem, K.M.; MacGregor, G.A. Gradual reduction of sugar in soft drinks without substitution as a strategy to reduce overweight, obesity, and type 2 diabetes: A modelling study. *Lancet Diabetes Endocrinol.* **2016**, *4*, 105–114. [CrossRef]
77. MacGregor, G.A.; Hashem, K.M. Action on sugar—Lessons from UK salt reduction programme. *Lancet* **2014**, *383*, 929–931. [CrossRef]
78. Greenfield, H.; Maples, J.; Wills, R.B.H. Salting of food—A function of hole size and location of shakers. *Nature* **1983**, *301*, 331–332. [CrossRef] [PubMed]
79. Greenfield, H.; Smith, A.M.; Wills, R. Influence of multi-holed shakers on salting on food. *Hum. Nutr. Appl. Nutr.* **1984**, *38*, 199–201. [PubMed]
80. Farleigh, C.; Shepherd, R.; Wharf, S. The effect of manipulation of salt pot hole size on table salt use. *Food Qual. Prefer* **1990**, *2*, 13–20. [CrossRef]
81. Lioe, H.N.; Selamat, J.; Yasuda, M. Soy sauce and its umami taste: A link from the past to current situation. *J. Food Sci.* **2010**, *75*, R71–R76. [CrossRef] [PubMed]
82. Fukutome, N. Soy Sauces of Asia: Soy Sauce Usage in the Philippines, Thailand and Vietnam. Food Culture. Available online: https://www.kikkoman.co.jp/kiifc/foodculture/pdf_25/e_014_021.pdf (accessed on 21 August 2019).
83. dos Santos, B.A.; Campagnol, P.C.B.; Morgano, M.A.; Pollonio, M.A.R. Monosodium glutamate, disodium inosinate, disodium guanylate, lysine and taurine improve the sensory quality of fermented cooked sausages with 50% and 75% replacement of NaCl with KCl. *Meat Sci.* **2014**, *96*, 509–513. [CrossRef] [PubMed]
84. Colmenero, F.J.; Ayo, M.; Carballo, J. Physicochemical properties of low sodium frankfurter with added walnut: Effect of transglutaminase combined with caseinate, KCl and dietary fibre as salt replacers. *Meat Sci.* **2005**, *69*, 781–788. [CrossRef]
85. John, S.K.; Rangan, Y.; Block, C.A.; Koff, M.D. Life-threatening hyperkalemia from nutritional supplements: Uncommon or undiagnosed? *Am. J. Emerg. Med.* **2011**, *29*, 1237.e1–1237.e2. [CrossRef] [PubMed]

86. Halim, J.; Bouzari, A.; Felder, D.; Guinard, J. The *Salt Flip*: Sensory mitigation of salt (and sodium) reduction with monosodium glutamate (MSG) in "Better-for-You" foods. *J. Food Sci.* **2020**, *85*, 2902–2914. [CrossRef]
87. Yamaguchi, S.; Kimizuka, A. Psychometric studies on the taste of monosodium glutamate. In *Glutamic Acid: Advances in Biochemistry and Physiology*; Filler, L.J., Garattini, S., Kare, M.R., Reynolds, A.R., Wurtman, R.J., Eds.; Raven Press: New York, NY, USA, 1979; pp. 35–54.
88. Dijksterhuis, G.; Boucon, C.; Le Berre, E. Increasing saltiness perception through perceptual constancy created by expectation. *Food Qual. Prefer* **2014**, *34*, 24–28. [CrossRef]
89. Emorine, M.; Septier, C.; Thomas-Danguin, T.; Salles, C. Heterogeneous salt distribution in hot snacks enhances saltiness without loss of acceptability. *Food Res. Int.* **2013**, *51*, 641–647. [CrossRef]
90. Dijksterhuis, G.B.; Le Berre, E.; Woods, A.T. Food products with improved taste. EP2451289 A1, 2010.
91. Spence, C. Analysing stereotypical food consumption behaviours: 'This way up?' Is there really a 'right' way to eat a biscuit? *Int. J. Food Des.* **2021**, *6*, 213–231. [CrossRef]
92. Spence, C. The tongue map and the spatial modulation of taste perception. *Curr. Res. Food Sci.* **2022**, *5*, 598–610. [CrossRef] [PubMed]
93. Spence, C. Factors affecting odour-induced taste enhancement. *Food Qual. Prefer* **2021**, *96*, 104393. [CrossRef]
94. Schifferstein, H.N.J.; Verlegh, P.W.J. The role of congruency and pleasantness in odor-induced taste enhancement. *Acta Psychol.* **1996**, *94*, 87–105. [CrossRef]
95. Stevenson, R.J.; Boakes, R.A. Sweet and sour smells: Learned synaesthesia between the senses of taste and smell. In *The Handbook of Multisensory Processing*; Calvert, G.A., Spence, C., Stein, B.E., Eds.; MIT Press: Cambridge, MA, USA, 2004; pp. 69–83.
96. Spence, C. Searching for perceptual similarity within, and between, the (chemical) senses. *i-Perception* **2022**, *13*, 20416695221124154. [CrossRef]
97. Linscott, T.D.; Lim, J. Retronasal odor enhancement by salty and umami tastes. *Food Qual. Prefer.* **2016**, *48*, 1–10. [CrossRef]
98. Dalton, P.; Doolittle, N.; Nagata, H.; Breslin, P. The merging of the senses: Integration of subthreshold taste and smell. *Nat. Neurosci.* **2000**, *3*, 431–432. [CrossRef]
99. Lim, J.; Fujimaru, T.; Linscott, T.D. The role of congruency in taste–odor interactions. *Food Qual. Prefer.* **2014**, *34*, 5–13. [CrossRef]
100. Murphy, C.; Cain, W.S. Taste and olfaction: Independence vs interaction. *Physiol. Behav.* **1980**, *24*, 601–605. [CrossRef]
101. Frank, R.A.; Byram, J. Taste–smell interactions are tastant and odorant dependent. *Chem. Senses* **1988**, *13*, 445–455. [CrossRef]
102. Djordjevic, J.; Zatorre, R.J.; Jones-Gotman, M. Odor-induced changes in taste perception. *Exp. Brain Res.* **2004**, *159*, 405–408. [CrossRef] [PubMed]
103. Lawrence, G.; Salles, C.; Septier, C.; Busch, J.; Thomas-Danguin, T. Odour–taste interactions: A way to enhance saltiness in low-salt content solutions. *Food Qual. Prefer.* **2009**, *20*, 241–248. [CrossRef]
104. Batenburg, M.; van der Velden, R. Saltiness enhancement by savory aroma compounds. *J. Food Sci.* **2011**, *76*, S280–S288. [CrossRef] [PubMed]
105. Lawrence, G.; Salles, C.; Palicki, O.; Septier, C.; Busch, J.; Thomas-Danguin, T. Using cross-modal interactions to counterbalance salt reduction in solid foods. *Int. Dairy J.* **2011**, *21*, 103–110. [CrossRef]
106. Nasri, N.; Beno, N.; Septier, C.; Salles, C.; Thomas-Danguin, T. Cross-modal interactions between taste and smell: Odour-induced saltiness enhancement depends on salt level. *Food Qual. Prefer.* **2011**, *22*, 678–682. [CrossRef]
107. Nasri, N.; Septier, C.; Beno, N.; Salles, C.; Thomas-Danguin, T. Enhancing salty taste through odour–taste–taste interactions: Influence of odour intensity and salty tastants' nature. *Food Qual. Prefer.* **2013**, *28*, 134–140. [CrossRef]
108. Seo, H.-S.; Iannilli, E.; Hummel, C.; Okazaki, Y.; Buschhüter, D.; Gerber, J.; Krammer, G.E.; van Lengerich, B.; Hummel, T. A salty-congruent odor enhances saltiness: Functional magnetic resonance imaging study. *Hum. Brain Mapp.* **2011**, *34*, 62–76. [CrossRef]
109. Manabe, M.; Ishizaki, S.; Yamagishi, U.; Yoshioka, T.; Oginome, N. Retronasal odor of dried bonito stock induces umami taste and improves the palatability of saltiness. *J. Food Sci.* **2014**, *79*, S1769–S1775. [CrossRef]
110. Niimi, J.; Eddy, A.I.; Overington, A.R.; Heenan, S.P.; Silcock, P.; Bremer, P.J.; Delahunty, C.M. Aroma–taste interactions between a model cheese aroma and five basic tastes in solution. *Food Qual. Prefer.* **2013**, *31*, 1–9. [CrossRef]
111. Chokumnoyporn, N.; Sriwattana, S.; Phimolsiripol, Y.; Torrico, D.D.; Prinyawiwatkul, W. Soy sauce odour induces and enhances saltiness perception. *Int. J. Food Sci. Technol.* **2015**, *50*, 2215–2221. [CrossRef]
112. Emorine, M.; Septier, C.; Andriot, I.; Martin, C.; Salles, C.; Thomas-Danguin, T. Combined heterogeneous distribution of salt and aroma in food enhances salt perception. *Food Funct.* **2015**, *6*, 1449–1459. [CrossRef] [PubMed]
113. Syarifuddin, A.; Septier, C.; Salles, C.; Thomas-Danguin, T. Reducing salt and fat while maintaining taste: An approach on a model food system. *Food Qual. Prefer.* **2016**, *48*, 59–69. [CrossRef]
114. Kakutani, Y.; Narumi, T.; Kobayakawa, T.; Kawai, T.; Kusakabe, Y.; Kuneida, S.; Wada, Y. Saltiness intensity enhancements by odor of soy sauce synchronized with breathing. *Trans. Virtual Real. Soc. Jpn.* **2019**, *24*, 77–82.
115. Onuma, T.; Maruyama, H.; Sakai, N. Enhancement of saltiness perception by monosodium glutamate taste and soy sauce odor: A near-infrared spectroscopy study. *Chem. Senses* **2018**, *43*, 151–167. [CrossRef] [PubMed]
116. Manabe, M.; Sakaue, R.; Obata, A. Contribution of the retronasal odor of soy sauce to salt reduction. *J. Food Sci.* **2020**, *85*, 2523–2529. [CrossRef]

117. Sinding, C.; Thibault, H.; Hummel, T.; Thomas-Danguin, T. Odor-induced saltiness enhancement: Insights into the brain chronometry of flavor perception. *Neuroscience* **2021**, *452*, 126–137. [CrossRef]
118. Emorine, M.; Septier, C.; Martin, C.; Cordelle, S.; Sémon, E.; Thomas-Danguin, T.; Salles, C. Salt and aroma compound distributions influence flavour release and temporal perception while eating hot-served flans. *Molecules* **2021**, *26*, 1300. [CrossRef]
119. Wongthahan, P.; Sae-Eaw, A.; Prinyawiwatkul, W. Sensory lexicon and relationships among brown colour, saltiness perception and sensory liking evaluated by regular users and culinary chefs: A case of soy sauces. *Int. J. Food Sci. Technol.* **2020**, *55*, 2841–2850. [CrossRef]
120. Boakes, R.; Hemberger, H. Odour-modulation of taste ratings by chefs. *Food Qual. Prefer.* **2012**, *25*, 81–86. [CrossRef]
121. Flaherty, T.J.; Lim, J. Individual differences in retronasal odor responsiveness: Effects of aging and concurrent taste. *Chemosens. Percept.* **2017**, *10*, 91–103. [CrossRef]
122. Lim, J.; Johnson, M.B. The role of congruency in retronasal odor referral to the mouth. *Chem. Senses* **2012**, *37*, 515–522. [CrossRef] [PubMed]
123. Gifford, S.R.; Clydesdale, F.M. The psychophysical relationship between color and sodium chloride concentrations in model systems. *J. Food Prot.* **1986**, *49*, 977–982. [CrossRef] [PubMed]
124. Gifford, S.R.; Clydesdale, F.M.; Damon, R.A. The psychophysical relationship between color and salt concentrations in chicken flavored broths. *J. Sens. Stud.* **1987**, *2*, 137–147. [CrossRef]
125. Chan, M.M.; Kane-Martinelli, C. The effect of color on perceived flavor intensity and acceptance of foods by young adults and elderly adults. *J. Am. Diet. Assoc.* **1997**, *97*, 657–659. [CrossRef]
126. Maga, J.A. Influence of color on taste thresholds. *Chem. Senses* **1974**, *1*, 115–119. [CrossRef]
127. Wang, Q.J.; Spence, C. Drinking through rosé-coloured glasses: Influence of wine colour on the perception of aroma and flavour in wine experts and novices. *Food Res. Int.* **2019**, *126*, 108678. [CrossRef]
128. Spence, C.; Levitan, C.; Shankar, M.U.; Zampini, M. Does food color influence taste and flavor perception in humans? *Chemosens. Percept.* **2010**, *3*, 68–84. [CrossRef]
129. Sukkwai, S.; Chonpracha, P.; Kijroongrojana, K.; Prinyawiwatkul, W. Influences of a natural colourant on colour and salty taste perception, liking, emotion and purchase intent: A case of mayonnaise-based dipping sauces. *Int. J. Food Sci. Technol.* **2017**, *52*, 2256–2264. [CrossRef]
130. Sukkwai, S.; Kijroongrojana, K.; Chonpracha, P.; Pujols, K.D.; Alonso-Marenco, J.R.; Ardoin, R.; Prinyawiwatkul, W. Effects of colorant concentration and 'natural colour' or 'sodium content' claim on saltiness perception, consumer liking and emotion, and purchase intent of dipping sauces. *Int. J. Food Sci. Technol.* **2018**, *53*, 1246–1254. [CrossRef]
131. Spence, C. Multisensory sweetness enhancement: Comparing olfaction and vision. In *Smell, Taste, Eat: The Role of the Chemical Senses in Eating Behaviour*; Stafford, L., Ed.; Palgrave Macmillan: London, UK, *in press*.
132. Wang, Q.J.; Mielby, L.; Thybo, A.K.; Bertelsen, A.S.; Kidmose, U.; Spence, C.; Byrne, D.V. Sweeter together? Assessing the combined influence of product-related and contextual factors on perceived sweetness of fruit beverages. *J. Sens. Stud.* **2019**, *34*, e12492. [CrossRef]
133. Sugimori, E.; Kawasaki, Y. Cross-modal correspondence between visual information and taste perception of bitter foods and drinks. *Food Qual. Prefer.* **2022**, *98*, 104539. [CrossRef]
134. Ikeda, K. New seasonings. *Chem. Senses* **2002**, *27*, 847–849. [CrossRef]
135. Fateminia, M.; Ghotbabadi, T.D.; Azad, K.M. Perceptions of the taste of colors in children and adults. *Color Res. Appl.* **2020**, *45*, 743–753. [CrossRef]
136. Wan, X.; Woods, A.T.; van Den Bosch, J.J.F.; McKenzie, K.J.; Velasco, C.; Spence, C. Cross-cultural differences in crossmodal correspondences between basic tastes and visual features. *Front. Psychol.* **2014**, *5*, 151–158. [CrossRef] [PubMed]
137. Spence, C.; Wan, X.; Woods, A.; Velasco, C.; Deng, J.; Youssef, J.; DeRoy, O. On tasty colours and colourful tastes? Assessing, explaining, and utilizing crossmodal correspondences between colours and basic tastes. *Flavour* **2015**, *4*, 23. [CrossRef]
138. Woods, A.T.; Marmolejo-Ramos, F.; Velasco, C.; Spence, C. Using single colors and color pairs to communicate basic tastes II: Foreground–background color combinations. *i-Perception* **2016**, *7*, 2041669516663750. [CrossRef]
139. Woods, A.T.; Spence, C. Using single colors and color pairs to communicate basic tastes. *i-Perception* **2016**, *7*, 2041669516658817. [CrossRef]
140. Spence, C.; Levitan, C.A. Explaining crossmodal correspondences between colours and tastes. *i-Perception* **2021**, *12*, 20416695211018223. [CrossRef]
141. Chonpracha, P.; Gao, Y.; Tuuri, G.; Prinyawiwatkul, W. Possible sugar and calorie reduction by visual cues: A case of syrup added to brewed coffee. *J. Food Sci.* **2019**, *84*, 3784–3792. [CrossRef]
142. Wang, Q.J.; Mielby, L.A.; Junge, J.Y.; Bertelsen, A.S.; Kidmose, U.; Spence, C.; Byrne, D.V. The Role of Intrinsic and Extrinsic Sensory Factors in Sweetness Perception of Food and Beverages: A Review. *Foods* **2019**, *8*, 211. [CrossRef]
143. Ferber, C.; Cabanac, M. Influence of noise on gustatory affective ratings and preference for sweet or salt. *Appetite* **1987**, *8*, 229–235. [CrossRef]
144. Woods, A.T.; Poliakoff, E.; Lloyd, D.M.; Kuenzel, J.; Hodson, R.; Gonda, H.; Batchelor, J.; Dijksterhuis, G.B.; Thomas, A. Effect of background noise on food perception. *Food Qual. Prefer.* **2011**, *22*, 42–47. [CrossRef]
145. Yan, K.S.; Dando, R. A crossmodal role for audition in taste perception. *J. Exp. Psychol. Hum. Percept. Perform.* **2015**, *41*, 590–596. [CrossRef]

146. Seo, H.-S.; Hummel, T. Auditory–olfactory integration: Congruent or pleasant sounds amplify odor pleasantness. *Chem. Senses* **2011**, *36*, 301–309. [CrossRef]
147. Guetta, R.; Loui, P. When music is salty: The crossmodal associations between sound and taste. *PLoS ONE* **2017**, *12*, e0173366. [CrossRef]
148. Knoeferle, K.M.; Woods, A.; Käppler, F.; Spence, C. That sounds sweet: Using cross-modal correspondences to communicate gustatory attributes. *Psychol. Mark.* **2015**, *32*, 107–120. [CrossRef]
149. Knöferle, K.; Spence, C.; Knoeferle, K. Crossmodal correspondences between sounds and tastes. *Psychon. Bull. Rev.* **2012**, *19*, 992–1006. [CrossRef] [PubMed]
150. Wang, Q.; Woods, A.; Spence, C. "What's your taste in music?" A comparison of the effectiveness of various soundscapes in evoking specific tastes. *i-Perception* **2015**, *6*, 2041669515622001. [CrossRef]
151. Wang, Q.J.; Keller, S.; Spence, C. Metacognition and crossmodal correspondences between auditory attributes and saltiness in a large sample study. *Multisens. Res.* **2021**, *34*, 785–805. [CrossRef]
152. Blecken, D. Hold the Sugar: A Chinese Café Brand is Offering Audio Sweeteners. *Campaign*, 13 February 2017.
153. Christensen, C.M. Effects of solution viscosity on perceived saltiness and sweetness. *Percept. Psychophys.* **1980**, *28*, 347–353. [CrossRef]
154. van Rompay, T.J.; Groothedde, S. The taste of touch: Enhancing saltiness impressions through surface texture design. *Food Qual. Prefer.* **2019**, *73*, 248–254. [CrossRef]
155. Escobar, F.B.; Wang, Q.J.; Corredor, A.; Velasco, C. The taste of visual textures. *Food Qual. Prefer.* **2022**, *100*, 104602. [CrossRef]
156. Seo, H.-S.; Iannilli, E.; Hummel, C.; Okazaki, Y.; Buschhüter, D.; Gerber, J.; Krammer, G.E.; Seo, H.-S.; Arshamian, A.; Schemmer, K.; et al. Cross-modal integration between odors and abstract symbols. *Neurosci. Lett.* **2010**, *478*, 175–178. [CrossRef] [PubMed]
157. Juravle, G.; Olari, E.-L.; Spence, C. A taste for beauty: On the expected taste, hardness, texture, and temperature of geometric shapes. *i-Perception* **2022**, *13*, 20416695221120948. [CrossRef] [PubMed]
158. Spence, C.; Corujo, A.; Youssef, J. Cotton candy: A gastrophysical investigation. *Int. J. Gastron. Food Sci.* **2019**, *16*, 100146. [CrossRef]
159. Pikler, J. *Schriften zur Anspassungstheorie des Empfindungsvorganges*; Leipzig: Leipzig, Germany, 1922.
160. Bandura, A. *Social Learning Theory*; Prentice Hall: Englewood Cliffs, NJ, USA, 1977.
161. Stibe, A.; Krüger, N.; Behne, A. Knowledge behavior gap model: An application for technology acceptance. In *Mobile Web and Intelligent Information Systems. MobiWIS 2022. Lecture Notes in Computer Science, 13475*; Awan, I., Younas, M., Poniszewska-Marańda, A., Eds.; Springer: Cham, Switzerland, 2022.
162. Maibach, E.; Murphy, D.A. Self-efficacy in health promotion research and practice: Conceptualization and measurement. *Health Educ. Res.* **1995**, *10*, 37–50. [CrossRef]
163. National Research Council. Committee on Diet and Health. *Diet and Health: Implications for Reducing Chronic Disease Risk*; National Academy Press: Washington, DC, USA, 1989.
164. Rimal, R.N. Closing the knowledge-behavior gap in health promotion: The mediating role of self-efficacy. *Health Commun.* **2000**, *12*, 219–237. [CrossRef]
165. Tinsley, B.J. Multiple influences on the acquisition and socialization of children's health attitudes and behavior: An integrative review. *Child Dev.* **1992**, *63*, 1043–1069. [CrossRef]
166. Farquhar, J.W.; Fortmann, S.P.; Flora, J.A.; Taylor, B.; Haskell, W.L.; Williams, P.T.; Maccoby, N.; Wood, P.D. Effects of community-wide education on cardiovascular disease risk factors: The Stanford five-city project. *J. Am. Med. Assoc.* **1990**, *262*, 359–365. [CrossRef]
167. Cheung, J.; Neyle, D.; Chow, P.P.K. Current knowledge and behavior towards salt reduction among Hong Kong citizens: A cross-sectional survey. *Int. J. Environ. Res. Public Health* **2021**, *18*, 9572. [CrossRef]
168. Marakis, G.; Tsigarida, E.; Mila, S.; Panagiotakos, D.B. Knowledge, attitudes and behavior of Greek adults towards salt consumption: A Hellenic food authority project. *Public Health Nutr.* **2014**, *17*, 1877–1893. [CrossRef] [PubMed]
169. Grimes, C.A.; Kelley, S.J.; Stanley, S.; Bolam, B.; Webster, J.; Khokhar, D.; Nowson, C.A. Knowledge, attitudes and behaviors related to dietary salt among adults in the state of Victoria, Australia 2015. *BMC Public Health* **2017**, *17*, 532. [CrossRef] [PubMed]
170. Bhana, N.; Utter, J.; Eyles, H. Knowledge, attitudes and behaviours related to dietary salt intake in high-income countries: A systematic review. *Curr. Nutr. Rep.* **2018**, *7*, 183–197. [CrossRef] [PubMed]
171. Du, X.; Fang, L.; Xu, J.; Chen, X.; Bai, Y.; Wu, J.; Wu, L.; Zhong, J. The association of knowledge, attitudes and behaviors related to salt with 24-h urinary sodium, potassium excretion and hypertensive status. *Sci. Rep.* **2022**, *12*, 13901. [CrossRef]
172. Zandstra, E.H.; Lion, R.; Newson, R.S. Salt reduction: Moving from consumer awareness to action. *Food Qual. Prefer.* **2016**, *48*, 376–381. [CrossRef]
173. Santos, J.A.; Sparks, E.; Thout, S.R.; McKenzie, B.; Trieu, K.; Hoek, A.; Johnson, C.; McLean, R.; Arcand, J.; Campbell, N.R.C.; et al. The science of salt: A global review on changes in sodium levels in foods. *J. Clin. Hypertens.* **2019**, *21*, 1043–1056. [CrossRef]
174. Powles, J.; Fahimi, S.; Micha, R.; Khatibzadeh, S.; Shi, P.; Ezzati, M.; Engell, R.E.; Lim, S.S.; Danaei, G.; Mozaffarian, D.; et al. Global, regional and national sodium intakes in 1990 and 2010: A systematic analysis of 24 h urinary sodium excretion and dietary surveys worldwide. *BMJ Open* **2013**, *3*, e003733. [CrossRef]
175. World Health Organization. *Guideline: Sodium Intake for Adults and Children*; WHO: Geneva, Switzerland, 2012. Available online: https://www.who.int/nutrition/publications/guidelines/sodium_intake_printversion.pdf (accessed on 28 August 2022).

176. Oria, M.; Harrison, M.; Stallings, V.A. *Dietary Reference Intakes for Sodium and Potassium*; National Academy of Sciences, Engineering, and Medicine and The National Academies Press: Washington, DC, USA, 2019.
177. He, F.J.; Tan, M.; Ma, Y.; MacGregor, G.A. Salt reduction to prevent hypertension and cardiovascular disease: JACC state-of-the-art review. *J. Am. Coll. Cardiol.* **2020**, *75*, 632–647. [CrossRef]
178. He, F.; Li, J.; MacGregor, G. Effect of longer term modest salt reduction on blood pressure: Cochrane systematic review and meta-analysis of randomised trials. *BMJ* **2013**, *346*, f1325. [CrossRef]
179. He, F.J.; Brinsden, H.C.; MacGregor, G.A. Salt reduction in the United Kingdom: A successful experiment in public health. *J. Hum. Hypertens.* **2014**, *28*, 345–352. [CrossRef]
180. Winkler, J.T. Nutritional reformulation: The unobtrusive strategy. *Food Sci. Technol.* **2014**, *28*, 20–24.
181. Johnson, C.M.; Angell, S.Y.; Lederer, A.; Dumanovsky, T.; Huang, C.; Bassett, M.T.; Silver, L.D. Sodium content of lunchtime fast food purchases at major US chains. *Arch. Intern. Med.* **2010**, *170*, 732–734. [CrossRef] [PubMed]
182. James, W.; Ralph, A.; Sanchez-Castillo, C. The dominance of salt in manufactured food in the sodium intake of affluent societies. *Lancet* **1987**, *329*, 426–429. [CrossRef]
183. Beauchamp, G.K.; Bertino, M.; Moran, M. Sodium regulation: Sensory aspects. *J. Am. Diet. Assoc.* **1982**, *80*, 40–45. [CrossRef]
184. Hoppu, U.; Hopia, A.; Pohjanheimo, T.; Rotola-Pukkila, M.; Mäkinen, S.; Pihlanto, A.; Sandell, M. Effect of salt reduction on consumer acceptance and sensory quality of food. *Foods* **2017**, *6*, 103. [CrossRef]
185. Liem, D.G.; Miremadi, F.; Keast, R.S.J. Reducing Sodium in Foods: The Effect on Flavor. *Nutrients* **2011**, *3*, 694–711. [CrossRef]
186. Busch, J.; Feunekes, G.; Hauer, B.; den Hoed, W. Salt reduction and the consumer perspective. *New Food* **2010**, *2*, 36–39.
187. Goffe, L.; Wrieden, W.; Penn, L.; Hillier-Brown, F.; Lake, A.A.; Araújo-Soares, V.; Summerbell, C.; White, M.; Adamson, A.J.; Adams, J. Reducing the Salt Added to Takeaway Food: Within-Subjects Comparison of Salt Delivered by Five and 17 Holed Salt Shakers in Controlled Conditions. *PLoS ONE* **2016**, *11*, e0163093. [CrossRef]
188. Goffe, L.; Hillier-Brown, F.; Doherty, A.; Wrieden, W.; Lake, A.A.; Araujo-Soares, V.; Summerbell, C.; White, M.; Adamson, A.J.; Adams, J. Comparison of sodium content of meals served by independent takeaways using standard versus reduced holed salt shakers: Cross-sectional study. *Int. J. Behav. Nutr. Phys. Act.* **2016**, *13*, 102. [CrossRef]
189. Villinger, K.; Wahl, D.R.; Engel, K.; Renner, B. Nudging sugar portions: A real-world experiment. *BMC Nutr.* **2021**, *7*, 65. [CrossRef]
190. Spence, C. Why should vanilla be the most liked smell? *Nature Food* **2022**.
191. Nakamura, Y.; Goto, T.K.; Tokumori, K.; Yoshiura, T.; Kobayashi, K.; Nakamura, Y.; Honda, H.; Ninomiya, Y.; Yoshiura, K. The temporal change in the cortical activations due to salty and sweet tastes in humans: fMRI and time-intensity sensory evaluation. *NeuroReport* **2012**, *23*, 400–404. [CrossRef] [PubMed]
192. Woods, A.T.; Lloyd, D.M.; Kuenzel, J.; Poliakoff, E.; Dijksterhuis, G.B.; Thomas, A. Expected taste intensity affects response to sweet drinks in primary taste cortex. *NeuroReport* **2011**, *22*, 365–369. [CrossRef] [PubMed]
193. Nitschke, J.B.; Dixon, G.E.; Sarinopoulos, I.; Short, S.; Cohen, J.D.; E Smith, E.; Kosslyn, S.M.; Rose, R.M.; Davidson, R. Altering expectancy dampens neural response to aversive taste in primary taste cortex. *Nat. Neurosci.* **2006**, *9*, 435–442. [CrossRef]
194. Liang, P.; Roy, S.; Chen, M.-L.; Zhang, G.-H. Visual influence of shapes and semantic familiarity on human sweet sensitivity. *Behav. Brain Res.* **2013**, *253*, 42–47. [CrossRef]
195. Bolton, A. No Salt, No Problem! Japanese Electro Fork Zaps Flavour into Your Mouth. *CNet*, 31 March 2016.
196. Spence, C. Gastrophysics: The psychology of herbs and spices. In Proceedings of the Oxford Symposium on Food and Cookery, Oxford, UK, 9 July–1 August 2021.
197. Hartley, I.E.; Liem, D.G.; Keast, R.S. Females' ability to discriminate MSG from NaCl influences perceived intensity but not liking of MSG added vegetable broths. *J. Food Sci.* **2020**, *85*, 3934–3942. [CrossRef]
198. Chi, S.; Chen, T. Predicting optimum monosodium glutamate and sodium chloride concentrations in chicken broth as affected by spice addition. *J. Food Process. Preserv.* **1992**, *16*, 313–326. [CrossRef]
199. Maluly, H.D.B.; Arisseto-Bragotto, A.P.; Reyes, F.G.R. Monosodium glutamate as a tool to reduce sodium in foodstuffs: Technological and safety aspects. *Food Sci. Nutr.* **2017**, *5*, 1039–1048. [CrossRef]
200. Wang, S.; Adhikari, K. Consumer perceptions and other influencing factors about monosodium glutamate in the United States. *J. Sens. Stud.* **2018**, *33*, e12437. [CrossRef]
201. Schaumburg, H.H.; Byck, R.; Gerstl, R.; Mashman, J.H. Monosodium L-Glutamate: Its Pharmacology and Role in the Chinese Restaurant Syndrome. *Science* **1969**, *163*, 826–828. [CrossRef]
202. Prescott, J.; Young, A. Does information about MSG (monosodium glutamate) content influence consumer ratings of soups with and without added MSG? *Appetite* **2002**, *39*, 25–33. [CrossRef] [PubMed]
203. Halpern, B.P. What's in a name? Are MSG and umami the same? *Chem. Senses* **2002**, *27*, 845–846. [CrossRef] [PubMed]
204. Zautcke, J.L.; Schwartz, J.A.; Mueller, E.J. Chinese Restaurant Syndrome: A review. *Ann. Emerg. Med.* **1986**, *15*, 1210–1213. [CrossRef]
205. Nierenberg, A. The campaign to redefine 'Chinese Restaurant Syndrome. *The New York Times*, 16 January 2020.
206. McPhee, S. Colonel Sanders' top secret: Uproar after the crucial ingredient in KFC seasoning is exposed—And it's left fried-chicken lovers SHOCKED. *Daily Mail Online*, 14 September 2022.
207. Kobayashi, C.; Kennedy, L.M.; Halpern, B.P. Experience-induced changes in taste identification of monosodium glutamate (MSG) are reversible. *Chem. Senses* **2006**, *31*, 301–306. [CrossRef] [PubMed]

208. Kurihara, K. Umami the fifth basic taste: History of studies on receptor mechanisms and role as a food flavor. *BioMed Res. Int.* **2015**, *2015*, 189402. [CrossRef]
209. Sitwell, W. *The Restaurant: A history of Eating Out*; Simon & Schuster: London, UK, 2020.
210. Schulkin, J. *Sodium Hunger: The Search for a Salty Taste*; Cambridge University Press: New York, NY, USA, 1991.
211. Harnack, L.J.; Cogswell, M.E.; Shikany, J.M.; Gardner, C.D.; Gillespie, C.; Loria, C.M.; Zhou, X.; Yuan, K.; Steffen, L.M. Sources of Sodium in US Adults From 3 Geographic Regions. *Circulation* **2017**, *135*, 1775–1783. [CrossRef]
212. Scourboutakos, M.J.; Semnani-Azad, Z.; L'Abbe, M.R. Restaurant meals: Almost a full day's worth of calories, fats, and sodium. *JAMA Int. Med.* **2013**, *173*(14), 1373–1374. [CrossRef]
213. Kolata, G. Why everything we know about salt may be wrong. *The New York Times*, 8 May 2017.

foods

MDPI

Article

The Crick-Eatery: A Novel Approach to Evaluate Cricket (*Acheta domesticus*) Powder Replacement in Food Products through Product Eating Experience and Emotional Response

Isaac Ho [1], Adelynn Peterson [1], Jack Madden [1], Kylie Wai [1], Ruta Lesniauskas [2], Jeff Garza [3], Attila Gere [4], Samir Amin [1] and Amy Lammert [1,*]

1 Department of Food Science and Nutrition, California Polytechnic State University, San Luis Obispo, CA 93407, USA
2 Garza Consulting, Evanston, IL 60201, USA
3 Garza Consulting, Grand Rapids, MI 49501, USA
4 Institute of Food Science and Technology, Hungarian University of Agriculture and Life Sciences, H-1118 Budapest, Hungary
* Correspondence: alammert@calpoly.edu

Abstract: This study was conducted to evaluate three different food products containing cricket powder for consumer acceptability, emotional response, satiety, and plate waste. US untrained consumers (n = 108), from the San Luis Obispo, CA area, were recruited to evaluate three food products (sausage, pasta, and brownies) as components in a three-course meal that either contain cricket powder (CP) or not (Control). The CP sausage was found to have lower liking scores than the Control for the attributes tested ($p < 0.05$). The CP pasta was found to be higher in overall liking than the Control ($p < 0.05$). The CP Brownies were rated highly across the attributes, except for texture and aftertaste ($p < 0.05$). Though the CP products were found to be as acceptable as the Controls, the use of cricket powder may have affected the texture and flavor profile of both the CP sausage and brownies. The participants selected more positive emotions terms for both the CP and Control products than negative emotions. Negative terms selected, such as worried, decreased once the products were consumed ($p < 0.05$). Plate waste and subjective satiety may also be indicators of consumer acceptability. Significant correlations were found between appearance liking and satiety as well as taste liking and plate waste for both the Control and CP products/dishes ($p < 0.05$). Based on this work, future acceptance of insect-based products may be encouraged by evaluating the products throughout an eating experience.

Keywords: cricket powder; sausage; pasta; chocolate brownies; plate waste; emotions; subjective satiety; acceptability; three-course meal

Citation: Ho, I.; Peterson, A.; Madden, J.; Wai, K.; Lesniauskas, R.; Garza, J.; Gere, A.; Amin, S.; Lammert, A. The Crick-Eatery: A Novel Approach to Evaluate Cricket (*Acheta domesticus*) Powder Replacement in Food Products through Product Eating Experience and Emotional Response. *Foods* **2022**, *11*, 4115. https://doi.org/10.3390/foods11244115

Academic Editor: Witoon Prinyawiwatkul

Received: 2 November 2022
Accepted: 10 December 2022
Published: 19 December 2022

1. Introduction

By 2050, the human population is expected to reach 9 billion [1]. To feed the growing population, the United Nations found that food production will need to double, including protein [2]. The global protein market was found to be worth 38 billion USD in 2019 and is expected to grow at a rate of 9.1% from 2020 to 2027 [3]. Currently, the protein market is mainly comprised of animal protein production, which continues to grow as developing countries adopt Western practices such as meat consumption [4]. However, traditional animal protein production, or livestock farming, contributes to the overconsumption of limited natural resources, such as land and water, and the production of greenhouse gas emissions [4,5]. As a result, sustainable methods may need to develop to mitigate the effects of livestock farming, but also provide protein-rich food sources.

Alternative protein sources, such as plant-based proteins and cultured meat, have been developed for commercial sale and may offset the demand for animal proteins [3].

However, when replacing or reducing animal protein consumption, not all protein alternatives may meet every consumer need. Plant-based proteins may not be as comparable to animal proteins due to food functionalities, which may affect the sensorial aspects of plant-based products [6,7]. Cultured meat, or lab-grown meat, may deter consumers due to its "unnatural" perception by some, but also its higher cost of production than conventional animal meat may reduce its accessibility to some consumers [8,9]. Therefore, a variety of alternative proteins may need to be developed, as certain consumer groups may value one alternative over another [10].

Entomophagy, the practice of consuming insects, may be one of those alternatives. Insect consumption has historically, and continues to be, a food source for two billion people across countries in Asia, Latin America, and Africa [4,11–14]. For regions where livestock farming may be limited, insects may be a viable option. Compared to conventional animal protein, insects can provide all the essential amino acids necessary for human nutrition but require fewer resources, such as lower feed and land requirements, than livestock [5,15–17]. Therefore, insects may offer new opportunities in the food industry to develop protein-based products.

Currently, there has been investment in producing insects for human consumption. There are over 2000 edible insect species that are consumed across certain regions around the world [18]. However, selecting which insect species to farm may need to be considered, as each species may vary in nutritional profile [16,19]. In China, 34 edible insect species, which include silkworm larvae (*Bombyx mori*), yellow mealworm (*Tenebrio molitor*), and Orthoptera species (locusts, grasshoppers, and mole crickets), have been developed and assessed for production for human consumption [20]. In Western countries, there has been a focus on developing methods to farm the house cricket (*Acheta domesticus*) and yellow mealworms for human consumption [18,21]. Western governments, such as the EU, have even started to recognize and regulate insect species, such as yellow mealworms, as food for humans [22].

Insect production may raise concerns, as it is a new and developing method of food production. High levels of pathogens or other contaminants, such as heavy metals found in the feed used in insect farming, may be the main vector of contamination for insects intended for human consumption [4,23]. Not to mention, insects may be considered an allergen concern. Insects have a similar protein profile to shellfish/crustaceans that may cause allergic responses in certain individuals [24,25]. However, these concerns may be managed through regulations on agricultural feed, and preventative controls such as HACCP procedures and required labeling on insect-based foods [4,23]. Requirements for high-quality feed for insects intended for human food may lead to higher production costs when compared to conventional animal meat production [26]. As a result, investment into developing insect farming may be necessary to increase the productivity and profitability of the insect market. In Europe, nearly 90% of the investment costs of insect farming are covered by venture capital firms in Europe [21]. In North America, startup companies, such as Aspire Food Group and All Things Bugs (GrioPro®), have been funded by both government and private bodies for cricket farming, for both the food and agriculture industries [18]. However, when introducing insects into the Western market, insects may not be as accepted as a food source.

The future acceptance of Insects as food may be influenced by psychological factors that may deter consumers from eating them. Disgust, a culturally induced form of rejection, is a way for humans to avoid foods that may cause illness and disease [27,28]. In Western culture, insects are considered vectors of disease as some consumers view them as filthy or associate them with contamination and spoilage [27,29–31]. Food neophobia, the avoidance of trying new, unfamiliar foods, may also influence consumers with the intention to consume insects [32,33]. Lombardi et al. [34] found that food neophobia negatively affected the willingness to pay for insect-based products among Italian consumers. For Western consumers, insects may be considered a novel food as conventional livestock are considered the predominant protein source. This may be the result of insects in Western regions being smaller than insects found in tropical regions, which may have led to insects in Western regions being seen as an insubstantial food source [35]. As a result, both factors may affect influence the future adoption of insects by Western consumers [36].

Attitudes towards insects may also predict the willingness/intent to consume insects [33,37,38]. La Barbera et al. [31] found that positive attitudes (low disgust and high interest) towards insects, as well as individuals' openness to new food experiences among Danish and Italian consumers, may encourage the consumption of insects. In China, consumers who have previously consumed insects have been found to have positive attitudes towards them, with less self-reported disgust, and would be more willing to eat them than those who have not [39,40]. In Latin American countries, regions that do not typically incorporate insects into their food may be unfamiliar with the practice; however, consumers have been found to be more willing to eat insects if prepared in familiar ways, such as being fried or roasted [41–43]. To better predict the acceptance of insects, methods to change the attitudes of consumers and increase their familiarity with entomophagy may help to move past the barriers of disgust.

Previous research has focused on identifying interest to consume insects, which may increase their acceptance. One strategy may be to identify early adopters, the first group of consumers who gain interest and accept a new product [44,45]. Once accepted by this group, widespread adoption may occur within the market [46–49]. Another way may be to inform consumers of the sustainable benefits of consuming insects. The addition of information regarding entomophagy, when evaluating insect-based products, has been found to either increase interest/intent to consume insects or increase positive emotions during testing [34,47,50–52]. Serpico et al. [50] suggested that increasing positive emotions, such as curiosity and interest, while decreasing negative emotions, such as disgust and worry, may encourage consumers to eat insects. As a result, identifying consumers with a high interest in insect-based products may be a viable way to introduce insects as a food source.

To increase interest to consume insects, determining the delivery method, or the product appropriateness, may be a consideration. Product appropriateness, or the perception of what is deemed acceptable by consumers, such as when incorporating an ingredient like insects, may increase acceptance by consumers [53–55]. When incorporating insects into food products, nonvisible forms, such as powders and ground meals, may be a viable option. Western consumers may be more willing to try insect-based products if they contained nonvisible insects than visible forms (whole insects) [30,33,56,57]. Cricket powder or powders from other Orthoptera species may be a viable insect powder for Western consumers as it has been found to contain a high protein content (42.0–65.5% dry matter) [58–60]. Not to mention, products that contain cricket powder are available in the Western market, including protein bars (EXO), pasta (Bugsolutely), and snack foods (Chirps) [61]. However, when incorporating insects into conventional food products, consumers may have certain expectations for the sensory characteristics of these products. Cricket powder has been found to decrease overall consumer acceptability, such as in terms of textural properties, of products and may lead to rejection by consumers [52,62].

When developing food products, especially novel foods such as insects, predicting their future acceptance may go beyond consumer liking when insect-based products are compared to their conventional counterparts. In a preliminary study, Ho et al. [63] developed three different food products containing cricket powder: sausage, pasta, and chocolate brownies. These products were evaluated for nutritional content and quality parameters. To further determine if these products are comparable to conventional food products, understanding how consumers may eat these products in normal eating situations may be a consideration. Previous studies have focused on the food-eating experience and emotions as predictors of consumer acceptance and food-choice behavior [64–67]. Consumer liking and positive emotions were found to increase when consumers deemed food products appropriate [66–68]. Insects, which may invoke negative emotions such as disgust by consumers, may benefit from an evaluation of their products under these conditions, by increasing more positive emotions [68]. Not to mention, there is limited research on complete dishes involving insects, as prior studies have evaluated mainly snack foods (chips, crackers, cookies, etc.), which is a method of easily incorporating insects into food products [33,62,69,70].

As a result, the three food products (sausage, pasta, and brownies) were developed into a three-course meal that US consumers would evaluate. This study was conducted (1) to evaluate different product forms containing insect powders, (2) to understand and identify psychological factors and attitudes that may influence the consumption of insects among US consumers, (3) to evaluate consumer acceptability of insect-based foods with emotional response, and (4) to determine if the product eating experience (food delivery method, plate waste, and satiety) affects the future acceptance of insects as food.

2. Materials and Methods

A two-week sensory test was conducted to evaluate the three different products containing cricket (*Acheta domesticus*) powder. A mild Italian-style fresh sausage, dried pasta, and chocolate brownie were developed to either NOT contain cricket powder (Control) or contain cricket powder (CP) by Ho et al. [63]. The three products were evaluated by utilizing them as major components in a three-course meal that included an appetizer (sausage), entrée (pasta), and dessert (brownie). The CP sausage, pasta, and brownies were developed to contain optimum usage levels of cricket powder (6.23%, 5% and 7%, respectively), to maximize cricket powder replacement without affecting quality parameters [63]. Participants were asked to evaluate the products used in their respective dishes. Figure 1 describes how the sensory test was conducted and the variables (consumer acceptability, emotions, subjective satiety, and plate waste) that were measured throughout the test. All online questionnaires for recruitment and sensory testing were developed and completed using Red Jade sensory software (Martinez, CA, USA). During Week 1, participants were asked to evaluate the three-course meal containing Control products while Week 2 had CP products. Participants were pre-screened to be willing to consume products containing insects over the course of the study; however, for the purpose of this study, they were not informed which products either contained insects or not during each week. Each week was conducted as blind taste testing, where the participants were asked to evaluate each dish without being given information regarding whether the products contained insects or not. After the completion of both weeks of testing, the participants were informed which dishes from each week contained insects (CP) or not (Control). Previous research, which is further explained later, has found that this may affect how consumers perceive the product, which can affect evaluation [51,71].

2.1. Products Tested and Sample Preparation

The products were incorporated into dishes for the three-course meal described in Appendix A. The formulations for both the Control and CP versions of the sausage, pasta, and brownies, and how they were prepared, are described by Ho et al. [63]. The same cricket powder (JR Unique Foods, Thani, Thailand) used in the previous study was used for this study [72]. The three dishes used for the three-course meal were the following: a sausage bruschetta with an onion and pepper topping (Figure A1), pasta with charred-tomato sauce (Figure A2), and brownie trifle dessert with berry sauce (Figure A3). Products and dish components that were not made the day of were made exactly one week before each testing day. All ingredients were purchased from local suppliers/grocery stores within the San Luis Obispo, CA area and stored properly before use.

Since plate waste, later discussed, was a factor in this study, each participant was given substantial amounts of each product/dish that consumers may typically be served in food-eating situations. Serving sizes were determined by increasing the reference weights (g), described by the FDA guidelines for the Reference Amounts Customarily Consumed (RACC) [73], for appetizers, dried pasta, and brownies by 20%. The increased serving sizes based on the RACC for the sausage appetizer, dried pasta, and brownies were targeted to be approximately 105 g, 66 g, and 48 g, respectively. The RACC for dried pasta and brownies was selected as there is no reference amount, in weight, for an entrée or dessert. However, components, aside from the main product components for each dish, were measured to limit differences in weight when served to participants. The average weights of the appetizer, entrée, and dessert dishes (excluding the weight of the serveware) were 119.1 g, 201.3 g, and 78.3 g, respectively.

Products were prepared and served in a manner similar to how consumers may eat them in normal dining situations (Figures A1–A3). Dishes were prepared on premium disposable plates (Munfix, Mezlex Corp, Monroe, NY, USA) and dessert cups (Eupako, Eupako Co., Ltd., Shenzhen, China) for presentation purposes.

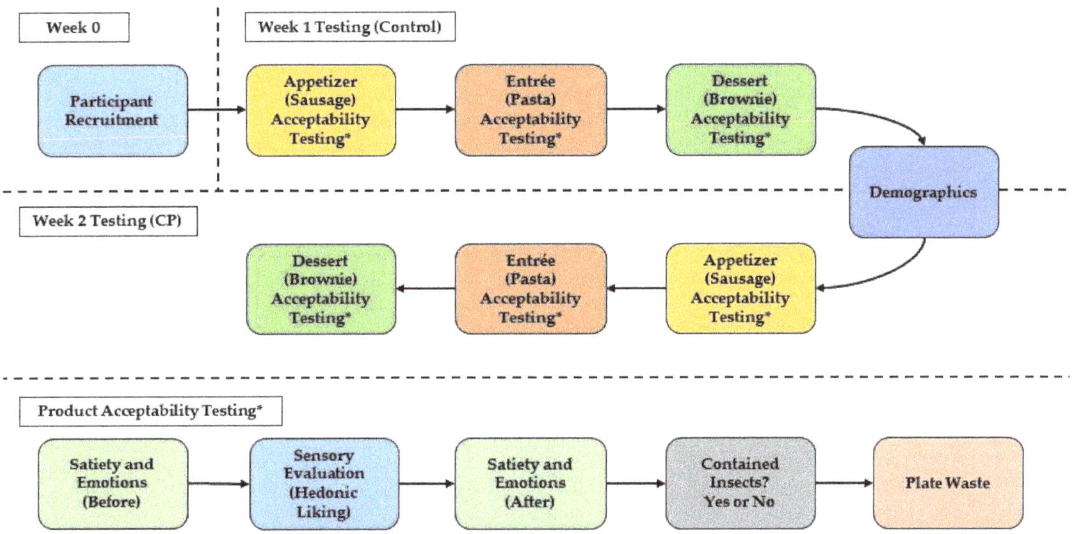

Figure 1. Flow diagram of the sensory test conducted. The Product Acceptability Testing flow diagram describes the factors measured during each product evaluation. * For week 1 and week 2 of acceptability testing for all products evaluated, subjects answered four questionnaires, subjective satiety and emotions before and after consumption, hedonic questionnaire, and answered whether or not the product contained insects. Upon completion of testing, all products were weighed to determine plate waste.

2.1.1. Appetizer Preparation

The appetizer components were developed using the formulations found in Table A1. Prior to testing day, the sausage was prepared and weighed into 40 g patties and vacuum-packed in bags. The sausage patties were placed into a walk-in freezer set at −26.1 °C.

During the day of testing, for the onions and pepper topping, the olive oil and sliced onions were cooked over a stovetop until the onions were caramelized. The roasted red peppers were sliced and added to the caramelized onions and mixed until homogenous. The topping was kept in a hot water bath at 60 °C before serving.

For dish preparation, the vacuum-packed sausages were placed into a hot water bath set at 71.1 °C, using an Anova Precision Cooker (Anova Culinary, San Francisco, CA, USA), for at least 1 h before preparation. The bread was sliced at an angle to be about 1.5 cm in thickness. The bread slices were topped with approximately 1 g of parmesan cheese on each slice and toasted in a kitchen oven using the low broiler setting for 5 min or until the top was golden brown in color. The cooked sausages were removed from the hot water bath and placed into an Avantco sandwich grill (Model P70S, Meridian, ID, USA) set at 176.7 °C (350°F) for 1 1/2 min to sear. The sausages were removed from the grill and each patty was sliced into 6 pieces. To assemble, two slices of toasted bread were placed on a plate, followed by one sausage patty (3 pieces per slice), and 59.1 mL (using a 1/4 US measuring cup) of the onions and peppers topping was placed over it. The dish was served immediately.

2.1.2. Entrée Preparation

Entrée components were prepared using the formulations found in Table A2. Prior to testing day, the pasta was prepared and allowed to dry. The dried pasta was pre-weighed to be approximately 66 g and placed into individual plastic bags for storage. For the tomato sauce, the canned diced tomatoes, garlic, salt, pepper, and the first addition of olive oil were mixed in a large container until homogenous. The tomato mixture was poured into baking sheets lined with aluminum foil prior to baking. The sheets were placed into a Combi Oven (Rational Model CPC 102, Landsberg am Lech, Germany) set at 218.3 °C (425°F) for 30 min, or until the tops of the tomatoes were charred. The sheets were removed from the oven, and the sauce was poured into a large container. The second addition of olive oil was added to this and mixed thoroughly, and the tomato sauce was allowed to cool at ambient temperature until ready to handle. The tomato sauce was sealed in vacuum pack bags and placed into a walk-in freezer set at −26.1 °C prior to testing day.

During sensory testing, the tomato sauce bags were placed into a hot water bath using an Anova Precision Cooker (Anova Culinary, San Francisco, CA, USA) set at 73.9 °C (165°F) for at least 1 h prior to serving. The tomato sauce was then poured into hotel pans that were in a hot water bath set at 73.9 °C (165°F). The pasta was boiled for 4 min, drained, and placed into mixing bowls. Then 59.1 mL (1/4 US measuring cup) of tomato sauce, using a 2-ounce (59.1 mL) ladle, was poured onto the pasta and mixed using a pair of tongs. The dressed pasta was immediately placed onto a plate and topped with the residual tomato sauce from the bowl, 1 g of parmesan cheese, and 1 g of fresh basil. The pasta was served immediately.

2.1.3. Dessert Preparation

Formulations for each dessert component can be found in Table A3. Prior to testing, the brownies were prepared and baked in 18 in by 13 in (45.7 cm by 33.0 cm) sheet pans in a Combi Oven set at 162.8 °C (325°F) for 25 min. The brownies were removed from the oven and allowed to equilibrate to room temperature before storage. The brownies were wrapped in aluminum foil and placed in a walk-in freezer set at −26.1 °C. For the berry sauce, the frozen berries, sugar, and water were mixed in a pot and heated over a stovetop. The berry mixture was allowed to be heated until the sauce reached a Brix of 40 using a refractometer (Atago Model 3810, Atago USA, Inc., Bellevue, WA, USA). The sauce was immediately removed from the stovetop and strained to remove the pulp and seeds. The strained sauce was poured into a 32 oz deli container and placed into a walk-in freezer for storage. The berry sauce was placed into a fridge set at 4 °C a day before testing.

During testing, the brownies were removed from the freezer and allowed to equilibrate to room temperature for at least 2 h. The brownies were cut into 1.5 cm by 1.5 cm squares (brownie bites) for preparation. Heavy cream, sugar, and vanilla extract were mixed using a stand mixer (Model KitchenAid Pro 600 KP26M1X, KitchenAid, Benton Harbor, MI, USA) for the whipped topping. The mixture was whipped until stiff peaks were formed. The whipped topping was poured into Ziploc resealable bags that would be used to pipe the topping onto the dessert.

For assembly, the brownie bites were pre-weighed to be 48 g for each serving. For the berry sauce, 8 mL of sauce was measured using a 10 mL syringe (Care Touch, Westminster, CO, USA) for each serving. Half of the berry sauce (4 mL) was used to line the inside of the dessert cup followed by half of the 48 g of brownie bites. About 1 US tablespoon (14.8 mL) of whipped topping was piped onto the brownie bites followed by the rest of the brownie bites. Another 1 tablespoon of whipped topping was piped onto the second layer of brownies and the remaining 4 mL of berry sauce was drizzled over the whipped topping.

The brownie dessert cups were placed into the refrigerator and/or served immediately when needed.

2.2. Ethics Statement

This research was reviewed and approved by the Cal Poly San Luis Obispo Institutional Review Board (2022-037-CP).

2.3. Participants and Recruitment

A recruitment email was sent through the Cal Poly SLO Sensory consumer database in Red Jade to untrained consumers. The email contained a website link to answer a screener questionnaire. Consumers would be qualified for participation if they had the following: age 18 and over, no declared food allergies and/or intolerances, had a consumption frequency of at least once a month for two of the three products tested (sausage, pasta, and brownies), be willing to consume food products containing insects and non-visible insect powders, and able to come in during the days and times offered for the sensory test. After screening, 164 people were qualified for the 2-week sensory test and were sent a follow-up email with instructions on how to participate. All 164 qualified consumers were recruited, expecting absences over the course of study. Testing occurred on Monday and Wednesday from 1:30 pm to 4:30 pm PST, due to facility availabilities, with six 30-min sessions with 10 people per session. There was a one-week gap between the weeks of testing to act as a washout period for the participants. Qualified participants were informed that the sensory evaluation would involve consuming a three-course meal and were asked to sign up for at least one session offered either Monday or Wednesday each week. Participants would receive a $25 Amazon gift card after the completion of each week. By the end, 108 subjects (n = 108) were used in the study, with 106 (n = 106) of the 108 participating in Week 1 (Control) and 103 of the 108 (n = 103) participating in Week 2 (CP). Overall, 102 of the 108 subjects (n = 102) participated in both weeks of testing.

2.4. Data Collection and Study Design

2.4.1. Testing Room Setting and Procedure

The three-course meal was prepared in the culinary lab kitchen, ready-to-serve, and given to participants who were waiting in the classroom attached to the culinary lab at Cal Poly, San Luis Obispo (Cal Poly SLO). Due to the facilities at Cal Poly SLO, there is a teaching classroom adjacent to the culinary kitchen. Testing was conducted there for food preparation reasons to ensure that the optimum temperatures of the products/dishes were similar to how consumers may be served in a restaurant setting and/or normal eating situations. Though this study did not focus on a restaurant setting, the products were served restaurant-style due to the variables, later explained, that were being measured. The tables in the classroom were arranged at the perimeter of the room so that participants would be facing away from each other during testing. The tables were lined with white plastic table covers (Prestee, Perch HQ, Boston, MA, USA) and participants were given silver-lined plastic silverware (Munfix, Mezlex Corp, Monroe, NY, USA) to eat with. To simulate a food-eating situation when consuming a three-course meal, no saltine crackers were provided over the course of the study and the participants were instructed to drink the water provided between each dish.

When participants checked in for their session, they were given their participant number attached to a table card holder (Hohiya, Hohiya Enterprise Co., Ltd., New Taipei City, Taiwan), which is typically used at some restaurants to indicate their order number. Participants were asked to be seated with only one person per table and with at least one empty seat between each other. Participants were given iPads with iOS 15.4.1 software (Model MP2F2LL/A, Apple, Cupertino, CA, USA) which gave them access to Red Jade website to answer questions for product evaluation. Before testing, participants were informed that they would be evaluating a three-course and that they could eat as much as they wanted of each dish. To indicate that they had finished eating each dish, they were asked to move the cardholder with their participant number from the left to the right side of their table. This would indicate that they were finished eating, and that their plate could be removed from their table and replaced with the next dish item. After all product evaluation was completed and they had finished consuming the last item (dessert), they were allowed to leave the room and had completed that week of testing.

2.4.2. Consumer Acceptability Testing

After being seated and ready to begin, participants were first given the sausage bruschetta appetizer, followed by the pasta entrée, and lastly, the brownie dessert. Each dish was labeled with a 4-digit code that participants had to enter into the Red Jade questionnaire; this would lead them to the next questionnaire page that contained a set of questions pertaining to the dish/product. Once the sample code was entered, participants responded to the questions that evaluated the products for the following attributes on a 9-point hedonic scale (1 = Dislike Extremely to 9 = Like Extremely): appearance, aroma, overall liking, taste, texture, flavor, and aftertaste.

Due to the sausage bruschetta appetizer being a more complex dish than the pasta entrée and brownie dessert, consumers evaluated the sausage and the whole bruschetta for the attributes tested. Consumers only evaluated the pasta product and brownie product from their respective dishes.

After evaluating each product/dish, participants were asked on a binomial scale if the dish contained insects (1 = Yes) or did not contain insects (0 = No). This was to determine if the participants were able to discern which products contained insects over the course of the study.

2.4.3. Emotional Response

Emotional responses were evaluated before and after consuming each product/dish. A modified version of the EsSense25 questionnaire [74], a check-all-that-apply (CATA) list of emotion terms, developed by Serpico et al. [50], was used to evaluate the emotions participants felt. The list of emotion terms was selected to reduce panelist fatigue when answering all of the questions. The original EsSense25 questionnaire includes a list of 25 emotion terms. The modified version used in the study included a 26th emotion term, "daring", from the EsSense Profile® [75], which may be associated with the interest to consume insects [50]. When answering the CATA, participants were asked to select any emotion terms they were feeling during evaluation. If the participants were to select the term "disgusted", the questionnaire would ask them a follow-up question to rate their feeling of disgust (Disgust Rating) [50] on a 5-point scale (1 = Not Disgusted at All; 5 = Extremely Disgusted). Participants were asked to answer the CATA questions before and after the hedonic questions when consuming each of the three dishes tested.

2.4.4. Plate Waste and Satiety

Since the products were served at amounts consumers may typically consume them, plate waste was measured over the course of the study. Before a dish was served to a participant, the dish was labeled with the participant's code, hidden from view, to keep track of which plate belonged to which participant. The dish was weighed as the initial weight using a digital balance scale (Model Adventurer Pro AV3102, Ohaus Corporation, Parsippany, NJ, USA) before being served to the participants. Scales were calibrated with a 1 kg and 2 kg standard stainless-steel weight (Models S-20879 and S-15293, Uline, Pleasant Prairie, WI, USA) before each of testing. Once the participants indicated they had finished done consuming the dish, it was removed and weighed as the final weight. Plate waste was determined using the following equation:

$$\text{Plate Waste }(\%) = \frac{\text{Final Weight of Dish} - \text{Average Plate Weight}}{\text{Initial Weight of Dish} - \text{Average Plate Weight}} \times 100\%$$

where, the average plate weight was determined by weighing 10 plates used for each respective dish. This was performed in triplicate.

Another factor to consider may be the level of satiety participants may have experienced before testing and after consuming each dish. Participants rated their subjective satiety on the Satiety Labeled Intensity Magnitude (SLIM) scale [76], which rated from 0 = Greatest Imaginable Hunger to 100 = Greatest Imaginable Fullness. Participants were asked to rate their satiety before and after the sensory evaluation/consumption of each product/dish.

2.4.5. Sociodemographics, Attitudes and Beliefs towards Entomophagy, Food Neophobia

To limit the fatigue participants may have had while evaluating the products, a website link to access the demographics questionnaire was emailed through Red Jade after the first week of testing. Participants were asked to answer sociodemographic questions which included the following: age group, gender, ethnic background, highest level of education, diet with respect to animal products, and insect consumption familiarity.

To evaluate the participants used in this study and their attitudes toward entomophagy, they were asked to answer the Entomophagy Attitudes Questionnaire (EAQ) [41]. The EAQ evaluated three factors that included five statements regarding the disgust towards insects (EAQ-D), three statements about the interest to consume insects (EAQ-I), and two statements regarding the use of insects as animal feed (EAQ-F). Participants were asked to rate their agreement with these statements on a 7-point scale (1 = Completely disagree; 7 = Completely agree). To the authors' knowledge, the EAQ has not been used to evaluate US consumers [39,41]. This study would also be the first to evaluate the attitudes of consumers who would be consuming insect-based products.

The demographics questionnaire included other questions regarding the consumption of insects. A set of questions developed by Rovai et al. [47] was used to determine which cluster groups (No-thank-yous, Hideaways, Daredevils, and Peekaboos) the participants aligned towards. The set of questions included questions that evaluated the participants' beliefs about consuming insects, lifestyle, and purchasing of food on a 5-point agreement scale (1 = Disagree completely; 5 = Agree completely), and liking of products containing nonvisible insects on a 5-point Likert scale (1 = Dislike very much; 5 = Like very much). Previous research using this set of questions was conducted through online surveys that evaluated the concept of insects as food [46,47]. Therefore, this study wanted to evaluate consumers who were willing to consume insect-based products in person.

Another factor that was measured was food neophobia. Participants were asked to answer the Food Neophobia Scale (FNS) [77] by rating their agreement with ten statements on a 7-point scale (1 = Completely disagree; 7 = Completely agree). By the end of the study, 106 subjects (n = 106) of the 108 responded to the demographics questionnaire.

2.5. Data Analysis

Analyses were conducted using R-project (version 4.1.3, R Foundation, Vienna, Austria) and RStudio (Version 2022.07.2, RStudio, PBC, Boston, MA, USA) to compare differences between the Control and CP product/dishes [78]. All responses collected as binary responses, Yes/No if it contained insects or emotional responses, were rescaled as percentages of participant selection to better analyze the results. The multcomp, multcompView, emmeans R packages and the aov function were used for means being compared, using a two-way ANOVA. The multcomp and multcompView R packages and the prop.test function were used to compare differences in percentages using a two-tailed z-test. A 95% confidence level (α = 0.05) was used for all statistical analyses.

2.5.1. Sociodemographics, Attitudes and Beliefs towards Entomophagy, Food Neophobia

The proportions and means for the sociodemographics, EAQ, and FNS were determined based on the participants' responses. Scores of the food neophilia statements from the FNS were reversed [32,77]. Participants were assigned to a cluster group (No-Thank-

Yous, Hideaways, Daredevils, Peek-a-boos, or Flat) using a typing tool that evaluated the responses to the set of questions that were developed by Rovai et al. [47]. Participants who were not able to be determined to be part of any cluster group were assigned as "Flat". From the cluster assignment, the Hideaways group (57.5%) was found to be the most predominant among the participants used for this study. The mean scores regarding the Hideaways' attitudes and beliefs towards entomophagy and food neophobia were also determined.

2.5.2. Consumer Acceptability, Plate Waste, and Insect Identification

The mean hedonic liking scores and plate waste were analyzed using a two-way ANOVA blocked on Treatment (Control and CP) and Participant. A two-tailed z-test was also conducted on the percentages of participants who answered "Yes" to whether the product/dish contained insects or did not contain insects between the Control and CP versions. For this study, participants were not informed in the beginning which dishes contained insects. All dishes in Week 2 contained insects (CP). Participants who were able to discern that the Week 2 dishes (CP) contained insects and selected "Yes" were identified as the Insect ID subgroup for each dish. Liking scores and plate waste were analyzed for the subgroup using the tests mentioned.

2.5.3. Emotional Responses

Percentages of emotion term selection were determined both before and after consumption of each product/dish. Emotional responses before and after consumption were compared for differences in selection using two-tailed z-tests. Analyses were conducted independently for each week (Control and CP).

The change in emotion term selection (ΔEmotion) was also determined by subtracting the percentage after by the before emotion term selection (After %–Before %) for each participant. The ΔEmotion was analyzed using a two-way ANOVA, blocked on Treatment (Control and CP) and Participant, by evaluating differences in change in emotion between the Control and CP product/dishes. Analysis was also conducted to compare changes in emotion across the three-course meal (Appetizer, Entrée, and Dessert) using a two-way ANOVA, blocked on Product (sausage, pasta, and brownie) and Participant. This was followed by Tukey's means comparisons tests to compare changes in emotions between the three different dishes.

2.5.4. Relationship between Consumer Acceptability, Plate Waste, and Satiety

Mean subjective satiety was analyzed per product/dish to determine changes in satiety before and after the consumption of each dish. Analyses were conducted independently on each dish and each week (Control and CP) using a two-way ANOVA blocked on Product Evaluation (Product Before and After) and Participant.

Pearson's r correlation coefficients were determined to assess a relationship between liking scores, subjective satiety, and plate waste using the Analysis ToolPak from Excel for Microsoft 365 MSO (Version 2210 Build 16.0.15726.20188 Microsoft Corp., Redmond, WA, USA). Correlations were ran using the participant-level data as aggregate scores. Correlation coefficients were determined between the liking scores of the bruschetta, sausage, pasta, and brownies and the following: satiety before consumption, satiety after consumption, ΔSatiety, and plate waste. For each participant, ΔSatiety was determined by subtracting the subjective satiety after score by the subject satiety before score (Satiety After–Satiety Before) for each product/dish evaluated. Satiety variables and plate waste collected from the appetizer were used as aggregate scores for both the bruschetta and sausage attribute liking scores.

Correlation coefficients were determined to be significant ($p < 0.05$) using a table of critical values for Pearson's r correlations using an $\alpha = 0.05$ [79]. Degrees of freedom were determined by subtracting the number of pairings (n) by 2. For the Control and CP independent correlations, there were 4 pairings ($n = 4$) for the bruschetta, sausage, pasta, and brownie, and attribute liking scores to satiety before consumption, satiety after consumption, ΔSatiety, and plate waste. For the Control and CP (combined) correlations there were 8 pairings (n = 8).

3. Results and Discussion

3.1. Sociodemographics

Table 1 displays the sociodemographic information of the 106 of the 108 participants (n = 106) who responded to the demographics questionnaire. It also shows the participants' familiarity with consuming insects, as well as which cluster group they were assigned to. There was a higher proportion of female participants (n = 72; 67.9%) than male participants (n = 33; 31.1%). The 35–44 Age Group was the most predominant (28.3%). The educational background varied among the participants with 77.4% having a bachelor's degree or higher. This may be expected as recruitment was conducted through the university and the local community. The participants were familiar with the consumption of insects with 84.0% stating their familiarity with it. From the cluster group assignments, the Hideaways were found to be the most predominant (57.5%), followed by the Peekaboos group (11.3%), then the Daredevils (10.4%), and lastly the No-Thank-Yous (6.6%). The Hideaways group had the highest proportion, which may be an indicator that the participants recruited may be willing to consume insects in nonvisible forms [47]. This also supports the findings from Rovai et al. [47], that the No-Thank-Yous may reject occasions to consume insects, due to the low representation of this group, since a requirement during recruitment was the willingness to eat visible and nonvisible insects.

Table 1. Sociodemographics, insect familiarity, and cluster assignments of the participants (n = 106) who responded to the demographics questionnaire. Proportions are rounded to the nearest tenth place.

	Proportions (%)	**Base Size n**
Total	-	106
Gender Group		
Male	31.1	33
Female	67.9	72
Prefer not to say	0.9	1
Age Group		
18–24	16.0	17
25–34	18.9	20
35–44	28.3	30
45–54	17.9	19
55–64	17.0	18
65 and Over	1.9	2
Ethnicity		
White or Caucasian	70.8	75
Hispanic or Latino	11.3	12
Black or African American	0	0
Native American or American Indian	0.9	1
Asian or Pacific Islander	15.1	16
Other	1.9	2

Table 1. *Cont.*

	Proportions (%)	Base Size *n*
Highest Education Level		
Some High School	0	0
High School Graduate or Equivalent	0	0
Some College	16.0	17
Trade, Technical or Vocational Training	0.9	1
Associate Degree	5.7	6
Bachelor's Degree	40.6	43
Master's Degree	27.4	29
Professional Degree	0	0
Doctorate Degree	9.4	10
Diet		
No restrictions	91.5	97
Limited animal products	7.5	8
Vegetarian	0.9	1
Vegan	0	0
Other	0	0
Insect Familiarity Statements		
Yes, I have heard of the eating of insects and I know what it means	84.0	89
I have heard of the eating of insects but actually don't know what it means	15.1	16
No, I have never heard of the eating of insects	0.9	1
Cluster Group Assignments		
No-Thank-Yous	6.6	7
Hideaways	57.5	61
Daredevils	10.4	11
Peek-a-boos	11.3	12
Flat	14.2	15

3.2. *Attitudes and Beliefs towards Insects and Food Neophobia*

Table 2 shows the mean scores for the EAQ factors, food neophobia scale, lifestyle assessment, purchasing of food considerations, and liking of products containing nonvisible insects of all participants, as well as the subscores for the Hideaways.

Table 2. Mean scores for the EAQ factors, FNS, belief about consuming insects, purchasing considerations, and liking of products containing nonvisible insects for all participants and the Hideaways subgroup.

Statements/Items	Mean Score	
EAQ—Disgust (EAQ-D) [7A]	**All (*n* = 106)**	**Hideaways (*n* = 61)**
I would be disgusted to eat any dish with insects	2.26	2.38
Thinking about the flavor that a bug might have sickens me	2.79	3.03
If I ate a dish and then came to know that there were insects among the ingredients, I would be disgusted	2.11	2.23
I would avoid eating a dish with insects among the ingredients: even if it was cooked by a famous chef	1.80	1.84
I would be bothered by finding dishes cooked with insects on a restaurant menu	2.03	2.15

Table 2. *Cont.*

Statements/Items	Mean Score	
EAQ—Interest (EAQ-I) [7A]		
I'd be curious to taste a dish with insects, if cooked well	6.13	6.07
In special circumstances: I might try to eat a dish of insects	5.75	5.59
At a dinner with friends I would try new foods prepared with insect flour	6.14	6.15
EAQ—Animal Feed (EAQ-F) [7A]		
Using insects as feed is a good way of producing meat	5.47	5.36
I think it is fine to give insect-based to fish that are farmed for human consumption	6.14	6.25
Food Neophobia [7A]		
At dinner parties, I will try a new food (R)	1.25	1.25
Ethnic food looks too weird to eat	1.65	1.61
I am afraid to eat things I have never had before	2.14	2.18
I am constantly sampling new and different foods (R)	2.24	2.44
I am very particular about the foods I will eat (R)	3.03	3.21
I do not trust new foods	2.18	2.28
I like foods from different countries (R)	1.58	1.70
I like to try new ethnic restaurants (R)	1.73	1.82
I will eat almost anything (R)	2.29	2.56
If I do not know what is in a food, I won't try it	2.73	2.84
Beliefs about Consuming Insects [5A,*]		
I would consume a live insect	1.96	1.66
I would consume a whole roasted insect	2.99	2.46
I would consume a whole insect incorporated into a dish	3.38	2.89
I would consume a packaged food that contains WHOLE insects	2.81	2.28
I would consume a packaged food that contains NONVISIBLE insects	4.42	4.38
I would support my family members or children eating insects	4.42	4.39
I think more people should consume insects	3.88	3.79
I think that insects should be used as livestock feed	4.28	4.26
I would consume livestock that was fed insects	4.64	4.64
Lifestyle Assessment [5A,*]		
I exercise regularly	3.99	4.02
I enjoy eating foods and ingredients from other cultures	4.65	4.64
I and or other members of my household have food allergies that prevent us from eating certain foods	1.72	1.80
I am willing to pay more for all natural or organic foods	3.52	3.56
I am concerned about the environmental impact of the foods I eat	3.82	3.74
I mostly eat a plant-based diet	2.32	2.28
I am willing to modify my diet in order to be healthier	4.22	4.26
I am always looking for new foods to help me live a healthier life	4.11	4.07
It is difficult for me to understand what products are truly nutritious and or healthy for me and my family	2.43	2.48
I am loyal to brands that I like	3.67	3.66
I eat raw fish sushi	4.26	4.10

Table 2. *Cont.*

Statements/Items	Mean Score	
Purchasing of Food Considerations [5A,*]		
Cost	4.38	4.39
Dietary restrictions	2.24	2.30
Convenience	3.90	3.85
Sustainable ingredients	3.54	3.49
Organic	3.11	3.15
Non-GMO	2.95	2.95
Taste	4.75	4.80
Nutritional value	4.15	4.16
Sustainably produced	3.47	3.52
Local	3.73	3.70
Fresh	4.29	4.31
Minimally processed	3.68	3.72
Natural	3.62	3.54
Preservative free	3.37	3.36
Brand reputation	3.38	3.33
Products Containing Nonvisible Insects [5L,*]		
Chip, cracker, or puffed snack	4.28	4.28
Pasta product	4.24	4.30
Dessert product, cookie or cake	4.08	4.00
Beef jerky type product	3.94	3.87
Protein shake or powder	4.18	4.20
Cereal granola	4.07	3.98
Baked product, bread	4.16	4.05
Meat analog, sausage or ground meat	4.20	4.15
Protein bar	4.25	4.28

[7A] Participants were asked to rate on a 7-point agreement scale (1 = Completely Disagree; 7 = Completely Agree). [5A] Participants were asked to rate on a 5-point agreement scale (1 = Completely Disagree; 5 = Completely Agree). [5L] Participants were asked to rate on a 5-point Likert scale (1 = Dislike very much; 5 = Like very much). (R) Food neophilia statement scores from the FNS were reversed. * The set of statements were developed by Rovai et al. [47].

3.2.1. Entomophagy Attitudes Questionnaire and Food Neophobia

For the EAQ scores, the participants self-reported, on a 7-point scale, low disgust (mean scores of 1.80–2.79 for EAQ-D statements), high interest to consume insects (mean scores of 5.75–6.14 for EAQ-I statements), and high interest to use insects as animal feed (mean scores of 5.47 and 6.14 for EAQ-F statements). When compared to the Hideaways' subscores, the mean scores for each statement were comparable. The largest difference of 0.24 was found for the "Thinking about the flavor that a bug might have sickens me (EAQ-D)" agreement score between all participants and the Hideaways. For the FNS scores, all participants and the Hideaways subgroup self-reported low food neophobia with the highest rated statement being "I am very particular about the foods I will eat (reversed)" at 3.03 and 3.21 (on a 7-point scale), respectively.

Disgust, food neophobia, and interest/intention toward insects may play key roles in the future acceptance of insect-based foods in the Western market. Both disgust and food neophobia have been heavily studied as negative predictors of the intention to consume insects [33,36,37,70,80]. In Western countries, disgust and/or food neophobia may lead to an aversion to consuming insects. To reduce the negative perception of consumers, exposure and familiarization to the consumption of insects may increase interest. Previous studies have found that providing information about the benefits of insects or consumers tasting insects for the first time may increase the intention and/or willingness to consume them in the future [47,50,52,56,57,81–83]. Evaluating these three factors may provide a scope of how certain consumer groups may perceive insect-based products once introduced into the market.

The EAQ was used in this study as it has been validated to relate disgust, food neophobia, and interest across consumer groups [31]. Verneau et al. [39] had previously implemented the instrument in China. They found that consumers with a history of consuming insects self-reported low disgust with the intent to consume insects. La Barbera et al. [41] had also used the same instrument on Chilean consumers, which predicted their intent to consume insects as well as incorporate insects into their diets. This study was conducted to evaluate the use of the EAQ on US consumers. From the results, US consumers, from the San Luis Obispo, California area, display low disgust and food neophobia but high interest to consume insects. As a result, this suggests that the participants in this study may be potential consumers of insects and/or insect-based products in the near future.

3.2.2. Beliefs, Lifestyle, Purchasing Considerations, and Product Liking

For the beliefs about consuming insects, the mean agreement scores for the statements varied between 1.96 to 4.64, for all participants, on a 5-point scale. However, the highest mean scores (above 4 mean score) were for the statements "I would consume a packaged food that contains NONVISIBLE insects", "I would support my family members or children eating insects", "I think that insects should be used as livestock feed", and "I would consume livestock that was fed insects".

This may imply that the participants would be more willing to consume insects in nonvisible forms (powders) than visible forms (whole). This further supports previous findings that some consumers may be more willing to consume nonvisible insects integrated into food products than whole insects [30,56,57,84]. This can be further supported as there was a difference in mean scores between all participants and the Hideaways for the statements "I would consume a whole roasted insect" and " ... whole insect incorporated into a dish" of 0.53 and 0.49, respectively. The mean score from all participants may be influenced by the Peek-a-boos and Daredevils from the study. Rovai et al. [47] suggested that the Peek-a-boos and Daredevils groups may be more willing to adopt insect-based food products, if this included nonvisible forms, whole insects, and/or foods containing whole insects, than the Hideaways.

The mean scores regarding feeding livestock with insects further support the high mean scores for the EAQ-F statements suggesting that insects may be an alternative animal feed to supply livestock production. Indirect entomophagy, the use of insects for animal rearing, may be viable to incorporate insects into the food supply [85]. Consumers who have self-reported interest to use insects for animal meat production may suggest an intention to consume food products using indirect entomophagy in the future [31,39,41].

From the lifestyle assessment, the participants self-reported high openness to trying new foods from other cultures (4.65), high value for healthy practices including "exercising regularly", "modifying diets", and "looking for new foods" (mean scores 3.99, 4.22, and 4.11, respectively) and high consumption of raw fish sushi (4.26). Ruby and Rozin [82] found that high sushi intake may be an indicator of the future acceptance of insects among US and Indian consumers. This study also evaluated this factor, which may predict that US consumers, in the San Luis Obispo area, may be more willing to consume insects in the near future.

For purchasing food considerations, the participants highly rated (above 4 mean score) "cost", "taste", "nutritional value", and "fresh" on a 5-point scale. "Convenience" may also be a consideration due to its high mean score (3.90). When compared to the Hideaways, the largest difference in mean scores was 0.08 between all participants and the subgroup. Pollard et al. [86], when determining factors that affect food choice of vegetable and fruit intake, suggested that food choice is a complex process that involves sensory appeal, cost, familiarity and habit, time constraints, and other factors. Placentino et al. [87] found that Italian athletes were willing to taste cricket-enriched energy bars due to the curiosity in their protein content and its textural properties. As a result, insect-based food products may need to meet these considerations to encourage their acceptance.

When evaluating products containing nonvisible insects, the participants rated the products with high liking, with "beef jerky type product" rated the lowest (3.94) and "chip, cracker, or puffed snack" rated the highest (4.28) on a 5-point scale. The mean liking scores between all participants and the Hideaways were comparable with a mean score difference of 0.11. Food appropriateness may be a consideration when incorporating insects into food products. Ardoin and Prinyawiwatkul [55] found that bakery/cereal and snack products, which include bread, bakery products such as cookies, and protein bars were deemed appropriate for the incorporation of cricket powder among US consumers. However, Lombardi et al. [34] found that pasta, a staple food, was deemed more appropriate by Italian students when compared to cookies. Meat products containing insects have also been evaluated for whether consumers are willing to consume them or if they are deemed appropriate [54,83,88,89]. Overall, food appropriateness, while not measured in this study, may be a factor when developing insect-based food products, as positive food sensory experiences with each associated product may encourage future acceptance [53,54]. From the results, the high liking across all the products evaluated by the participants may suggest their openness to incorporating nonvisible insects into these products.

3.3. Consumer Acceptability, Plate Waste, Insect Identification

Table 3 shows the sensory evaluation results for all participants and the Insect ID subgroup along with plate waste and the percentage of participants who selected "Yes" if the dish contained insects.

Table 3. Mean attribute liking scores, percentage of participants who selected "Yes" the dish contained insects, and plate waste for the dishes/products evaluated (bruschetta, sausage, pasta, and brownies) for all participants and the Insect ID subgroup. Base sizes (n) of participants that evaluated each product/dish are also shown. Parameters colored (blue for ALL participants, yellow for the Insect ID subgroup) and denoted were found to be significantly different ($p < 0.05$) using a two-way ANOVA, blocked on Treatment and Participants, for hedonic liking and plate waste. A two-tailed z-test was used for the "Yes, it contained insects" percentage.

Product and Attributes	All Participants		Insect ID Subgroup	
	Control	CP	Control	CP
Bruschetta	n = 106	n = 103	n = 48	
Appearance [2]	6.80	6.67	6.83	6.40
Aroma [1,2]	7.42	7.13	7.52	6.98
Overall [1,2]	7.57	7.14	7.65	6.96
Taste [1,2]	7.56	7.09	7.58	6.83
Texture [2]	7.20	6.97	7.31	6.62
Flavor [1,2]	7.49	7.08	7.56	6.87
Aftertaste [1,2]	7.11	6.50	6.98	6.29
Contained Insects? Yes (%) [1]	22	47	-	-
Plate Waste (%)	31.0	32.7	31.1	33.4
Sausage			n = 48	
Appearance [1,2]	6.53	5.96	6.67	5.90
Aroma [1,2]	7.20	6.76	7.33	6.69
Overall [1,2]	7.33	6.50	7.42	6.21
Taste [1,2]	7.31	6.71	7.42	6.40
Texture [1,2]	7.18	6.34	7.23	6.00
Flavor [1,2]	7.29	6.56	7.19	6.23
Aftertaste [1,2]	6.80	6.11	6.75	5.83

Table 3. *Cont.*

Product and Attributes	All Participants		Insect ID Subgroup	
	Control	CP	Control	CP
Pasta			$n = 32$	
Appearance	6.97	6.98	6.88	6.72
Aroma	7.01	7.05	6.69	6.56
Overall [1]	6.60	6.90	6.34	6.50
Taste	6.55	6.74	6.34	6.34
Texture	6.67	6.70	6.41	6.25
Flavor	6.69	6.82	6.69	6.56
Aftertaste	6.34	6.44	6.32	6.22
Contained Insects? Yes (%)	29	31	-	-
Plate Waste (%) [2]	40.9	44.2	32	39.4
Brownie			$n = 43$	
Appearance	7.45	7.42	7.54	7.44
Aroma [1]	7.42	7.67	7.69	7.81
Overall [2]	7.87	7.65	7.96	7.30
Taste [1,2]	7.78	7.41	7.92	7.14
Texture [1,2]	7.33	6.98	7.43	6.77
Flavor [2]	7.75	7.50	7.90	7.05
Aftertaste [1,2]	7.36	6.89	7.34	6.58
Contained Insects? Yes (%)	29	42	-	-
Plate Waste (%) [2]	27.5	30.7	23.1	30.7

[1] A significant difference was found between the Control and CP versions of that product by ALL participants from the study ($p < 0.05$). [2] A significant difference was found between the Control and CP versions of that product by participants from the Insect ID subgroup per dish/product ($p < 0.05$).

3.3.1. Sausage Bruschetta Appetizer

For all participants, the Control bruschetta was rated significantly higher in liking than the CP version for aroma, overall liking, taste, flavor, and aftertaste ($p < 0.05$). No differences were found in liking for appearance and texture ($p > 0.05$). For the sausage product, there was a significant difference in liking between the Control and CP sausages for all attributes ($p < 0.05$). A higher percentage of participants indicated that the CP appetizer contained insects in the dish ($p < 0.05$). For plate waste, no differences were found when comparing the Control and CP appetizer ($p > 0.05$). Overall, the mean liking scores for the Control and CP bruschetta were from 6.80 to 7.57 and 6.50 to 7.14, respectively, across all attributes. The mean liking scores across all attributes for the Control and CP sausage were from 6.53 to 7.33 and 5.96 to 6.76, respectively.

Previous studies have found that the incorporation of insect powders in meat/sausage products reduced consumer liking. Caparros Megido et al. [83] found that beef burger patties containing mealworms (*Tenebrio molitor*) were rated lower in overall liking and other attributes by Belgian students, when compared to all-beef patties; however, the mealworm beef patties were rated higher than legume-based patties. Choi et al. [90] also found that consumer acceptance decreased as mealworm (*Tenebrio molitor*) replacement increased in pork frankfurters. For Orthoptera species, Cruz-López et al. [91] found that increasing grasshopper (*Sphenarium purpurascens*) flour decreased the overall liking of pork sausages. However, another study found that sensory characteristics did not change between different levels of mealworm replacement in pork patties except for juiciness [92].

When incorporating nonmeat ingredients into meat products such as sausages, previous studies have found that consumer acceptance may be affected [93]. Plant byproducts, such as okara, have been found to decrease sensory scores in beef patties and burgers [94,95]. However, other studies using different ingredients, such as pulses and mushroom, either maintain sensory scores or even increased them [93,96]. As a result, considerations when utilizing a nonmeat ingredient in meat products may include how it affects the overall product.

The decrease in attribute liking of the CP sausage may be attributed to the changes in the quality of the sausage product. Ho et al. [63] found that the CP sausage became darker than the Control and textural changes were found for firmness and springiness. When using cricket powder, Smarzyński et al. [97] found that the addition of cricket powder decreased the lightness of pork pâtés, and shifted the color more towards blue. Previous research using Orthoptera powders (*Acheta domesticus* and *Sphenarium purpurascens*) changed the textural properties in cased sausages [91,98]. A decrease in attributes such as taste, flavor, and texture may be the result of the cricket powder. The flavor of cricket powder has been described as woody, nutty, and even similar to pet food [89,99,100]. Not to mention, the exoskeleton of insects may lead to a gritty texture [98,99]. However, Cruz-López et al. [91] and Keto et al. [101] suggest that the negative effects on these attributes may be offset with the addition of spices and flavorings that may improve the acceptance of meat products containing insects. From the results, the sausage and bruschetta were acceptable for the participants (6.50 and 7.14, respectively, in overall liking). In addition, the sausage product was served as a component in the bruschetta appetizer and may have had an effect as no differences were found in liking scores for appearance and texture ($p > 0.05$). Therefore, when using insect powders in meat products, the negative changes in sensory characteristics may be mitigated through formulation and product delivery methods.

3.3.2. Pasta Entrée

The CP pasta entrée rated higher ($p < 0.05$) in overall liking than the Control by all participants. No differences were found for the other hedonic attributes, "Yes, it contained insects" percentage, and plate waste ($p > 0.05$). Overall, the mean liking scores for the Control and CP pasta were from 6.34 to 7.01 and 6.44 to 7.05, respectively, across all attributes.

Previous studies have evaluated consumer liking of pasta containing insect ingredients [102–105]. Duda et al. [102] found that 5% usage of cricket powder in pasta was met with high acceptance by consumers. Results from this study may also support this as the participants scored the CP pasta, which contains 5% cricket powder (Table A2), higher in overall liking than the Control ($p < 0.05$). However, results from Duda et al. [102] found that insect powders may decrease acceptance as insect powder usage increases. Çabuk et al. [104] and Jakab et al. [105] support this, as the sensory characteristics were reduced with the addition of insect powders in egg pasta and millet pasta, respectively. Though sensory liking may decrease, the use of insect powders in pasta, especially in healthier alternatives, may provide new market opportunities. Biro et al. [103] found that Hungarian students preferred buckwheat pasta containing silkworm (*Bombyx mori*) powder more than the control buckwheat pasta.

To further support this, previous research has found that insect-based pasta has been found to resemble whole wheat pasta which is considered a healthier option to conventional pasta [102,106]. This may be a result of the cricket powder shifting the color of pasta to be dark in appearance [102,106]. To mitigate this, Ho et al. [63] developed the Control and CP pasta formulations with caramel coloring to limit noticeable differences in appearance that the participants may detect. Lombardi et al. [34] mentioned that pasta is considered a staple food with a low hedonic value. However, this can be mitigated through pasta preparation such as the addition of dressings and sauces [34,83,107]. From the results, both the Control and CP pasta were rated as acceptable by the participants. Therefore, the use of insects may provide opportunities to develop health-oriented pasta products.

3.3.3. Brownie Dessert

For all participants, the CP brownie was rated significantly higher in liking for aroma than the Control ($p < 0.05$). However, the CP dessert was rated significantly lower in taste, texture, and aftertaste than the Control ($p < 0.05$). No differences were found for the other attributes (appearance and flavor), "Yes, it contained insects" percentage, and plate waste

($p > 0.05$). Overall, the mean liking scores for the Control and CP brownies were from 7.33 to 7.87 and 6.89 to 7.67, respectively, across all attributes evaluated.

The use of cricket powder in brownies and other sweet bakery products may be a viable application. As mentioned, cricket powder has been found to have a nutty and woody flavor [99,100]. From a product development perspective, using cricket powder in chocolate-based products, such as brownies, may be a method to utilize the sensory characteristics of cricket and other insect powders. Castro Delgado et al. [108] found that cricket powder in chocolate chip cookies had no effect on liking when compared to their control among US consumers. The use of chocolate may mitigate the negative effects of insect powders on bakery products, as it may reduce the shift in color and possibly lead to a decrease in liking in appearance/color [109,110].

Gurdian et al. [52] evaluated the cricket powder replacement in brownies, and the brownies containing cricket powder were rated lower in liking than the brownies that did not. However, Gurdian et al. [52] suggested that this may be due to the effect of the formulation of the brownies. From the results, although there were differences in attribute liking, the CP brownie may be as acceptable as the Control; this may be due to the formulation as well as the product delivery method. Due to the gritty texture of cricket powder [98,99], Ho et al. [63] further ground the cricket powder used in the formulation. Cricket powder has been found to change the textural properties of bakery products which may be due to the prevention of gluten formation and/or changes in amylose/amylopectin proportions [109–111]. As a result, methods to mitigate the effects of insect powders on bakery products may be a consideration, including formulation.

3.3.4. Insect ID Subgroup

For the bruschetta and sausage, the Insect ID subgroup ($n = 48$) rated the Control higher in liking than the CP version for all attributes tested ($p < 0.05$). There was no difference for plate waste ($p > 0.05$). For the pasta, no differences in attribute liking were found ($p > 0.05$), but there was a higher amount of plate waste ($p < 0.05$) for the CP pasta than the Control by the subgroup ($n = 32$). The Control brownie was rated higher in liking ($p < 0.05$) for overall liking, taste, texture, flavor, and aftertaste by the Insect ID subgroup ($n = 43$). No differences in liking were found for appearance and aroma ($p > 0.05$). There was also higher plate waste ($p < 0.05$) collected from the CP brownie than the Control by the subgroup.

There were more differences in attributes evaluated between the bruschetta appetizer and brownies within the Insect ID subgroup ($p < 0.05$). Not to mention, the differences in mean liking scores of the Control and CP bruschetta, sausage, and brownies for the Insect ID subgroup were comparatively higher than by all participants. The largest difference in mean liking scores for the bruschetta, when evaluated by all participants and the Insect ID subgroup, was 0.47 to 0.75, respectively, for taste. For the sausage, the largest difference was 0.84 (all participants) to 1.23 (Insect ID subgroup) for texture. For the brownie, the largest difference between all participants and the Insect ID subgroup was 0.25 to 0.85, respectively, for flavor. However, there was no difference found in liking for flavor when evaluated by all participants ($p > 0.05$).

This study did not inform the participants which products/dishes did or did not contain insects. Previous research has found that informing consumers that products that contained insects may lead to lower expected liking scores prior to tasting [51,52,54,89]; however, expected liking may not necessarily affect future acceptance. Schouteten et al. [89] found that informed conditions did not affect the sensory profile between beef burgers and insect-based burgers when evaluated, and the low acceptance of the insect-based burger was due to low sensory quality. Although there were more differences in attribute liking for the Insect ID subgroups, the subsamples may be too small ($n = 48$ for the bruschetta and sausage; $n = 43$ for the brownie). Not to mention, the Insect ID subgroup that selected "Yes, it contained insects" may have selected the correct answer by chance. For the purposes of this study, the authors wanted to determine whether the participants were able to detect if the products contained insects. From the results, most of the participants were unaware that the products/dishes evaluated had even contained insects. At the end of the study, some of the participants commented that they did not believe any of the products tested even contained any insects.

3.4. *Emotional Response*

3.4.1. Emotional Response before and after Evaluation

Figure 2 shows the emotion term selection before and after evaluating each dish. For the Control appetizer, there was a significant increase ($p < 0.05$) in the selection of positive emotions by participants for happy, pleasant, satisfied, and warm after consumption. There was a significant decrease ($p < 0.05$) in emotion term selection by participants for active (positive) and worried (negative) after consumption. For the CP appetizer, there was an increase ($p < 0.05$) in the selection of the satisfied and warm (positive) as well as understanding (neutral) after consumption. There was a significant decrease ($p < 0.05$) in the selection for bored (neutral) for the CP (from 7% to 0%); however, there was also a decrease in the selection of bored for the Control appetizer (from 7% to 1%), though no significance was found ($p > 0.05$).

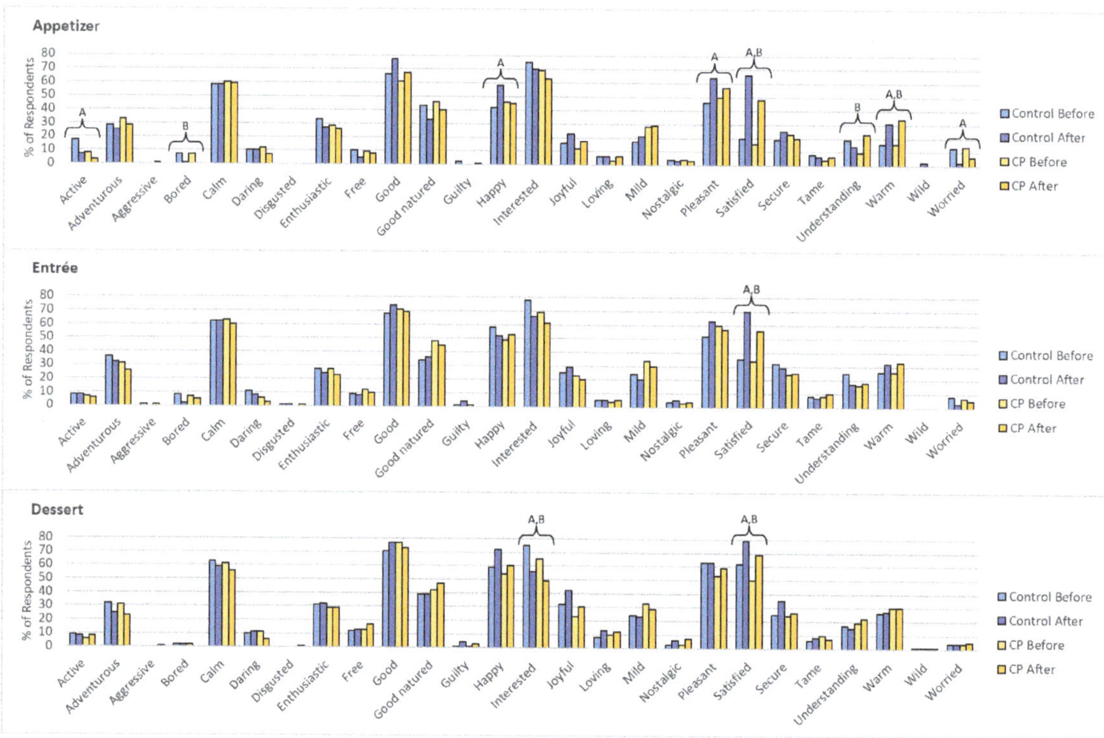

Figure 2. The percentage of participants who selected an emotion term before and after each dish evaluation. Bar graphs were colored blue for the Control dishes ($n = 106$) and yellow for the CP dishes ($n = 103$). Emotions that were denoted with a number were found to have a significant difference between the percentages of participants who selected the emotion term before and after consumption, using a two-tailed z-test ($p < 0.05$), for (A) the Control dish and (B) the CP dish.

For the entrée, there was a significant increase ($p < 0.05$) in the selection of the term satisfied (positive) after consumption for both the Control and CP versions. No change in selection was found for the other emotion terms ($p > 0.05$).

When evaluating the dessert, there was an increase ($p < 0.05$) in selection for the positive terms interested and satisfied. No change in selection was found for the other emotion terms for the dessert ($p > 0.05$).

Measuring emotional responses elicited by food may expand findings during sensory evaluation to better predict the future acceptance of food products [112–115]. The use of emotion terms/lexicons has suggested that the selection of positive (happy, active, interested, satisfied, warm, etc.) and neutral/unclassified (understanding, daring, wild) emotion terms were associated with highly liked products, while lowly liked products were associated with negative emotions (disgusted, bored, and worried) [74,115].

From the results, the participant selected positive emotions such as calm, good, happy, interested, pleasant, satisfied, and warm. The products/dishes may have elicited more positive emotions than negative emotions after consumption of each dish. This can be supported as the selection of satisfied increased ($p < 0.05$) after the consumption/evaluation of all dishes. This emotion may relate to the food satisfaction participants may perceive in certain eating situations, which may relate to food appropriateness and acceptance [64]. To the authors' knowledge, the relation between emotions such as satisfied and food satisfaction has not been studied, something which may give more insight in the future acceptance of new food products such as insects. However, the selection of the term interested decreased after the evaluation of all the dishes, which was significant for the dessert ($p < 0.05$) but not for the appetizer and entrée ($p > 0.05$). Overall, positive emotions were elicited throughout the study which may support the high liking scores of the products/dishes.

Regarding insects as food, it has been found that insects elicit negative emotions such as disgust when people are aware that the product contains insects [84]. From this study, the products/dishes evaluated showed limited negative emotions throughout the courses served in this study. There was little to no selection of negative emotion terms, including disgusted and bored by participants when evaluating all products/dishes. At most, 1% of participants self-reported the term disgusted for the Control and CP entrée and CP dessert. Schouteten et al. [89] found that when participants were informed that the insect-based burgers contained insects, negative emotions, such as disgust, were reported. Le Goff and Delarue [71] found that participants expressed more negative facial expressions when they were informed that the chip products were "enriched with cricket protein" than when they were not. This may be due to information regarding food products that may elicit negative emotions associated with that product, such as insects [84,113,114]. However, negative emotions consumers may experience have been found to be reduced once they taste the product [71,89]. This can be further supported as there was a decrease in the selection of the term worried ($p < 0.05$) after the consumption of the Control appetizer. Emotion terms, such as worried, have been found to be associated with novel foods due to concerns of safety and feelings of distrust [70]. The participants may have experienced worry when evaluating the Control appetizer as this was the first product/dish being evaluated in the overall study with the participants unaware of which products may contain insects.

3.4.2. Changes in Emotional Response

To better understand the emotions the participants self-reported, Figure 3 shows the change in emotion term selection throughout the three-course meal for each week (Control and CP). For the Control dishes, there was a significant difference between the changes in emotion selection for active, happy, pleasant, pleasant, satisfied, secure, understanding, warm, and worried ($p < 0.05$). With the CP dishes, changes in emotion selection for the terms active, understanding, warm, and worried were significantly different between the dishes ($p < 0.05$). For the appetizer, the change in the selection of happy was positive for the Control, while it was negative for the CP version ($p < 0.05$). The change in selection of satisfied was greater for the Control appetizer than the CP appetizer ($p < 0.05$). For the entrée, the change in the selection of pleasant was positive for the Control, while it was negative for the CP entrée ($p < 0.05$). For both the appetizer and entrée, the change in selection for understanding was more positive for both CP versions, while it was negative for both Controls.

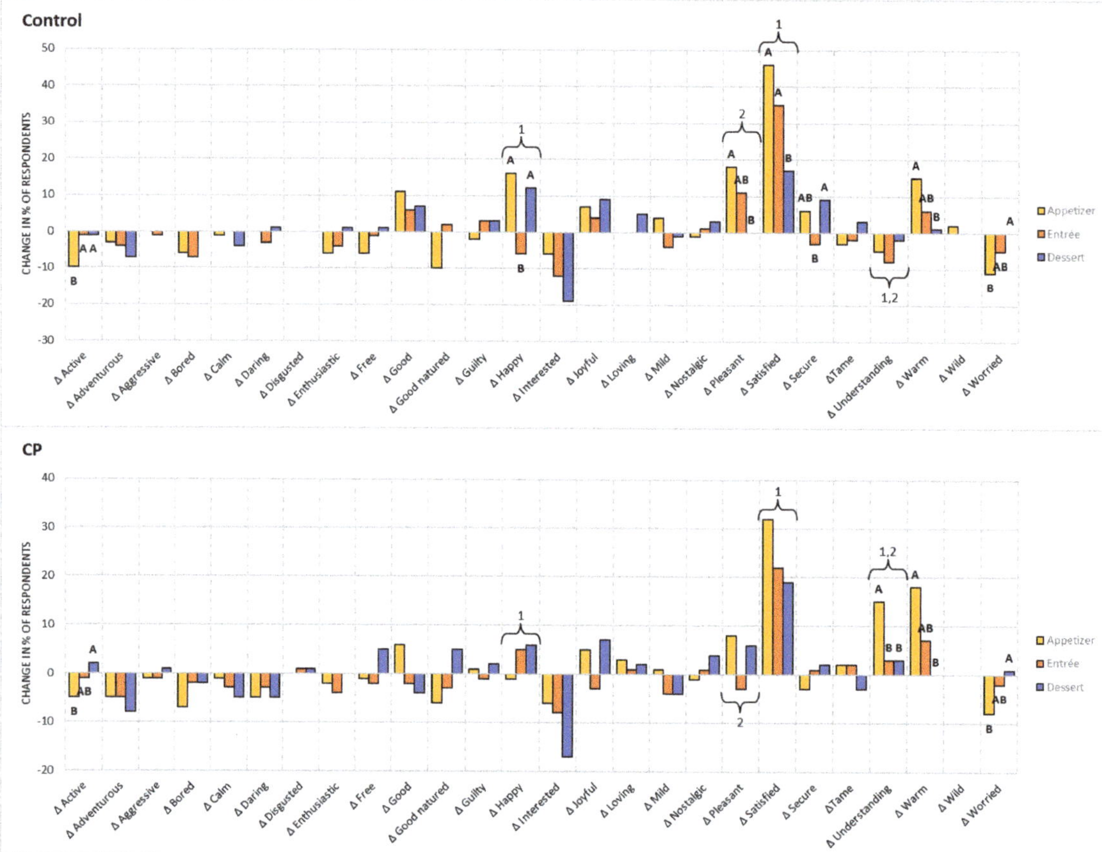

Figure 3. ΔEmotion for each emotion term before and after evaluating a dish for both the Control and CP versions. Emotions terms denoted by a number (shown on both versions) were found to have a significant difference ($p < 0.05$) in the change in emotion term selection between the Control and CP, with (1) being the Appetizer and (2) the Entrée, using a two-way ANOVA blocked on Treatment (Control and CP) and Participant. No differences were found for the change in emotion selection between the Control and CP ($p > 0.05$). Control and/or CP dishes with differentiating letters, for each emotion term, were found to be significantly different ($p < 0.05$) using a two-way ANOVA, blocked on Product (sausage, pasta, and brownie) and Participant, followed by Tukey's means comparisons test.

ΔEmotion was analyzed to understand and visualize emotional responses before and after the consumption of each dish. The selection of positive emotions increased for good, happy, pleasant, satisfied, secure, and warm for the Control and CP dishes. However, the change in the selection of positive emotions was less prominent, with a decrease in the selection of emotions terms, for the CP when compared to Control. This can be supported due to a difference between the changes in the selection of happy and satisfied ($p < 0.05$). King et al. [116] suggested that the number of samples tested may affect the intensity of certain emotion terms including interested and satisfied. Not to mention, when evaluating facial expressions, de Wijk et al. [117] found that repeated exposure when evaluating the same products may affect emotional responses over time. Limitations of using emotional response measurements, such as EsSense Profile® or other emotion term lists, may include heavy cognitive involvement by the participants, which may lead to fatigue during product evaluation. Although there was no difference ($p > 0.05$), there was a negative change in the selection of interested throughout the three-course meal for both the Control and CP versions. The feeling of interest may be associated with the curiosity the consumers experienced once the dish was served to them. Once the participants had consumed and evaluated the product, the loss of interest in the dish may be due to that associated curiosity being fulfilled.

For the CP dishes, there was an increase in the selection of understanding while there was a decrease for the Control dishes. This occurrence was the highest for the CP appetizer ($p < 0.05$), which may support that the participants were able to detect that the dish did contain insects. However, the increase in understanding, an unclassified/neutral emotion, may be an indicator that the CP dishes may have not elicited negative emotions. This can be further supported, as there was a decrease in the selection of worried by the participants.

The emotion of satisfied may be of interest as the selection of this term was the most prominent among all the terms selected. In a food service setting, previous studies have found that feelings of satisfaction and contentment when consuming foods may be factors for food choice and consumption [65,118]. When evaluating food products through full-size servings, fulfilment-related emotions, such as satisfaction, may need to be further studied to understand the future acceptance of food products. From the results, there was a decrease in the change of satisfied over the three-course meals ($p < 0.05$). This may relate to the changes in perceived hunger/satiety, which is later explained, that consumers may experience when consuming multiple dishes in a normal eating situation. As a result, further research may be necessary to understand emotional responses during eating situations, such as a three-course meal.

Overall, throughout the three-course meal, the participants selected more positive emotion terms when compared to the negative terms provided for both the Control and CP dishes. Previous research supports that reducing negative emotions associated with insect-based food products but increasing positive emotions may increase future consumption of insects [27,50,84]. The frequency of positive emotions when evaluating the products may support that the CP products were as acceptable as the Control products.

3.5. Relationship between Consumer Acceptability, Plate Waste, and Satiety

3.5.1. Satiety

Figure 4 shows the subjective satiety reported by the participants for each week during product/dish testing during a three-course meal. Overall, there was a significant difference in reported mean satiety before and after the consumption/evaluation of each dish for both the Control and CP ($p < 0.05$). From the results, the participants' perceived satiety was increasing, suggesting that the participants' perceived fullness was increasing, or perceived hunger was decreasing, as they continued to consume each dish.

Satiety, or perceived/subjective satiety, has been studied as a driver of liking and food intake for food product development [119,120]. Satiety may be affected by the sensory properties of the product, including texture, as well as the serving size, and whether the product is large-sized or bite-size [121,122]. The change in perceived satiety after consumption has been found to be a predictor of purchase intent and acceptance of a product [119]. To the authors' knowledge, there is limited research on measuring the acceptability of products, especially in an eating situation where participants are served in a three-course meal. Overall, subjective/perceived satiety increased over the course of product/dish evaluation, which is to be expected as the participants were allowed to eat as much as they liked.

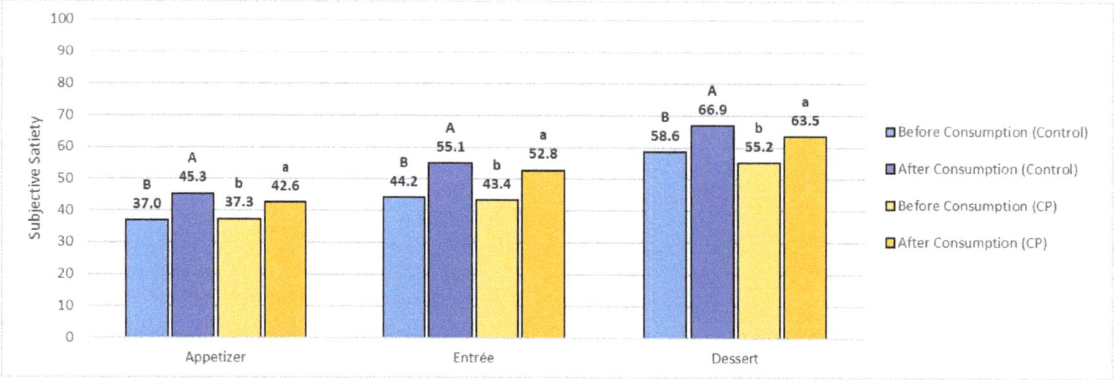

Figure 4. Mean subjective satiety reported by the participants, using the SLIM scale [76], before and after the consumption of each dish. Bar graphs were colored blue for the Control dishes and yellow for the CP dishes. Differentiating letters (uppercase for Control, lowercase for CP) shows a significant difference was found between the before and after subjective satiety using a two-way ANOVA, blocked on Product Evaluation (Product Before and After) and Participant, for each dish ($p < 0.05$).

3.5.2. Consumer Liking, Satiety, and Plate Waste

Table 4 shows the correlation coefficients between the attribute liking scores (including all products/dishes), satiety variables, and plate waste. For the Control, there were significant negative relationships between flavor liking and ΔSatiety, as well as attribute liking scores (overall liking, taste, and texture) and plate waste ($p < 0.05$). Although no significance was found ($p > 0.05$), attribute liking scores (aroma, flavor, aftertaste) were found to have a high negative correlation with ΔSatiety and plate waste ($-1 > r > -0.70$). There were significant positive correlations between appearance liking and satiety before and after ($p < 0.05$). A moderate positive correlation was found between texture liking and satiety before ($0.50 > r > 0.70$), though no significance ($p > 0.05$).

Table 4. Correlation coefficients between attribute liking scores, ΔSatiety, satiety before and after, and plate waste. Coefficients were either colored blue if $r > 0.50$ or yellow if $r < -0.50$. Correlation coefficients denoted were found to be significant ($p < 0.05$).

		Control		
Attributes	**Δsatiety**	**Satiety before**	**Satiety after**	**Plate Waste (%)**
Appearance	0.06	0.96 *	0.96 *	−0.22
Aroma	−0.85	0.26	0.15	−0.89
Overall	−0.91	0.37	0.25	−0.98 *
Taste	−0.92	0.32	0.20	−0.97 *
Texture	−0.83	0.55	0.44	−0.96 *
Flavor	−0.97 *	−0.08	−0.20	−0.91
Aftertaste	−0.85	0.43	0.32	−0.94

Table 4. *Cont.*

CP				
Attributes	**Δsatiety**	**Satiety before**	**Satiety after**	**Plate Waste (%)**
Appearance	0.75	0.85	0.88	0.11
Aroma	0.47	0.90	0.86	−0.32
Overall	0.40	0.81	0.77	−0.33
Taste	0.25	0.80	0.73	−0.52
Texture	0.40	0.79	0.76	−0.31
Flavor	−0.05	0.31	0.25	−0.40
Aftertaste	0.51	0.86	0.84	−0.22
Control and CP Combined				
Attributes	**Δsatiety**	**Satiety before**	**Satiety after**	**Plate Waste (%)**
Appearance	0.57	0.84 *	0.88 *	−0.06
Aroma	0.23	0.60	0.59	−0.51
Overall	0.06	0.54	0.51	−0.68
Taste	0.00	0.49	0.45	−0.77 *
Texture	0.11	0.63	0.60	−0.66
Flavor	−0.02	0.08	0.07	−0.68
Aftertaste	0.20	0.53	0.53	−0.63

* Correlation coefficients were found to be significant from the table of critical values for Pearson's r correlations ($p < 0.05$).

For the CP dishes, no significant relationships were found ($p > 0.05$). However, appearance liking had a high positive correlation with ΔSatiety ($0.70 > r > 1$), while aftertaste liking was found to moderate positive correlation with ΔSatiety ($0.50 > r > 0.70$). There was a high positive correlation between attribute liking (appearance, aroma, overall liking, taste, texture, and aftertaste) and subjective satiety for both before and after ($0.70 > r > 1$). There was also a moderate negative correlation between taste liking and plate waste ($-0.70 > r > -0.50$).

When Control and CP are combined, a significant positive correlation was found between appearance liking and satiety before and after ($p < 0.05$). A moderate positive correlation was found between appearance liking and ΔSatiety, as well as for attribute liking (aroma, overall liking, texture, and aftertaste) and satiety before and after consumption ($0.50 > r > 0.70$). There was a significant negative relationship between taste liking and plate waste ($p < 0.05$). However, no significance ($p > 0.05$) attribute liking for aroma, overall liking, texture, flavor, and aftertaste, were found to have a moderate negative correlation with plate waste ($-0.70 > r > -0.50$).

Plate waste has been used to assess the implementation of new food products in food service settings, especially for child nutrition programs [123–126]. Plate waste has also been determined to be a predictor of food choice regarding portion control [127,128]. From the results, a negative relationship between consumer liking scores and plate waste was found. As plate waste increased, the liking scores decreased across the products/dishes evaluated. When introducing novel foods in US school lunch programs, it was found that plant-based meals increased plate waste by students [129]. However, this may be due to the unfamiliarity of these foods. Martins et al. [130] suggested that the sensory characteristics that increase satisfaction with certain meals may influence the food waste produced by US students as well.

Regarding subjective satiety, the results may suggest that perceived satiety may be a factor in consumer acceptance. The perceived satiety consumers may experience when evaluating a product possibly affects their evaluation. As ΔSatiety increased, consumer liking of the attributes decreased, while satiety before and after positively affected consumer liking. However, Yeomans et al. [131] suggested that flavor and palatability of porridges may be a better predictor of the intent to purchase of a product than perceived satiety. Further research into perceived satiety over the course of product evaluation may be necessary.

Although this may be a novel approach, plate waste may give insight into the future consumer acceptance of new food products. Typically, product evaluation involved providing participants with small portions of the product to be tested. For the purposes of this study, the products were served in a manner that consumers may typically consume in certain eating situations, such as a three-course meal. From the results, plate waste may be an indicator of future acceptance as there was a moderate negative correlation between both the Control and CP dishes (combined) and plate waste.

4. Conclusions

This study was designed to evaluate potential products containing insect powders. As a result, this was a blind study where participants were not informed which products contained insects. The products were measured for consumer acceptability as well as measuring emotional response, satiety, and plate waste as factors in the future acceptance of these products. The US consumers recruited for this study were found to be potential early adopters of insect-based foods and who may be open to different food products, especially for the products tested, containing nonvisible insects. During consumer acceptability testing, though there was a difference in attribute liking, the CP products/dishes were still considered acceptable by the participants when compared to the Controls. These findings provide more insight into the products developed by Ho et al. [63], and that formulation and serving methods may increase the overall acceptability of food products containing insect powders. This can also be further supported by the fact that the emotional responses evaluated as more positive emotion terms, such as satisfied and interested, were selected more than negative emotions, such as disgusted and worried. Plate waste may also assess the performance of these in certain eating situations. Consumer liking may relate to how much plate waste is produced by consumers and predict acceptance of novel foods such as insect-based foods. The findings from this study may provide a novel approach to assessing new food products by measuring different factors during the product consumption experience.

Author Contributions: Conceptualization, I.H., R.L., J.G., A.G., S.A. and A.L.; methodology, I.H., J.G., A.G., S.A. and A.L.; software, I.H., R.L., J.G. and A.L.; validation, I.H., R.L., J.G. and A.L.; formal analysis, I.H., R.L., J.G. and A.L.; investigation, I.H., A.P., J.M., K.W., S.A. and A.L.; resources, S.A. and A.L.; data curation, I.H., A.P., J.M., K.W., S.A. and A.L.; writing—original draft preparation, I.H. and A.L.; writing—review and editing, I.H., A.P., J.M., K.W., R.L., J.G., A.G., S.A. and A.L.; visualization, I.H., R.L., J.G. and A.L.; supervision, S.A. and A.L.; project administration, A.L.; funding acquisition, S.A. and A.L. All authors have read and agreed to the published version of the manuscript.

Funding: This work was supported by the California State University Agricultural Research Institute (Award No. 21-03-105).

Institutional Review Board Statement: The study was conducted in accordance with the Declaration of Helsinki, and approved by the Institutional Review Board of Cal Poly San Luis Obispo (2022-037-CP).

Informed Consent Statement: Informed consent was obtained from all subjects involved in the study.

Data Availability Statement: The data presented in this study are available on request from the corresponding author. The data are not publicly available due to privacy and ethical restrictions.

Acknowledgments: The authors would like to thank the students from the Advanced Culinary Course at Cal Poly SLO during the Spring 2022 quarter. These students helped prepare the products and execute the study.

Conflicts of Interest: The authors declare no conflict of interest. J.G. and R.L. are from Garza Consulting which is a consumer, sensory, and statistical consulting company. Their role in this research was to provide guidance to the statistical design of the experiment, suggestions on the questionnaire, and data analysis. The two co-authors were not compensated, but were project collaborators.

Appendix A

Appetizer—Sausage bruschetta with an onion and pepper topping

(a) (b)

Figure A1. Sausage bruschetta appetizer presented to participants for product evaluation. (**a**) Sausage bruschetta contained the Control sausage used in the study and (**b**) the sausage bruschetta contained the CP sausage.

Table A1. Ingredients used and their sources for the Sausage with Peppers and Onions Bruschetta. The sausage component shows the formulations for both the Control and CP versions. Usage levels (%) shown are the percentages used for each dish component. Percentages were rounded to the nearest hundredths.

Dish Component	Ingredient	Usage Level (%)		Source
		Control	CP	
Sausage	Pork Shoulder	62.29	56.06	Cal Poly Meats, San Luis Obispo, CA, USA
	Pork Backfat	31.15	31.15	Cal Poly Meats, San Luis Obispo, CA, USA
	Cricket Powder	-	6.23	JR Unique Foods, Udon Thani, Thailand
	Fennel Seed, Ground	0.91	0.91	Private Selection, Kroger Co., Cincinnati, OH, USA
	Dried Basil	0.13	0.13	First Street, Amerifoods Trading Co., Los Angeles, CA, USA
	Dried Oregano	0.16	0.16	First Street, Amerifoods Trading Co., Los Angeles, CA, USA
	Crushed Red Chili Pepper	0.17	0.17	Kirkland Signature, Costco Wholesale Co., Seattle, WA, USA
	Garlic, Minced	3.74	3.74	Christopher Ranch, Gilroy, CA, USA
	Onion Powder	0.37	0.37	First Street, Amerifoods Trading Co., Los Angeles, CA, USA
	Black Pepper, Ground	0.09	0.09	First Street, Amerifoods Trading Co., Los Angeles, CA, USA
	Salt	0.76	0.76	First Street, Amerifoods Trading Co., Los Angeles, CA, USA
	Dried Thyme	0.14	0.14	First Street, Amerifoods Trading Co., Los Angeles, CA, USA
	Dried Rosemary	0.10	0.10	Private Selection, Kroger Co., Cincinnati, OH, USA

Table A1. *Cont.*

Dish Component	Ingredient	Usage Level (%)	Source
Bruschetta (Toasted Bread)	French Bread, Sliced	-	First Street, Amerifoods Trading Co., Los Angeles, CA, USA
	Grated Parmesan Cheese	-	First Street, Amerifoods Trading Co., Los Angeles, CA, USA
Onion and Pepper Topping	Yellow Onions, Sliced	54.50	Western Onion, Camarillo, CA, USA
	Extra Virgin Olive Oil	4.07	Kirkland Signature, Costco Wholesale Co., Seattle, WA, USA
	Canned Roasted Red Peppers, Sliced	41.42	First Street, Amerifoods Trading Co., Los Angeles, CA, USA

Entrée—Pasta with a charred tomato sauce

(a)

(b)

Figure A2. Pasta entrée presented to participants for product evaluation. (**a**) Pasta dish containing the Control pasta used in the study and (**b**) the pasta dish containing the CP pasta.

Table A2. Ingredients used and their sources for the pasta entrée. The pasta component shows the formulations for both the Control and CP versions. Usage levels (%) shown are the percentages used for each dish component. Percentages were rounded to the nearest hundredths.

Dish Component	Ingredient	Use Level (%)		Source
		Control	CP	
Pasta	Durum Wheat Semolina	61.41	73.87	Miller Milling Company, Fresno, CA, USA
	Whole Wheat Flour	14.41	-	King Arthur Flour, Norwich, VT, USA
	Cricket Powder	-	5.00	JR Unique Foods, Udon Thani, Thailand
	Powdered Caramel Coloring (P600)	0.10	0.075	Sethness Products Company, Clinton, IA, USA
	Water	24.08	21.05	-
Tomato Sauce	Canned Tomatoes, Diced	76.37		First Street, Amerifoods Trading Co., Los Angeles, CA, USA
	Extra Virgin Olive Oil, 1st Addition	5.81		Kirkland Signature, Costco Wholesale Co., Seattle, WA, USA
	Fresh Garlic, Minced	7.69		Christopher Ranch, Gilroy, CA, USA
	Salt	0.70		First Street, Amerifoods Trading Co., Los Angeles, CA, USA
	Black Pepper, Ground	0.17		First Street, Amerifoods Trading Co., Los Angeles, CA, USA
	Extra Virgin Olive Oil, 2nd Addition	9.27		Kirkland Signature, Costco Wholesale Co., Seattle, WA, USA

Table A2. *Cont.*

Dish Component	Ingredient	Use Level (%)	Source
Topping	Grated Parmesan Cheese	-	First Street, Amerifoods Trading Co., Los Angeles, CA, USA
	Fresh Basil, Chiffonade	-	Simple Truth Organic, Kroger Co., Cincinnati, OH, USA

Dessert—Brownie trifle dessert with a berry sauce

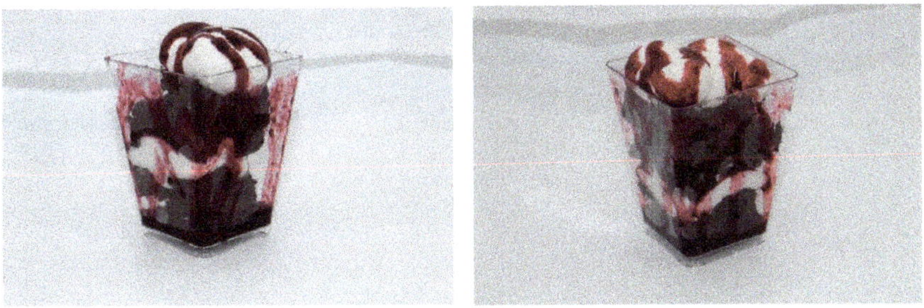

(a) (b)

Figure A3. Brownie dessert that was presented to participants for product evaluation. (**a**) Brownie dessert contained the Control brownie used in the study and (**b**) the brownie dessert contained the CP pasta.

Table A3. Ingredients used and their sources for the brownie dessert. The pasta component shows the formulations for both the Control and CP versions. Usage levels (%) shown are the percentages used for each dish component. Percentages were rounded to the nearest hundredths.

Dish Component	Ingredient	Usage Level (%)		Source
		Control	CP	
Brownie	Dutch Cocoa Powder	4.91	4.91	The Hershey Co., Hershey, PA, USA
	Bittersweet Chocolate	11.86	11.86	Puratos Chocolate USA, Kenosha, WI, USA
	Unsalted Butter	26.18	26.18	Kirkland Signature, Costco Wholesale Co., Seattle, WA, USA
	Eggs, Beaten	16.36	16.36	The Kroger Co., Cincinnati, OH, USA
	Granulated Sugar	21.27	21.27	Sysco Corp., Houston, TX, USA
	Salt	0.36	0.36	First Street, Smart & Final, Commerce, CA, USA
	Vanilla Extract	0.57	0.57	Cook Flavoring Co., Paso Robles, CA, USA
	Bread Flour	9.41	2.41	General Mills Operations Inc., Minneapolis, MN, USA
	Cricket Powder	-	7.00	JR Unique Foods, Udon Thani, Thailand
	Bittersweet Chocolate Chunks	9.07	9.07	Puratos Chocolate USA, Kenosha, WI, USA
Whipped Cream Topping	Heavy Whipping Cream	82.64		Producers, Fresno, CA, USA
	Granulated Sugar	16.53		Sysco Corp., Houston, TX, USA
	Vanilla Extract	0.83		Cook Flavoring Co., Paso Robles, CA, USA
Berry Sauce	Frozen Berry Blend (Raspberries, Blueberries, Blackberries)	64.10		Kirkland Signature, Costco Wholesale Co., Seattle, WA, USA
	Granulated Sugar	22.91		Sysco Corp., Houston, TX, USA
	Water	12.99		-

References

1. Van Huis, A.; van Itterbeeck, J.; Klunder, H.; Mertens, E.; Muir, G.; Vantomme, P. *Edible Insects—Future Prospects for Food and Feed Security*; FAO: Rome, Italy, 2013.
2. United Nations. *Food Production Must Double by 2050 to Meet Demand from World's Growing Population, Innovative Strategies Needed to Combat Hunger, Experts Tell Second Committee*; United Nations: New York, NY, USA, 2009.
3. Ismail, B.P.; Senaratne-Lenagala, L.; Stube, A.; Brackenridge, A. Protein Demand: Review of Plant and Animal Proteins Used in Alternative Protein Product Development and Production. *Anim. Front.* **2020**, *10*, 53–63. [CrossRef] [PubMed]
4. Baiano, A. Edible Insects: An Overview on Nutritional Characteristics, Safety, Farming, Production Technologies, Regulatory Framework, and Socio-Economic and Ethical Implications. *Trends Food Sci. Technol.* **2020**, *100*, 35–50. [CrossRef]
5. Govorushko, S. Global Status of Insects as Food and Feed Source: A Review. *Trends Food Sci. Technol.* **2019**, *91*, 436–445. [CrossRef]
6. Onwezen, M.C.; Bouwman, E.P.; Reinders, M.J.; Dagevos, H. A Systematic Review on Consumer Acceptance of Alternative Proteins: Pulses, Algae, Insects, Plant-Based Meat Alternatives, and Cultured Meat. *Appetite* **2021**, *159*, 105058. [CrossRef] [PubMed]
7. Sagis, L.M.; Yang, J. Protein-Stabilized Interfaces in Multiphase Food: Comparing Structure-Function Relations of Plant-Based and Animal-Based Proteins. *Curr. Opin. Food Sci.* **2022**, *43*, 53–60. [CrossRef]
8. Siegrist, M.; Sütterlin, B.; Hartmann, C. Perceived Naturalness and Evoked Disgust Influence Acceptance of Cultured Meat. *Meat. Sci.* **2018**, *139*, 213–219. [CrossRef]
9. Tucker, C.A. The Significance of Sensory Appeal for Reduced Meat Consumption. *Appetite* **2014**, *81*, 168–179. [CrossRef]
10. Gere, A.; Harizi, A.; Bellissimo, N.; Roberts, D.; Moskowitz, H. Creating a Mind Genomics Wiki for Non-Meat Analogs. *Sustainability* **2020**, *12*, 5352. [CrossRef]
11. Van Huis, A. Insects as Human Food. In *Ethnozoology*; Elsevier: Amsterdam, The Netherlands, 2018; pp. 195–213. [CrossRef]
12. Gómez-Luciano, C.A.; de Aguiar, L.K.; Vriesekoop, F.; Urbano, B. Consumers' Willingness to Purchase Three Alternatives to Meat Proteins in the United Kingdom, Spain, Brazil and the Dominican Republic. *Food Qual. Prefer* **2019**, *78*, 103732. [CrossRef]
13. Cerda, H.; Martinez, R.; Briceno, N.; Pizzoferrato, L.; Manzi, P.; Ponzetta, M.T.; Marin, O.; Paoletti, M.G. Palm Worm: (*Rhynchophorus palmarum*) Traditional Food in Amazonas, Venezuela—Nutritional Composition, Small Scale Production and Tourist Palatability. *Ecol. Food Nutr.* **2001**, *40*, 13–32. [CrossRef]
14. Yi, C.; He, Q.; Wang, L.; Kuang, R. The Utilization of Insect-Resources in Chinese Rural Area. *J. Agric. Sci.* **2010**, *2*, 146. [CrossRef]
15. Jantzen da Silva Lucas, A.; Menegon de Oliveira, L.; da Rocha, M.; Prentice, C. Edible Insects: An Alternative of Nutritional, Functional and Bioactive Compounds. *Food Chem.* **2020**, *311*, 126022. [CrossRef] [PubMed]
16. De Castro, R.J.S.; Ohara, A.; Aguilar, J.G.D.S.; Domingues, M.A.F. Nutritional, Functional and Biological Properties of Insect Proteins: Processes for Obtaining, Consumption and Future Challenges. *Trends Food Sci. Technol.* **2018**, *76*, 82–89. [CrossRef]
17. Jonas-Levi, A. The High Level of Protein Content Reported in Insects for Food and Feed Is Overestimated. *J. Food Compos. Anal.* **2017**, *62*, 184–188. [CrossRef]
18. Dossey, A.T.; Tatum, J.T.; McGill, W.L. Modern Insect-Based Food Industry: Current Status, Insect Processing Technology, and Recommendations Moving Forward. In *Insects as Sustainable Food Ingredients*; Elsevier: Amsterdam, The Netherlands, 2016; pp. 113–152. [CrossRef]
19. Gere, A.; Radványi, D.; Héberger, K. Which Insect Species Can Best Be Proposed for Human Consumption? *Innov. Food Sci. Emerg. Technol.* **2019**, *52*, 358–367. [CrossRef]
20. Gao, Y.; Wang, D.; Xu, M.-L.; Shi, S.-S.; Xiong, J.-F. Toxicological Characteristics of Edible Insects in China: A Historical Review. *Food Chem. Toxicol.* **2018**, *119*, 237–251. [CrossRef]
21. Thrastardottir, R.; Olafsdottir, H.T.; Thorarinsdottir, R.I. Yellow Mealworm and Black Soldier Fly Larvae for Feed and Food Production in Europe, with Emphasis on Iceland. *Foods* **2021**, *10*, 2744. [CrossRef]
22. European Commission Rules Yellow Mealworms as Safe to Eat in the EU. Available online: https://www.businessinsider.com/insects-bugs-edible-environment-sustainability-eco-friendly-health-sustainable-farming-2021-5 (accessed on 8 August 2021).
23. Żuk-Gołaszewska, K.; Gałęcki, R.; Obremski, K.; Smetana, S.; Figiel, S.; Gołaszewski, J. Edible Insect Farming in the Context of the EU Regulations and Marketing—An Overview. *Insects* **2022**, *13*, 446. [CrossRef]
24. Pali-Schöll, I.; Meinlschmidt, P.; Larenas-Linnemann, D.; Purschke, B.; Hofstetter, G.; Rodríguez-Monroy, F.A.; Einhorn, L.; Mothes-Luksch, N.; Jensen-Jarolim, E.; Jäger, H. Edible Insects: Cross-Recognition of IgE from Crustacean- and House Dust Mite Allergic Patients, and Reduction of Allergenicity by Food Processing. *World Allergy Organ. J.* **2019**, *12*, 100006. [CrossRef]
25. De Gier, S.; Verhoeckx, K. Insect (Food) Allergy and Allergens. *Mol. Immunol.* **2018**, *100*, 82–106. [CrossRef]
26. Niyonsaba, H.H.; Höhler, J.; Kooistra, J.; van der Fels-Klerx, H.J.; Meuwissen, M.P.M. Profitability of Insect Farms. *J. Insects Food Feed* **2021**, *7*, 923–934. [CrossRef]
27. Hartmann, C.; Bearth, A.; Hartmann, C.; Bearth, A. Bugs on the Menu: Drivers and Barriers of Consumer Acceptance of Insects as Food. In *Edible Insects in the Food Sector*; Springer International Publishing: Cham, Switzerland, 2019; pp. 45–55. [CrossRef]
28. Orsi, L.; Voege, L.L.; Stranieri, S. Eating Edible Insects as Sustainable Food? Exploring the Determinants of Consumer Acceptance in Germany. *Food Res. Int.* **2019**, *125*, 108573. [CrossRef] [PubMed]
29. Looy, H.; Dunkel, F.V.; Wood, J.R. How Then Shall We Eat? Insect-Eating Attitudes and Sustainable Foodways. *Agric. Hum. Values* **2014**, *31*, 131–141. [CrossRef]

30. Lammers, P.; Ullmann, L.M.; Fiebelkorn, F. Acceptance of Insects as Food in Germany: Is It about Sensation Seeking, Sustainability Consciousness, or Food Disgust? *Food Qual. Prefer.* **2019**, *77*, 78–88. [CrossRef]
31. La Barbera, F.; Verneau, F.; Videbæk, P.N.; Amato, M.; Grunert, K.G. A Self-Report Measure of Attitudes toward the Eating of Insects: Construction and Validation of the Entomophagy Attitude Questionnaire. *Food Qual. Prefer.* **2020**, *79*, 103757. [CrossRef]
32. Sogari, G.; Menozzi, D.; Mora, C. The Food Neophobia Scale and Young Adults' Intention to Eat Insect Products. *Int. J. Consum. Stud* **2019**, *43*, 68–76. [CrossRef]
33. Sogari, G.; Menozzi, D.; Hartmann, C.; Mora, C. How to Measure Consumers Acceptance Towards Edible Insects?—A Scoping Review About Methodological Approaches. In *Edible Insects in the Food Sector*; Springer International Publishing: Cham, Switzerland, 2019; pp. 27–44. [CrossRef]
34. Lombardi, A.; Vecchio, R.; Borrello, M.; Caracciolo, F.; Cembalo, L. Willingness to Pay for Insect-Based Food: The Role of Information and Carrier. *Food Qual. Prefer.* **2019**, *72*, 177–187. [CrossRef]
35. Kim, T.-K.; Yong, H.I.; Kim, Y.-B.; Kim, H.-W.; Choi, Y.-S. Edible Insects as a Protein Source: A Review of Public Perception, Processing Technology, and Research Trends. *Food Sci. Anim. Resour.* **2019**, *39*, 521–540. [CrossRef]
36. Mancini, S.; Moruzzo, R.; Riccioli, F.; Paci, G. European Consumers' Readiness to Adopt Insects as Food. A Review. *Food Res. Int.* **2019**, *122*, 661–678. [CrossRef]
37. La Barbera, F.; Verneau, F.; Amato, M.; Grunert, K. Understanding Westerners' Disgust for the Eating of Insects: The Role of Food Neophobia and Implicit Associations. *Food Qual. Prefer.* **2018**, *64*, 120–125. [CrossRef]
38. Mancini, S.; Sogari, G.; Menozzi, D.; Nuvoloni, R.; Torracca, B.; Moruzzo, R.; Paci, G. Factors Predicting the Intention of Eating an Insect-Based Product. *Foods* **2019**, *8*, 270. [CrossRef] [PubMed]
39. Verneau, F.; Zhou, Y.; Amato, M.; Grunert, K.G.; la Barbera, F. Cross-Validation of the Entomophagy Attitude Questionnaire (EAQ): A Study in China on Eaters and Non-Eaters. *Food Qual. Prefer.* **2021**, *87*, 104029. [CrossRef]
40. Liu, A.J.; Li, J.; Gómez, M.I. Factors Influencing Consumption of Edible Insects for Chinese Consumers. *Insects* **2019**, *11*, 10. [CrossRef] [PubMed]
41. La Barbera, F.; Verneau, F.; Amato, M.; Grunert, K.G.; Schnettler, B. Acceptance of Insect-Based Food in Chile: Evidence from a Survey Using the Entomophagy Attitude Questionnaire (EAQ). *Food Qual. Prefer.* **2021**, *93*, 104269. [CrossRef]
42. Bisconsin-Júnior, A.; Rodrigues, H.; Behrens, J.H.; da Silva, M.A.A.P.; Mariutti, L.R.B. "Food Made with Edible Insects": Exploring the Social Representation of Entomophagy Where It Is Unfamiliar. *Appetite* **2022**, *173*, 106001. [CrossRef]
43. Lucchese-Cheung, T.; de Aguiar, L.K.; da Silva, R.F.F.; Pereira, M.W. Determinants of the Intention to Consume Edible Insects in Brazil. *J. Food Prod. Mark.* **2020**, *26*, 297–316. [CrossRef]
44. House, J. Consumer Acceptance of Insect-Based Foods in the Netherlands: Academic and Commercial Implications. *Appetite* **2016**, *107*, 47–58. [CrossRef]
45. Frattini, F.; Bianchi, M.; de Massis, A.; Sikimic, U. The Role of Early Adopters in the Diffusion of New Products: Differences between Platform and Nonplatform Innovations. *J. Prod. Innov. Manag.* **2014**, *31*, 466–488. [CrossRef]
46. Rovai, D.; Amin, S.; Lesniauskas, R.; Wilke, K.; Garza, J.; Lammert, A. Are Early Adopters Willing to Accept Frozen, <scp>ready-to-cook</Scp> Mealworms as a Food Source? *J. Sens. Stud.* **2022**, *37*, e12774. [CrossRef]
47. Rovai, D.; Michniuk, E.; Roseman, E.; Amin, S.; Lesniauskas, R.; Wilke, K.; Garza, J.; Lammert, A. Insects as a Sustainable Food Ingredient: Identifying and Classifying Early Adopters of Edible Insects Based on Eating Behavior, Familiarity, and Hesitation. *J. Sens. Stud.* **2021**, *36*, e12681. [CrossRef]
48. Ho, I.; Gere, A.; Chy, C.; Lammert, A. Use of Preference Analysis to Identify Early Adopter Mind-Sets of Insect-Based Food Products. *Sustainability* **2022**, *14*, 1435. [CrossRef]
49. Rogers, E.M. *Diffusion of Innovations*; Simon and Schuster: New York, NY, USA, 2003.
50. Serpico, M.; Rovai, D.; Wilke, K.; Lesniauskas, R.; Garza, J.; Lammert, A. Studying the Emotional Response to Insects Food Products. *Foods* **2021**, *10*, 2404. [CrossRef] [PubMed]
51. Pambo, K.O.; Okello, J.J.; Mbeche, R.M.; Kinyuru, J.N.; Alemu, M.H. The Role of Product Information on Consumer Sensory Evaluation, Expectations, Experiences and Emotions of Cricket-Flour-Containing Buns. *Food Res. Int.* **2018**, *106*, 532–541. [CrossRef] [PubMed]
52. Gurdian, C.E.; Torrico, D.D.; Li, B.; Tuuri, G.; Prinyawiwatkul, W. Effect of Informed Conditions on Sensory Expectations and Actual Perceptions: A Case of Chocolate Brownies Containing Edible-Cricket Protein. *Foods* **2021**, *10*, 1480. [CrossRef] [PubMed]
53. Tan, H.S.G.; Verbaan, Y.T.; Stieger, M. How Will Better Products Improve the Sensory-Liking and Willingness to Buy Insect-Based Foods? *Food Res. Int.* **2017**, *92*, 95–105. [CrossRef]
54. Tan, H.S.G.; Fischer, A.R.H.; van Trijp, H.C.M.; Stieger, M. Tasty but Nasty? Exploring the Role of Sensory-Liking and Food Appropriateness in the Willingness to Eat Unusual Novel Foods like Insects. *Food Qual. Prefer.* **2016**, *48*, 293–302. [CrossRef]
55. Ardoin, R.; Prinyawiwatkul, W. Product Appropriateness, Willingness to Try and Perceived Risks of Foods Containing Insect Protein Powder: A Survey of U.S. Consumers. *Int. J. Food Sci. Technol.* **2020**, *55*, 3215–3226. [CrossRef]
56. Jensen, N.H.; Lieberoth, A. We Will Eat Disgusting Foods Together—Evidence of the Normative Basis of Western Entomophagy-Disgust from an Insect Tasting. *Food Qual. Prefer.* **2019**, *72*, 109–115. [CrossRef]
57. Woolf, E.; Zhu, Y.; Emory, K.; Zhao, J.; Liu, C. Willingness to Consume Insect-Containing Foods: A Survey in the United States. *LWT* **2019**, *102*, 100–105. [CrossRef]

58. Stone, A.K.; Tanaka, T.; Nickerson, M.T. Protein Quality and Physicochemical Properties of Commercial Cricket and Mealworm Powders. *J. Food Sci. Technol.* **2019**, *56*, 3355–3363. [CrossRef]
59. Montowska, M.; Kowalczewski, P.Ł.; Rybicka, I.; Fornal, E. Nutritional Value, Protein and Peptide Composition of Edible Cricket Powders. *Food Chem.* **2019**, *289*, 130–138. [CrossRef] [PubMed]
60. Rumpold, B.A.; Schlüter, O.K. Potential and Challenges of Insects as an Innovative Source for Food and Feed Production. *Innov. Food Sci. Emerg. Technol.* **2013**, *17*, 1–11. [CrossRef]
61. Food Products Made from Insects. Available online: https://www.insectgourmet.com/food-products-made-from-insects/ (accessed on 13 September 2022).
62. Ardoin, R.; Marx, B.D.; Boeneke, C.; Prinyawiwatkul, W. Effects of Cricket Powder on Selected Physical Properties and US Consumer Perceptions of Whole-Wheat Snack Crackers. *Int. J. Food Sci. Technol.* **2021**, *56*, 4070–4080. [CrossRef]
63. Ho, I.; Peterson, A.; Madden, J.; Huang, E.; Amin, S.; Lammert, A. Will It Cricket? Product Development and Evaluation of Cricket (*Acheta domesticus*) Powder Replacement in Sausage, Pasta, and Brownies. *Foods* **2022**, *11*, 3128. [CrossRef] [PubMed]
64. Cardello, A.V.; Schutz, H.; Snow, C.; Lesher, L. Predictors of Food Acceptance, Consumption and Satisfaction in Specific Eating Situations. *Food Qual. Prefer.* **2000**, *11*, 201–216. [CrossRef]
65. Hartwell, H.J.; Edwards, J.S.A.; Brown, L. The Relationship between Emotions and Food Consumption (Macronutrient) in a Foodservice College Setting—A Preliminary Study. *Int. J. Food Sci. Nutr.* **2013**, *64*, 261–268. [CrossRef] [PubMed]
66. Piqueras-Fiszman, B.; Jaeger, S.R. The Effect of Product–Context Appropriateness on Emotion Associations in Evoked Eating Occasions. *Food Qual. Prefer.* **2015**, *40*, 49–60. [CrossRef]
67. Cardello, A.V.; Schutz, H.G. Food Appropriateness Measures as an Adjunct to Consumer Preference/Acceptability Evaluation. *Food Qual. Prefer.* **1996**, *7*, 239–249. [CrossRef]
68. Motoki, K.; Park, J.; Spence, C.; Velasco, C. Contextual Acceptance of Novel and Unfamiliar Foods: Insects, Cultured Meat, Plant-Based Meat Alternatives, and 3D Printed Foods. *Food Qual. Prefer.* **2022**, *96*, 104368. [CrossRef]
69. Barton, A.; Richardson, C.D.; McSweeney, M.B. Consumer Attitudes toward Entomophagy before and after Evaluating Cricket (*Acheta domesticus*)-based Protein Powders. *J. Food Sci.* **2020**, *85*, 781–788. [CrossRef]
70. Ardoin, R.; Prinyawiwatkul, W. Consumer Perceptions of Insect Consumption: A Review of Western Research since 2015. *Int. J. Food Sci. Technol.* **2021**, *56*, 4942–4958. [CrossRef]
71. Le Goff, G.; Delarue, J. Non-Verbal Evaluation of Acceptance of Insect-Based Products Using a Simple and Holistic Analysis of Facial Expressions. *Food Qual Prefer* **2017**, *56*, 285–293. [CrossRef]
72. Cricket Powder (*Acheta domesticus*)—JR Unique Foods Ltd. Available online: https://jrunique.com/product/acheta-cricket-powder/ (accessed on 25 August 2022).
73. Guidance for Industry: Reference Amounts Customarily Consumed (List of Products for Each Product Category) | FDA. Available online: https://www.fda.gov/regulatory-information/search-fda-guidance-documents/guidance-industry-reference-amounts-customarily-consumed-list-products-each-product-category (accessed on 28 July 2022).
74. Nestrud, M.A.; Meiselman, H.L.; King, S.C.; Lesher, L.L.; Cardello, A.V. Development of EsSense25, a Shorter Version of the EsSense Profile®. *Food Qual. Prefer.* **2016**, *48*, 107–117. [CrossRef]
75. King, S.C.; Meiselman, H.L. Development of a Method to Measure Consumer Emotions Associated with Foods. *Food Qual. Prefer.* **2010**, *21*, 168–177. [CrossRef]
76. Cardello, A.V.; Schutz, H.G.; Lesher, L.L.; Merrill, E. Development and Testing of a Labeled Magnitude Scale of Perceived Satiety. *Appetite* **2005**, *44*, 1–13. [CrossRef]
77. Pliner, P.; Hobden, K. Development of a Scale to Measure the Trait of Food Neophobia in Humans. *Appetite* **1992**, *19*, 105–120. [CrossRef]
78. R Core Team. R: A Language and Environment for Statistical Computing. R Foundation for Statistical Computing. Available online: https://www.r-project.org/ (accessed on 31 October 2022).
79. Aspelmeier, J. Table of Critical Values for Pearson's R. Available online: https://pdf4pro.com/amp/view/table-of-critical-values-for-pearson-s-r-59198f.html (accessed on 31 October 2022).
80. Menozzi, D.; Sogari, G.; Veneziani, M.; Simoni, E.; Mora, C. Eating Novel Foods: An Application of the Theory of Planned Behaviour to Predict the Consumption of an Insect-Based Product. *Food Qual. Prefer.* **2017**, *59*, 27–34. [CrossRef]
81. Tan, H.S.G.; Fischer, A.R.H.; Tinchan, P.; Stieger, M.; Steenbekkers, L.P.A.; van Trijp, H.C.M. Insects as Food: Exploring Cultural Exposure and Individual Experience as Determinants of Acceptance. *Food Qual. Prefer.* **2015**, *42*, 78–89. [CrossRef]
82. Ruby, M.B.; Rozin, P. Disgust, Sushi Consumption, and Other Predictors of Acceptance of Insects as Food by Americans and Indians. *Food Qual. Prefer.* **2019**, *74*, 155–162. [CrossRef]
83. Caparros Megido, R.; Gierts, C.; Blecker, C.; Brostaux, Y.; Haubruge, É.; Alabi, T.; Francis, F. Consumer Acceptance of Insect-Based Alternative Meat Products in Western Countries. *Food Qual. Prefer.* **2016**, *52*, 237–243. [CrossRef]
84. Gmuer, A.; Nuessli Guth, J.; Hartmann, C.; Siegrist, M. Effects of the Degree of Processing of Insect Ingredients in Snacks on Expected Emotional Experiences and Willingness to Eat. *Food Qual. Prefer.* **2016**, *54*, 117–127. [CrossRef]
85. Van Huis, A. Edible Insects Contributing to Food Security? *Agric. Food Secur.* **2015**, *4*, 20. [CrossRef]
86. Pollard, J.; Kirk, S.F.L.; Cade, J.E. Factors Affecting Food Choice in Relation to Fruit and Vegetable Intake: A Review. *Nutr. Res. Rev.* **2002**, *15*, 373–387. [CrossRef]

87. Placentino, U.; Sogari, G.; Viscecchia, R.; de Devitiis, B.; Monacis, L. The New Challenge of Sports Nutrition: Accepting Insect Food as Dietary Supplements in Professional Athletes. *Foods* **2021**, *10*, 1117. [CrossRef] [PubMed]
88. Verbeke, W. Profiling Consumers Who Are Ready to Adopt Insects as a Meat Substitute in a Western Society. *Food Qual. Prefer.* **2015**, *39*, 147–155. [CrossRef]
89. Schouteten, J.J.; de Steur, H.; de Pelsmaeker, S.; Lagast, S.; Juvinal, J.G.; de Bourdeaudhuij, I.; Verbeke, W.; Gellynck, X. Emotional and Sensory Profiling of Insect-, Plant- and Meat-Based Burgers under Blind, Expected and Informed Conditions. *Food Qual. Prefer.* **2016**, *52*, 27–31. [CrossRef]
90. Choi, Y.-S.; Kim, T.-K.; Choi, H.-D.; Park, J.-D.; Sung, J.-M.; Jeon, K.-H.; Paik, H.-D.; Kim, Y.-B. Optimization of Replacing Pork Meat with Yellow Worm (*Tenebrio molitor* L.) for Frankfurters. *Korean J. Food Sci. Anim. Resour.* **2017**, *37*, 617–625. [CrossRef]
91. Cruz-López, S.O.; Álvarez-Cisneros, Y.M.; Domínguez-Soberanes, J.; Escalona-Buendía, H.B.; Sánchez, C.N. Physicochemical and Sensory Characteristics of Sausages Made with Grasshopper (*Sphenarium purpurascens*) Flour. *Foods* **2022**, *11*, 704. [CrossRef]
92. Choi, J.-H.; Yong, H.I.; Ku, S.-K.; Kim, T.-K.; Choi, Y.-S. The Quality Characteristics of Pork Patties According to the Replacement of Mealworm (*Tenebrio molitor* L.). *Korean J. Food Cook. Sci.* **2019**, *35*, 441–449. [CrossRef]
93. Pintado, T.; Delgado-Pando, G. Towards More Sustainable Meat Products: Extenders as a Way of Reducing Meat Content. *Foods* **2020**, *9*, 1044. [CrossRef]
94. Su, S.I.T.; Yoshida, C.M.P.; Contreras-Castillo, C.J.; Quiñones, E.M.; Venturini, A.C. Okara, a Soymilk Industry by-Product, as a Non-Meat Protein Source in Reduced Fat Beef Burgers. *Ciência E Tecnol. De Aliment.* **2013**, *33*, 52–56. [CrossRef]
95. Turhan, S.; Temiz, H.; Sagir, I. Utilization of wet okara in low-fat beef patties. *J. Muscle Foods* **2007**, *18*, 226–235. [CrossRef]
96. Wang, L.; Guo, H.; Liu, X.; Jiang, G.; Li, C.; Li, X.; Li, Y. Roles of Lentinula Edodes as the Pork Lean Meat Replacer in Production of the Sausage. *Meat. Sci.* **2019**, *156*, 44–51. [CrossRef] [PubMed]
97. Smarzyński, K.; Sarbak, P.; Musiał, S.; Jezowski, P.; Piatek, M.; Kowalczewski, P.T. Nutritional Analysis and Evaluation of the Consumer Acceptance of Pork Pâté Enriched with Cricket Powder-Preliminary Study. *Open Agric.* **2019**, *4*, 159–163. [CrossRef]
98. Kim, H.W.; Setyabrata, D.; Lee, Y.J.; Jones, O.G.; Kim, Y.H.B. Effect of House Cricket (*Acheta domesticus*) Flour Addition on Physicochemical and Textural Properties of Meat Emulsion Under Various Formulations. *J. Food Sci.* **2017**, *82*, 2787–2793. [CrossRef]
99. Dion-Poulin, A.; Turcotte, M.; Lee-Blouin, S.; Perreault, V.; Provencher, V.; Doyen, A.; Turgeon, S.L. Acceptability of Insect Ingredients by Innovative Student Chefs: An Exploratory Study. *Int. J. Gastron. Food Sci.* **2021**, *24*, 100362. [CrossRef]
100. Sogari, G.; Menozzi, D.; Mora, C. Sensory-Liking Expectations and Perceptions of Processed and Unprocessed Insect Products. *Int. J. Food Syst. Dyn.* **2018**, *9*, 314–320. [CrossRef]
101. Keto, L.; Stefanski, T.; Isokangas, A.; Rautio, P.; Rokka, S.; Pihlanto, A. Risk Profile Related to Production and Consumption of Insects as Food and Feed. *EFSA J.* **2015**, *13*, 4257. [CrossRef]
102. Duda, A.; Adamczak, J.; Chełmińska, P.; Juszkiewicz, J.; Kowalczewski, P. Quality and Nutritional/Textural Properties of Durum Wheat Pasta Enriched with Cricket Powder. *Foods* **2019**, *8*, 46. [CrossRef]
103. Biró, B.; Fodor, R.; Szedljak, I.; Pásztor-Huszár, K.; Gere, A. Buckwheat-Pasta Enriched with Silkworm Powder: Technological Analysis and Sensory Evaluation. *LWT* **2019**, *116*, 108542. [CrossRef]
104. Çabuk, B.; Yılmaz, B. Fortification of Traditional Egg Pasta (Erişte) with Edible Insects: Nutritional Quality, Cooking Properties and Sensory Characteristics Evaluation. *J. Food Sci. Technol.* **2020**, *57*, 2750–2757. [CrossRef]
105. Jakab, I.; Tormási, J.; Dhaygude, V.; Mednyánszky, Z.S.; Sipos, L.; Szedljak, I. Cricket Flour-Laden Millet Flour Blends' Physical and Chemical Composition and Adaptation in Dried Pasta Products. *Acta Aliment.* **2020**, *49*, 4–12. [CrossRef]
106. Pasini, G.; Cullere, M.; Vegro, M.; Simonato, B.; Dalle Zotte, A. Potentiality of Protein Fractions from the House Cricket (*Acheta domesticus*) and Yellow Mealworm (*Tenebrio molitor*) for Pasta Formulation. *LWT-Food Sci. Technol.* **2022**, *164*, 113638. [CrossRef]
107. Hartmann, C.; Shi, J.; Giusto, A.; Siegrist, M. The Psychology of Eating Insects: A Cross-Cultural Comparison between Germany and China. *Food Qual. Prefer.* **2015**, *44*, 148–156. [CrossRef]
108. Castro Delgado, M.; Chambers, E.; Carbonell-Barrachina, A.; Noguera Artiaga, L.; Vidal Quintanar, R.; Burgos Hernandez, A. Consumer Acceptability in the USA, Mexico, and Spain of Chocolate Chip Cookies Made with Partial Insect Powder Replacement. *J. Food Sci.* **2020**, *85*, 1621–1628. [CrossRef]
109. Pauter, P.; Różańska, M.; Wiza, P.; Dworczak, S.; Grobelna, N.; Sarbak, P.; Kowalczewski, P. Effects of the Replacement of Wheat Flour with Cricket Powder on the Characteristics of Muffins. *Acta Sci. Pol. Technol. Aliment.* **2018**, *17*, 227–233. [CrossRef]
110. Bawa, M.; Songsermpong, S.; Kaewtapee, C.; Chanput, W. Nutritional, Sensory, and Texture Quality of Bread and Cookie Enriched with House Cricket (*Acheta domesticus*) Powder. *J. Food Process. Preserv.* **2020**, *44*, e14601. [CrossRef]
111. De Oliveira, L.M.; da Silva Lucas, A.J.; Cadaval, C.L.; Mellado, M.S. Bread Enriched with Flour from Cinereous Cockroach (*Nauphoeta cinerea*). *Innov. Food Sci. Emerg. Technol.* **2017**, *44*, 30–35. [CrossRef]
112. Jiang, Y.; King, J.M.; Prinyawiwatkul, W. A Review of Measurement and Relationships between Food, Eating Behavior and Emotion. *Trends Food Sci. Technol.* **2014**, *36*, 15–28. [CrossRef]
113. Spinelli, S.; Masi, C.; Zoboli, G.P.; Prescott, J.; Monteleone, E. Emotional Responses to Branded and Unbranded Foods. *Food Qual. Prefer.* **2015**, *42*, 1–11. [CrossRef]
114. Spinelli, S.; Masi, C.; Dinnella, C.; Zoboli, G.P.; Monteleone, E. How Does It Make You Feel? A New Approach to Measuring Emotions in Food Product Experience. *Food Qual. Prefer.* **2014**, *37*, 109–122. [CrossRef]

115. Ng, M.; Chaya, C.; Hort, J. Beyond Liking: Comparing the Measurement of Emotional Response Using EsSense Profile and Consumer Defined Check-All-That-Apply Methodologies. *Food Qual. Prefer.* **2013**, *28*, 193–205. [CrossRef]
116. King, S.C.; Meiselman, H.L.; Thomas Carr, B. Measuring Emotions Associated with Foods: Important Elements of Questionnaire and Test Design. *Food Qual. Prefer.* **2013**, *28*, 8–16. [CrossRef]
117. De Wijk, R.A.; Kaneko, D.; Dijksterhuis, G.B.; van Zoggel, M.; Schiona, I.; Visalli, M.; Zandstra, E.H. Food Perception and Emotion Measured over Time In-Lab and in-Home. *Food Qual. Prefer.* **2019**, *75*, 170–178. [CrossRef]
118. Andaleeb, S.S.; Caskey, A. Satisfaction with Food Services. *J. Foodserv. Bus. Res.* **2007**, *10*, 51–65. [CrossRef]
119. Li, X.E.; Jervis, S.M.; Drake, M.A. Examining Extrinsic Factors That Influence Product Acceptance: A Review. *J. Food Sci.* **2015**, *80*, R901–R909. [CrossRef]
120. Hetherington, M.; Havermans, R.C. Sensory-Specific Satiation and Satiety. In *Satiation Satiety Control Food Intake*; Woodhead Publishing: Sawston, UK, 2013; pp. 253–269. [CrossRef]
121. Weijzen, P.L.G.; Liem, D.G.; Zandstra, E.H.; de Graaf, C. Sensory Specific Satiety and Intake: The Difference between Nibble- and Bar-Size Snacks. *Appetite* **2008**, *50*, 435–442. [CrossRef]
122. Nguyen, Q.C.; Varela, P. Identifying Temporal Drivers of Liking and Satiation Based on Temporal Sensory Descriptions and Consumer Ratings. *Food Qual. Prefer.* **2021**, *89*, 104143. [CrossRef]
123. Hamdi, N.; Ellison, B.; McCaffrey, J.; Metcalfe, J.J.; Hoffman, A.; Haywood, P.; Prescott, M.P. Implementation of a Multi-Component School Lunch Environmental Change Intervention to Improve Child Fruit and Vegetable Intake: A Mixed-Methods Study. *Int. J. Environ. Res. Public Health* **2020**, *17*, 3971. [CrossRef]
124. Toma, A.; Omary, M.B.; Marquart, L.F.; Arndt, E.A.; Rosentrater, K.A.; Burns-Whitmore, B.; Kessler, L.; Hwan, K.; Sandoval, A.; Sung, A. Children's Acceptance, Nutritional, and Instrumental Evaluations of Whole Grain and Soluble Fiber Enriched Foods. *J. Food Sci.* **2009**, *74*, H139–H146. [CrossRef]
125. Buzby, J.C.; Guthrie, J. *Plate Waste in School Nutrition Programs: Final Report to Congress*; Economic Research Service, U.S. Dept. of Agriculture: Washington, DC, USA, 2002.
126. Chawner, L.R.; Blundell-Birtill, P.; Hetherington, M.M. An Online Study Examining Children's Selection of Vegetables at Mealtimes: The Role of Meal Contexts, Variety and Liking. *Appetite* **2022**, *169*, 105803. [CrossRef] [PubMed]
127. Vanhatalo, S.; Liedes, H.; Pennanen, K. Nature Ambience in a Lunch Restaurant Has the Potential to Evoke Positive Emotions, Reduce Stress, and Support Healthy Food Choices and Sustainable Behavior: A Field Experiment among Finnish Customers. *Foods* **2022**, *11*, 964. [CrossRef] [PubMed]
128. Vermote, M.; Versele, V.; Stok, M.; Mullie, P.; D'Hondt, E.; Deforche, B.; Clarys, P.; Deliens, T. The Effect of a Portion Size Intervention on French Fries Consumption, Plate Waste, Satiety and Compensatory Caloric Intake: An on-Campus Restaurant Experiment. *Nutr. J.* **2018**, *17*, 43. [CrossRef] [PubMed]
129. Lindke, A.R.; Smith, T.A.; Cotwright, C.J.; Morris, D.; Cox, G.O. Plate Waste Evaluation of Plant-Based Protein Entrees in National School Lunch Program. *J. Nutr. Educ. Behav.* **2022**, *54*, 12–19. [CrossRef] [PubMed]
130. Liz Martins, M.; Rodrigues, S.S.P.; Cunha, L.M.; Rocha, A. Factors Influencing Food Waste during Lunch of Fourth-Grade School Children. *Waste Manag.* **2020**, *113*, 439–446. [CrossRef]
131. Yeomans, M.; Weinberg, L.; James, S. Effects of Palatability and Learned Satiety on Energy Density Influences on Breakfast Intake in Humans. *Physiol. Behav.* **2005**, *86*, 487–499. [CrossRef]

foods

Article

Green-Labelled Rice versus Conventional Rice: Perception and Emotion of Chinese Consumers Based on Review Mining

Heng Xu [1,*], Mengyun Xiao [1], Jun Zeng [1] and Huihui Hao [2]

[1] School of Management, Henan University of Technology, Zhengzhou 450001, China
[2] Faculty of Agribusiness and Commerce, Lincoln University, Lincoln 7647, New Zealand
* Correspondence: xuheng@haut.edu.cn

Abstract: The COVID-19 pandemic increased public health awareness, changing consumers' sensitivity and beliefs about food health. Food anxiety and health scares turn consumers toward safe and healthy foods to strengthen their immunity, which makes green food more popular. However, it remains unclear how to understand the gap between consumer intention to purchase green food and their actual purchasing behaviour. Taking rice as an object of study, comparing differences in consumer perceptions and emotions towards green-labelled rice and conventional rice is beneficial for understanding the components and psychological characteristics of consumer perceptions of green food. Therefore, we used topic modelling and sentiment analysis to explore consumers' focus of attention, attitudinal preferences, and sentiment tendencies based on the review (n = 77,429) from JD.com. The findings revealed that (1) consumers' concerns about green-labelled rice are increasing rapidly, and most have a positive attitude; (2) consumers of green-labelled rice are more concerned about origin, aroma, and taste than conventional rice; (3) consumers of conventional rice are more concerned about the cost-performance ratio, while consumers of green-labelled rice are also price-sensitive; (4) green label mistrust and packaging breakage during logistics are the leading causes of negative emotions among consumers of green-labelled rice. This study provides a comparative analysis of consumer perceptions and emotions between the two types of rice, thus revealing the main influencing factors of the intention-behaviour gap and providing valuable consumer insights for the promotion of green consumption and the sustainable development of the green food industry.

Keywords: green food; consumer perception; topic modeling; sentiment analysis

Citation: Xu, H.; Xiao, M.; Zeng, J.; Hao, H. Green-Labelled Rice versus Conventional Rice: Perception and Emotion of Chinese Consumers Based on Review Mining. *Foods* **2023**, *12*, 87. https://doi.org/10.3390/foods12010087

Academic Editor: Witoon Prinyawiwatkul

Received: 25 October 2022
Revised: 15 December 2022
Accepted: 22 December 2022
Published: 24 December 2022

1. Introduction

The popularity of green consumption has led to a growing interest in green food [1]. In particular, the pandemic has increased public awareness of food safety and environmental awareness, and strongly influenced consumers' habits and behaviours [2], leading to healthier lifestyles [3]. Due to personal concerns about safety and health, consumers are eager to strengthen their immunity through healthy eating [4]. With its specific attributes, green food meets the current demand [5]. Research demonstrated that consumers' perceived severity of the epidemic and their own safety awareness contribute to their intention to purchase green food [6]. Qi et al. [5] also concluded that the outbreak of the COVID-19 pandemic has strongly influenced consumers' habits and behaviours, creating a more sustainable and healthier era of consumption. In this context, there is a high interest in the sustainability of food choices [7,8]. However, understanding the gap between consumers' willingness to purchase green food and their purchasing behaviour remains a pressing issue [9]. In order to identify the causes of the "intention-behaviour" gap, it is essential to explore consumers' attitudinal preferences and emotional dispositions towards green-labelled food and conventional food.

Green food is produced in an excellent ecological environment and meets technical standards of safety, nutrition, and non-pollution [10]. Due to the strict control of the ecological environment where agricultural products are produced, green food not only prevents people from food-borne hazards caused by food contamination but also plays a crucial role in reducing carbon emissions and protecting the natural ecological environment. Besides, it is widely believed that green food is produced in an environmentally friendly way and is healthier than food produced traditionally [11].

Nearly half of the world's population relies on rice as a staple food. Although rice grown in a green sustainable environment sells for a higher price in the global market, it is increasingly attracting consumer attention [12,13]. Moreover, people are willing to pay for rice with environmental attributes (i.e., green-labelled rice) rather than rice without environmental attributes (i.e., conventional rice) [14]. The study showed that consumer attitudes toward green-labelled rice are positively correlated with concerns about food safety and the perceived importance of environmental consequences [15]. In addition to health and environmental protection, a traceable label certification system is a significant factor in consumers' positive attitude towards green-labelled rice, even among those with little knowledge of eco-labelling standards [13,16].

Product reviews on e-commerce platforms generally require consumers to confirm receipt before they post anonymously. These reliably reflect consumers' authentic evaluations and opinions on the quality, price, and experience of the purchased products, providing a large amount of data to support our study of consumer attitudes and behaviour from the user perspective [17,18]. Therefore, through text mining we can understand the potential focus of consumers, explore the actual trend of consumers' green consumption during the COVID-19 pandemic, and analyse the difference between consumers' focus and attitude when buying green-labelled rice and conventional rice. Based on the above results, through further use of the sentiment analysis method, we can observe the changes in the emotional characteristics of consumers towards different food products [19] and analyse the thematic preferences of positive and negative emotions of consumers towards two types of rice. These results will help us to compare the differences in consumers' perceptions and emotions towards different food products and explore consumers' purchasing behaviour.

Taking green-labelled rice and conventional rice as the research object, this study selects user-generated content (i.e., product reviews) from JD.com in China as the data source. Then, the topic modelling and sentiment analysis are adopted to extract consumers' attention and compare consumer attitudes and sentiments towards the two types of rice, so as to gain insight into the components and psychological characteristics of consumers' perception of green food.

2. Literature Review

2.1. Consumer Attitude towards Green-Labelled Rice

Consumer attitudes are the main drivers behind consumption behaviour [20], and consumer attitudes towards green-labelled rice significantly influence their choices [21]. Green-labelled rice is significantly better than conventional rice in terms of nutritional quality and has fewer heavy metals than conventional rice, making it more compliant with safety and environmental standards [22]. Most studies have shown consumers' positive attitudes towards green-labelled rice rich in vitamins and other nutrients [13]. In China, consumers have shown a favourable preference for green-labelled rice due to health benefits and environmental protection considerations [16].

Previous studies have focused on the factors influencing the purchase intention of green-labelled rice. Firstly, health and food safety are the essential attributes for consumers to choose green-labelled rice [22], as people generally believe green-labelled rice is healthier and safer than conventional rice [23]. Secondly, in the context of a low-carbon society, consumers' subjective environmental knowledge has a significant effect on the purchase of green-labelled rice with environmental attributes [24]. This is because the production of green products is less polluting to the environment, and it is a prosocial and environmentally friendly behaviour [25]. Finally, consumer trust in green labels is an essential factor influencing consumers' purchasing behaviour [26]. It contributes to the purchase and attitudinal loyalty of green products, and a nutritional label or brief claims will improve consumers' attitudes towards rice [26]. Samant and Seo [27] also concluded that the effects of sustainability-related label claims on quality perception and acceptability of food become significantly more pronounced when consumers understand and trust the label claims.

Emotions are considered to be important drivers of food-related cognition and behaviour [28]. According to the technology acceptance theory, the acceptance of new technology or models demand positive attention and sufficient understanding from consumers [29]. By paying attention to the new things and perceiving their characteristics, consumers generate sentimental preference for these new things and then produce a behavioural intention [30]. The traditional approach of consumer sentiment research is mainly through verbal self-reporting questionnaires [31] which is easy to apply, cost-effective, and discriminatory, but may not capture the full range of emotions individuals might experience in response to food, and therefore may not properly measure food-evoked emotions [32,33]. Alternative methods based on big data for sentiment analysis are becoming popular [34], and sentiment classification methods using machine learning and deep learning are widely used in lots of studies. Compared to traditional self-reporting questionnaire methods, sentiment analysis, through online reviews, allows us to capture the emotions evoked by consumers' direct experience with food [28], enabling a more objective determination of consumers' emotions and attitudes toward green foods. It is worth noting that although some scholars have proposed the main reasons that affect consumers' negative attitudes towards green-labelled rice as the distrust and lack of knowledge about green labels [18], no study has yet illustrated consumers' positive and negative attitudes towards buying green-labeled rice by comparison with conventional rice.

Consumer attitudes are linked to a complex set of ideas, motivations, and experiences [35]. The sustainability of the green-labelled rice industry depends on consumers' concerns and attitudes toward green-labelled rice [36]. Therefore, a comparative analysis of consumers' attitudinal preferences and emotional dispositions towards the two types of rice based on user-generated content provides a new perspective for the green food sector and helps to understand the gap between consumers' willingness to buy green food and their actual purchasing behaviour.

2.2. Text Mining in Green Labeled Food

Green food in China refers to a wide array of primary and processed agricultural products that are safe, nutritious and of high quality for human consumption. Hassan et al. [37] argue that the term "green food" considers not only the culture of food production, but also the protection of the environment, economics of stakeholders, and social relationships. Organic food, which also has safety and environmental attributes, is not accepted widely among the population due to the most stringent certification standards. In contrast, green food has higher consumer awareness and is widely followed in China [38], and researchers are increasingly interested in green food consumption. Therefore, this study also focuses on green food.

Most previous studies have used traditional questionnaires or interviews to explore consumers' preferred motivations for purchasing green-labelled rice [16,39], as well as their purchase intentions and influencing factors [24]. However, regional surveys are limited to the themes covered by research questions and usually only capture small or moderate sample sizes (Table 1), making it challenging to explore the perceptions and attitudinal preferences of multi-faceted populations on health issues [40].

Table 1. Summary of Literature in the Field of Green-Labelled Rice.

Author	Method	Sample size	Conclusion
Qi & Ploeger [5]	Interview	28	The COVID-19 pandemic has increased consumers' willingness to buy green food. The high price of green food, the problem of unavailability, the problem of mistrust, and limited knowledge are the main factors that trigger the IBG (Intentional Behavioural Gap).
Tong et al. [24]	Questionnaire	622	Subjective environmental knowledge and concerns about food quality have a significant impact on consumers' willingness to purchase green-labelled rice, and individual socio-demographic characteristics affect consumers' choice of green-labelled rice, including age, education, health status, and income.
My et al. [41]	Experiment	199	Consumers willing to pay premiums for sustainably produced rice are more health-conscious; have better knowledge and greater trust in food quality certification for rice; and tend to be more environmentally conscious. Enhancing consumers' understanding and trust in food quality certification can help increase consumers' acceptance of sustainably produced rice.
Anang et al. [42]	Interview	100	The most preferred attributes of rice are taste, cooking quality, cooking time, and aroma, and consumers are willing to pay higher premiums for the aroma and origin of rice. In contrast, the least preferred attributes are price, impurities, and product origin.
Hao et al. [43]	Text mining	25,000	Package integrity, delivery timeliness, door-to-door delivery, and service responsiveness are the most important logistical factors for consumers when purchasing rice and produce.
Hu et al. [14]	Content analysis	142,158	Green-labelled rice outperformed conventional rice in terms of appearance and cooking quality. However, in terms of protein content, there was no obvious difference between green-labelled rice and conventional rice.

Considering the anonymity of the Internet, consumers tend to express their opinions and tastes freely on review sites and share genuine reviews and experiences of products [44,45]. Online reviews thus convey emotional information about consumers and reflect their attitudes towards products [46]. The data from online reviews can be mined to obtain more practical and objective information, as it is not influenced by the subjective will of the sender [43].

Text mining approaches are gaining popularity in research in the field of food science [40]. We can use the LDA model to extract the focus of consumers' attention on green-labelled food from a large amount of review content, and sentiment analysis methods can be used to analyse consumers' attitudinal preferences towards green-labelled food. In recent studies, Hao et al. [43] analysed the logistical factors affecting consumer satisfaction when purchasing rice and produce through text mining techniques based on consumer comments on the JD.com. Singh et al. [18] analysed public attitudes towards green-labelled foods by using thematic clustering and the sentiment lexicon approach based on Twitter comments. Thus, text mining methods can offer more objective insights into consumer attitudes towards green-labelled rice to provide deeper insights and more diverse recommendations [47], which can help promote sustainable development in the green food sector.

3. Methodology

As shown in Figure 1, firstly, we used web crawling techniques to obtain consumer review data on green-labelled rice and conventional rice in JD.com and preprocessed them accordingly. Secondly, the LDA topic model was used to extract topics from the review text in order to uncover the focus of consumer attention and to conduct a significant difference analysis. Finally, the SnowNLP was used for sentiment classification to identify consumers' sentiment tendencies and analyse the causes of different emotions through word frequency statistics.

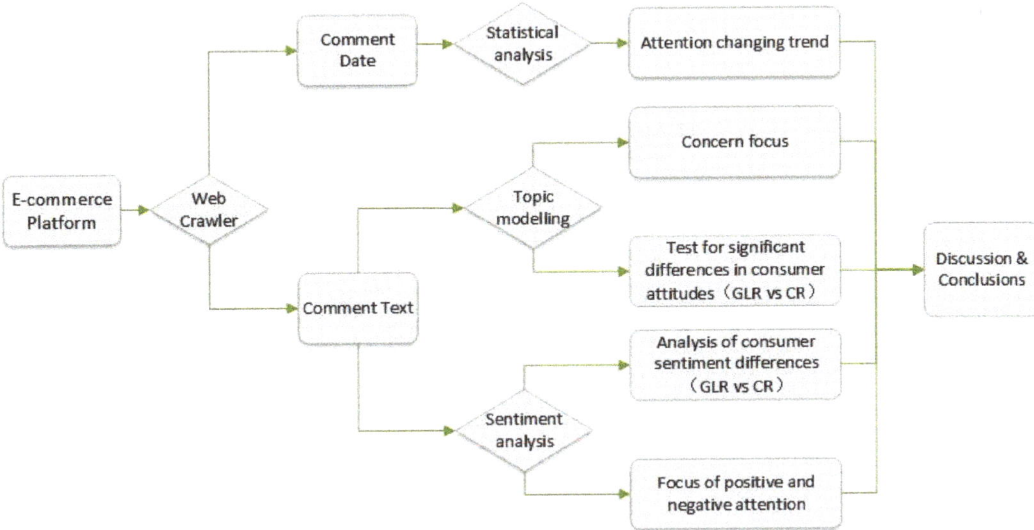

Figure 1. The analysis framework of review mining.

3.1. Chinese Green Food Label

The green food label (Figure 2.) is a product quality certification mark officially registered by the Green Food Development Center of China with the Trademark Office of the State Administration for Industry and Commerce. Green food regulations and standards were developed and established following the Codex Alimentarius programmed by the Codex Alimentarius Commission (CAC) [38], which is a unique labelling certification in China [16]. It consists of three parts: the sun above; the leaf below and the bud in the center; and the logo is a square circle, meaning protection and safety. The logo indicates that the environmental quality of the place where the product is made conforms to the Environmental Quality Standards for Green Food Production Areas. The production process is strictly under the guidelines for the use of green food production materials and the requirements of the production operation procedures, meeting the technical standards of safety, nutrition, and non-pollution. The logo is awarded by the Green Food Development Center and the Green Food Certification Review Committee after a systematic assessment and certification. Green food follows a "from farm to fork" control principle similar to the Hazard Analysis Critical Control Point (HACCP) system [38].

Figure 2. Chinese green food label.

3.2. Data Collection

This study chose JD.com as our data source. JD.com is one of the most popular B2C platforms in China's e-commerce sector, with 570 million active annual users and a well-established consumer post-purchase evaluation mechanism [48]. According to the sales ranking of the products on JD.com and the fan base of the store where the product belongs, the rice brands that are well-known and popular among consumers were selected as follows: October Paddy (fans: 15,168,000), San Ren Xing (fans: 1,226,000), Tian Yuan Dao (fans: 2,159,000), Chai Huo Da Yuan (fans: 10,107,000), Qiao Fu Da Yuan (fans: 299,000), and National Treasure Qiaomi (fans: 1,050,000). In addition, the number of product comments of these brands is over 100,000, providing us with rich data from which to extract valuable information using text mining methods [49].

In April 2022, we used the Python crawler to obtain product reviews for the above brands of green-labelled rice and conventional rice products, including user ID, comment posting times, and comment texts. A total of 82,398 raw comments were obtained, of which 41,394 were for green-labelled rice and 41,004 were for conventional rice. Furthermore, we pre-processed the raw comment data as follows.

(1) Data denoising. Remove interfering information from the original comments, including duplicate comments, advertisements, comments that are irrelevant to the research, and meaningless words and symbols.

(2) The stopword dictionary is added. In order to improve the efficiency of word separation, a total of 1550 deactivated words were aggregated from the Chinese deactivation table of HIT and the Chinese deactivation table of Baidu [49].

(3) A customized word segmentation dictionary is constructed. Considering the complexity of online content, some particular words could not be identified in the word separation process. For example, the specialized words of rice types such as "long-grain rice" and "pearl rice"; the emotional words from evaluation such as "not good" and "don't like"; and the e-commerce words such as "home delivery" and "logistics speed". In order to improve the accuracy of word separation, we developed a custom dictionary through manual supervision, and the specific process is as follows: ① obtain the subject words with high word frequency in the product titles and the high-frequency words expressed by consumers; ② add the above words to the custom dictionary and carry out word separation; ③ according to the word frequency statistics of word separation, add the words with unreasonable word separation to the custom dictionary through manual checking, a total of 30 words (Table 2).

Table 2. Custom dictionaries.

No.	Word (Chinese)	Word (English)	No.	Word (Chinese)	Word (English)
1	不好吃	Not good	16	送货上门	Home delivery
2	不香	Not fragrant	17	物流速度	Logistics speed
3	不喜欢	Don't like	18	发货速度	Shipping speed
4	不划算	Not a good deal	19	京东快递	JDL express
5	米香	Rice fragrant	20	真空包装	Vacuum packed
6	新米	New rice	21	包装破损	Packaging broken
7	陈米	Stale rice	22	漏气	Leaking
8	长粒香	Long-grain rice	23	原产地	Origin
9	珍珠米	Pearl rice	24	绿色食品	Green food
10	旧米	Old rice	25	生产日期	Production date
11	五常大米	Wuchang rice	26	煮粥	Cooked porridge
12	颗粒均匀	Uniform grains	27	降价	Price reduced
13	颗粒饱满	Full of grains	28	性价比	Value for money
14	晶莹剔透	Crystal clear	29	发霉	Mouldy
15	软糯	Soft and sticky	30	国家地理标志	National geographical indication

In the end, we obtained 77,429 valid reviews, of which 38,656 and 38,772 were for green-labelled rice and conventional rice, respectively.

3.3. Topic Modeling with LDA

This study uses the LDA model for thematic clustering of product reviews to discover the focus and attitudes of consumers towards green-labelled rice and conventional rice. The LDA is an unsupervised Bayesian probabilistic model for discovering the hidden semantics of textual data and contains a three-layer structure of words, topics, and documents [50]. The model is not only applicable to short Chinese texts but also has good reliability [51,52].

By training the model, the optimal number of topics can be found based on the topic confusion, and lower confusion values represent a better model fit [53]. However, the simplicity and interpretability of the textual content should be taken into account when selecting the number of topics [50]. By setting the different numbers of topics for the experiment, we obtained (num_topics, perplexity) = $\{\cdots(9, 49{,}891.04), (10, 49{,}671.81), (11, 50{,}148.18)\cdots\}$. When num_topics = 10 we get the minimum value of perplexity and therefore, the number of topics is determined as 10 in this research.

3.4. Sentiment Analysis

This study uses SnowNLP, which is more established in the Chinese language domain, to perform sentiment analysis on product reviews [54]. The model is based on a machine learning Bayesian algorithm to train and predict the data, normalizing the sentiment score of each text to be between 0 and 1. The results of the model indicate the probability of the sentiment, with close to 1 being positive and close to 0 being negative [55]. The original training data of this model are different from the text data of this study. Therefore, in order to improve the accuracy of the sentiment calculation, the model needs to be retrained.

(1) We conducted a random sampling based on 10% of the number of valid comments. A total of 7743 comments were sampled to build the corpus.

(2) We invited two PhD level researches in food science and customer relationship management to individually annotate each comment in the corpus with positive and negative sentiment polarity within five working days. Comments with consistent annotation results were formally admitted into the corpus. For comments with inconsistent results, we invited a professor of management to perform a secondary annotation, taking the majority (2:1) as the final sentiment polarity according to the annotation results and eliminating the controversial comments. We obtained 7564 validly annotated comments in the end.

(3) In total, 80% of the valid annotated comments were used as the training set and 20% as the test set. The Sentiment module of SnowNLP was called for training and testing, and the original model was compared with the trained model. The results showed that the accuracy of the original model was 80.37%, the precision was 79.38%, and the recall rate was 79.28%; the accuracy of the trained model was 88.71%, the precision was 93.82%, and the recall rate was 91.08%. Therefore, the sentiment analysis model established in this study can effectively handle review texts in the food domain.

(4) The sentiment value (between 0 and 1) was calculated for the remaining 90% of valid comments using the retrained model. Sentiment values ≥ 0.6 were taken as positive sentiment and ≤ 0.4 as negative sentiment and comments with insignificant sentiment between 0.4 and 0.6 were excluded [56].

3.5. Statistical Tests

(1) We used SPSS to conduct a one-way ANOVA to determine whether there was a significant difference between the topics of green-labelled rice and conventional rice by comparing the mean weights of each topic.

(2) We used independent sample T-Tests to assess which topics differed significantly in mean weights. When the sample size is large, the significance level should be set at a lower level [57], and therefore based on $\alpha = 0.05 \times \sqrt{100/N}$ [58], we set the significance level at 0.0005.

(3) We calculated the absolute effect size of Cohen's d by dividing the mean difference by the combined standard deviation [57]. Referring to the extended Monte Carlo study [59], the effect sizes were interpreted as follows: trivial ($0 \leq d < 0.2$), small ($0.2 \leq d < 0.5$), medium ($0.5 \leq d < 0.8$), and large ($0.8 \leq d \leq 1$) [60].

4. Results

4.1. Consumer Concerns

We counted the number of consumer comments on green-labelled rice and conventional rice between January 2016 and April 2022. Figure 3 shows the fitted curves and the curves in the level of consumer attention to green-labelled rice and conventional rice. These two curves show the nearly same changing trends in consumer attention. The R^2 of the two types of rice converges to 1, indicating a good fit. Overall, consumer attention in both types of rice has increased year after year. The introduction of national food safety policies between 2017 and 2018 has led consumers to focus on safer and more environmentally friendly green foods [61], with consumer concern in green-labelled rice overtaking conventional rice for the first time between 2017 and 2018, and consumer concern for green-labelled rice has proliferated since then. The estimated trend of the fitted curve shows that in the future, consumers will pay more attention to green-labelled rice than conventional rice.

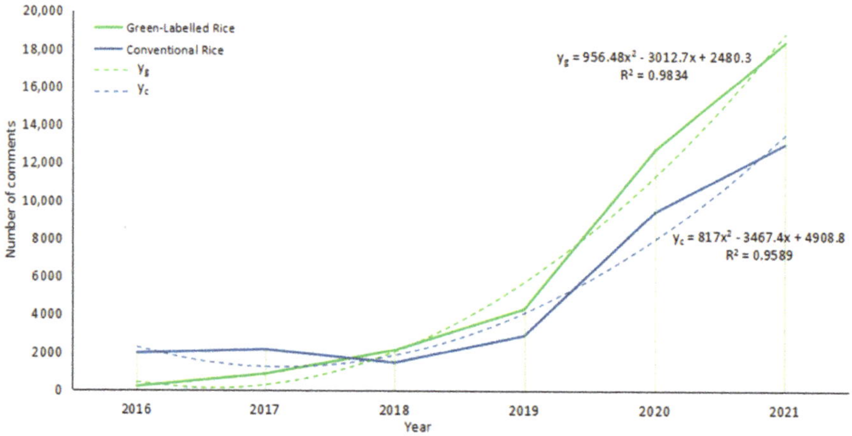

Figure 3. Level of attention on green-labelled rice and conventional rice.

Based on the LDA model, we extracted ten meaningful and relevant topics. These topics cover a range of issues, such as consumers' perception of product features, the experience of the purchase process, and subjective evaluation of the product. We provided a name for each topic based on the high-frequency feature words extracted, as shown in Table 3. They are logistics speed, origin, taste and flavour, appearance characteristics, price, aroma, product packaging, impurity content, quality evaluation, and production date. The intensity of topics is obtained by calculating the probability distribution of documents representing users' degree of attention regarding a topic [62]. Furthermore, the probabilities of each document belonging to different topics are the topics to calculate.

The extracted topics are potential influence factors that consumers were concerned about when purchasing rice [17], and they are based on the sensory experience of consumers experiencing the food. The human-food interaction is a multi-module experience in which consumers rarely separate individual senses independently [63]. When confronted with food stimuli, people perceive and integrate information from all senses through unconscious neurophysiological processes [28]. In the case of rice, this sensory information includes visual (appearance characteristics, product packaging, and impurity content), olfactory (aroma), and gustatory (taste and flavour) sensations. Appearance characteristics, impurity content, aroma, taste, and flavour reflect the quality of rice to a certain extent, while product packaging is a concern of consumers about the integrity of packaging and food safety during food transportation. At the same time, the essential attributes of rice, such as origin, price, and production date, are also factors of concern to consumers.

In addition, for the e-commerce platform, as a channel for consumers to shop, logistics speed is the most important factor for consumers. Its transportation efficiency and service quality will also have a certain degree of influence on consumers' experience of buying green-labelled rice.

Table 3. Topic Identification and Sorting of Intensity.

No.	Topic Identification	Intensity of Topic	Feature Words
1	Logistics speed	0.139834305	Logistics, Jingdong, speed, express, soon, service, home delivery, satisfaction, service attitude, epidemic
2	Origin	0.12553394	Northeast, Wuchang, ecological, Heilongjiang, South, Hubei, origin, trust, quality, Jingshan
3	Taste and flavour	0.121185015	Taste, flavour, tasty, delicious, soft, fluffy, sweet, fragrant, delicate, palatable
4	Appearance characteristics	0.115346825	Rice, full, grainy, crystal clear, clean, evenly grained, fresh, colour, translucent, broken rice
5	Price	0.08884901	Price, activity, cheap, value for money, discount, bargain, affordable, support, supermarket, guarantee
6	Aroma	0.088206845	Aroma, smell, fragrant, rice, delicious, worth, trust, brand, fragrant rice, aromatic
7	Product packaging	0.08736981	Vacuum packed, leaky, outer packaging, broken, epidemic, tight, shipping, complete, sealed, fine, intact, sturdy
8	Impurity content	0.08170846	Impurities, insects, rubbish, stale rice, lousy review, disappointment, broken rice, mouldy, yellowing, white spots
9	Quality evaluation	0.07233044	Quality, satisfied, loved, great, joyous, recommended, poor, new rice, affordable, five stars
10	Production date	0.070206845	Rice, fresh, date, taste, date of production, colour, old rice, month, new rice, local

4.2. Differences in Consumer Attitudes

Based on the LDA model, we portrayed the weight distribution of green-labelled rice and conventional rice on different topics (Figure 4). For green-labelled rice, origin, aroma, taste, and flavour are most important to consumers. For conventional rice, appearance characteristics, taste, and flavour are most important to consumers. In addition, the speed of logistics in e-shopping experience is of great concern to consumers of both types of rice.

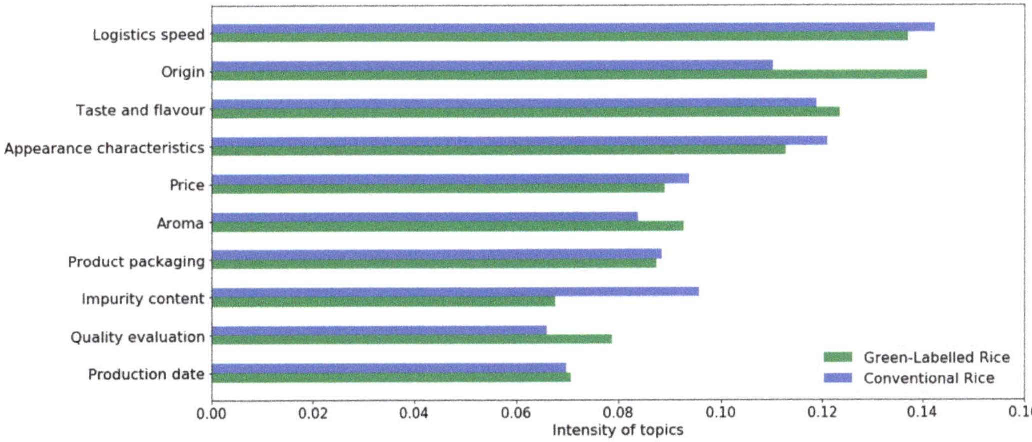

Figure 4. Comparison of topic weights for green-labelled rice and conventional rice.

Consumer attitudes towards the two types of rice differed under the same topic. Based on the topic weights, we conducted a one-way ANOVA and T-Test and used Cohen's d

Mean to indicate the magnitude of the differences (Table 4). Prior to this, we performed a normality analysis of the data. Due to the large sample size, we chose a K-S test. The results showed that the p-value > α = 0.0005, which indicated that the data were approximately normally distributed and could be subjected to subsequent statistical analysis. The results showed significant differences in consumers' attitudes towards green-labelled rice and conventional rice across eight topics (intensity of topics = 0.8424), except for the two topics, which are the production date and the product packaging (intensity of topics = 0.1576).

Table 4. Comparison of Green-Labelled Rice (GLR) and Conventional Rice (CR).

No.	Topic	ANOVA (F-Value)	T-Test Result	Cohen's d Mean	Effect Size
1	Logistics speed	13.114 *	*GLR < CR	0.1	Trivial
2	Origin	561.437 *	*GLR > CR	0.2	Small
3	Taste and flavour	11.983 *	*GLR > CR	0.1	Trivial
4	Appearance characteristics	113.309 *	*GLR < CR	0.1	Trivial
5	Price	95.328 *	*GLR < CR	0.1	Trivial
6	Aroma	78.356 *	*GLR > CR	0.2	Small
7	Product packaging	0.005	NS	NS	NS
8	Impurity content	776.172 *	*GLR < CR	0.2	Small
9	Quality evaluation	211.099 *	*GLR > CR	0.2	Small
10	Production date	1.040	NS	NS	NS

Not significant (NS): adjusted p-value > 0.0005, significant (*): adjusted p-value $\leq$ 0.0005.

According to the T-Test results, consumers who buy green-labelled rice are more concerned with origin, taste and flavour, aroma, and quality evaluation than conventional rice. Consumers who buy conventional rice are more concerned with logistics speed, appearance characteristics, price, and impurity content than green-labelled rice. Furthermore, according to Cohen's d Mean value and Effect Size results, there is a trivial (d = 0.1) difference in consumer concerns about logistics speed, taste and flavour, appearance characteristics and price between green-labelled rice and conventional rice. However, the difference between concerns is less than 0.2 standard deviations, and the difference is insignificant and not meaningful [60,64]. In addition, there is a small (d = 0.2) difference in consumer concerns about the origin, aroma, impurity content, and quality evaluation.

Consumers who buy green-labelled rice are more aware of the origin and pay more attention to the quality of the rice. Consumers who are concerned about food safety are also usually concerned about origin information [16], as origin will match quality ratings [65] and the origin information on product packaging will influence consumer perceptions of product quality [66]. The green label means that the environment of origin meets environmental standards, so the quality and safety of green-labelled rice are more assured.

The aroma is a more distinctive feature of green-labelled rice, and this is what consumers look for. Green-labelled rice has higher culinary quality than conventional rice [14], and its overall quality and taste are better [65]. However, the results showed that the difference in consumer attitudes towards taste and flavour between the two types of rice was trivial (d = 0.1). The sense of smell affects consumers' enjoyment and perception of food [67], and the difference in consumer attitudes towards rice aroma between the two types of rice is also relatively large (d = 0.2).

Foods' overall liking and purchase intent are influenced by visual inputs [68]. Consumers of conventional rice were more concerned about impurities in rice (d = 0.2). Combining the characteristic words extracted from the impurity content topic, such as: "worm", "mouldy", and "yellowing", indicates that significantly more problems with impurities occur in conventional rice than in green-labelled rice, which also means that there is still more room for improvement in quality control of conventional rice.

It is worth noting that price is an essential consideration for consumers when purchasing goods. The price of green-labelled rice is also generally higher than conventional rice. Previous research [21] has suggested that consumers who purchase green-labelled rice are not sensitive to price. However, we found a trivial (d = 0.1) difference in consumer attitudes towards price between the two types of rice, such that this difference was not practically significant. It shows that consumers who buy green-labelled rice are equally sensitive to price.

4.3. Differences in Consumer Sentiment

Food-elicited emotion is increasingly becoming critical for product differentiation [69]. Based on the sentiment values calculated by the SnowNLP model, we counted the number of positive and negative comments on each topic for both types of rice. As shown in Table 5, except for the topic of impurity content, the number of positive comments on each of the remaining nine topics for green-labelled rice was more significant than the number of negative comments. In total, 82.65% of consumers had positive sentiments towards green-labelled rice, 14.82% had negative sentiments, and 2.53% had little sentiment, indicating that consumers were relatively more satisfied with green-labelled rice. The non-parametric test results show that consumers have higher positive emotions about the logistics speed, origin, taste and flavour, and aroma of green-labelled rice, and more negative emotions about price, product packaging, impurity content and production date. In contrast, they had a more negative sentiment towards the impurity content of the rice and product packaging. In comparison, 63.62% of consumers had positive sentiments towards conventional rice, 33.50% had negative sentiments, and 2.87% had little sentiments. The results of the comparison of the topics of positive and negative emotions show that the number of positive emotions is higher in logistics speed, origin, appearance characteristics, and aroma. Whereas, the negative sentiment towards conventional rice mainly focused on rice quality, which shows that conventional rice has an apparent disadvantage in terms of quality.

Table 5. Distribution of the number of positive and negative comments for green-labelled rice and conventional rice and comparison of positive (Pos) and negative (Neg) topics.

No.	Topic	Green-Labelled Rice			Conventional Rice		
		Positive	Negative	Pos vs. Neg	Positive	Negative	Pos vs. Neg
1	Logistics speed	5296	448	*Pos > Neg	5122	1040	*Pos > Neg
2	Origin	5247	338	*Pos > Neg	3616	594	*Pos > Neg
3	Taste and flavour	4365	549	*Pos > Neg	2986	1394	*Pos < Neg
4	Appearance characteristics	3581	455	NS	3928	981	*Pos > Neg
5	Price	2068	486	*Pos < Neg	1691	1262	*Pos < Neg
6	Aroma	2740	156	*Pos > Neg	1841	370	*Pos > Neg
7	Product packaging	1843	901	*Pos < Neg	869	1940	*Pos < Neg
8	Impurity content	559	1253	*Pos < Neg	376	2965	*Pos < Neg
9	Quality evaluation	1884	237	NS	1004	473	*Pos < Neg
10	Production date	1215	341	*Pos < Neg	735	655	*Pos < Neg
	Total	28,798	5164		22,168	11,674	

Not significant (NS): adjusted p-value > 0.0005, significant (*): adjusted p-value $\leq$ 0.0005.

The results of the T-test (Table 6) show a significant difference in the sentiment values between the two types of rice on 10 topics and that the probability of positive sentiment is greater for green-labelled rice than for conventional rice. In particular, in terms of product packaging, the difference in sentiment between the two types of rice is significant (d = 0.8), with consumers having an overall negative attitude towards the product packaging of conventional rice (mean (CR) = 0.3283 $\leq$ 0.4). In terms of rice impurity content, consumers have a low mean positive sentiment towards both types of rice, with both being negative. Consumers who purchase conventional rice have a higher probability of negative sentiment towards impurity content (1 − Mean (CR) = 1 − 0.1394 = 0.8606). In addition, the difference in the mean values of consumer sentiment towards the two types of rice in terms of speed of logistics, origin, and appearance characteristics is small, and both have a high probability of positive sentiment.

Table 6. Comparison of positive emotion probability between green-labelled rice (GLR) and conventional rice (CR).

No.	Topic	T-Test Result			Cohen's d Mean	Effect Size
		Mean (GLR)	Mean (CR)	GLR vs. CR		
1	Logistics speed	0.9074	0.8188	*GLR > CR	0.3	Small
2	Origin	0.9123	0.8414	*GLR > CR	0.3	Small
3	Taste and flavour	0.8623	0.6704	*GLR > CR	0.5	Medium
4	Appearance characteristics	0.8685	0.7923	*GLR > CR	0.2	Small
5	Price	0.7887	0.5690	*GLR > CR	0.5	Medium
6	Aroma	0.9257	0.8175	*GLR > CR	0.4	Small
7	Product packaging	0.6494	0.3283	*GLR > CR	0.8	Large
8	Impurity content	0.3149	0.1394	*GLR > CR	0.5	Medium
9	Quality evaluation	0.8658	0.6709	*GLR > CR	0.5	Medium
10	Production date	0.7576	0.5307	*GLR > CR	0.6	Medium

Not significant (NS): adjusted p-value > 0.0005, significant (*): adjusted p-value $\leq$ 0.0005.

Further, we explored the specific reasons for each sentiment by analysing the frequency of words in comments of green-labelled rice and conventional rice [70].

As shown in Figure 5a, the positive words of green-labelled rice mainly include "delicious", "Wuchang", "worthwhile", "new rice", "full", " fragrant", "reassuring", "family", and "green food". This shows that consumers have a positive attitude towards the quality, taste, and safety of green-labelled rice. As shown in Figure 5b, the positive terms of conventional rice include "logistics", "speed", "flavor", "affordable", "porridge", "fresh", "discount", "service", and "round". This shows that consumers are more concerned about the value of consumption and service experience and have a positive attitude towards the cost-performance ratio of conventional rice. In particular, there is more positive feedback from consumers on the logistics speed and appearance characteristics of traditional rice.

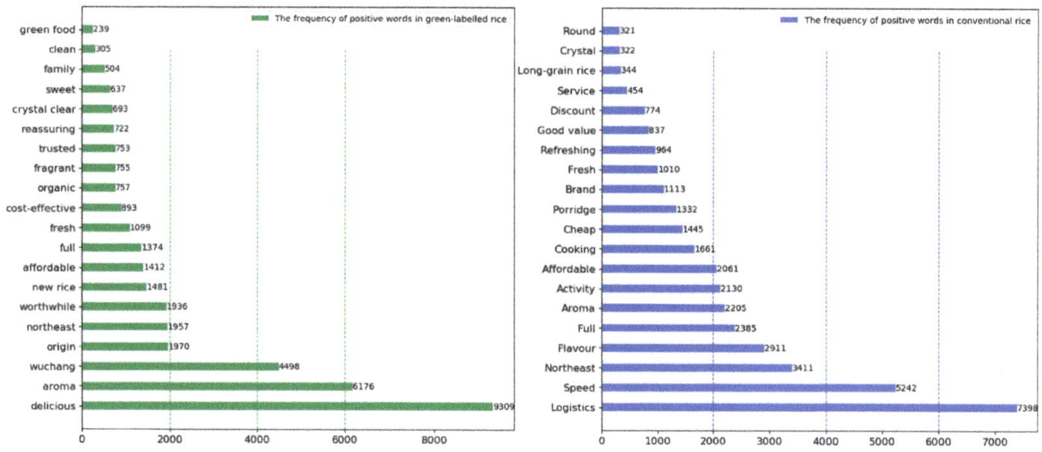

(a) **(b)**

Figure 5. (**a**) Frequency of positive emotion-related words for green-labelled rice; (**b**) frequency of positive emotion-related words for conventional rice.

As shown in Figure 6a, the negative terms of green-labelled rice mainly include "stale rice", "insects", "leaking", "disappointing", "musty", "broken", "unworthy", and "doubtful". This indicates that consumers' frustration with green-labelled rice is mainly reflected in distrust of the green food label and broken packaging. As shown in Figure 6b, the negative terms of conventional rice include "unpalatable", "leaky", "stale rice", "broken", "rubbish", "bugs", "unpleasant", and "sketchy". It shows that poor quality and broken packaging are the main reasons that consumers have negative sentiment, also, long storage time and excessive impurities impact consumer sentiment.

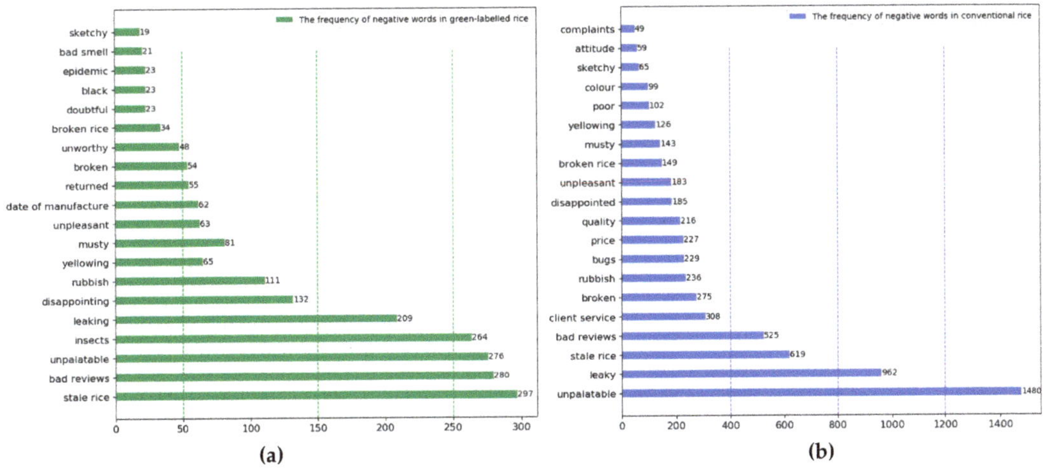

Figure 6. (**a**) Frequency of negative emotion-related words for green-labelled rice; (**b**) frequency of negative emotion-related words for conventional rice.

5. Discussion

Firstly, the pandemic is essential in raising consumer concerns about green-labelled rice. Public health incidents have increased consumer awareness of food safety, and health scares have prompted consumers to turn to healthier foods perceived as more natural [3]. Driven by egoism, consumers are more willing to purchase green food with quality assurance for their health [6,71]. Meanwhile, the pandemic has also triggered a sense of environmental concern and responsibility among consumers, leading to a greater awareness of personal social responsibility [24]. Environmental awareness has become a significant incentive for environmentally friendly food consumption [72]. Driven by altruism, consumers perceive the purchase of green food as an environmentally friendly act that demonstrates personal ethics and thus, holds positive attitudes towards green-labelled food. Consumers with high levels of altruistic environmental values are more likely to purchase green products [73]. This study supports Qi et al.'s [5] conclusion that consumers' ethical attitudes influence the willingness to purchase green-labelled food. In addition, the strong government support for environmental protection and green consumption under the COVID-19 pandemic are driving factors for consumer interest in green label rice [74].

Secondly, price is an essential factor that consumers consider when making purchase decisions [75,76]. Stolz et al. [21] indicated that consumers who prefer green-labelled foods are not price-sensitive, and those who are price-sensitive prefer conventional products. Baudry et al. [7] also concluded that consumers who buy green-labelled foods are less concerned about price. However, we found that consumers who buy green-labelled rice are more concerned about price, and price-sensitive consumers also buy green-labelled rice. On the one hand, people's income constraints or lower income expectations due to the pandemic could lead to spending less on green food [77] and caring about the price. Several studies have also confirmed that household income and expenditure affect consumers' consumption of green foods [8,78]. In addition, the level of health and safety concerns motivates consumers to pay a premium for green foods, even when constrained by income [79]. On the other hand, the anonymous commenting mechanism on e-commerce platforms does not merely directly facilitate the expression of opinions by usually reticent consumers [80], it also facilitates the expression of genuine opinions and attitudes towards the purchased product [80,81] for effectively providing consumers with emotional safety [44]. Thus, authentic and massive consumer reviews bring objective consumer insights and reflect rational consumer behaviour during the COVID-19 pandemic.

Finally, mistrust of the green label is the leading cause of negative sentiment. On the one hand, although most consumers have a positive attitude toward green-labelled food, some remain skeptical of the safety, nutritional, and environmental quality characteristics due to the lack of proper and objective understanding of the green labels [18]. On the other hand, the presence of a few counterfeit products and low-quality green-labelled rice in the market has led to a bias in consumer expectations, triggering skepticism towards green-labelled rice, including distrust of third-party labelling verification [26]. Lack of label trust reduces consumer expectations of the benefits of green-labelled foods and has been a barrier to developing the green-labelled food market [82]. In addition, several studies have demonstrated that green food label certification has a significant impact on consumer attitudes toward green foods [83,84]. Yin et al. [85] found that the trust of consumers with high-risk perceptions in food safety has been reduced to a very low level, which also negatively affects their trust in Chinese certification labels. Wang et al. [86] also concluded that consumer trust in green labels and certification organizations significantly affects consumers' willingness to purchase green labelled foods. This was also verified in our study from the sentiment analysis of consumers' online reviews. Therefore, the government needs to promote knowledge about green labelling and strengthen the regulation of the green food market.

6. Conclusions

Understanding consumers' perceptions and emotions towards green-labelled food is of genuine value for developing the green food industry and promoting green consumption, but also of strategic significance for building a low-carbon society and achieving green sustainable development. This study overcomes the limitations of traditional research in terms of data acquisition. It uses product reviews on e-commerce platforms to explore consumers' different concerns and emotional tendencies toward green-labelled rice and conventional rice. The results show that consumers' concerns about green-labelled rice have shown a significant upward trend since the COVID-19 out-break; consumers are more concerned about the origin, aroma, and taste of green-labelled rice than conventional rice; consumers of conventional rice are more concerned about the cost-performance ratio, while consumers of green-labelled rice are also price-sensitive; and that most consumers have a positive attitude towards green-labelled rice, while mistrust of green labels and broken packaging during logistics are the main causes of negative emotions.

6.1. Theoretical Contributions

Firstly, we applied the methods of text mining, sentiment analysis, and statistical tests to study consumer perceptions and sentiments towards green-labelled rice and conventional rice by collecting online reviews from the e-commerce platform. By using these methods, we can obtain more realistic and objective information, which is a big improvement compared with the previous literature that adopted interviews or questionnaires to survey. This approach suggests that review data from e-commerce platforms are useful for research in the field of green food, providing a new perspective for the study of public perceptions and emotions about green food, and contributing to the study of consumer behaviour in the field of green food.

Secondly, this study found that label mistrust was the leading cause of negative consumer sentiment towards green-labelled rice based on consumer review data, as consumers' distrust of the green label directly led to questions about the quality of the product. It also further validates that the issue of label mistrust is critical in discouraging consumers from purchasing green-labelled foods [77].

Thirdly, price is an important influencing factor for consumers when purchasing goods. This study found that consumers who purchase green-labelled rice are also consistently concerned about price, in contrast to Stolz et al.'s [21] finding that consumers who purchase green-labelled rice are not price-sensitive. In addition, price-sensitive consumers also purchase green-labelled rice. In the context of the pandemic, increased concern about food safety and environmental health has directly contributed to greater consumer acceptance of green-labelled rice.

6.2. Implications for Practice

Firstly, to improve consumers' trust in green labelling, the government should strictly supervise the origin of rice that meets the national environmental quality standards for green food and strengthen the governance of online and offline markets. Furthermore, the government should promote a unified green product certification and labelling system, while actively spreading and popularising knowledge of green food labelling.

Secondly, as the integrity of food packaging is a considerable and important factor for logistics [43], online retailers in the green food industry should reinforce the packaging quality of green-labelled rice to reduce the problem of packaging breakage during the shipping and courier process. At the same time, they should work closely with courier companies to improve transportation efficiency and provide optimal delivery solutions for consumers in different and remote regions.

6.3. Limitations and Future Research Directions

Due to the limitations of publicly crawling the product review on JD.com, we could not obtain the statistical characteristics of users who posted reviews, such as age, gender, education level, and income. In future research, assuming factors of consumers' statistical characteristics as control variables, questionnaires and empirical analysis can further explore the differences in consumers' perceptions and emotional characteristics of green-labelled rice and conventional rice under different sample dimensions. In addition, the level of consumers' perception and trust in green labels is a crucial factor influencing consumers' purchase intentions and decisions. However, existing research has not identified the psychological mechanisms. Therefore, we will adopt the social value theory and social trust theory to explain the mechanism of consumers' trust in green labels.

Author Contributions: H.X. contributed to conceptualization, methodology, and supervision; M.X. contributed to the data acquisition, formal analysis, and writing and editing of the original draft; J.Z. contributed to validation; H.H. contributed to review and editing. All authors have read and agreed to the published version of the manuscript.

Funding: This research received no external funding.

Institutional Review Board Statement: Not applicable.

Informed Consent Statement: Not applicable.

Data Availability Statement: The data are available from the corresponding author.

Acknowledgments: The authors would like to thank the anonymous reviewers for their valuable comments, and Sicong Wang for the proofreading of this manuscript.

Conflicts of Interest: The authors declare no conflict of interest.

References

1. Eyinade, G.A.; Mushunje, A.; Yusuf, S.F.G. The willingness to consume organic food: A review. *Food Agric. Immunol.* **2021**, *32*, 78–104. [CrossRef]
2. Wojciechowska-Solis, J.; Kowalska, A.; Bieniek, M.; Ratajczyk, M.; Manning, L. Comparison of the Purchasing Behaviour of Polish and United Kingdom Consumers in the Organic Food Market during the COVID-19 Pandemic. *Int. J. Environ. Res. Public Health* **2022**, *19*, 1137. [CrossRef]
3. Xie, X.; Huang, L.; Li, J.J.; Zhu, H. Generational differences in perceptions of food health/risk and attitudes toward organic food and game meat: The case of the COVID-19 crisis in China. *Int. J. Environ. Res. Public Health* **2020**, *17*, 3148. [CrossRef] [PubMed]
4. Ghufran, M.; Ali, S.; Ariyesti, F.R.; Nawaz, M.A.; Aldieri, L.; Xiaobao, P. Impact of COVID-19 to customers switching intention in the food segments: The push, pull and mooring effects in consumer migration towards organic food. *Food. Qual. Prefer.* **2022**, *99*, 104561. [CrossRef]
5. Qi, X.; Ploeger, A. Explaining Chinese consumers' green food purchase intentions during the COVID-19 pandemic: An extended Theory of Planned Behaviour. *Foods* **2021**, *10*, 1200. [CrossRef] [PubMed]
6. Wang, H.; Ma, B.; Cudjoe, D.; Bai, R.; Farrukh, M. How does perceived severity of COVID-19 influence purchase intention of organic food? *Br. Food J.* **2021**, *124*, 3353–3367. [CrossRef]
7. Baudry, J.; Péneau, S.; Allès, B.; Touvier, M.; Hercberg, S.; Galan, P.; Amiot, M.J.; Lairon, D.; Méjean, C.; Kesse-Guyot, E. Food choice motives when purchasing in organic and conventional consumer clusters: Focus on sustainable concerns (The NutriNet-Santé Cohort Study). *Nutrients* **2017**, *9*, 88. [CrossRef]
8. Yin, S.; Wu, L.; Du, L.; Chen, M. Consumers' purchase intention of organic food in China. *J. Sci. Food Agric.* **2010**, *90*, 1361–1367. [CrossRef]
9. Tezer, A.; Bodur, H.O. The green consumption effect: How using green products improves consumption experience. *J. Consum. Res.* **2020**, *47*, 25–39. [CrossRef]
10. Zhu, Q.; Li, Y.; Geng, Y.; Qi, Y. Green food consumption intention, behaviors and influencing factors among Chinese consumers. *Food. Qual. Prefer.* **2013**, *28*, 279–286. [CrossRef]
11. Woese, K.; Lange, D.; Boess, C.; Bögl, K.W. A comparison of organically and conventionally grown foods—Results of a review of the relevant literature. *J. Sci. Food Agric.* **1997**, *74*, 281–293. [CrossRef]
12. Liu, Z.; Yuan, Y.; Xie, T.; Zhang, Y.; Shao, S.; Nie, J.; Xia, W.; Rogers, K.M.; Zhang, W. Long-term agricultural effects on the authentication accuracy of organic, green, and conventional rice using isotopic and elemental chemometric analyses. *J. Agric. Food Chem.* **2020**, *68*, 1213–1225. [CrossRef] [PubMed]
13. My, N.H.D.; Van Loo, E.J.; Rutsaert, P.; Tuan, T.H.; Verbeke, W. Consumer valuation of quality rice attributes in a developing economy: Evidence from a choice experiment in Vietnam. *Br. Food J.* **2018**, *120*, 1059–1072. [CrossRef]
14. Hu, X.; Zhang, W.; Lu, L.; Shao, Y.; Chen, M.; Zhu, Z.; Mou, R. Comparison of quality of appearance, cooking quality, and protein content of green-labeled rice and conventional rice. *Cereal Chem.* **2022**, *99*, 873–883. [CrossRef]
15. My, N.H.D.; Rutsaert, P.; Van Loo, E.J.; Verbeke, W. Consumers' familiarity with and attitudes towards food quality certifications for rice and vegetables in Vietnam. *Food Control* **2017**, *82*, 74–82. [CrossRef]
16. Liu, Q.; Yan, Z.; Zhou, J. Consumer choices and motives for eco-labeled products in China: An empirical analysis based on the choice experiment. *Sustainability* **2017**, *9*, 331. [CrossRef]
17. Xu, H.; Zhang, M.; Zeng, J.; Hao, H.; Lin, H.-C.K.; Xiao, M. Use of Latent Dirichlet Allocation and Structural Equation Modeling in Determining the Factors for Continuance Intention of Knowledge Payment Platform. *Sustainability* **2022**, *14*, 8992. [CrossRef]
18. Singh, A.; Glińska-Neweś, A. Modeling the public attitude towards organic foods: A big data and text mining approach. *J. Big Data* **2022**, *9*, 2. [CrossRef]
19. Rajeswari, B.; Madhavan, S.; Venkatesakumar, R.; Riasudeen, S. Sentiment analysis of consumer reviews–a comparison of organic and regular food products usage. *Rajagiri Manag. J.* **2020**, *14*, 155–167. [CrossRef]
20. Lee, H.J.; Yun, Z.S. Consumers' perceptions of organic food attributes and cognitive and affective attitudes as determinants of their purchase intentions toward organic food. *Food. Qual. Prefer.* **2015**, *39*, 259–267. [CrossRef]
21. Stolz, H.; Stolze, M.; Hamm, U.; Janssen, M.; Ruto, E. Consumer attitudes towards organic versus conventional food with specific quality attributes. *NJAS-Wagen. J. Life Sci.* **2011**, *58*, 67–72. [CrossRef]
22. Xu, J.; Zhong, J.; Zhang, B.; Li, X. Green labelled rice shows a higher nutritional and physiochemical quality than conventional rice in China. *Foods* **2021**, *10*, 915. [CrossRef] [PubMed]
23. Magkos, F.; Arvaniti, F.; Zampelas, A. Organic food: Buying more safety or just peace of mind? A critical review of the literature. *Crit. Rev. Food Sci. Nutr.* **2006**, *46*, 23–56. [CrossRef] [PubMed]
24. Tong, Q.; Anders, S.; Zhang, J.; Zhang, L. The roles of pollution concerns and environmental knowledge in making green food choices: Evidence from Chinese consumers. *Food Res. Int.* **2020**, *130*, 108881. [CrossRef]
25. Kareklas, I.; Carlson, J.R.; Muehling, D.D. "I eat organic for my benefit and yours": Egoistic and altruistic considerations for purchasing organic food and their implications for advertising strategists. *J. Advert.* **2014**, *43*, 18–32. [CrossRef]
26. Gorton, M.; Tocco, B.; Yeh, C.H.; Hartmann, M. What determines consumers' use of eco-labels? Taking a close look at label trust. *Ecol. Econ.* **2021**, *189*, 107173. [CrossRef]
27. Samant, S.S.; Seo, H.S. Quality perception and acceptability of chicken breast meat labeled with sustainability claims vary as a function of consumers' label-understanding level. *Food. Qual. Prefer.* **2016**, *49*, 151–160. [CrossRef]

28. Kaneko, D.; Toet, A.; Brouwer, A.M.; Kallen, V.; Van Erp, J.B.F. Methods for evaluating emotions evoked by food experiences: A literature review. *Front. Psychol.* **2018**, *9*, 911. [CrossRef]
29. Taherdoost, H. A review of technology acceptance and adoption models and theories. *Procedia Manuf.* **2018**, *22*, 960–967. [CrossRef]
30. Kouparis, M. Applying the technology acceptance model and flow theory to online consumer behaviour. *Inf. Syst. Res.* **2002**, *13*, 205–223. [CrossRef]
31. Churchill, A.; Behan, J. Comparison of methods used to study consumer emotions associated with fragrance. *Food. Qual. Prefer.* **2010**, *21*, 1108–1113. [CrossRef]
32. Jaeger, S.R.; Cardello, A.V.; Schutz, H.G. Emotion questionnaires: A consumer-centric perspective. *Food. Qual. Prefer.* **2013**, *30*, 229–241. [CrossRef]
33. Venkatraman, V.; Dimoka, A.; Pavlou, P.A.; Vo, K.; Hampton, W.; Bollinger, B.; Hershfield, H.E.; Ishihara, M.; Winer, R.S. Predicting advertising success beyond traditional measures: New insights from neurophysiological methods and market response modeling. *J. Mark. Res.* **2015**, *52*, 436–452. [CrossRef]
34. Tian, G.; Lu, L.; McIntosh, C. What factors affect consumers' dining sentiments and their ratings: Evidence from restaurant online review data. *Food. Qual. Prefer.* **2021**, *88*, 104060. [CrossRef]
35. Yiridoe, E.K.; Bonti-Ankomah, S.; Martin, R.C. Comparison of consumer perceptions and preference toward organic versus conventionally produced foods: A review and update of the literature. *Renew. Agric. Food Syst.* **2005**, *20*, 193–205. [CrossRef]
36. Bourn, D.; Prescott, J. A comparison of the nutritional value, sensory qualities, and food safety of organically and conventionally produced foods. *Crit. Rev. Food Sci. Nutr.* **2002**, *42*, 1–34. [CrossRef]
37. Hassan, M.; Wen, X.; Xu, J.; Zhong, J.; Li, X. Development and challenges of green food in China. *Front. Agric. Sci. Eng.* **2020**, *7*, 56–66. [CrossRef]
38. Xu, J.; Zhang, Z.; Zhang, X.; Ishfaq, M.; Zhong, J.; Li, W.; Zhang, F.; Li, X. Green Food Development in China: Experiences and Challenges. *Agriculture* **2020**, *10*, 614. [CrossRef]
39. Naseem, A.; Mhlanga, S.; Diagne, A.; Adegbola, Y.P.; Midingoyi, S. Economic analysis of consumer choices based on rice attributes in the food markets of West Africa—The case of Benin. *Food Secur.* **2013**, *5*, 575–589. [CrossRef]
40. Kim, A.Y.; Ha, J.G.; Choi, H.; Moon, H. Automated text analysis based on skip-gram model for food evaluation in predicting consumer acceptance. *Comput. Intell. Neurosci.* **2018**, *2018*, 9293437. [CrossRef]
41. My, N.H.D.; Demont, M.; Van Loo, E.J.; Guia, A.; Rutsaert, P.; Tuan, T.H.; Verbeke, W. What is the value of sustainably-produced rice? Consumer evidence from experimental auctions in Vietnam. *Food Policy* **2018**, *79*, 283–296. [CrossRef]
42. Anang, B.T.; Adjetey, S.N.A.; Abiriwe, S.A. Consumer preferences for rice quality characteristics and the effects on price in the Tamale Metropolis, Northern Region, Ghana. *Int. J. AgriSci.* **2011**, *1*, 67–74. Available online: https://www.researchgate.net/publication/215591227 (accessed on 15 December 2022).
43. Hao, H.; Guo, J.; Xin, Z.; Qiao, J. Research on e-Commerce Distribution Optimization of Rice Agricultural Products Based on Consumer Satisfaction. *IEEE Access* **2021**, *9*, 135304–135315. [CrossRef]
44. Whitlock, J.L.; Powers, J.L.; Eckenrode, J. The virtual cutting edge: The internet and adolescent self-injury. *Dev. Psychol.* **2006**, *42*, 407–417. [CrossRef] [PubMed]
45. Bae, S.; Lee, T. Product type and consumers' perception of online consumer reviews. *Electron. Mark.* **2011**, *21*, 255–266. [CrossRef]
46. Fan, Z.P.; Che, Y.J.; Chen, Z.Y. Product sales forecasting using online reviews and historical sales data: A method combining the Bass model and sentiment analysis. *J. Bus. Res.* **2017**, *74*, 90–100. [CrossRef]
47. Zhao, Y.; Xu, X.; Wang, M. Predicting overall customer satisfaction: Big data evidence from hotel online textual reviews. *Int. J. Hosp. Manag.* **2019**, *76*, 111–121. [CrossRef]
48. Sun, X.; Han, M.; Feng, J. Helpfulness of online reviews: Examining review informativeness and classification thresholds by search products and experience products. *Decis. Support Syst.* **2019**, *124*, 113099. [CrossRef]
49. Huang, H.; Long, R.; Chen, H.; Sun, K.; Li, Q. Exploring public attention about green consumption on Sina Weibo: Using text mining and deep learning. *Sustain. Prod. Consump.* **2022**, *30*, 674–685. [CrossRef]
50. Chen, L.; Wang, P.; Ma, X.; Wang, X. Cancer communication and user engagement on Chinese social media: Content analysis and topic modeling study. *J. Med. Internet Res.* **2021**, *23*, e26310. [CrossRef]
51. Wu, X.; Wu, H.; Xie, X.; Xu, J.; Zhang, T. Short text topic recognition and optimization method for university online community. In Proceedings of the Artificial Intelligence and Security, New York, NY, USA, 26–28 July 2019.
52. Bastani, K.; Namavari, H.; Shaffer, J. Latent Dirichlet allocation (LDA) for topic modeling of the CFPB consumer complaints. *Expert Syst. Appl.* **2019**, *127*, 256–271. [CrossRef]
53. Blei, D.M.; Ng, A.Y.; Jordan, M.I. Latent dirichlet allocation. *J. Mach. Learn. Res.* **2003**, *3*, 993–1022.
54. Liu, J.; Zhou, Y.; Jiang, X.; Zhang, W. Consumers' satisfaction factors mining and sentiment analysis of B2C online pharmacy reviews. *BMC Med. Inform. Decis. Mak.* **2020**, *20*, 194. [CrossRef]
55. Yu, S.; Eisenman, D.; Han, Z. Temporal Dynamics of Emotions during the COVID-19 Pandemic at the Center of Outbreak: A Sentimental Analysis of Weibo Tweets from Wuhan. *J. Med. Internet Res.* **2021**, *23*, e27078. [CrossRef]
56. Zhang, B.; Xiao, P.; Yu, X. The Influence of Prosocial and Antisocial Emotions on the Spread of Weibo Posts: A Study of the COVID-19 Pandemic. *Discret. Dyn. Nat. Soc.* **2021**, *2021*, 8462264. [CrossRef]

57. Karami, A.; Dahl, A.A.; Shaw, G.J.; Valappil, S.P.; Turner-McGrievy, G.; Kharrazi, H.; Bozorgi, P. Analysis of social media discussions on (#) diet by blue, red, and swing states in the US. *Healthcare* **2021**, *9*, 518. [CrossRef]
58. Good, I.J. C140. Standardized tail-area prosabilities. *J. Stat. Comput. Simul.* **1982**, *16*, 65–66. [CrossRef]
59. Sawilowsky, S.S. New effect size rules of thumb. *J. Mod. Appl. Stat. Meth.* **2009**, *8*, 26. [CrossRef]
60. Beisland, E.G.; Gjeilo, K.H.; Andersen, J.R.; Bratås, O.; Bø, B.; Haraldstad, K.; Hjelmeland, I.H.H.; Iversen, M.M.; Løyland, B.; Norekvål, T.M. Quality of life and fear of COVID-19 in 2600 baccalaureate nursing students at five universities: A cross-sectional study. *Health Qual. Life Outcomes* **2021**, *19*, 198. [CrossRef]
61. Notice of the State Council Municipality on Printing and Distributing the 13th Five-Year National Food Safety Plan and the 13th Five-Year National Drug Safety Plan. Available online: http://www.gov.cn/zhengce/content/2017-02/21/content_5169755.htm (accessed on 15 December 2022).
62. Jian, F.; Yajiao, W.; Yuanyuan, D. Microblog topic evolution computing based on LDA algorithm. *Open Phys.* **2018**, *16*, 509–516. [CrossRef]
63. Beekman, T.L.; Crandall, P.G.; Seo, H.S. You Eat How You Think: A Review on the Impact of Cognitive Styles on Food Perception and Behavior. *Foods* **2022**, *11*, 1886. [CrossRef]
64. Braga, A.A.; Weisburd, D.L. Does hot spots policing have meaningful impacts on crime? Findings from an alternative approach to estimating effect sizes from place-based program evaluations. *J. Quant. Criminol.* **2020**, *38*, 1–22. [CrossRef]
65. Fiamohe, R.; Nakelse, T.; Diagne, A.; Seck, P.A. Assessing the effect of consumer purchasing criteria for types of rice in Togo: A choice modeling approach. *Agribusiness* **2015**, *31*, 433–452. [CrossRef]
66. Jarma Arroyo, S.E.; Hogan, V.; Ahrent Wisdom, D.; Moldenhauer, K.A.K.; Seo, H.S. Effect of geographical indication information on consumer acceptability of cooked aromatic rice. *Foods* **2020**, *9*, 1843. [CrossRef]
67. Seo, H.S.; Pramudya, R.C.; Singh, A.; Hummel, T. Recent evidence for the impacts of olfactory disorders on food enjoyment and ingestive behavior. *Curr. Opin. Food Sci.* **2021**, *42*, 187–194. [CrossRef]
68. Gurdian, C.E.; Torrico, D.D.; Li, B.; Prinyawiwatkul, W. Effect of serving plate types and color cues on liking and purchase intent of cheese-flavored tortilla chips. *Foods* **2021**, *10*, 886. [CrossRef]
69. Jiang, Y.; King, J.M.; Prinyawiwatkul, W. A review of measurement and relationships between food, eating behavior and emotion. *Trends Food Sci. Technol.* **2014**, *36*, 15–28. [CrossRef]
70. Li, L.; Liu, X.; Zhang, X. Public attention and sentiment of recycled water: Evidence from social media text mining in China. *J. Clean. Prod.* **2021**, *303*, 126814. [CrossRef]
71. Ben Khadda, Z.; Ezrari, S.; Radouane, N.; Boutagayout, A.; El Housni, Z.; Lahmamsi, H.; Zahri, A.; Houssaini, T.; El Ghadraoui, L.; Elamine, Y.; et al. Organic food consumption and eating habit in Morocco, Algeria, and Tunisia during the COVID-19 pandemic lockdown. *Open Agric.* **2022**, *7*, 21–29. [CrossRef]
72. Li, G.; Yang, L.; Zhang, B.; Li, X.; Chen, F. How do environmental values impact green product purchase intention? The moderating role of green trust. *Environ. Sci. Pollut. Res.* **2021**, *28*, 46020–46034. [CrossRef]
73. Loebnitz, N.; Aschemann-Witzel, J. Communicating organic food quality in China: Consumer perceptions of organic products and the effect of environmental value priming. *Food. Qual. Prefer.* **2016**, *50*, 102–108. [CrossRef]
74. Lin, L.; Zhou, D.; Ma, C. Green food industry in China: Development, problems and policies. *Renew. Agric. Food Syst.* **2010**, *25*, 69–80. [CrossRef]
75. Liang, R.D. Predicting intentions to purchase organic food: The moderating effects of organic food prices. *Br. Food J.* **2016**, *118*, 183–199. [CrossRef]
76. Al-Swidi, A.; Huque, M.R.S.; Hafeez, M.H.; Shariff, M.N.M. The role of subjective norms in theory of planned behavior in the context of organic food consumption. *Br. Food J.* **2014**, *116*, 1561–1580. [CrossRef]
77. Qi, X.; Yu, H.; Ploeger, A. Exploring influential factors including COVID-19 on green food purchase intentions and the intention–behaviour gap: A qualitative study among consumers in a Chinese context. *Int. J. Environ. Res. Public Health* **2020**, *17*, 7106. [CrossRef]
78. Zhang, B.; Fu, Z.; Huang, J.; Wang, J.; Xu, S.; Zhang, L. Consumers' perceptions, purchase intention, and willingness to pay a premium price for safe vegetables: A case study of Beijing, China. *J. Clean. Prod.* **2018**, *197*, 1498–1507. [CrossRef]
79. Rizzo, G.; Borrello, M.; Dara Guccione, G.; Schifani, G.; Cembalo, L. Organic food consumption: The relevance of the health attribute. *Sustainability* **2020**, *12*, 595. [CrossRef]
80. Haines, R.; Hough, J.; Cao, L.; Haines, D. Anonymity in computer-mediated communication: More contrarian ideas with less influence. *Group Decis. Negot.* **2014**, *23*, 765–786. [CrossRef]
81. Askari, A.; Jalilvand, A.; Neshati, M. On Anonymous Commenting: A Greedy Approach to Balance Utilization and Anonymity for Instagram Users. In Proceedings of the 42nd International ACM SIGIR Conference, Paris, France, 21–25 July 2019.
82. Nuttavuthisit, K.; Thøgersen, J. The importance of consumer trust for the emergence of a market for green products: The case of organic food. *Group Decis. Negot.* **2017**, *140*, 323–337. [CrossRef]
83. Song, L.; Lim, Y.; Chang, P.; Guo, Y.; Zhang, M.; Wang, X.; Yu, X.; Lehto, M.R.; Cai, H. Ecolabel's role in informing sustainable consumption: A naturalistic decision making study using eye tracking glasses. *J. Clean. Prod.* **2019**, *218*, 685–695. [CrossRef]
84. Grankvist, G.; Biel, A. Predictors of purchase of eco-labelled food products: A panel study. *Food. Qual. Prefer.* **2007**, *18*, 701–708. [CrossRef]

85. Yin, S.; Chen, M.; Xu, Y.; Chen, Y. Chinese consumers' willingness-to-pay for safety label on tomato: Evidence from choice experiments. *China Agric. Econ. Rev.* **2017**, *9*, 141–155. [CrossRef]
86. Wang, J.; Tao, J.; Chu, M. Behind the label: Chinese consumers' trust in food certification and the effect of perceived quality on purchase intention. *Food Control* **2020**, *108*, 106825. [CrossRef]

foods

MDPI

Article

Factors Influencing Consumers' Willingness-to-Try Seafood Byproducts

Silvia Murillo [1], Ryan Ardoin [2] and Witoon Prinyawiwatkul [1,*]

1 Agricultural Center, School of Nutrition and Food Sciences, Louisiana State University, Baton Rouge, LA 70803, USA; smurillomiguez1@lsu.edu
2 Food Processing and Sensory Quality Research Unit, Southern Regional Research Center, USDA-ARS, New Orleans, LA 70124, USA; ryan.ardoin@usda.gov
* Correspondence: wprinya@lsu.edu

Abstract: With increasing global demand for seafood, seafood byproducts (SB) utilization can contribute to a more sustainable food supply chain through waste-to-value food product development. However, consumer perceptions of SB (e.g., fish skin and bones) are underexplored. Therefore, this study aims to evaluate some factors influencing consumers' willingness-to-try seafood byproducts. An online survey was conducted in the USA regarding intervention of SB informational cues with N = 904 adult seafood consumers internationally. The proportion of consumers willing to try SB increased significantly (McNemar's test, $\alpha = 0.05$) from 47% to 68% after SB safety and health claims had been presented in the questionnaire. Gender, race, SB knowledge, and previous SB consumption were significant predictors of trial intent (based on logistic regression), as were emotional baseline scores during the COVID-19 pandemic. Males were more open to SB consumption than females, and racial identity was associated with differential responsiveness to SB information. Higher levels of "bored" and "unsafe" feelings, and lower levels of "free" were associated with increased SB trial intent. Potential SB consumers identified fish products (82% willingness-to-try); seasoning mix, sauces, and dressing (71% willingness-to-try); and soup and gravy products (62% willingness-to-try) as most appropriate for SB incorporation. Predominant reasons for SB avoidance were concerns about sensory quality, safety, and nutrition. These consumer-driven data could guide SB product development concepts to encourage trial and overcome aversions through new consumption experience.

Keywords: byproduct; seafood; willingness to try; product appropriateness; emotions; safety claim; health benefit claim

Citation: Murillo, S.; Ardoin, R.; Prinyawiwatkul, W. Factors Influencing Consumers' Willingness-to-Try Seafood Byproducts. *Foods* **2023**, *12*, 1313. https://doi.org/10.3390/foods12061313

Academic Editor: Begoña Panea

Received: 8 March 2023
Revised: 15 March 2023
Accepted: 16 March 2023
Published: 20 March 2023

1. Introduction

Reducing food loss and waste has been among the most significant challenges facing sustainable food systems and global food security [1]. It has been estimated that almost one-third of food produced for human consumption is wasted every year, and the amount of food loss is expected to increase with population growth and consumption [2]. Meat, poultry, and fish have accounted for approximately 30% of total food value lost [2], with the global demand for seafood increasing faster than that of terrestrial meat [3]. Global per capita seafood consumption has grown by a rate of 122%, from 0.9 kg in 1961 to 20.5 kg in 2019 [4]. Generation of seafood loss, byproducts, and waste can occur throughout the supply chain, from harvesting to processing, distribution, and consumption [5]. During fish processing, for example, almost 55% of biomass can be diverted to non-food production streams [2] such as cosmetics, pharmaceuticals, fishmeal, and animal feed, or be discarded. This includes potentially edible byproducts with nutritional value [6]. The entire seafood industry generates about 60% of byproducts which are mainly skin, viscera, bones, fins, belly flaps, heads, exoskeletons, and blood. For example, fish filleting generates 200 to 400 kg of byproducts per ton. The crustacean industry follows the same fashion as the fish industry, generating around 80% of byproducts of the total weight of prawns, shrimps, crabs, and lobsters [7,8].

In the hierarchy of biomass value, pharmaceutical applications yield a higher economic value than human food applications [9]. However, societal value of keeping valuable nutrients in the food supply chain should also be considered. With current and projected rates of seafood production (and subsequent byproduct generation) and the variety of SB sources and composition [10], food applications of SB could work alongside other production streams to reduce waste rather than be in the competition. Strategies to progress toward more sustainable food systems have included related concepts of circular economy, byproduct valorization, and waste-to-value production [11,12]. These approaches aim to minimize food loss by keeping edible components of dietary or nutritional value in the food production system. This often involves new food applications of ingredients which may be unfamiliar to consumers [11,12]. Commercialization of these products needs in-depth research on consumers to overcome barriers such as consumer perception, unfamiliarity, negative emotions, and potential risk that can decline the willingness-to-try a new product.

Prior research found that appropriate product concepts could encourage consumers' willingness-to-try novel food sources and emphasized that appropriateness of an ingredient within a product is best determined by the potential users/consumers [13]. The willingness-to-try response is a measure of consumers' intent or "willingness" to engage in a hypothetical behavior (for instance, seafood byproduct consumption). Such metrics are common in survey-based studies of novel foods, where actual behaviors cannot be observed, and/or past opportunities may not have been presented [14]. The most common drivers of consumers' willingness-to-try novel food products are positive emotions, sensory attributes, product appropriateness, familiarity (previous positive experiences), knowledge (product information), and education. In contrast, some factors that inhibit consumers' willingness-to-try products include unfamiliarity with the products or raw materials, cultural preferences, negative emotions, and food neophobia [13–16]. Some food applications of SB have shown promise with consumer segments (e.g., catfish bone powder in a seafood breading mixture [17]), others have demonstrated disconfirmed sensory expectancy on the perceived product performance (e.g., dark orange color in dipping sauces containing shrimp head pigment [18]). With global demand for the production of seafood rising [3], there is increasing need for consumer-driven product development concepts to keep valuable seafood components in the human food supply.

Some consumers have demonstrated negative psychological reactions to consumption of byproducts, despite potential benefits of sustainable food sources [11]. Barriers to adoption of foods containing unusual ingredients such as insects [13] and animal byproducts [15] have been noted, but consumers' openness to SB consumption has been underexplored on a large scale. Along with appropriate products, communication of ingredient safety and potential benefits have been suggested to overcome negative attitudes and emotions which may hinder trial of unfamiliar foods derived from byproducts [11,15]. This current study will provide essential data such as appropriateness, perceived risk, that can be applied by product developers to create novel products. The objectives of this research were to (1) evaluate consumers' openness to SB consumption under different [increasingly] informed conditions (definition only, safety informational cue, and health benefit informational cue), (2) reveal both emotional and demographic predictors of SB consumption, (3) determine appropriate food product categories for SB application, and (4) identify potential risks associated with the unwillingness to try SB.

2. Materials and Methods

2.1. Online Survey and Data Collection

An online survey was designed and administered using Qualtrics online software (Qualtrics Provo, UT, USA). Weblinks and QR Codes were generated to distribute the survey through email, online platforms, social media, and a consumer database ("Tiger Tasters", Baton Rouge, LA, USA). Potential respondents were screened for age (at least

18 years old) and seafood consumption ("Do you consume seafood or fish products?" = yes). Participation was entirely voluntary, and informed consent was collected. The questionnaire and research protocol were approved by Louisiana State University Agricultural Center Institutional Review Board (Baton Rouge, LA, USA; IRBAG-21-0063). Data were collected online from April to June of 2020, the months immediately after COVID-19 was declared a global pandemic [19]. Figure 1 exhibits the schematic overview of the survey, which was completed by a total of 904 adult seafood consumers. To achieve the study goal, the survey was designed to allow each step to collect essential data to predict the willingness-to-try (GWWT: General willingness-to-try [with seafood byproduct definition only]; GWTTS: General willingness-to-try [with definition and safety information]; GWTTSH: General willingness-to-try [with definition, safety information, and health benefit information]) based on demographics, emotions, prior consumption experience, and informational cues, to identify the to p products appropriate for incorporation of seafood byproducts, and to identify factors (risks) inhibiting consumers' willingness-to-try seafood byproducts. Details for the methods in each step are given below.

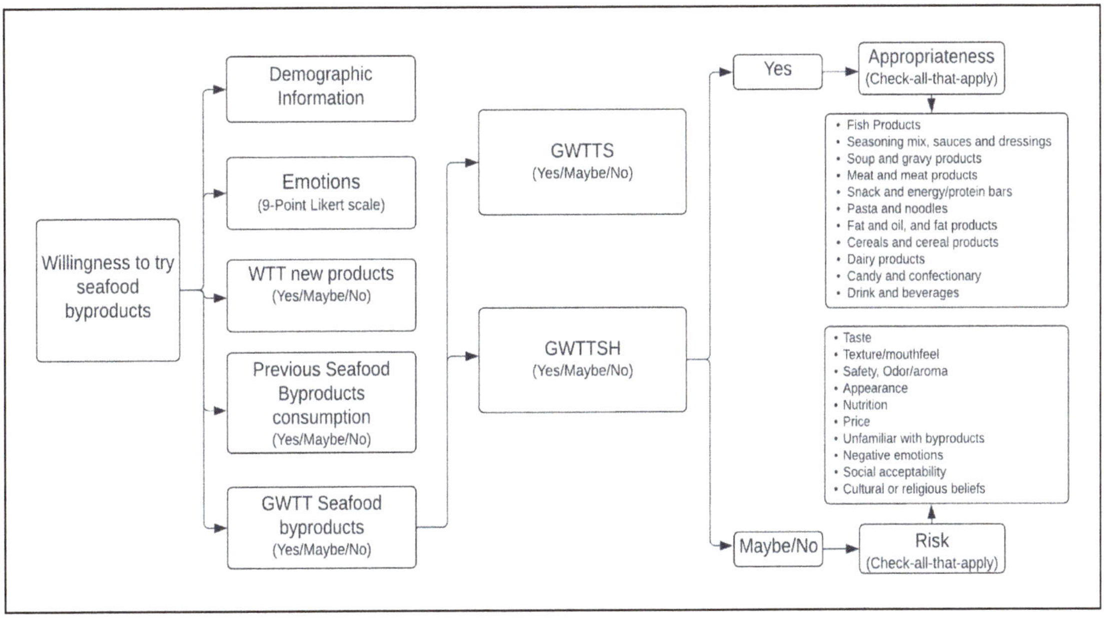

Figure 1. Schematic overview of the online survey work. WTT: Willingness-to-try, GWWT: General willingness-to-try [with seafood byproduct definition only], GWTTS: General willingness-to-try [with definition and safety information], GWTTSH: General willingness-to-try [with definition, safety information, and health benefit information].

2.2. Demographic Information

Over half (62%) of the respondents were considered US consumers ("Have you lived in US for the last three years?" = yes) and non-US consumers were not asked to report their nationality, so their locations were unknown. Sixty-three percent of respondents were female. Proportions of the population sample by race were as follows: 39% Caucasian, 37% Hispanic/Latin/Spanish origin, 19% Asian, 3% Black/African American, and 2% other races. The largest age segment was 26- to 35-years-olds (37%), followed by ages 19–25 (17%), 36–45 (16%), 46–55 (15%), 56–65 (9%), and 65 or older (6%). The majority of the respondents (94%) were college-educated, and about 76% were employed either full-time or part-time at the time this survey was administered (Table 1).

Table 1. Demographic and socioeconomic information of participants (N = 904).

Demographic and Socioeconomic Data			
	Age (%)		
19-25 years		17.04	
26-35 years		36.50	
36-45 years		15.82	
46-55 years		15.49	
56-65 years		8.96	
>65 years		6.19	
	Race (%)		
White		38.61	
Hispanic/Latin/Spanish origin		37.06	
Asian		19.47	
Black/African American		2.99	
Other		1.88	
	Education Level (%)		
Graduate degree (MS, Ph.D., ED)		58.52	
College or equivalent		35.62	
High School diploma or below		5.86	
	Gender (%)		
Female		62.94	
Male		37.06	
Employment status (%)	Before COVID-19		During COVID-19 *
Employed Full time, on-site	64.05		17.37
Employed Full-time, work from home	6.75		45.35
Employed Part-time on-site	8.74		3.32
Employed Part-time, work from home	3.98		10.4
Retired	7.19		7.3
Unemployed	9.29		16.26

* Data were collected online from April to June of 2020.

Eighty-three percent of consumers responded "yes" to the question "Are you typically willing to try new foods?" An additional 14% responded "not sure." While this was not a measure of food neophobia per se, it indicated that most respondents were at least open to the idea of new food experiences. Seventy-six percent of consumers claimed to know what seafood byproducts were. Regardless of their responses, all participants were subsequently presented with the following SB definition: "Related to seafood and aquacultural products, a byproduct is something (for example, bone, skin, gut) which is produced during the manufacturing or processing of another product (for example, catfish fillet)." After presentation of this SB definition, 46% of participants responded "not sure" to the question, "Have you ever eaten food products that contain or are fortified with seafood or aquacultural (farm-raised) byproducts such as head, bones, gut, or skin?"; 39% responded "yes."

2.3. Emotional Profiles

Consumers rated 21 baseline emotions in response to the question "How have you been feeling during the COVID-19 pandemic?" Twenty-one emotion terms (selected from the EsSense Profile® [20]) were presented in a randomized order and rated using a bipolar 9-point Likert scale (disagree extremely, disagree very much, disagree moderately, disagree slightly, neither disagree or agree, agree slightly, agree moderately, agree very much, agree extremely) with a tenth "not applicable" option. The list included the following emotions: active, adventurous, aggressive, bored, calm, eager, energetic, enthusiastic, free, friendly, good, glad, happy, healthy, loving, nostalgic, pleased, peaceful, satisfied, unsafe, and worried. Baseline emotional profiles obtained from the data were treated as psychographic predictors of SB consumption.

2.4. Willingness-to-Try Seafood Byproduct, Perceived Risks, and Product Appropriateness

In this experiment, consumers' general willingness-to-try foods made with SB was measured under three informed conditions. Participants were first asked, "Would you be willing to try new products containing a small portion of seafood byproducts such as bone and skin?" (yes/maybe/no). This response was subsequently referred to as the general willingness-to-try (GWTT) response and was obtained given only the SB definition provided earlier. Next, participants answered the question with an additional safety informational cue: "Would you be willing to try new products containing a small portion of seafood byproducts such as bone and skin, knowing it is safe for consumption?" (yes/maybe/no). This willingness-to-try response under the informed condition of SB safety was later referred to as (GWTTS). Finally, consumers answered the question, "Would you be willing to try new products fortified with safe seafood byproducts that are claimed to provide health benefits?" (yes/maybe/no). This response, with an added informational cue related to potential SB health benefits, was referred to as GWTTSH.

Consumers who responded "maybe" or "no" to the GWTTSH question were considered reluctant or averse consumers, respectively, and directed to answer the question "Why?" and reported the reasons for their response from a list of perceived risks [13,20] in a check-all-that-apply (CATA) format. The list of risks included: taste, appearance, odor/aroma, texture/mouthfeel, safety, nutrition, negative emotions (boredom, disgust, fear, guilty, worry, etc.,), social acceptability, cultural or religious beliefs, unfamiliar with byproducts, and price. This list was meant to identify potential negative consequences consumers may anticipate from SB consumption (and thus avoid it), and to encompass the four dimensions of risk perception presented by Baker et al. [16]: functional, physical, social, and psychological risks.

Consumers who responded "yes" to the GWTTSH question were considered potential users of SB and were directed to a product appropriateness question: "Which group of products would you be willing to try if they contained seafood byproducts?" Using the CATA method, consumers selected all applicable groups from the list of 12 product categories [21]: dairy products; bakery products; meat and meat products including poultry and game; fat and oil, and fat products; drink and beverages; snack and energy/protein bars; candy and confectionery, cereals and cereal products, seasoning mix, sauces and dressings; soup and gravy products (canned and frozen); fish and fish products including mollusks and crustaceans; pasta and noodles.

2.5. Statistical Analysis

Willingness-to-try responses were reported as percentage [of the total population sample] "yes" responses to GWTT, GWTTS, and GWTTSH questions, and as percentage [of potential SB consumers only] selection for each product category in the CATA list. Effects of informational cues were evaluated by marginal homogeneity tests. The Stuart-Maxwell test was first performed to demonstrate significant overall shifts in responses (yes/maybe/no) from one informed condition to the next, and the McNemar's test evaluated increases in "yes" responses from GWTT to GWTTS, and from GWTTS to GWTTSH. Demographic data were reported as percentages of the total population sample and were used as predictors (regressors) of the GWTT, GWTTS, and GWTTSH responses in separate logistic regression models. Likewise, consumers' emotional profiles were used as predictors of the GWTTS, GWTTS, and GWTTSH responses in separate logistic regression models, with each emotion treated as a numerical variable (from 1-9 on the 9-point scale, or 0 for "not applicable"). Therefore, two logistic regression models were fitted: one with the full set of demographics, and another with the full set of emotions as predictors. The variable inflation factor (VIF) diagnostic did not show any evidence of multicollinearity which could have skewed effect estimates. Relative appropriateness of each product category for incorporation of SB was determined by frequency of selection (% willingness-to-try) from the CATA list. Perceived risks were also reported as percentage [of reluctant/averse consumers] selection. Data were analyzed by Microsoft® Excel® (Version 2206 Build 16.0.15330.20260; Microsoft Corporation,

Redmond, WA, USA) and SAS (Copyright© 2016 SAS Institute Inc., Cary, NC, USA). A significance level of $\alpha = 0.05$ was used throughout.

3. Results and Discussion

3.1. Effects of Informational Cues on Willingness-to-Try Seafood Byproducts

Initially, slightly less than half (47%) of the respondents (N = 904) reported "yes" for GWTT for food made with SB (Figure 2). Responses to the GWTT question were provided after presentation of a SB definition only, and it was hypothesized that additional SB informational cues could increase willingness-to-try.

Figure 2. Willingness-to-try food containing seafood byproduct under different informed conditions. [1] Percentage of 904 consumer survey respondents. [2] GWTT: General willingness-to-try [with seafood byproduct definition only], GWTTS: General willingness-to-try [with definition and safety information], GWTTSH: General willingness-to-try [with definition, safety information, and health benefit information]. * There was a significant increase in "yes" responses (McNemar's test) with additional information from GWTT to GWTTS ($p < 0.001$) and from GWTTS to GWTTSH ($p < 0.01$).

With additional informational cue "... knowing it is safe for consumption," the percentage of consumers willing to try food made with SB (GWTTS) increased from 47% to 64% (Figure 2). Concerns about the safety of novel foods (especially animal products) are documented reasons for avoidance, and for consumption to occur, assurances of safety precede sensory appeal [22]. Effectiveness of food safety information relies on credibility of the source and consumers' experiences [23], and results (Figure 2) indicated that a significant proportion of respondents gave credence to the SB safety claim (Stuart Maxwell test, $p < 0.001$; McNemar's test, $p < 0.001$). For these consumers, safety of SB was likely unknown, uncertain, or not salient before the given informational cue. This study demonstrated that consumers' knowledge and trust in SB safety may be an important factor in product adoption (see later discussion of perceived safety risk).

Consumers' positive GWTTSH significantly rose to 68% (Stuart Maxwell test, $p < 0.001$; McNemar's test, $p = 0.007$), with the additional informational cue "... claimed to provide health benefits" (Figure 2). Other authors [24] found that information about public health increased openness to SB, whereas in this study the message was intended to appeal to individual health benefits. Various nutrients and health-promoting compounds have been identified in SB (see [10] for a review), and health-benefit claims have shown to be effective in increasing purchase intent for specific foods containing SB [17,25]. However, the magnitude of the effect may depend on overall product acceptability [25]. Since no actual food stimuli were presented in this research, the observed effects demonstrated attitudinal shifts in favor of SB consumption.

For this research, informational cues were defined as extrinsic messaging [23] which prompted decision-making based on limited information processing [26]. The observed positive impact of SB information cues (a safety claim for GWTTS, and a health benefit claim for GWTTSH) on SB trial intent suggested that they triggered interest and attention [21], were considered relevant and credible [23], and thus evoked more favorable judgements of SB among a significant proportion of consumers (Figure 2). When consumers are unfamiliar with a product, they often seek more information [16]. However, as the risk perception data will show, extrinsic information alone was not sufficient to overcome all the cognitive barriers to SB consumption.

3.2. Individual Predictors of Willingness-to-Try Seafood Byproducts

3.2.1. Demographics

Based on logistic regression modeling, gender was found to be a consistently significant demographic predictor of general willingness-to-try responses across information levels (GWTT, GWTTS, and GWTTSH; Table 2). In each case, males were more likely to exhibit SB trial intent than females. However, the discrepancy between genders lessened with each additional informational cue, indicated by decreasing odds ratio (OR) estimates from 2.4 (GWTT) to 1.6 (GWTTS) to 1.4 (GWTTSH). While SB information increased "yes" responses for both genders, data suggested that it was more effective in changing the minds of female consumers. These findings supported suppositions that males are more open to novel foods, while females are more health and food-safety conscious [12].

Table 2. Demographic predictors of willingness-to-try [1] food products containing seafood byproducts, based on N = 904 consumer responses.

	GWTT		GWTTS		GWTTSH	
Parameter	**95% CI [2]**	**OR [3]**	**95% CI**	**OR**	**95% CI**	**OR**
Gender (M:F)	1.74–3.22	2.4	1.18–2.20	1.6	1.047–1.99	1.4
US consumer	0.77–1.65	-	-	-	-	-
Race [4]	-	-	-		-	
A:B	-	-	1.39–9.32	3.6	1.08–7.25	2.8
H:B	-	-	1.56–9.15	3.8	1.71–9.99	4.1
W:B	-	-	1.01–5.74	2.4	-	-
O:B	-	-	-	-	1.12–19.64	4.7
H:W	-	-	1.06–2.32	1.6	1.28–2.86	1.9
Previous [5]	3.52–6.74	4.9	2.27–4.50	3.2	2.30–4.73	3.3
Knowledge [6]	1.14–2.29	1.6	1.21–4.49	1.7	1.44–2.84	2.0

[1] Based on separate logistic regression models of responses: GWTT = General willingness-to-try [with seafood byproduct definition only], GWTTS = General willingness-to-try [with definition and safety information], and GWTTSH = General willingness-to-try [with definition, safety information, and health benefit information]. [2] 95% Confidence intervals. [3] Odds ratio (OR; relative odds of a "yes" response) estimates were only presented and interpreted for predictors found to be significant to the model (p-value $\leq$ 0.05). [4] Significant between-group effects are presented for self-identified Asian:Black/African American (A:B), Hispanic/Latin/Spanish origin:Black/African American (H:B), White:Black/African American (W:B), Other race:Black/African American (O:B), and Hispanic/Latin/Spanish origin:White consumers (H:B). [5] Previous consumption of seafood byproducts (yes:no).[6] Self-reported initial knowledge of seafood byproducts (yes:no).

For the purposes of this study, self-reported race was meant to reflect consumers' social, cultural, and/or ethnic identity, rather than biological or genetic information [27]. All race categories exhibited a significant increase in willingness-to-try SB after informational cues had been given to them, yet the race variable revealed significant differences in frequency of GWTTS and GWTTSH responses (Table 2). As race was not a significant predictor of the initial GWTT response, its significance in subsequent models indicated that race was associated with differential responsiveness to informational cues regarding SB consumption.

Asian consumers exhibited the highest SB trial intent at each information level (62% GWTT, 76% GWTTS, and 77% GWTTSH), and Black/African American consumers reported the lowest (22% GWTT, 37% GWTTS, and 41% GWTTSH) despite being more responsive to informational cues (higher % increase in "yes" responses). As such, odds ratio (OR) estimates between Asian and Black/African American consumers decreased with increasing information, from GWTTS to GWTTSH (Table 2). It is possible that raw differences in WTT frequencies were due to relative novelty (vs. familiarity) of the food source. In Asian countries, seafood byproducts and coproducts have been more commonly available for human consumption than in the west [28]. In this study, US consumer status was not a significant predictor of WTT responses (Table 2), but it was worth noting that 85% of Black/African American respondents were US consumers, and 81% of Asian respondents were not.

Based on percent increase in "yes" responses to willingness-to-try questions, Hispanic/Latin/Spanish origin consumers were most positively responsive to informational cues regarding SB, increasing from 49% GWTT to 74% GWTTSH. After safety and health benefit claims, respectively, odds of GWTTS (OR = 1.6) and GWTTSH (OR = 1.9) were significantly higher for Hispanic/Latin/Spanish origin consumers than White consumers. Compared to Black/African American consumers, estimated odds of a "yes" response were more than twice as high for White (OR = 2.4 for GWTTS) and up to four times higher for Hispanic/Latin/Spanish origin consumers (OR=4.1 for GWTTSH; Table 2) during the period of this study. In previous research [17], combined messages about safety and health benefits of catfish byproduct had a greater impact on Hispanic/Latin Americans than US Americans. In the present study, self-reported race added resolution to statistical models, and evidenced that racial identity (or associated factors) can play a part in adoption of new foods such as SB, along with other individual factors.

Both reported knowledge of SB and previous SB consumption were associated with increased odds of trial intent across all informed conditions (Table 1). These effects were independent of race and gender (no significant interaction effects during model selection). It has been theorized that consumers' level of familiarity with a product moderates their responsiveness to extrinsic cues [24]. In the present study, existing knowledge about SB had a cumulatively positive effect on trial intent along with informational cues, with OR estimates increasing from 1.6 (GWTT) to 1.7 (GWTTS) to 2.0 (GWTTSH; Table 2).

High OR estimates for previous SB consumers (OR ranging from 3.2-4.9) added to existing evidence that trial of a novel food can encourage future instances [13]. Positive impacts of SB knowledge and previous consumption further motivated the need for appropriate products containing SB which encourage initial trial. Through consumption behaviors, familiarity with SB can be increased, providing consumers with more complex schemas to evaluate actual product quality [29] and guide further development.

3.2.2. Emotions

Food-related emotions can be studied from the perspective of how emotions affect food choice, or from the perspective of how food choice affects emotions [30]. This research was concerned with the former, using consumers' emotional baseline scores (Table S1) during an early stage of the COVID-19 pandemic in 2020 as predictors of willingness-to-try foods containing SB. The COVID-19 outbreak impacted consumers' emotions and food-related behaviors [31]. These included purchase and consumption habits [31] as well as food waste behavior [32]. The timing of similar online studies was considered a limitation to determine consumer perception and behavioral partners [15], but the current research was interested in how these unique conditions affected perceptions of waste-to-value foods made with SB.

Still, a follow-up post COVID-19 pandemic online study would be warranted to investigate how emotional predictors of SB consumptions have changed, if at all.

Initially, only the emotion *bored* was significant in modeling the GWTT response (Table 3; full models including non-significant emotions are presented in supplementary Tables S2–S4). Strict interpretation of this relationship would imply that for every one-unit increase in *bored* (on a 9-point scale), estimated odds of GWTT = "yes" would increase by 10% (OR = 1.1). It was noticed that Asian consumers reported higher mean *bored* scores (6.7 on 9-point scale) during COVID-19 than each other group, while also reporting the highest GWTT intent. A study conducted in China in March, 2020 described associations between consumers' boredom (due to limited activity during the COVID-19 "lockdown"), sensations seeking, and consumption willingness [33]. Thus, it is possible that increased boredom induced by the COVID-19 lockdown (Table S4) may have left consumers seeking new food experiences, and increased openness to SB consumption.

Table 3. Significant emotional predictors [1] of willingness-to-try [2] foods containing seafood byproducts [2].

Response	Emotion	95% CI [3]	Odds Ratio
GWTT	Bored	1.01–1.14	1.1
GWTTS	Enthusiastic	1.01–1.21	1.1
	Free	0.83–0.96	0.89
	Unsafe	1.00–1.16	1.1
GWTTSH	Free	0.84–0.92	0.88
	Loving	0.84–1.00	0.92
	Unsafe	1.04–1.03	1.1

[1] Based on logistic regression models with 21 emotional predictors. Significance of terms was determined by Type III Wald tests (p-value ≤ 0.05). [2] GWTT = general willingness-to-try [with seafood byproduct definition only], GWTTS = general willingness-to-try [with definition and safety information], and GWTTSH = general willingness-to-try [with definition, safety information, and health benefit information]. [3] 95% Confidence intervals.

The baseline emotions, enthusiastic, free, and unsafe, became significant predictors of the GWTTS response, and free, loving, and unsafe were significant predictors of the GWTTSH response. Perhaps more relevant than the magnitude of these effects (all OR estimates between 0.89 and 1.1; Table 3) were their directions. In some instances, positive emotions would be expected to increase (expected OR > 1) and negative emotions expected to decrease (expected OR < 1) the motivation for consumption [30]. However, in the present case, feeling increasingly free or loving (positive emotions) was associated with reduced odds of willingness-to-try SB (OR < 1; Table 3), and feeling more unsafe (a negative emotion) was associated with higher willingness-to-try SB odds (OR > 1). These findings were interpreted in the context of pandemic-elicited emotions.

The significance of new terms after presentation of informational cues was indicative of the emotion profiles of consumers who were responsive to the safety and health claims. Consumers feeling more unsafe during the pandemic may have been reluctant to report positive GWTT and thus representative of those influenced by the safety claim. Grunert [23] posed that during non-crisis times, the food supply is considered safe, so safety has less influence on food choice. During a pandemic, however, many consumers took extra measures to ensure personal safety [31]. The SB safety information cue may have reassured some of these consumers, making unsafe a significant predictor of both GWTTS and GWTTSH responses.

Similar to the *bored* emotion, activity restrictions imposed by the pandemic may have left consumers feeling less *free* and lacking in variety of experience [33]. These consumers with lower baseline *free* scores were seemingly influenced by informational cues, perhaps seeking new (but safe) food experiences. *Enthusiastic* has been called a sensation-seeking emotion [14] when presenting as a high-arousal motivator of new food experiences, and high states of arousal have been associated with attention to informational cues elsewhere [16]. Based on the GWTTS logistic regression model (Table 3), consumers feeling less *free* and/or more *enthusiastic* during the pandemic had higher odds of positive

SB trial intent, perhaps related to the desire for variety and sensation under otherwise restricted conditions.

Objective explanations of lay persons' self-reported emotions suffer from certain limitations [34]. First, there may not be consensus among scientists on definitions of emotional descriptors or terms. Perhaps more concerning is that naïve consumers with different linguistic backgrounds may use words and verbal scales differently; this can be exacerbated when translating from one language to another [34]. In the present internationally distributed online survey, it was unclear why the positive emotion *loving* became a significant predictor of the GWTTSH response with an OR estimate < 1, although Asian consumers reported significantly lower mean baseline *loving* scores along with the directionally highest proportion of GWTTSH = "yes" responses. Despite inherent limitations in psychological self-reporting, use of emotions to predict food choice does provide information beyond typical demographics and hedonic ratings [30]. In this study, openness to SB consumption and responsiveness to informational cues were affected by consumer emotions during the COVID-19 pandemic.

3.3. Product Appropriateness for Seafood Byproducts

Relative appropriateness of a product category was measured by frequency of selection from the CATA WTT list and reported as percentage of the 611 consumers who responded "yes" to the GWTTSH question. Among these potential SB consumers, 82% reported willingness-to-try *fish and fish products including mollusks, and crustaceans* if they contained SB, making it the most appropriate food product category among the twelve categories presented (Figure 3). After fish and fish products, the most appropriate categories were *seasoning mix, sauces and dressings* (71% WTT), *soup and gravy products (canned and frozen)* (62% WTT), *meat and meat products including poultry and game* (61% WTT), and *snack and energy/protein bars* (58% WTT). The median number of categories selected was 5, and the mode was 3, indicating that SB was considered appropriate for versatile applications by potential consumers.

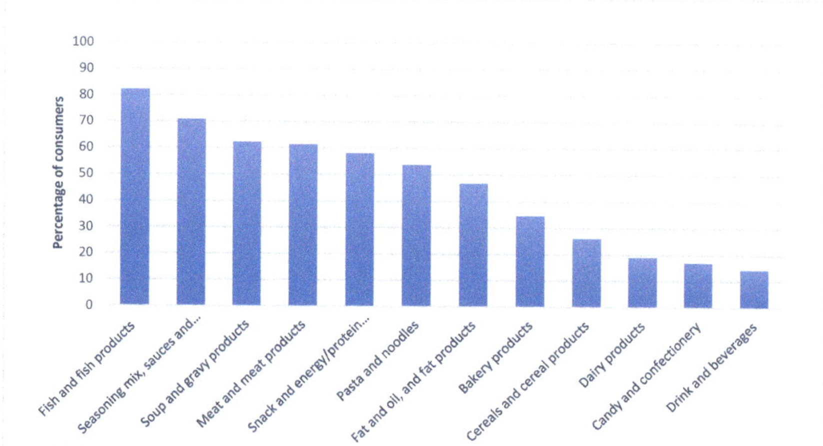

Figure 3. Product appropriateness of foods containing seafood byproduct based on willingness-to-try responses from 611 consumers.

Appropriateness has been called a situational aspect of acceptance and represents a cognitive dimension of food choice [35] separate from pure hedonic assessment. No actual products were evaluated in this study, so WTT responses were expected to be made based on mental constructs of hypothetical products, and appropriateness was intended to indicate a perceived fit of the ingredient (i.e., SB) within a final food product concept.

In these situations, consumers may rely on expectations and/or past experiences when making their judgments [23].

For the sake of this discussion, the term seafood will be used to include finfish, crustaceans, mollusks, and other aquatic animals reared or captured for human consumption [3,4]. In this case, an edible seafood byproduct would still be inherently considered seafood, and potential consumers identified its most appropriate "fit" within its own product category (fish and fish products including mollusks, and crustaceans). In our previous research, catfish bone powder has found favorable acceptance when incorporated into breading mixes used to dredge fried catfish, and knowledge of its use increased purchase intent [9]. In another study [36], tilapia skins were made into ready-to-eat puffed snacks with only the addition of oil and seasonings, and in some parts of the world small fish bones have been eaten as snacks [3]. Within this product category, potential SB applications can include introduction into other seafood products [17], either overlapping with other categories (e.g., both seafood and snack) and/or standing alone as a new seafood product unto itself [36,37].

The seasoning mix, sauces, and dressings category was deemed second most appropriate for SB usage (Figure 3). Fish sauces, as an example, have been well-known and consumed in Asia and can be made using fish heads, bones, and entrails. Garum is thought to itself have originated as a byproduct of fish salting [37]. Beyond fermentation, various processing technologies have been used to transform would-be seafood waste into potential food ingredients, including oils which may lend to sauces/dressings, although sensory quality remains a consideration [10]. However, to determine the product appropriateness of SB, product developers should consider the market niche based on the consumer culture and demographic. Cross cultural studies with actual products containing SB should be conducted to provide clarity of the appropriateness of each category.

Among the top five product categories reported (Figure 3), examples of each can be found in literature (i.e., seafood, sauces, soups, sausage, snacks [3]). However, more widespread acceptance of foods made with SB will likely depend on a combination of perceived quality dimensions, such as appropriateness of usage (a search characteristic [11]), sensory quality (an experience characteristic [11]), and perceived product benefits (a credence characteristic [11]). Therefore, product development efforts should target the wants and needs of intended consumers, whether it be snacks fortified with powdered fish byproduct for school children in Ghana [38] or pâté made with fish roe, milt, and liver among teenagers in Ireland [39]. The present consumer-driven appropriateness data were intended as a starting point to guide effective SB valorization concepts which will encourage consumption.

3.4. Perceived Risks of Seafood Byproduct Consumption

When consumers make choices in the face of uncertainty (e.g., whether or not to try an unfamiliar food), they must judge the likelihood of negative consequences such as physical harm (safety), undesirable sensory experiences (taste, texture, etc.,), or negative social impacts and the severity of such outcomes, or perceived risks [16]. These subjective evaluations can lead consumers to avoid certain foods when perceived risk is high. Based on this theoretical approach to risk perception and consumer decision-making, the term risk has been used to describe barriers to novel food consumption [13].

Consumers who did not report willingness-to-try a SB-containing product after a safety claim and health benefit claim ("maybe" or "no" to the GWTTSH question) identified perceived risks, or reasons for their SB aversion from a CATA list of 11 items. Of these 293 consumers, 91% were considered SB-reluctant (GWTTSH = "maybe") and 9% were considered SB-averse (GWTTSH = "no"). The frequency of risk selection was reported as percentages of 293 SB-reluctant/averse consumers (Figure 4). The median number of risks selected was 4, and the mode was 3.

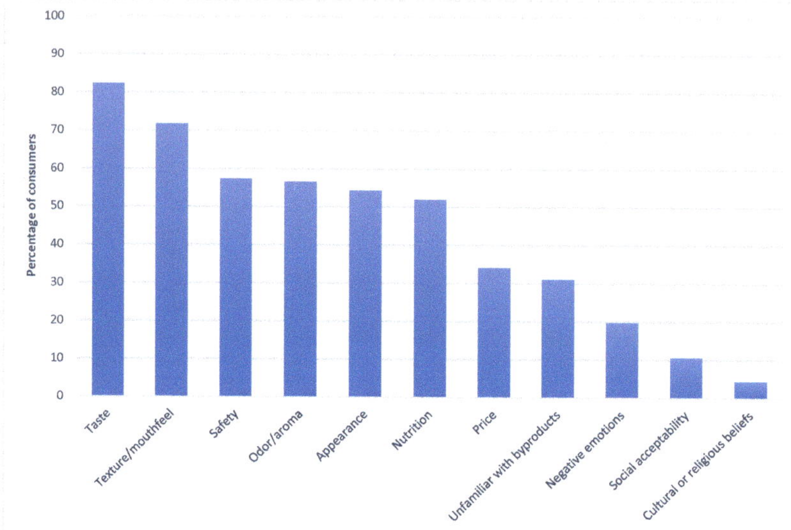

Figure 4. Perceived risks associated with seafood byproduct trial intent, based on responses from 293 consumers who were unwilling to try foods containing seafood byproducts.

Ninety-seven percent of reluctant/averse SB consumers cited at least one sensory attribute (taste, texture/mouthfeel, odor/aroma, or appearance) as a reason for SB avoidance. Taste is consistently a predominant driver of food choice, and was the most common reason for seafood byproduct avoidance (82% selection; Figure 4). Taste was followed by texture/mouthfeel (72% selection), odor/aroma (tied with safety at 57% selection), and appearance (54% selection; Figure 4) as the top cited SB risks. Lower sensory expectations of waste-to-value foods are not unique to SB, and have been shown for both plant [11] and animal [15] sourced byproducts. As only 60 of 293 SB-reluctant/averse consumers reported previous SB consumption, it was inferred that concerns about sensory quality were largely based on expectations rather than experience, and that SB was expected to be of inferior quality to more commonly consumed seafood portions. However, actual trials (Catfish nuggets/Fish skin snack) indicated that with the additional product information consumers accepted the product and these novel products have potential to be introduced to the market [14,36].

Despite an informational cue claiming that products containing SB are safe for consumption, safety was still the third most common reason for trial aversion (57% selection; Figure 4). Interestingly, consumers citing safety as a reason for SB avoidance reported lower unsafe emotion scores (a mean of 5.7 on a 9-point scale) than their counterparts who did not cite safety as a perceived risk (a mean unsafe score of 5.9). It was inferred that SB were considered less safe than typical seafood products by these consumers.

Food safety can be considered a credence characteristic, since the safety of a food is not directly experienced but must rely on trust [23]. As new technologies and uses for SB emerge, safety and risk management are priorities for industry [10,40] and nature of the material, production practices and processing methods, as well as appropriate testing [40] should be considered. Regulatory bodies also play a role in appropriate usage, labeling, and inspection of byproducts to ensure safety [41]. Labeling implications include identification of allergens [42] (i.e., fish or shellfish allergy) and any ingredient processing identification requirements (such as those existing for "mechanically separated" meat in the US) [41]. Reluctant/averse consumers' trust in SB safety ultimately may depend on attitudes toward producers and regulations [23] as well as exposure, rather than simple claims. Concerns about nutrition of SB also remained an issue for 52% of reluctant/averse consumers, even after a health claim. As opposed to sensory quality (which is more subjective), content and recovery of nutrients and health-promoting compounds in SB has been empirically documented [10]. Fish and seafood have been considered healthy food choices [6], but based on the present data, SB has yet to be elevated to the same perceptions of healthfulness.

Remaining concerns of price, unfamiliarity, negative emotions, social acceptability, and cultural or religious beliefs were each cited less than 50% of the time, but are common barriers to acceptance of food from unusual sources. Some of these issues have been related to individual traits, such as food neophobia (aversion to new or unusual foods) [12,14], disgust sensitivity [14], or socio-cultural characteristics [15] elsewhere. While personal traits may be more stable, unfamiliarity can be overcome through exposure [14]. Additionally, education and proper communication regarding the SB benefits may help minimize misconceptions about risk related to SB consumption [13]. In all but three instances, each of these lesser cited risks was selected alongside more prevalent concerns of sensory quality, safety and/or nutrition of SB.

4. Concluding Remarks and Future Prospects

Not all, but most, seafood consumers participated in this study were willing to try foods made with SB. Much reluctance was overcome through simple informational cues, and potential consumers identified fish products as the best cognitive "fit" for seafood byproduct utilization. Initial openness to SB consumption may be best predicted by previous experience, while responsiveness to SB safety and health information was associated with individual factors (e.g., gender, race). Additionally, salient emotions can help predict seafood byproduct trial intent, which may have been motivated by a desire for varied experience during the COVID-19 pandemic. Barriers to SB trial intent can largely be overcome through cognitive strategies, but some consumers remain unconvinced of sensory appeal (mainly taste) and safety.

This experiment provided meaningful insight for product developers to create new products with SB guided by consumer-by-consumer data. However, results should not be generalized outside of the current population samples. Participants were unpaid volunteers, and the data were self-reported. The research was performed during the COVID-19 pandemic when participants may have faced many restrictions. No actual products were used to evaluate the WTT (GWTT, GWTTS, and GWTTSH), risks, or appropriateness in this study. Currently, seafood byproducts do not face any special labeling regulations other than reporting the ingredient and allergenicity. Agencies may work within each country to establish appropriate requirements to help the seafood industry commercialize new products containing SB. Future studies should be focused on studying consumer perception in actual trials with a specific food matrix for each byproduct.

Supplementary Materials: The following supporting information can be downloaded at: https://www.mdpi.com/article/10.3390/foods12061313/s1. Table S1: Mean and standard deviation of emotions by race on a scale of 9-points. Table S2: Logistic regression model of general willingness-to-try (GWTT) foods containing seafood byproducts. Table S3: Logistic regression model of general willingness-to-try with safety claim (GWTTS) foods containing seafood byproducts. Table S4: Logistic regression model of general willingness-to-try with safety and health benefit claim (GWTTSH), foods containing seafood byproducts. All authors have read and agreed to the published version of the manuscript.

Author Contributions: S.M., conceptualization; methodology; formal analysis; investigation; data curation; writing—original draft preparation. R.A., conceptualization; methodology; validation; data curation; writing—original draft preparation and editing. W.P.; conceptualization; methodology; validation; investigation; resources; writing—review and editing; supervision; project administration and resource/funding acquisition. All authors have read and agreed to the published version of the manuscript.

Funding: This work was partially supported by the USDA National Institute of Food and Agriculture Hatch project (Accession No. 1022270; Hatch Project No. LAB94473) and the LSU Agricultural Center.

Institutional Review Board Statement: The study was conducted according to the guidelines of the Declaration of Helsinki and approved by the Institutional Review Board of Louisiana State University Agricultural Center (IRBAG-21-0063).

Data Availability Statement: The data that support the findings of this study are available from the corresponding author upon reasonable request.

Acknowledgments: We acknowledge the graduate students under the direct supervision of the corresponding author in the Sensory Services Lab, LSU Agricultural Center for assistance with this work.

Conflicts of Interest: The authors declare no conflict of interest. The funders had no role in the design of the study; in the collection, analyses, or interpretation of data; in the writing of the manuscript, or in the decision to publish the results.

References

1. Muth, M.K.; Birney, C.; Cuéllar, A.; Finn, S.M.; Freeman, M.; Galloway, J.N.; Gee, I.; Gephart, J.; Jones, K.; Low, L.; et al. A systems approach to assessing environmental and economic effects of food loss and waste interventions in the United States. *Sci. Total Environ.* **2019**, *685*, 1240–1254. [CrossRef]
2. FAO. *The State of Food and Agriculture 2019. Moving Forward on Food Loss and Waste Reduction*; Food and Agriculture Organization of the United Nations: Rome, Italy, 2019. [CrossRef]
3. FAO. *The State of World Fisheries and Aquaculture 2020*; Food and Agriculture Organization of the United Nations: Rome, Italy, 2020. [CrossRef]
4. FAO. *The State of World Fisheries and Aquaculture 2022*; Food and Agriculture Organization of the United Nations: Rome, Italy, 2022. [CrossRef]
5. Love, D.C.; Fry, J.P.; Milli, M.C.; Neff, R.A. Wasted seafood in the United States: Quantifying loss from production to consumption and moving toward solutions. *Glob. Environ. Chang.* **2015**, *35*, 116–124. [CrossRef]
6. Šimat, V. Valorization of seafood processing byproducts. In *Valorization of Agri-Food Wastes and Byproducts*; Bhat, R., Ed.; Academic Press: London, UK, 2021; pp. 515–536.
7. Ioannis, S.; Persefoni, T. *Seafood Processing Technology, Quality and Safety-Fish and Waste Management*; Wiley Press: Oxford, UK, 2014; pp. 2063–2309.
8. Ozogul, F.; Hamed, I.; Yesim, O.; Regenstein, J. Crustacean By-Products. *J. Food Chem.* **2019**, 33–38. [CrossRef]
9. Hoang, D.L.; Davis, C.; Moll, H.C.; Nonhebel, S. Can multiple uses of biomass limit the feedstock availability for future biogas production? An overview of biogas feedstocks and their alternative uses. *Energies* **2020**, *13*, 2747. [CrossRef]
10. Ozogul, F.; Cagalj, M.; Šimat, V.; Ozogul, Y.; Tkaczewska, J.; Hassoun, A.; Kaddoure, A.; Kuley, E.; Rathod, N.; Phadke, G. Recent developments in valorisation of bioactive ingredients in discard/seafood processing by-products. *Trends Food Sci. Technol.* **2021**, *116*, 559–582. [CrossRef]
11. Aschemann-Witzel, J.; Peschel, A.O. How circular will you eat? The sustainability challenge in food and consumer reaction to either waste-to-value or yet underused novel ingredients in food. *Food Qual. Prefer.* **2019**, *77*, 15–20. [CrossRef]
12. Coderoni, S.; Perito, M.A. Sustainable consumption in the circular economy. An analysis of consumers' purchase intentions for waste-to-value food. *J. Clean. Prod.* **2020**, *252*, 119870. [CrossRef]
13. Ardoin, R.; Prinyawiwatkul, W. Product appropriateness, willingness to try and perceived risks of foods containing insect protein powder: A survey of U.S. consumers. *Int. J. Food Sci. Technol.* **2020**, *55*, 3215–3226. [CrossRef]
14. Ardoin, R.; Prinyawiwatkul, W. Consumer perceptions of insect consumption: A review of western research since 2015. Int. J. *Food Sci. Technol.* **2021**, *56*, 4942–4958. [CrossRef]
15. Bearth, A.; Khunnutchanart, K.; Gasser, O.; Hasler, N. The whole beast: Consumers' perceptions of and willingness-to-eat animal by-products. *Food Qual. Prefer.* **2021**, *89*, 104144. [CrossRef]
16. Baker, M.A.; Shin, J.T.; Kim, Y.W. An exploration and investigation of edible insect consumption: The impacts of image and description on risk perceptions and purchase intent. *Psychol. Mark.* **2016**, *33*, 94–112. [CrossRef]
17. Murillo, S.; Ardoin, R.; Watts, E.; Prinyawiwatkul, W. Effects of catfish (*Ictalurus punctatus*) bone powder on consumers' liking, emotions, and purchase intent of fried catfish strips. *Foods* **2022**, *11*, 540. [CrossRef]
18. Sukkwai, S.; Kijroongrojana, K.; Chonpracha, P.; Pujols, K.D.; Alonso-Marenco, J.R.; Ardoin, R.; Prinyawiwatkul, W. Effects of colorant concentration and 'natural colour' or 'sodium content' claim on saltiness perception, consumer liking and emotion, and purchase intent of dipping sauces. *Int. J. Food Sci. Technol.* **2018**, *53*, 1246–1254. [CrossRef]
19. World Health Organization. WHO Director-General's Opening Remarks at the Media Briefing on COVID-19. Geneva, Switzerland. 11 March 2020. Available online: https://www.who.int/director-general/speeches/detail/who-directorgeneral-s-opening-remarks-at-the-media-briefing-on-covid-19---11-march-2020 (accessed on 20 December 2022).
20. King, S.C.; Meiselman, H.L. Development of a method to measure consumer emotions associated with foods. *Food Qual. Prefer.* **2010**, *21*, 168–177. [CrossRef]
21. United States Department of Agriculture. What We Eat in America Food Categories. Available online: https://www.ars.usda.gov/ARSUserFiles/80400530/pdf/1720/Food_Category_List_2017-March%202020.pdf (accessed on 20 December 2022).

22. Henchion, M.; McCarthy, M.; O'Callaghan, J. Transforming beef by-products into valuable ingredients: Which spell/recipe to use? *Front. Nutr.* **2016**, *3*, 53. [CrossRef]
23. Grunert, K.G. Current issues in the understanding of consumer food choice. *Trends Food Sci. Technol.* **2022**, *13*, 275–285. [CrossRef]
24. Altintzoglou, T.; Honkanen, P.; Whitaker, R.D. Influence of the involvement in food waste reduction on attitudes towards sustainable products containing seafood by-products. *J. Clean. Prod.* **2021**, *285*, 125487. [CrossRef]
25. Sukkhown, P.; Pirak, T.; Chonpracha, P.; Ardoin, R.; Prinyawiwatkul, W. Seafood flavor perception, liking, emotion, and purchase intent of coated peanuts as affected by coating color and hydrolyzed squid peptide powder. *J. Food Sci.* **2019**, *84*, 1570–1576. [CrossRef]
26. Lin, Y.; Boh, W.F. Informational cues or content? Examining project funding decisions by crowdfunders. *Inf. Manag.* **2021**, *58*, 103499. [CrossRef]
27. United States Census Bureau. About the Topic of Race. Available online: https://www.census.gov/topics/population/race/about.html (accessed on 20 December 2022).
28. Newton, R.; Telfer, T.; Little, D. Perspectives on the utilization of aquaculture coproduct in Europe and Asia: Prospects for value addition and improved resource efficiency. *Crit. Rev. Food Sci. Nutr.* **2014**, *54*, 495–510. [CrossRef] [PubMed]
29. Rao, A.R.; Monroe, K.B. The moderating effect of prior knowledge on cue utilization in product evaluations. *J. Consum. Res.* **1988**, *15*, 253–264. [CrossRef]
30. Desmet, P.M.; Schifferstein, H.N. Sources of positive and negative emotions in food experience. *Appetite 2008. 50*, 290–301. [CrossRef]
31. Szymkowiak, A.; Gaczek, P.; Jeganathan, K.; Kulawik, P. The impact of emotions on shopping behavior during epidemic. What a business can do to protect customers. *J. Consum. Behav.* **2021**, *20*, 48–60. [CrossRef]
32. Brizi, A.; Biraglia, A. "Do I have enough food?" How need for cognitive closure and gender impact stockpiling and food waste during the COVID-19 pandemic: A cross-national study in India and the United States of America. *Pers. Indiv. Differ.* **2021**, *168*, 110396. [CrossRef] [PubMed]
33. Deng, S.; Wang, W.; Xie, P.; Chao, Y.; Zhu, J. Perceived severity of COVID-19 and post-pandemic consumption willingness: The roles of boredom and sensation-seeking. *Front. Psychol.* **2020**, *11*, 567784. [CrossRef]
34. Scherer, K.R. What are emotions? And how can they be measured? *Soc. Sci. Inf.* **2005**, *44*, 695–729. [CrossRef]
35. Schutz, H.G.; Martens, M. Appropriateness as a cognitive-contextual measure of food attitudes. In *Food, People and Society*; Frewer, L.J., Risvik, E., Schifferstein, H., Eds.; Springer: Berlin, Heidelberg, 2001; pp. 247–266.
36. Wang, L.; Chen, W.; Zhou, R.; Ren, Y.; Wang, K.; Jiang, N.; Gao, R. Physicochemical and sensory properties of a tilapia skin-based ready-to-eat snack prepared by infrared drying and air frying. *Appl. Food Res.* **2022**, *2*, 100155. [CrossRef]
37. Mouritsen, O.G.; Duelund, L.; Calleja, G.; Frøst, M.B. Flavour of fermented fish, insect, game, and pea sauces: Garum revisited. *Int. J. Gastron. Food Sci.* **2017**, *9*, 16–28. [CrossRef]
38. Abbey, L.; Glover-Amengor, M.; Hagan, L.; Mboom, F.P. Physicochemical properties and sensory attributes of local snacks fortified with powdered fish processing by-products and an underutilised fish species. *Ghana J. Agric. Sci.* **2019**, *54*, 26–35. [CrossRef]
39. Furey, A.E.; Hoeche, U.; McLaughlin, C.; Noci, F. Incorporation of roe, milt and liver from plaice (*Pleuronectes platessa*), herring (*Clupea harengus*) and cod (*Gadus morhua*) in newly developed seafood pâtés: Sensory evaluation by teenage consumers in Ireland and their attitudes to seafood. *Int. J. Gastron. Food Sci.* **2022**, *28*, 100524. [CrossRef]
40. Ong, K.J.; Johnston, J.; Datar, I.; Sewalt, V.; Holmes, D.; Shatkin, J.A. Food safety considerations and research priorities for the cultured meat and seafood industry. *Compr. Rev. Food Sci. Food Saf.* **2021**, *20*, 5421–5448. [CrossRef] [PubMed]
41. Jayathilakan, K.; Sultana, K.; Radhakrishna, K.; Bawa, A.S. Utilization of byproducts and waste materials from meat, poultry and fish processing industries: A review. *J. Food Sci. Technol.* **2012**, *49*, 278–293. [CrossRef] [PubMed]
42. Tahergorabi, R.; Hosseini, S.V.; Jaczynski, J. Seafood proteins. In *Handbook of Food Proteins*; Woodhead Publishing: Sawston, UK, 2011; pp. 116–149.

MDPI
St. Alban-Anlage 66
4052 Basel
Switzerland
www.mdpi.com

Foods Editorial Office
E-mail: foods@mdpi.com
www.mdpi.com/journal/foods

www.ingramcontent.com/pod-product-compliance
Lightning Source LLC
Chambersburg PA
CBHW061554220926
51850CB00005B/331